Palgrave Studies in Environmental Sustainability

Series Editor

Robert Brinkmann, Hofstra University, Hempstead, USA

The series will take advantage of the growing interest in a number of environmental issues in sustainability, particularly in those that focus on interdisciplinary approaches to solving real-world problems. Unfortunately, many of the themes of sustainability address short- and long-term survivability of our planet and there is a pressing need for access to information to those who seek to solve sustainability problems. Thus, the scope of the book series seeks to be comprehensive within an environmental sustainability framework. The topics covered in the series range from climate change to public land management. While no series can ever be truly comprehensive, the series will provide one of the most definitive surveys of knowledge in the area of environmental sustainability.

Kousik Das Malakar · Kajal Kumar Mandal ·
Jenia Mukherjee · Supriya Roy

Coastal Ecologies at Risk

Climate Disasters, Innovative Risk Reduction, and Socio-ecological Sustainability

palgrave
macmillan

Kousik Das Malakar
Department of Geography
GDC Memorial College, Chaudhary
Bansi Lal University
Bahal, Haryana, India

Kajal Kumar Mandal
Department of Geography
School of Basic Sciences, Central
University of Haryana
Mahendragarh, Haryana, India

Jenia Mukherjee
Department of Humanities and Social
Sciences
Indian Institute of Technology
Kharagpur, West Bengal, India

Supriya Roy
Department of Geography
Institute of Humanities and Social
Sciences, Visva-Bharati University
Santiniketan, West Bengal, India

ISSN 2946-5311 ISSN 2946-532X (electronic)
Palgrave Studies in Environmental Sustainability
ISBN 978-3-032-26080-2 ISBN 978-3-032-26081-9 (eBook)
https://doi.org/10.1007/978-3-032-26081-9

Cover illustration: Sean Tilden/alamy

This Palgrave Macmillan imprint is published by the registered company Springer Nature Switzerland AG
The registered company address is: Gewerbestrasse 11, 6330 Cham, Switzerland

Dedicated to
Young Scholars in the Field of Coastal Science, Climate Studies, Geography, Hazard and Disaster Studies, Environmental Science, Sustainability Science, Social Sciences and Policy Makers

Foreword

The proposed title, *Coastal Ecologies at Risk,* by Malakar et al., offers a timely and well-grounded contribution to contemporary debates on climate change, disaster risk, and sustainability in coastal regions. As climate-induced hazards intensify across littoral landscapes, the book frames coastal socio-ecological systems not merely as spaces of exposure and vulnerability, but as dynamic environments shaped by long-standing and evolving human–environment relationships. Central to the volume is the argument that sustainability in coastal regions cannot be secured through technocratic solutions alone; rather, it demands an integrated understanding of ecological processes, social institutions, local knowledge systems, and governance arrangements.

The book is organised around a clear and well-structured intellectual progression. Beginning with the fundamentals of climate disasters and coastal ecologies, it establishes a conceptual foundation for interpreting coastal socio-ecological systems through the lenses of resilience, adaptation, and management. Subsequent sections critically examine risk reduction practices within policy and institutional contexts, engaging with both existing governance responses and emerging future challenges. A distinctive strength of the book lies in its sustained engagement with participatory approaches, risk perceptions, and field-based narratives, which add empirical richness and social depth to the analysis. The integration of GIScience with participatory methods further enhances the work, illustrating how spatial technologies can be effectively combined with community-based knowledge to support more inclusive, grounded, and context-sensitive approaches to coastal disaster risk reduction.

Overall, this volume makes a substantial scholarly and practical contribution by bringing together theory, policy analysis, and applied case studies within a unified socio-ecological framework. It is my hope that the work will be widely recognised and appreciated by researchers, planners, policy professionals, decision-makers, and advanced students working in the fields of coastal sustainability, disaster risk reduction, and climate adaptation. Published by *Palgrave Macmillan Cham*, the book stands as an important reference for understanding how coastal regions can respond to growing climate risks while advancing pathways toward socio-ecological sustainability.

Bahal, India

Dr. S. K. Sinha
Director, BRCM Education Society

Preface

Coastal socio-ecological systems occupy a *paradoxical position* within the contemporary climate crisis. They are among the most productive, densely populated, and culturally diverse regions of the world, while also being highly exposed to climate-induced hazards such as cyclones, storm surges, coastal flooding, sea-level rise, salinization, shoreline erosion, and a wide range of interconnected socio-environmental challenges. Over recent decades, both the frequency and intensity of these hazards have increased, alongside an expansion in their spatial reach. Events that were once episodic now appear as persistent and cumulative pressures. These changes have deepened existing social inequalities, disrupted local livelihoods, accelerated ecosystem degradation, and placed mounting pressure on governance capacities across many coastal regions, particularly in the Global South. In this setting, coastal ecologies can no longer be understood only as biophysical entities or treated as passive recipients of climate impacts. They are better seen as dynamic socio-ecological systems in which environmental processes, physical infrastructures, institutional arrangements, knowledge systems, and everyday community practices remain closely connected. Responses to coastal risk therefore call for integrative frameworks that cross disciplinary boundaries and draw on climate science, disaster studies, ecology, geography, planning, governance, and participatory forms of knowledge. This book addresses this need by moving beyond fragmented and sector-specific perspectives to present a holistic, empirically grounded, and practice-oriented understanding of *Coastal Ecologies at Risk*.

The book centers on the growing precarity of coastal socio-ecological systems under climate stress. The subtitle, *Climate Disasters, Innovative Risk Reduction, and Socio-ecological Sustainability*, reflects the book's dual focus on examining the drivers of coastal risk and identifying pathways toward resilience and sustainability. Disasters are approached not as isolated or purely natural events, but as outcomes shaped by interacting climatic, social, institutional, and ecological processes. At the same time, the book gives attention to innovation in disaster risk reduction through ecosystem-based, community-based, technological, and participatory approaches. The book aligns closely with the intellectual vision of the *Palgrave Studies in Environmental Sustainability* series. It engages with sustainability as a dynamic, contested, and place-specific process rather than a fixed or universal goal. By bringing disaster risk reduction, climate adaptation, governance, and socio-ecological perspectives into conversation, the volume contributes to sustainability scholarship in contexts marked by rapid environmental change, uncertainty, and deeply rooted social vulnerability. Its focus on coastal regions adds an important spatial and thematic dimension, highlighting areas that function both as global climate frontiers and as everyday living spaces for millions of people.

A key strength of this book lies in its integrative and participatory orientation. It brings theory and practice into dialogue, links global frameworks with local realities, and connects technological tools with lived experiences. The chapters offer an interdisciplinary view of coastal ecologies that links science, governance, and climate disaster responses. They examine climate impacts on densely populated coasts while drawing lessons that speak to the protection of vulnerable coastal ecosystems across different regions. The volume advances field-based interdisciplinary research on coastal socio-ecological sustainability and brings research practice and indigenous knowledge into discussions of coastal resilience and policy-making. By incorporating community perceptions, indigenous and local knowledge, and participatory cartography into spatial and policy analyses, the book moves beyond technocratic approaches to risk assessment. It shows how socially grounded and spatially explicit methods can support disaster risk reduction and sustainability across diverse coastal settings. The book is organized into five interrelated parts comprising fourteen chapters that build a narrative from foundational concepts to applied practices.

Part I, *Fundamentals of Climate Disasters and Coastal Ecologies*, sets out the conceptual and analytical foundations of the volume. It introduces key climate disaster processes affecting coastal systems and outlines approaches to assessing and mapping socio-ecological vulnerability. This section provides a baseline for understanding how hazards, exposure, and vulnerability intersect in coastal contexts. ***Part II***, *Understanding Coastal Socio-ecological Systems: Climate Disasters, Coastal Resilience, and Management*, extends the discussion by focusing on processes of resilience-building. It examines coastal infrastructural resilience, risk perception and communication, early warning systems, ecosystem-based adaptation, community-based and participatory approaches, and integrated coastal zone management. The chapters illustrate multiple ways through which resilience can be developed across scales. ***Part III***, *Risk Reduction Practices and Socio-ecological Sustainability: Policy Frameworks and Future Challenges*, shifts attention to governance, policy, and long-term sustainability. It engages with policy frameworks, coastal governance, sustainable development, and the blue economy, while reflecting on emerging challenges and opportunities for disaster risk reduction under accelerating climate change. ***Part IV***, *Participatory Approaches in Coastal Disaster Risk Reduction and Sustainability: Risk Perceptions and Field Narratives*, brings voices from the coast into focus. Through participatory insights and detailed field narratives, this section examines how local perceptions, lived experiences, and social relations shape disaster risk and adaptation practices. It highlights the role of co-produced knowledge in designing inclusive and context-sensitive interventions. ***Part V***, *Applications of GIScience and Participatory Approaches in Coastal Disaster Risk Reduction and Sustainability: Case Studies*, presents empirical case studies, mainly from the Indian coast. These chapters illustrate the use of geospatial science, participatory GIS, and community mapping in analyzing multi-hazard vulnerability, tracking coastal ecosystem change, integrating indigenous knowledge, and supporting participatory policy planning for socio-ecological sustainability.

The interdisciplinary scope of this book is one of its defining features. Drawing on geography, environmental science, disaster studies, ecology, planning, governance, and the social sciences, the book offers a multi-scalar and multi-method perspective on coastal risk and sustainability. Its contribution lies in the integration of spatial technologies with participatory methodologies and in its sustained focus on socio-ecological systems rather than hazard-specific or sectoral approaches. By situating empirical insights from the Global South within broader theoretical and policy debates, the volume speaks to both academic inquiry and applied practice. This book is intended for scholars, practitioners, policymakers, students, and all those engaged in understanding and addressing climate risks in coastal regions. It seeks not only to inform but also to encourage more inclusive, reflective, and context-sensitive approaches to coastal disaster risk reduction and socio-ecological sustainability. At a time when coastal futures are increasingly uncertain, integrated knowledge, participatory practice, and ethical engagement with vulnerable communities remain central to shaping resilient and just coastal pathways.

Bahal, India — Kousik Das Malakar
Mahendragarh, India — Kajal Kumar Mandal
Kharagpur, India — Jenia Mukherjee
Santiniketan, India — Supriya Roy

Acknowledgements With sincere appreciation, we extend our gratitude to the survey participants from the coastal regions of India, whose knowledge, lived experiences, and critical perspectives have been central to the intellectual development of this book, *Coastal Ecologies at Risk*. Their active participation has substantially enriched the empirical depth of the case study sections, providing grateful insights into the complex and dynamic interactions between coastal communities and their surrounding environments. We also acknowledge the extensive body of scholarly literature that has informed and guided this work, providing a solid theoretical and empirical foundation upon which the arguments of this book are built.

The publication of this volume as part of the *Palgrave Studies in Environmental Sustainability (Palgrave Macmillan, Cham)* series is a distinct honor. We express our sincere thanks to the editors and the publisher for their commitment to advancing carefully developed scholarship on coastal socio-ecological systems and sustainability. Their sustained efforts in fostering interdisciplinary academic inquiry have played a vital role in the dissemination of knowledge within this field. We further extend our deep gratitude to mentors, colleagues, and family members whose unwavering support, encouragement, and intellectual engagement have been invaluable throughout the course of this project.

We also wish to acknowledge the anonymous reviewers of this project, whose thoughtful and critical suggestions contributed significantly to improving the scope, clarity, and scholarly quality of the work.

Finally, we offer our heartfelt appreciation to the series editor, Robert Brinkmann, and the acquisitions editor, Rachael Ballard, for their insightful evaluations, professional guidance, and sustained support in bringing this book project to completion. Their engagement has been central to refining the manuscript, and their facilitation of this opportunity with Palgrave Macmillan is gratefully acknowledged.

New Delhi, India — Kousik Das Malakar
Mahendergarh, India — Kajal Kumar Mandal
Kharagpur, India — Jenia Mukherjee
Santiniketan, India — Supriya Roy

Competing Interests The authors declare no potential conflicts of interest.

Funding This book project did not receive financial support from any public, private, or non-profit organization.

Author Contributions

Monographs' conceptualization: K.D.M.; Project proposal develops, correspondence and management: K.D.M.; Lead author and supervision: K.D.M.; Co-author: K.K.M., J.M., and S.R.; Field investigation and data calculation: K.D.M., S.R., and K.K.M.; Field narrative thoughts: S.R., K.D.M., and J.M.; Methodologies: K.D.M., K.K.M., J.M., and S.R.; Software and applications: K.D.M., S.R., K.K.M., and J.M.; Tables, figures, photo plates, and mapping: K.D.M., K.K.M., and S.R.; Resources, research and writing: K.D.M., S.R., K.K.M., and J.M.; Reviewing roles: K.D.M., J.M., S.R., K.K.M., and other anonymous reviewers; Editing: K.D.M., and S.R.; Original draft design and preparation: K.D.M.; Proofing and others: K.D.M., K.K.M., S.R., and J.M.

K.D.M.: Kousik Das Malakar; K.K.M.: Kajal Kumar Mandal; J.M.: Jenia Mukherjee and S.R.: Supriya Roy

Contents

About the Authors

Dr. Kousik Das Malakar is a socio-ecological geographer, GIS analyst, and researcher with a specialisation in climate change and policy. He currently holds the position of Assistant Professor in the Department of Geography, GDC Memorial College (affiliated with the Chaudhary Bansi Lal University, Bhiwani), BRCM Vidyagram Campus, Haryana. Dr. Malakar earned his Ph.D. from the Department of Geography, Central University of Haryana, Mahendergarh. He also possesses a master's degree in social science geography from Jawaharlal Nehru University (JNU), New Delhi, and a bachelor's degree from Vidyasagar University, Medinipur, West Bengal. His research interests encompass the climate crisis, coastal socio-ecological systems, water governance, environments and societies, sustainability, disaster studies, participatory policy planning, and the application of geospatial technologies. In his field of research, Dr. Malakar has participated in over 150 academic events and has authored and edited more than 07 books published by Springer Nature, Emerald Publishing, and other reputed publishers. Dr. Malakar has published more than 50 research papers, conference proceedings, book chapters, general articles, and reports in nationally and internationally reputed publications. He has supervised more than 20 postgraduate project reports. He has received numerous certifications from the Indian Institute of Remote Sensing (IIRS) in Dehradun, the Geological Survey of India (GSI) in Hyderabad, and the National Institute of Disaster Management (NIDM) in New Delhi, recognising his ability to engage in and acquire research ideas and knowledge. Additionally, he actively serves as a reviewer for academic journals and books and is a member of several academic and professional organisations, including Academic Counsellor at IGNOU New Delhi, Human Development and Capability Association (USA), Unequal World (USA), the International Society for Urban Health (USA), the International Centre of Excellence in Engineering & Technology (Portugal), and the International Centre of Excellence in Humanities, Social Sciences, and Interdisciplinary Studies (Portugal).

Dr. Kajal Kumar Mandal is an Assistant Professor (Guest) in the Department of Geography at the Central University of Haryana, Mahendragarh, where he has been contributing to teaching and research since September 2023. He holds a Ph.D. in disaster management from Pondicherry University (a central university) and a master's degree in geography from Banaras Hindu University. He also holds advanced certifications in Geospatial Technologies from NRSC-ISRO and Geoinformatics from CED, India, and is a UGC-NET-qualified Junior Research Fellow (JRF) in Geography. Dr. Mandal's research spans flood hazard and risk assessment, disaster risk reduction, climate change adaptation, groundwater potential delineation, cyclone risk assessment, and hydrological and social vulnerability analysis. He is proficient in ArcGIS, ERDAS Imagine, Google Earth Engine, SNAP, HecRAS, and PCI-Geomatica, with hands-on expertise in Sentinel-1A SAR data processing, Multi-Criteria Decision Analysis (MCDA), and field-based geophysical surveys. He has an impressive publication record of 21 peer-reviewed research articles in reputed international journals, including Natural Hazards Research, Journal of Earth System Science, and Remote Sensing Applications. His research has addressed critical themes such as micro-level flood inundation mapping, social vulnerability assessment, groundwater delineation, drought hazard mapping, and cyclone risk modelling across diverse regions of India. Beyond academia, Dr. Mandal has applied his geospatial expertise in the industry, having served as a GIS specialist on the Yanbu Waterfront Marina Development Project in Saudi Arabia, where he conducted satellite-based mapping of mangroves and seagrasses along the Red Sea coast. He has also held teaching positions at Adamas University and JS University. Dr. Mandal has presented his work at international conferences and delivered guest lectures at multiple universities across India.

Dr. Jenia Mukherjee is a transdisciplinary researcher deploying environmental humanities as an *'engaged praxis'*-evident in her action projects in urban deltas and coastal environments of South and South-East Asia and Europe. She is investigating several global partnership grants, generously funded by DAAD-Germany; Social Science and Humanities Research Council, Canada; Swiss National Science Foundation, Switzerland; Water Development Partnership Programme, Netherlands, etc. The SOR4D, Switzerland funded ENGAGE4Sundarbans project in transboundary Sundarbans, spanning India and Bangladesh, is a transdisciplinary initiative committed to set up real-world laboratories in remote island villages of the largest mangrove delta in the world. Closely working with artists, Jenia uses a wide range of ethno-visual techniques and curates outreach activities such as exhibitions, field schools, etc. to reach out to a wider audience. Every year she conducts winter and summer field schools in the Sundarbans and the Chilika Lagoon, India. She has published in several journals including Environment and Planning E, Wires WATER, Environment and Urbanization, Urban Research and Practice, Marine Policy, Maritime Studies, Frontiers in Water, Hydrological Sciences Journal, etc. and is also the editor of Coastal Studies and Society (Sage). Her book on the social biography of Kolkata's canal and wetlands system Blue Infrastructures (Springer Nature 2020) won the Carson Writing Fellowship, and is included in the curricula of water studies and political ecology courses in Europe and the United States. She is an Associate Professor at the Department of Humanities and Social Sciences, Indian Institute of Technology Kharagpur, India.

Ms. Supriya Roy is a GIS analyst and socio-environmental geographer. She holds a master's degree in social science geography from the Department of Geography, Institute of Humanities & Social Sciences, Visva-Bharati University, Santiniketan, and a double bachelor's degree in social science geography from the University of Gour Banga and Vidyasagar College of Education, West Bengal (India). In her field of research, she has participated in more than 100 workshops/seminars/conferences/webinars and has contributed research monograph and articles to both national and international journals. She has been awarded multiple certifications from NIDM (New Delhi) and IIRS (Dehradun) for her active participation in research and the acquisition of knowledge and research ideas. Her research interests encompass a diverse range of topics, including coastal indigenous communities, anthropo-environmental conflicts, climate justice, climate migration, society and politics, participatory policy planning, and applications of geographic information science.

Abbreviations

CbA	Community-based Adaptation
CbM	Community-based Monitoring
CC	Climate Change
CCA	Climate Change Adaptation
CD	Climate Disasters
CDRR	Coastal Disaster Risk Reduction
CED	Coastal Ecosystem Development
CRR	Coastal Risk Reduction
CSA	Climate Smart Adaptation
DEM	Digital Elevation Model
DRM	Disaster Risk Management
DRR	Disaster Risk Reduction
EbA	Ecosystem-based Adaptation
EE	Environmental Education
EIA	Environmental Impact Assessment
EWS	Early Warning System
FGDs	Focus Group Discussions
GHG	Greenhouse Gases
GIS	Geographic Information System
GTI	Green Technology and Innovation

IbA	Infrastructure-based Adaptation
ICT	Information and Communication Technology
ICZM	Integrated Coastal Zone Management
IKS	Indigenous Knowledge Systems
IPCC	Intergovernmental Panel on Climate Change
LULC	Land Use and Land Cover
NbS	Nature-based Solutions
NDVI	Normalized Difference Vegetation Index
NGO	Non-governmental Organization
PC	Participatory Cartography
PCDRR	Participatory Coastal Disaster Risk Reduction
PGIS	Participatory Geographic Information System
PM&E	Participatory Monitoring and Evaluation
RP	Risk Perception
RS	Remote Sensing
SDGs	Sustainable Development Goals
SES	Social-ecological System
SEV	Socio-ecological Vulnerability
SFA	Sustainable Fisheries and Aquaculture

List of Figures

List of Tables

Part I

Fundamentals of Climate Disasters and Coastal Ecologies

1

Introduction to Climate Disasters and Coastal Ecologies

This chapter introduces the growing challenge of climate disasters and their close connection with coastal ecologies. It aims to build a clear understanding of how climate change is steadily reshaping coastal environments and affecting the people who depend on them for their lives and livelihoods. Climate disasters are presented not as rare or isolated events, but as the result of long-term changes in climate interacting with sensitive ecosystems and social systems. Rising sea levels, warming oceans, changing weather patterns, and human activities together increase the exposure and vulnerability of coastal regions. The chapter also highlights the importance of coastal ecologies such as mangroves, coral reefs, fisheries, and salt marshes. These ecosystems support biodiversity, protect shorelines, and provide natural defenses against storms and flooding. When these systems are damaged, the impacts of climate disasters become more severe, creating a cycle of ecological loss and social risk. Attention is given to the idea that coastal vulnerability is shaped not only by natural forces but also by social and economic conditions. The chapter introduces the need for sustainable and inclusive approaches to coastal risk reduction, stressing that

K. D. Malakar et al., *Coastal Ecologies at Risk*, Palgrave Studies in Environmental Sustainability, https://doi.org/10.1007/978-3-032-26081-9_1

protecting ecosystems and strengthening community resilience must go hand in hand in responding to climate change.

Key Points of the Chapter

- Understanding the concept and key drivers of climate disasters
- Explaining coastal socio-ecological systems in the context of climate change
- Discussing the impacts of climate disasters on coastal ecologies
- Describing coastal risk reduction and management practices

1.1 Climate Disasters

1.1.1 Concept of Climate Disasters

Climate disasters refer to severe and damaging environmental events that arise from the increasing intensity of climate processes in a changing global environment. These events are not isolated incidents but are closely connected to long-term changes in climate systems, including rising temperatures, altered rainfall patterns, and sea-level rise. Phenomena such as cyclones, floods, droughts, heatwaves, storm surges, coastal erosion, and saline intrusion are becoming more frequent and intense due to the combined effects of greenhouse gas emissions, changes in land use, and the degradation of natural ecosystems. In coastal regions, these disasters extend beyond physical damage and affect livelihoods, ecological stability, and patterns of human settlement (Siddiqui & Sahay, 2022; Yildiz & Shaw, 2024). Climate disasters are best understood as the result of interactions between environmental hazards and social conditions. A climatic event becomes a disaster not only because of its physical strength but because it affects people who are exposed and vulnerable. For example, the impact of a cyclone depends on factors such as housing quality, access to early warning systems, the condition of surrounding ecosystems, and existing social and economic inequalities. Coastal regions clearly illustrate this relationship. The destruction of mangroves, unplanned urban growth, and excessive use of coastal and marine resources weaken natural protection against storms and flooding. As a result, climatic events often turn into large-scale social and ecological crises. This shows that climate disasters are not purely natural events but are shaped by human actions that influence risk and resilience (Irshad, 2024; Siddiqui & Sahay, 2022). From a coastal ecological perspective, climate disasters also represent disruptions in the close relationship between land, sea, and living systems. Rising sea levels lead to the flooding of wetlands and low-lying areas, storm surges increase shoreline erosion, and warmer oceans affect coral reefs and fish populations. These changes weaken the ecosystem services that coastal environments provide, including food security, storm protection, carbon storage, and culturally important livelihoods. When these ecological functions decline, coastal communities become more exposed to future hazards. This creates a cycle in which environmental degradation increases disaster risk, and repeated disasters further damage ecosystems (Dutta et al., 2026; Rauf et al., 2026).

Another important aspect of climate disasters is their slow and cumulative nature. While sudden events such as cyclones and floods often receive immediate attention, gradual processes like soil salinization, loss of coastal vegetation, and changes in monsoon patterns can be equally damaging over time. In coastal areas, these slow changes reduce agricultural productivity, affect freshwater availability, and alter settlement patterns. In many cases, people are forced to relocate without experiencing a single dramatic disaster event. Recovery becomes increasingly difficult when communities face continuous environmental stress rather than short-term shocks. Understanding climate disasters therefore requires moving beyond a focus on individual events and recognizing them as ongoing conditions within stressed coastal environments (Kim, 2024; Malakar et al., 2023). Placing climate disasters within the broader context of coastal ecologies highlights the need for coordinated and long-term responses. It emphasizes the role of healthy ecosystems in reducing disaster impacts, the importance of sustainable coastal management, and the need to address social inequalities that increase vulnerability. As climate change continues to intensify, climate disasters provide a crucial framework for understanding how environmental change, ecological fragility, and human development interact, particularly in densely populated coastal regions (Sandifer et al., 2012).

1.1.2 Types of Climate Disasters

Climate disasters can be grouped into several related types based on the main climate processes through which they occur (Fig. 1.1). In fragile coastal ecologies, these types often overlap and reinforce one another, increasing overall risk. One major category is hydro-meteorological disasters, which include cyclones, storm surges, coastal flooding, and periods of extreme rainfall. These events are becoming more intense due to warming ocean waters and rising sea levels. As a result, coastal regions face growing threats of inundation, shoreline erosion, and damage to both infrastructure and natural ecosystems. Storms that were once occasional are now occurring more frequently, reshaping coastlines and disrupting human settlements as well as marine habitats. A second category consists of slow-onset climate disasters. These include sea-level rise, long-term coastal erosion, saltwater intrusion into freshwater sources and agricultural land, and the gradual loss of wetlands and mangrove forests. Unlike sudden disasters, these changes develop slowly over many years. Their impacts often remain unnoticed until livelihoods and ecological systems are already under severe stress. Soil salinization reduces agricultural productivity, freshwater becomes scarce, and low-lying islands and delta regions gradually become unsuitable for habitation. In many cases, communities are forced to relocate without experiencing a single dramatic disaster event.

Thermal and atmospheric extremes form another type of climate disaster that increasingly affects coastal and nearby regions. Heatwaves, rising sea surface temperatures, and changes in wind patterns and monsoon systems influence marine ecosystems, fisheries, and human health. Coral bleaching is one clear example of climate-related ecological damage. When coral reefs weaken, they lose their ability to protect shorelines from waves and storms. This increases exposure to erosion and flooding and intensifies the effects of other climate hazards. Another

Fig. 1.1 Types of climate disasters (Prepared by the authors)

important category is ecological climate disasters, which occur when climate stress leads to widespread disruption of biological systems. These include the decline of fisheries, loss of mangrove forests, degradation of coastal wetlands, and the spread of invasive species. Such changes reduce biodiversity and weaken the ecosystem services that support coastal livelihoods and protect shorelines. These ecological disruptions often occur alongside physical climate events, increasing their combined impact on both human and natural systems (Mohapatra, 2024; Schmidt et al., 2020; Tschumi & Zscheischler, 2020).

Finally, socio-economic climate disasters arise when climatic hazards interact with poverty, inequality, and weak systems of governance. In coastal regions, fishing communities, informal settlements, and small island populations are often the most affected by storms, flooding, and environmental decline. In these settings, the disaster extends beyond the immediate climatic event to include long-term damage to livelihoods, food security, and social stability. Understanding these different types of climate disasters highlights their layered and interconnected nature and points to the need for integrated approaches that link climate processes, ecological protection, and social resilience in coastal regions.

1.1.3 Causes and Drivers of Climate Disasters

Climate disasters develop through the combined influence of natural climate processes and human activities that have altered atmospheric, oceanic, and ecological systems (Fig. 1.2). One of the most important underlying causes is the steady increase in greenhouse gases such as carbon dioxide, methane, and nitrous oxide. These gases trap heat in the atmosphere and lead to global warming. As temperatures rise, both the air and oceans hold more energy, which strengthens weather systems. This results in more intense cyclones, heavier rainfall, longer heatwaves, and more severe droughts. In coastal ecologies, warming oceans also raise sea levels through thermal expansion, increasing the impact of storm surges, tidal flooding, and coastal erosion. Changes in oceanic and hydrological systems further contribute to climate disasters, particularly along coastlines. The melting of glaciers and polar ice sheets adds large volumes of water to the oceans, while warmer seawater expands, together causing sea levels to rise. At the same time, shifts in rainfall patterns and changes in monsoon systems affect river flows, groundwater recharge, and the movement of sediments toward coastal zones. These processes weaken the natural stability of deltas, estuaries, and wetlands, making them more vulnerable to flooding, land subsidence, and saltwater intrusion into freshwater sources.

Fig. 1.2 Causes and drivers of climate disasters (Prepared by the authors)

Human activities on land and in coastal areas greatly increase disaster risk by damaging ecosystems that once offered natural protection. Deforestation, drainage of wetlands, clearing of mangroves, and poorly planned coastal development reduce the ability of ecosystems to absorb wave energy, stabilize shorelines, and regulate water movement. When these natural buffers are lost, even moderate storms or high tides can cause serious damage. Rapid urban growth in low-lying coastal areas also places people, homes, and infrastructure in locations that are highly exposed to climate-related hazards (Abbass et al., 2022; Ismail-Zadeh, 2022).

Social and economic conditions play a major role in shaping the impacts of climate disasters. Poverty, inequality, weak infrastructure, and limited access to early warning systems reduce the ability of communities to prepare for and respond to climate-related events. In many coastal regions, livelihoods depend on activities such as fishing, farming, and tourism, which are highly sensitive to environmental change. Disruptions to these activities increase economic stress and social insecurity. Weak governance, poor land-use planning, and limited investment in resilient infrastructure further increase vulnerability and slow recovery after disasters. Overall, climate disasters arise from the interaction of climate change, environmental degradation, and social vulnerability. In coastal ecologies, these forces come together with particular intensity, as rising sea levels, fragile ecosystems, and dense human populations coexist in limited spaces. Therefore, these connected causes and drivers are essential for developing responses that not only address immediate disaster impacts but also reduce long-term risk by improving environmental management and social resilience in coastal regions (Ohlendorf, 2019; Roukounis & Tsihrintzis, 2022).

1.2 Coastal Ecologies and Ecosystems with Climate Disasters

1.2.1 Concept and Importance of Coastal Ecologies and Ecosystems

Coastal ecologies and ecosystems refer to the natural systems found where land and sea meet. These areas include mangroves, salt marshes, coral reefs, seagrass beds, estuaries, beaches, and deltas. They are shaped by the regular movement of tides, waves, river water, sediments, and living organisms. Coastal environments are not fixed landscapes. They are living systems that change over time in response to natural forces and human activities. At the same time, they provide the environmental base on which millions of people depend for housing, livelihoods, and cultural practices. One of the most important functions of coastal ecosystems is their support for biodiversity. Mangroves and estuaries serve as breeding and nursery areas for many fish and shellfish species. Coral reefs support some of the richest biological communities in the world, while wetlands offer habitats for birds and a wide range of aquatic organisms. These ecosystems are closely linked and sustain food chains that extend from coastal waters to the open ocean. When coastal ecosystems are healthy, they help maintain food security, ecological balance, and the overall productivity of marine environments (Jana & Singh, 2022).

Coastal ecologies also play a vital role in protecting shorelines and nearby human settlements from climate-related hazards. Mangrove forests, sand dunes, coral reefs, and wetlands reduce wave energy, limit storm surges, and slow coastal erosion. By stabilizing sediments and regulating water flow, these ecosystems help reduce flooding during storms and high tides. As climate change leads to rising sea levels and stronger storms, the protective role of coastal ecosystems becomes increasingly important for coastal safety. In addition to their ecological and protective roles, coastal ecosystems provide important economic and cultural benefits. Many livelihoods such as fishing, aquaculture, tourism, and coastal agriculture depend on healthy coastal environments. Beyond economic value, coastal areas hold strong cultural and social meaning for many communities. The sea and shoreline influence local traditions, beliefs, and everyday life. Damage to coastal ecosystems therefore affects not only the natural environment but also social well-being and cultural continuity (Singh et al., 2020). Within discussions of climate disasters, coastal ecologies hold a central position because they shape how environmental change affects human vulnerability. When these ecosystems are degraded, exposure to climate hazards increases and recovery becomes more difficult. In contrast, their protection and restoration can reduce risk and strengthen resilience. Understanding the nature and importance of coastal ecologies is therefore essential for addressing climate disasters and supporting sustainable futures in coastal regions.

1.2.2 Coastal Ecologies and Climate Disasters: An Interaction

Coastal ecologies and climate disasters are closely connected, and each influences the scale and effects of the other. Coastal ecosystems such as mangroves, coral reefs, wetlands, and sand dunes act as natural protection against climate hazards. They reduce the impact of cyclones, storm surges, coastal flooding, and erosion by absorbing wave energy, trapping sediments, and stabilizing shorelines. When these ecosystems are healthy, they can greatly reduce damage to homes, farmland, and infrastructure during extreme weather events. At the same time, climate disasters place strong pressure on coastal ecosystems. Powerful storms can uproot mangroves, damage coral reefs, and change the shape of coastlines. Prolonged flooding and changes in salinity affect wetland plants and soil conditions, reducing ecosystem health. Rising sea levels and warmer ocean temperatures add further stress, leading to coral bleaching, changes in fish populations, and the loss of sensitive habitats. These impacts weaken the ability of coastal ecosystems to recover and continue providing protection and resources (Ismail-Zadeh, 2022).

The relationship between coastal ecologies and climate disasters is reinforcing in nature. When ecosystems are degraded due to deforestation, pollution, or poorly planned development, their capacity to reduce climate impacts declines. This makes coastal regions more exposed to storms and flooding. In turn, repeated climate disasters cause further ecological damage, increasing long-term risk for both natural systems and human communities. This pattern is especially clear in low-lying deltas, small islands, and densely populated coastal areas, where environmental pressure and climate risks are already high. Human activities play a central role in shaping these interactions. The conversion of wetlands into urban or industrial land, the removal of mangroves for aquaculture, and excessive fishing reduce ecological resilience and increase vulnerability to climate hazards. In contrast, actions such as mangrove restoration, ecosystem-based adaptation, and sustainable coastal management can strengthen natural protection and reduce disaster risk. The way coastal ecosystems are used and managed therefore strongly influences how severely climate disasters affect coastal societies. Understanding the links between coastal ecologies and climate disasters is essential for developing effective responses to climate change in coastal regions. Coastal ecosystems should be seen not only as areas at risk but also as key contributors to resilience. Combining ecosystem conservation with disaster risk reduction can help coastal communities lower their vulnerability, recover more effectively, and build more sustainable and climate-resilient futures (Malakar et al., 2023; Roukounis & Tsihrintzis, 2022).

1.3 Coastal Hazards, Disasters, and Risks

1.3.1 Types and Factors of Coastal Hazards

Coastal hazards refer to natural and climate-related processes that pose risks to human life, property, ecosystems, and economic activities along coastal regions. These hazards emerge from the ongoing interaction of oceanic, atmospheric, geological, and ecological forces that shape the zone where land meets the sea (Fig. 1.3). With the progression of climate change, coastal hazards are becoming more frequent, more intense, and less predictable. Many coastlines are therefore turning into areas of heightened risk. A coastal hazard should be seen not only as a physical event but as a condition of potential harm that arises when environmental forces affect exposed and vulnerable coastal populations and ecosystems (Panda et al., 2026).

Coastal hazards can be grouped into several major categories based on the processes involved. Hydrodynamic hazards include storm surges, high waves, coastal flooding, and tidal inundation, often linked to cyclones and extreme weather events. Geomorphological hazards involve slower processes such as coastal erosion, shoreline retreat, and land subsidence, which gradually change the shape and stability of coastlines. Oceanic and climatic hazards include sea-level rise, warming ocean waters, and changes in ocean circulation, all of which increase the risk of flooding and habitat loss. Biological and ecological hazards also play an important role. The degradation of coral reefs, mangroves, and coastal wetlands weakens natural protective systems and increases the impact of physical hazards. A range of interconnected factors drives and intensifies coastal hazards. Climate change plays a central role by raising sea levels, strengthening storms, and altering rainfall and wind patterns. These changes increase wave energy, flooding frequency, and rates of shoreline erosion. Geological conditions also influence how coastlines respond to environmental stress. Factors such as sediment supply, coastal slope, and tectonic activity affect coastal stability. Low-lying deltas, sandy beaches, and estuarine zones are generally more vulnerable to erosion and inundation than rocky coastlines. Human activities strongly shape

Fig. 1.3 Types of coastal hazards (Prepared by the authors)

the severity of coastal hazards. Rapid and poorly planned urban development, coastal land reclamation, sand mining, deforestation, and the removal of wetlands and mangroves disrupt natural sediment movement and eliminate ecological buffers. As a result, coastlines become more fragile and less able to withstand oceanic and climatic forces. Social conditions further increase risk. High population density, poverty, and weak systems of governance often lead to the concentration of people and infrastructure in hazardous areas without sufficient protection or planning (Finkl, 2013; Rauf et al., 2026).

Therefore, coastal hazards reflect the combined influence of climate change, ecological conditions, and human actions. They are not simply natural threats but outcomes of long-term environmental change and development choices. This perspective highlights the need for integrated approaches that reduce risk, protect coastal ecosystems, and strengthen the resilience of communities living along the world's coastlines.

1.3.2 Types and Factors of Coastal Disasters

Coastal disasters refer to large-scale damaging events that occur in coastal areas when natural hazards interact with vulnerable human and ecological systems. Hazards such as storms, flooding, erosion, and sea-level rise are part of natural coastal processes, but they become disasters when communities, infrastructure, and ecosystems are unable to cope with or recover from their impacts (Fig. 1.4). A coastal disaster is therefore defined not only by the intensity of a climatic or ocean-related event, but by the extent of damage, displacement, ecological loss, and social and economic disruption it causes. With climate change increasing the frequency and strength of extreme events, coastal disasters are becoming more common, particularly in low-lying and densely populated coastal regions. Different forms of coastal disasters emerge depending on the dominant processes involved. Storm-related disasters include cyclones, hurricanes, storm surges, and extreme waves that lead to flooding, damage to buildings and infrastructure, and loss of life. Flooding and inundation disasters occur when high tides, heavy rainfall, and rising sea levels submerge coastal land, contaminate freshwater sources, and destroy crops and settlements. Erosion and land-loss disasters develop gradually as shorelines retreat, beaches disappear, and communities are forced to relocate over time. Ecological disasters also play a significant role. The degradation of coral reefs, mangroves, and wetlands reduces coastal protection, weakens fisheries, and leads to biodiversity loss, often creating long-term social and economic difficulties (Hoffmann, 2015; Hoffmann & Reicherter, 2016).

Fig. 1.4 Types of coastal disasters (Prepared by the authors)

A combination of natural and human-driven factors influences the occurrence and severity of coastal disasters. Climate change is a key factor, as warmer ocean temperatures strengthen storms and rising sea levels extend the reach of flooding and wave action. Changes in rainfall patterns and river flows affect the supply of sediments to coastal areas, influencing erosion and the stability of deltas. Geological conditions such as coastal slope, soil characteristics, and tectonic activity also shape how different coastlines respond to environmental stress. Human activities often intensify these natural pressures. Rapid and poorly planned urban expansion, the removal of mangroves and sand dunes, sand mining, and the conversion of wetlands for agriculture or industrial use weaken natural defenses and increase exposure to hazards. Social and economic conditions further shape vulnerability. Poverty, high population density, limited access to early warning systems, and weak disaster preparedness reduce the ability of coastal communities to respond effectively. As a result, similar climatic events can lead to very different outcomes depending on local ecological health and social capacity (Islam et al., 2025; Maharani et al., 2025). Coastal disasters reflect the interaction of climate processes, coastal ecosystems, and human societies. They emerge from long-term environmental change combined with patterns of development and resource use. This perspective points to the need for integrated approaches that reduce risk, protect coastal ecosystems, and strengthen the social capacity of communities living along vulnerable coastlines.

1.3.3 Types and Factors of Coastal Risks

Coastal risk refers to the likelihood and possible severity of harm that may occur when coastal hazards interact with exposed populations, infrastructure, and ecosystems. Unlike a disaster, which describes damage that has already taken place, risk focuses on future possibilities and potential outcomes. In coastal areas, risk includes the chance of flooding, erosion, storm damage, or ecosystem failure, as well as the extent to which human settlements, livelihoods, and natural systems may be affected. Coastal risk is therefore not determined by environmental forces alone but is shaped by social, economic, and ecological conditions. Several types of coastal risk can be identified based on the nature of exposure and potential loss (Fig. 1.5). Physical risk involves damage to land, housing, infrastructure, and coastal ecosystems caused by storms, waves, erosion, and rising sea levels. Economic risk refers to losses affecting fisheries, tourism, agriculture, ports, and other coastal activities that rely on stable environmental conditions. Social risk relates to threats to human life, health, housing security, and cultural heritage, with poor and marginalized communities often facing the greatest danger. Ecological risk includes the degradation or loss of mangroves, coral reefs, wetlands, and marine biodiversity that support coastal productivity and natural protection (Balica et al., 2012; Herreros-Cantis et al., 2020).

The level of coastal risk in any region is influenced by a combination of environmental and physical factors. Climate-related drivers such as sea-level rise, ocean warming, changing storm patterns, and shifting rainfall regimes increase the frequency and intensity of coastal hazards. Physical characteristics including coastal elevation, shoreline shape, sediment supply, and the presence of protective ecosystems affect how strongly these hazards are experienced. Low-lying deltas, sandy coastlines, and small islands are generally more exposed to flooding and erosion than elevated or rocky coasts. Human and institutional factors often play a

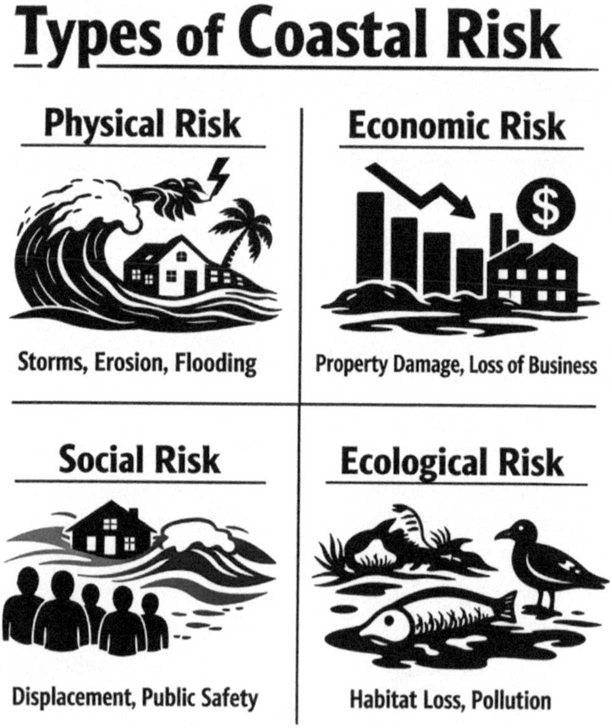

Fig. 1.5 Types of coastal risk (Prepared by the authors)

decisive role in shaping coastal risk. Rapid urban growth, poorly regulated construction, and the removal of mangroves, dunes, and wetlands reduce natural protection and increase exposure to hazards. Social and economic conditions such as poverty, limited access to information, weak governance, and inadequate disaster preparedness increase vulnerability and reduce the capacity of communities to adapt or recover. In such contexts, even moderate climatic events can result in serious impacts. A focus on coastal risk highlights the importance of planning and management that address vulnerability before disasters occur. Reducing risk requires attention to environmental protection, social equity, and long-term development choices, rather than relying solely on emergency response. This approach is especially important as climate change continues to intensify coastal hazards and place growing pressure on coastal societies and ecosystems (Hemmati et al., 2021; O'Hare & White, 2018).

1.4 Impacts of Climate Disasters on Coastal Ecologies

Climate disasters are among the strongest forces shaping coastal ecologies in the present period. Coastal environments lie at the meeting point of land, sea, and atmosphere, which makes them naturally dynamic but also highly sensitive to climatic stress. Events such as cyclones, storm surges, floods, marine heatwaves, droughts, and rising sea levels often occur together and interact with existing environmental pressures. These combined effects produce deep and sometimes lasting changes in coastal ecosystems (Fig. 1.6). As climate change increases the frequency and intensity of such events, many coastal regions are experiencing levels of disturbance that threaten both ecological diversity and human dependence on coastal resources. One of the most visible effects of climate disasters is the alteration of coastal landscapes. Storm surges and powerful waves erode beaches, breach barrier islands, and reshape shorelines. Floods can either deposit large amounts of sediment or wash it away from estuaries and deltas. These physical changes disturb habitats that depend on specific conditions of water depth, salinity, and sediment stability.

Mangroves, salt marshes, and tidal flats are especially vulnerable because they rely on relatively stable sediment and water conditions. When these foundational habitats are damaged or lost, the wider coastal ecosystem is affected, as many species depend on them for shelter, food, and breeding. Marine ecosystems are also highly sensitive to climate-related disturbances. Rising sea temperatures and marine heatwaves cause coral bleaching, weaken reef structures, and affect plankton and fish populations. Coral reefs support high levels of biodiversity and help reduce wave energy along coastlines. Damage to these reefs lowers their ecological productivity and reduces their ability to protect coastal areas. Changes in ocean temperature, currents, and oxygen levels can also alter fish migration patterns, disrupt spawning cycles, and reduce the reliability of fisheries that support both ecosystems and coastal livelihoods. Wetlands and estuaries experience a particular mix of pressures from climate disasters. Floods and storm surges raise salinity levels and introduce pollutants into estuarine systems, while droughts reduce freshwater inflow and concentrate salts and nutrients. These changes affect plant growth, soil conditions, and the survival of fish and invertebrates. Repeated exposure to such stress can gradually transform diverse wetlands into degraded ecosystems with limited ecological functions. As wetlands decline, their ability to store carbon, filter pollutants, and reduce flood impacts also decreases, which increases vulnerability to future disasters (Banik & Uddin, 2025).

The effects of climate disasters extend beyond immediate physical damage and drive long-term ecological change. Species that cannot adapt to altered temperatures, salinity levels, or disturbance patterns may decline or disappear. At the same time, more tolerant or invasive species may spread, leading to changes in species composition and food webs. Such ecological shifts often reduce biodiversity and are difficult to reverse. For example, the loss of reef-building corals can lead to the dominance of algae, changing the structure and function of reef ecosystems. Coastal ecosystems also play an important role in climate regulation through carbon storage in mangroves, seagrasses, and salt marshes. Damage to these ecosystems during climate disasters releases stored carbon into the atmosphere, adding to global warming. In this way, the degradation of coastal ecologies not only results from

Fig. 1.6 Major impacts of climate disasters on coastal ecologies (Prepared by the authors)

climate change but also contributes to it, reinforcing cycles of environmental stress. The social consequences of these ecological impacts are far-reaching. Coastal communities rely on healthy ecosystems for food, income, cultural practices, and protection from hazards. When fisheries decline, shorelines erode, or wetlands and reefs are destroyed, food security is threatened and livelihoods are disrupted. Indigenous groups and small-scale fishing communities are often among the most affected, as they depend closely on local ecosystems and have limited resources to adapt. Ecological damage and social vulnerability therefore become closely linked in coastal regions.

Climate disasters affect coastal ecologies in several key ways, including habitat destruction, loss of biodiversity, coral bleaching, increased erosion and land loss, changes in water quality and salinity, disruption of food webs and fisheries, decline in ecosystem services, release of stored carbon, and increased vulnerability of coastal communities. These impacts often accumulate over time, pushing ecosystems toward conditions where recovery becomes increasingly difficult. In some cases, repeated disturbances lead to major ecological shifts, such as the conversion of coral reefs to algae-dominated systems or vegetated wetlands to open water. These changes usually result in lower productivity and reduced resilience to future stress. An integrated view of climate disasters and coastal ecologies shows that environmental change, ecological processes, and human systems are closely connected. Coastal ecosystems are not passive elements within climate change but active components whose condition influences both vulnerability and resilience. Protecting and restoring these ecosystems supports disaster risk reduction, climate adaptation, and long-term sustainability. Future outcomes for coastal ecologies will depend largely on societal responses to climate stress. Actions such as mangrove restoration, reef conservation, sustainable fisheries management, and ecosystem-based coastal planning can strengthen resilience and support recovery. At the same time, limiting greenhouse gas emissions remains essential to reduce the scale of future damage. Acknowledging the central role of coastal ecologies in both climate impacts and climate solutions is an important step toward more secure and sustainable coastal futures (Das & Swain, 2024; Huang & Cheng, 2013).

1.5 Coastal Risk Reduction Practices: Society and Sustainability

Coastal risk reduction practices refer to the strategies used to limit harm from climate-related hazards such as storm surges, flooding, coastal erosion, and sea-level rise. Within the context of climate disasters and coastal ecologies, these practices focus on reducing risk before disasters occur rather than responding only after damage has been done. This approach recognizes that vulnerability is shaped not only by exposure to hazards but also by social conditions, governance structures, and the health of coastal ecosystems (Fig. 1.7). Reducing risk therefore requires attention to social preparedness, environmental protection, and long-term planning alongside technical measures. Social systems play a central role in coastal risk reduction. Strong institutions, clear policies, and effective coordination among authorities help reduce loss of life and property during extreme events. Measures such as early warning systems, evacuation planning, disaster education, and climate-resilient infrastructure are essential for protecting coastal populations. Community participation is equally important. Local knowledge, social networks, and collective action allow people to respond quickly to emergencies and adapt to changing risks. Because poorer and marginalized groups often face greater exposure to hazards, equity and social inclusion are critical for sustainable risk reduction (Maccarrone & Akhir, 2024; Uitto & Shaw, 2016).

Sustainability brings an ecological focus to coastal risk reduction by highlighting the protective role of natural systems. Ecosystem-based approaches such as mangrove restoration, wetland conservation, dune stabilization, and coral reef protection help reduce wave energy, slow erosion, and limit flooding. These natural systems also support biodiversity, fisheries, and coastal livelihoods. In many cases, nature-based solutions are more flexible and cost-effective than large engineered structures, especially under conditions of long-term climate change. Working with natural processes strengthens both ecological stability and human safety. Planning and development practices are most effective when social and ecological concerns are addressed together. Land-use regulations, climate-resilient housing, and the protection of critical coastal

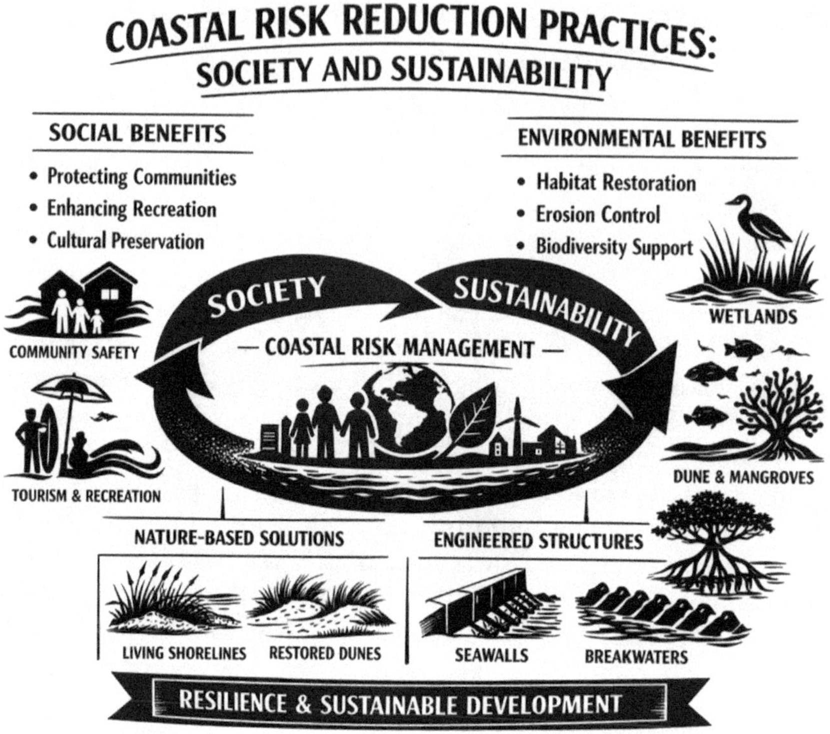

Fig. 1.7 Coastal risk reduction practices: Society and sustainability (Prepared by the authors)

ecosystems can reduce the concentration of people and infrastructure in high-risk areas. Promoting sustainable livelihoods such as responsible fishing, eco-tourism, and climate-adaptive agriculture reduces pressure on fragile ecosystems and improves economic security. These measures not only reduce disaster risk but also support long-term social and environmental well-being. Coastal risk reduction depends on the close connection between social resilience and ecological sustainability. Protecting communities from climate hazards requires the protection of the ecosystems that buffer coastlines and support livelihoods. Practices that combine community preparedness, inclusive governance, and ecosystem conservation provide a practical pathway toward safer and more adaptable coastal regions. As climate change continues to intensify coastal hazards, such integrated approaches are essential for building sustainable coastal futures (Almutairi et al., 2020; Durap & Balas, 2024).

1.6 Conclusion and Arguments

The discussion across earlier chapters shows that climate change, coastal ecologies, and human vulnerability are closely linked and must be considered together. Climate disasters are no longer rare or exceptional in coastal regions. Rising sea levels, warming oceans, extreme weather events, and long-term environmental degradation have made climatic stress a regular part of coastal life. These changes place growing pressure on both ecosystems and human communities, especially in low-lying and densely populated coastal areas. Coastal ecosystems such as mangroves, coral reefs, wetlands, and estuaries play a vital role in reducing the impacts of climate-related hazards. They help absorb wave energy, reduce flooding, stabilize shorelines, and support livelihoods and biodiversity. At the same time, these ecosystems are increasingly stressed by climate change and human activities such as deforestation, pollution, and unplanned development. As these natural systems weaken, coastal hazards intensify and risks increase. Environmental disturbance then extends beyond nature, creating social and economic problems that affect food security, housing, and livelihoods. Viewing coastal risk as the interaction of hazards, exposure, and vulnerability shifts attention away from short-term emergency response toward long-term resilience. This perspective highlights the importance of development choices, ecosystem health, and social conditions in shaping how climate impacts are experienced. Sustainable development, ecological protection, and social equity are closely connected, and progress in one area supports progress in the others.

Coastal risk reduction practices that combine community participation, ecosystem-based approaches, and climate-sensitive planning offer practical ways to reduce future losses. Community involvement strengthens preparedness and local response, while healthy ecosystems provide natural protection and support recovery. Planning that respects the changing nature of coastal environments helps limit settlement and infrastructure growth in high-risk areas. The long-term future of coastal regions depends largely on how societies value and manage their coastal ecologies. Protecting these environments is not only an environmental concern but also a foundation for resilience and sustainable living. Integrating ecological conservation into development and risk reduction strategies offers a pathway toward safer, more secure, and more resilient coastal societies under changing climate conditions. The chapter discussions lead to the following arguments:

(a) Climate disasters emerge from the interaction between natural climatic processes and human-driven environmental change. They should therefore be understood as socio-ecological phenomena shaped by development choices, land use practices, and governance systems, rather than as purely natural events.
(b) Coastal ecosystems such as mangroves, coral reefs, wetlands, and dunes function as critical natural buffers against climate hazards. When these ecosystems are degraded, their protective capacity declines, directly increasing exposure to storms, flooding, erosion, and other climate-related threats.
(c) Coastal hazards and disasters rarely occur in isolation. Storms, flooding, erosion, and sea-level rise interact with one another over time, producing cumulative impacts that affect ecosystems, infrastructure, and human livelihoods simultaneously.
(d) Coastal risk is influenced not only by physical exposure to hazards but also by social, economic, and institutional conditions. Poverty, inequality, weak governance, and limited access to information or resources play a major role in shaping vulnerability and disaster outcomes.
(e) Climate disasters and ecological degradation reinforce each other in a cyclical manner. Extreme events damage coastal ecosystems, while the loss of these ecosystems removes natural protection and increases the severity of future climate risks.
(f) Risk reduction practices that are based on ecosystems and natural processes provide more flexible and long-term protection than

approaches that rely solely on engineered structures. Such practices adapt better to changing climatic conditions while supporting biodiversity and livelihoods.

(g) Community participation and social equity are essential for effective and fair coastal risk management. Inclusive decision-making, recognition of local knowledge, and attention to vulnerable groups strengthen preparedness, response, and recovery.

References

Abbass, K., Qasim, M. Z., Song, H., et al. (2022). A review of the global climate change impacts, adaptation, and sustainable mitigation measures. *Environmental Science and Pollution Research, 29*, 42539–42559. https://doi.org/10.1007/s11356-022-19718-6

Almutairi, A., Mourshed, M., & Ameen, R. F. M. (2020). Coastal community resilience frameworks for disaster risk management. *Natural Hazards, 101*, 595–630. https://doi.org/10.1007/s11069-020-03875-3

Balica, S. F., Wright, N. G., & van der Meulen, F. (2012). A flood vulnerability index for coastal cities and its use in assessing climate change impacts. *Natural Hazards, 64*, 73–105. https://doi.org/10.1007/s11069-012-0234-1

Banik, M., & Uddin, M. (2025). Environmental degradation and human rights challenges for coastal populations in Bangladesh. *Discover Global Society, 3*, 137. https://doi.org/10.1007/s44282-025-00268-4

Das, A., & Swain, P. K. (2024). Navigating the sea level rise: Exploring the interplay of climate change, sea level rise, and coastal communities in India. *Environmental Monitoring and Assessment, 196*, 1010. https://doi.org/10.1007/s10661-024-13191-z

Durap, A., & Balas, C. E. (2024). Towards sustainable coastal management: A hybrid model for vulnerability and risk assessment. *Journal of Coastal Conservation, 28*, 66. https://doi.org/10.1007/s11852-024-01065-y

Dutta, S., Saha, P., Satapathy, S., Patel, S., & Debnath, B. (2026). Climate change induced natural hazards of agriculture in coastal zones. In K. K. Basheer Ahammed, A. C. Pandey, B. R. Parida, & I. W. G. A. Karang (Eds.), *Monitoring coastal marine ecosystems under climate change*. Springer. https://doi.org/10.1007/978-3-032-09137-6_13

Finkl, C. W. (2013). *Coastal Hazards*. Springer. https://doi.org/10.1007/978-94-007-5234-4.

Hemmati, M., Mahmoud, H. N., Ellingwood, B. R., et al. (2021). Unraveling the complexity of human behavior and urbanization on community vulnerability to floods. *Scientific Reports, 11*, 20085. https://doi.org/10.1038/s41598-021-99587-0

Herreros-Cantis, P., Olivotto, V., Grabowski, Z. J., et al. (2020). Shifting landscapes of coastal flood risk: Environmental (in)justice of urban change, sea level rise, and differential vulnerability in new York City. *Urban Transformations, 2*, 9. https://doi.org/10.1186/s42854-020-00014-w

Hoffmann, G. (2015). Coastal Disasters. In *Encyclopaedia of marine geosciences* (pp. 1–10). Springer-Verlag.

Hoffmann, G., & Reicherter, K. (2016). Geohazards: Coastal disasters. In J. Harff, M. Meschede, S. Petersen, & J. Thiede (Eds.), *Encyclopedia of marine geosciences. Encyclopedia of earth sciences series*. Springer. https://doi.org/10.1007/978-94-007-6238-1_169

Huang, Y., & Cheng, H. (2013). The impact of climate change on coastal geological disasters in southeastern China. *Natural Hazards, 65*, 377–390. https://doi.org/10.1007/s11069-012-0370-7

Irshad, S. M. (2024). *Economics of disasters and climate change: Risk and uncertainties*. Palgrave Macmillan Singapore. https://doi.org/10.1007/978-981-97-9415-7

Islam, M. K., Jahan, I., & Azad, A. (2025). Spatial assessment of coastal erosion vulnerability: Natural and anthropogenic factors in Chattogram, Bangladesh. *Natural Hazards Research, 6*(1), 50–67. https://doi.org/10.1016/j.nhres.2025.05.001

Ismail-Zadeh, A. (2022). Natural hazards and climate change are not drivers of disasters. *Natural Hazards, 111*, 2147–2154. https://doi.org/10.1007/s11069-021-05100-1

Jana, N. C., & Singh, R. B. (2022). *Climate, environment and disaster in developing countries*. Springer Singapore. https://doi.org/10.1007/978-981-16-6966-8.

Kim, K. G. (2024). Climate disaster risk management—Risk assessment and green recovery planning: Core concepts and identification of current issues. In K. G. Kim & C. Atkin (Eds.), *Cross-cultural perspectives on climate change adaptation. Cities and nature*. Springer. https://doi.org/10.1007/978-3-031-50365-8_2

Maccarrone, V., & Akhir, M. F. (Eds.). (2024). *Coastal sustainability: Insights from Southeast Asia and beyond*. Springer Cham. https://doi.org/10.1007/978-3-031-75749-5

Maharani, A., Sujarwoto, S., Susanti, H., et al. (2025). Association between climate related hazards and depression among coastal communities in Indonesia. *Scientific Reports, 15*, 6998. https://doi.org/10.1038/s41598-025-89298-1

Malakar, K. D., Kumar, M., Anand, S., & Kuzur, G. (2023). *Climate change and socio-ecological transformation: Vulnerability and sustainability*. Springer Singapore. https://doi.org/10.1007/978-981-99-4390-6

Mohapatra, M. (2024). Climate and weather forewarning Systems for Disaster Preparedness and Response. In A. K. Gupta, A. Gupta, & P. Acharya (Eds.), *Disaster risk and management under climate change. Disaster resilience and green growth*. Springer. https://doi.org/10.1007/978-981-99-4105-6_3

O'Hare, P., & White, I. (2018). Beyond 'just' flood risk management: The potential for—And limits to—Alleviating flood disadvantage. *Regional Environmental Change, 18*, 385–396. https://doi.org/10.1007/s10113-017-1216-3

Ohlendorf, L. M. (2019). *Drivers of climate change in urban India: Social values, lifestyles, and consumer dynamics in an emerging megacity*. Springer Cham. https://doi.org/10.1007/978-3-319-96670-0

Panda, J., Tom, G., Mukherjee, A., Bhasi, I., & Roy, D. (2026). Natural hazards in the coastal zones of India, associated vulnerability and risk assessment. In K. K. Basheer Ahammed, A. C. Pandey, B. R. Parida, & I. W. G. A. Karang (Eds.), *Monitoring coastal marine ecosystems under climate change*. Springer. https://doi.org/10.1007/978-3-032-09137-6_12

Rauf, A. U., Astuti, R. D. P., & Sarinah Basri, K. (2026). Natural hazards in coastal zones. In K. K. Basheer Ahammed, A. C. Pandey, B. R. Parida, & I. W. G. A. Karang (Eds.), *Monitoring coastal marine ecosystems under climate change*. Springer. https://doi.org/10.1007/978-3-032-09137-6_9

Roukounis, C. N., & Tsihrintzis, V. A. (2022). Indices of coastal vulnerability to climate change: A review. *Environmental Processes, 9*, 29. https://doi.org/10.1007/s40710-022-00577-9

Sandifer, P. A., Collier, T. K., & Trtanj, J. M. (2012). Coastal ecosystems, sustainable management. In R. A. Meyers (Ed.), *Encyclopedia of sustainability science and technology*. Springer. https://doi.org/10.1007/978-1-4419-0851-3_884

Schmidt, S., Nüsser, M., Baghel, R., et al. (2020). Cryosphere hazards in Ladakh: The 2014 Gya glacial lake outburst flood and its implications for risk assessment. *Natural Hazards, 104*, 2071–2095. https://doi.org/10.1007/s11069-020-04262-8

Siddiqui, A. R., & Sahay, A. (Eds.) (2022). *Climate change, disaster and adaptations: Contextualising human responses to ecological change*. Springer Cham. https://doi.org/10.1007/978-3-030-91010-5.

Singh, A., Fernando, R. L. S., & Haran, N. P. (2020). *Development in coastal zones and disaster management*. Palgrave Macmillan Singapore. https://doi.org/10.1007/978-981-15-4294-7

Tschumi, E., & Zscheischler, J. (2020). Countrywide climate features during recorded climate-related disasters. *Climatic Change, 158*, 593–609. https://doi.org/10.1007/s10584-019-02556-w

Uitto, J. I., & Shaw, R. (2016). *Sustainable development and disaster risk reduction*. Springer Tokyo. https://doi.org/10.1007/978-4-431-55078-5

Yildiz, A., & Shaw, R. (2024). Overview of the concept: Disaster and climate risk education. In A. Yildiz & R. Shaw (Eds.), *Disaster and climate risk education. Disaster risk reduction*. Springer. https://doi.org/10.1007/978-981-97-5987-3_1

2

Climate Disasters, Socio-Ecological Vulnerability, and Coastal Ecologies: Assessment and Mapping

This chapter introduces climate disasters as part of the wider social and ecological processes that shape coastal environments and the lives of people who depend on them. As climate change intensifies cyclones, sea-level rise, flooding, and coastal erosion, it becomes clear that disaster risk is not shaped by physical hazards alone. Risk grows from the close links between fragile ecosystems, social inequalities, and the ability of institutions to respond and adapt. Coastal regions, where livelihoods, settlements, and natural systems are closely connected, provide an important setting to examine these relationships. The chapter moves beyond a narrow focus on hazards and adopts a socio-ecological vulnerability perspective to explain how climate risks are created and unevenly experienced across coastal areas. It highlights how degraded ecosystems such as mangroves, wetlands, and coral reefs can increase disaster impacts, while social and governance conditions influence who faces the greatest losses and who recovers more slowly. By connecting vulnerability concepts with spatial assessment and sustainability concerns, the chapter lays the foundation for identifying high-risk coastal zones and supporting more inclusive and ecosystem-based approaches to resilience. It sets the stage

K. D. Malakar et al., *Coastal Ecologies at Risk*, Palgrave Studies in Environmental Sustainability, https://doi.org/10.1007/978-3-032-26081-9_2

for understanding climate disasters within the broader goal of sustainable coastal development.

Key Points of the Chapter

- Describes climate risk as the outcome of hazards, ecosystem degradation, and social inequality.
- Explains socio-ecological vulnerability through exposure, sensitivity, and adaptive capacity.
- Highlights how degraded mangroves and reefs intensify disaster impacts.
- Examines vulnerability hotspots using spatial mapping of ecological and social stress.
- Emphasizes the need for ecosystem-based adaptation and inclusive governance to build coastal resilience.

2.1 Socio-Ecological Vulnerability

2.1.1 Socio-Ecological Vulnerability Assessment: A Brief Review

Socio-ecological vulnerability assessment focuses on how closely linked human and environmental systems experience and respond to climate-related hazards. In coastal areas, people, livelihoods, ecosystems, and physical landscapes are deeply connected, and these links shape the impacts of climate disasters. Vulnerability cannot be explained only by physical conditions such as flooding or shoreline erosion, nor only by social conditions such as poverty or weak institutions. It develops through the interaction of fragile ecosystems, social inequality, and the ability of institutions to manage risk. Vulnerability therefore changes over time as social and environmental conditions evolve. Vulnerability is commonly described through three related components: exposure, sensitivity, and adaptive capacity (Fig. 2.1). Exposure refers to how often and how strongly coastal areas face hazards such as cyclones, storm surges, sea-level rise, and salinity intrusion. Sensitivity reflects how much communities and ecosystems are affected when hazards occur, shaped by factors such as ecosystem condition, dependence on natural resources, housing quality, and population characteristics. Adaptive capacity describes the ability of people, institutions, and ecosystems to prepare for, cope with, and recover from climate stresses, as well as to adjust to future changes. These components help explain why similar hazards can lead to very different outcomes across coastal regions. A key feature of socio-ecological vulnerability assessment is its focus on the interaction between social and ecological systems. The loss of mangroves, for example, increases exposure to storm damage, while poverty and insecure livelihoods limit the ability of households to reduce risk. At the same time, healthy ecosystems and effective institutions can reduce impacts and support recovery. This approach moves beyond listing risk factors and instead examines the processes that shape vulnerability over time. Vulnerability also varies across space and scale. Conditions that appear stable at regional levels may hide serious risks at the community or household level, particularly for marginalized groups. Different coastal ecosystems also respond in different ways to climatic and human pressures. Socio-ecological vulnerability assessment therefore provides a strong foundation for linking climate hazard analysis with adaptation planning and for supporting fair and sustainable responses to coastal climate risk (Jha et al., 2021; Kumar et al., 2024).

2.1.2 Assessment of Socio-Ecological Vulnerability: Purpose and Importance

Socio-ecological vulnerability assessment provides a strong foundation for examining how climate disasters affect connected human and environmental systems in coastal regions. The purpose of this approach is to move beyond narrow views that focus only on hazards or individual sectors and to present a broader picture of risk shaped by ecological conditions, social structures, and institutional capacity. In coastal areas, where livelihoods, settlement patterns, and ecosystem services are closely linked, this assessment helps explain why some communities and landscapes face greater losses even when exposed to similar climate events.

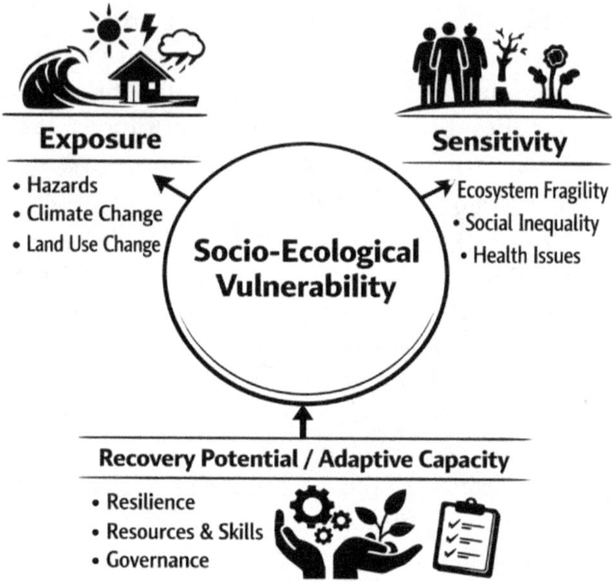

Fig. 2.1 Concept of socio-ecological vulnerability (Prepared by the authors)

An important aim of socio-ecological vulnerability assessment is to identify uneven patterns of risk across coastal spaces. Climate hazards such as cyclones, storm surges, flooding, and saline intrusion do not affect all populations in the same way. By combining social factors such as income, land ownership, education, gender, and livelihood dependence with ecological factors like mangrove cover, shoreline condition, and water quality, vulnerability assessment reveals how risk is distributed unevenly across people and places. This distinction helps separate simple exposure from deeper vulnerability caused by ecological degradation and limited capacity to respond (Lalyer et al., 2020).

Socio-ecological vulnerability assessment also plays a key role in supporting adaptation planning and disaster risk reduction. It helps guide decisions on where to invest resources and how to design responses that reflect local conditions. Vulnerability mapping can inform ecosystem restoration efforts, the placement of protective infrastructure, and targeted livelihood support. In this way, scientific analysis becomes more closely linked to practical action. This approach also supports long-term sustainability by examining how development choices, governance practices, and environmental change influence future climate risk. Activities such as wetland loss, unplanned urban growth, and poorly regulated coastal development can increase sensitivity to climate stress. By drawing attention to these processes, vulnerability assessment encourages development pathways that reduce risk and strengthen resilience over time. Socio-ecological vulnerability assessment connects climate risk with social equity and environmental sustainability. By linking ecosystem health with social conditions, it supports more balanced and inclusive responses to climate change in coastal regions (Kok et al., 2016; Sun et al., 2025).

Socio-ecological vulnerability assessment is important because it explains how climate disasters affect both coastal communities and the ecosystems that support them. In coastal regions, livelihoods, settlement patterns, and environmental processes are closely linked, so damage to one part often increases pressure on others. This approach treats vulnerability as a shared condition of social and ecological systems, which helps explain why the impacts of climate disasters and the pace of recovery vary across places and social groups. This assessment is especially useful for revealing uneven and often overlooked patterns of climate risk. Coastal populations do not face the same level of vulnerability. Social factors such as poverty, gender roles, dependence on natural resources, and access to institutions strongly influence how people prepare for and respond to hazards. Ecological conditions, including mangrove loss, shoreline erosion, and declining fisheries, further shape exposure and sensitivity to climate extremes. By linking social and ecological factors, socio-ecological vulnerability assessment helps identify areas where environmental degradation and social disadvantage occur together. Socio-ecological vulnerability assessment also supports planning, adaptation, and disaster risk reduction. It provides a clear basis for setting priorities, whether for ecosystem restoration, infrastructure development, or livelihood support. In a context of rising climate risk and limited resources, this kind of evidence is essential for fair and effective decision-making (Cheng, 2019; Maina et al., 2016; Malakar et al., 2023). The approach also contributes to long-term sustainability in coastal regions.

By connecting climate hazards with development choices and environmental management, it shows how present actions influence future vulnerability. In this way, socio-ecological vulnerability assessment helps guide coastal development toward resilience, ecological health, and social equity.

2.2 Conceptual Frameworks for Socio-Ecological Vulnerability: Models and Approaches

Conceptual frameworks for socio-ecological vulnerability assessment help explain how climate hazards interact with ecosystems and human systems in coastal regions. In coastal ecologies, vulnerability is shaped not only by exposure to storms, floods, or sea-level rise but also by ecosystem degradation, social inequality, dependence on natural resources, and the ability of institutions to respond. These frameworks organize complex relationships into clear structures that support assessment, spatial analysis, and policy planning. The exposure, sensitivity, and adaptive capacity framework is widely used to describe vulnerability in coastal systems. Exposure refers to the degree to which an area faces climate hazards, such as its location along cyclone-prone coastlines. Sensitivity reflects how strongly communities and ecosystems are affected, influenced by factors like reliance on fisheries or the condition of mangroves. Adaptive capacity describes the ability of people, institutions, and ecosystems to prepare for, respond to, and recover from climate stress. This framework is useful because it combines social and ecological factors in a single approach (Behnassi et al., 2021; Khromova et al., 2025; Zango-Palau et al., 2024).

The pressure, state, and response framework and its expanded form, DPSIR, focus on how human activities influence environmental conditions over time. In coastal regions, development, population growth, and tourism place pressure on ecosystems through activities such as wetland conversion or overfishing. These pressures change ecosystem conditions and lead to impacts like increased erosion or reduced protection from storms, which then raise social vulnerability. This framework is especially useful for linking vulnerability assessment with environmental management and policy decisions. The sustainable livelihoods framework offers insight into how vulnerability differs within coastal communities. It focuses on the resources people rely on to secure their livelihoods, including natural resources, housing and infrastructure, skills and health, social networks, and financial resources. These factors shape how households experience and respond to climate hazards and help explain why similar events produce different outcomes. The social and ecological systems framework highlights interactions and feedbacks across different scales. Coastal vulnerability is influenced not only by local conditions but also by wider governance systems, markets, and climate processes. This perspective supports the study of long-term change and resilience in rapidly changing coastal environments (Behnassi et al., 2021). Conceptual frameworks are often combined in applied vulnerability assessments. Using multiple approaches allows physical exposure, ecological sensitivity, and social conditions to be examined together. These frameworks play a central role in turning complex climate and environmental interactions into practical tools that support fair and sustainable responses to climate disasters in coastal regions (Fig. 2.2).

2.2.1 Socio-Ecological Vulnerability Assessment: IPCC Approaches

The Intergovernmental Panel on Climate Change has developed one of the most widely accepted frameworks for understanding vulnerability in the context of climate change. This framework is especially relevant for socio-ecological vulnerability assessment in coastal regions because it connects climate hazards with social conditions and environmental processes. Over successive assessment reports, the IPCC has gradually expanded its view of vulnerability, moving from a focus on physical impacts to a broader understanding of climate risk that includes social capacity, ecological conditions, and resilience within linked human and environmental systems. In earlier assessment reports, vulnerability was

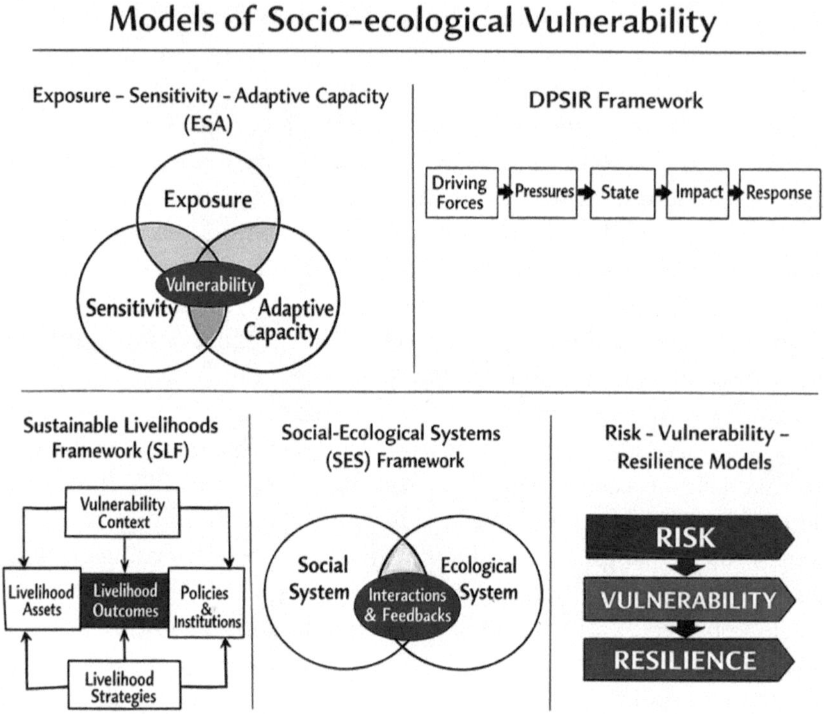

Fig. 2.2 Models of socio-ecological vulnerability (Prepared by the authors)

defined through three main components: exposure, sensitivity, and adaptive capacity. This approach made it possible to assess how coastal areas are exposed to hazards such as cyclones, storm surges, and sea-level rise, how sensitive ecosystems and communities are to these hazards, and how well they can respond and recover. This structure has been widely applied in coastal studies because it allows both environmental indicators, such as shoreline erosion and mangrove cover, and social indicators, such as income levels, infrastructure, and institutional access, to be examined together. In more recent assessment reports, the IPCC reorganized its framework around the concept of climate risk. Risk is understood as the result of the interaction between hazards, exposure, and vulnerability. In this formulation, vulnerability is shaped by sensitivity and adaptive capacity, while exposure refers to the presence of people, livelihoods, ecosystems, and assets in areas prone to climate hazards. This distinction is important for coastal regions because it separates physical location from the social and ecological factors that determine how severe the impacts will be (Das et al., 2020; Gupta et al., 2019; Lyngdoh et al., 2025).

The IPCC framework is valued for its flexibility and consistency across scales. It can be applied at the level of households, communities, or entire coastal regions while maintaining a clear structure. It also provides a shared language that links climate science, disaster risk reduction, ecosystem management, and development planning. In coastal contexts, the framework supports spatial vulnerability mapping by highlighting areas where high exposure overlaps with high sensitivity and low adaptive capacity. This makes it a strong foundation for identifying vulnerability hotspots and guiding targeted adaptation strategies in climate-affected coastal landscapes.

2.2.2 Socio-Ecological Vulnerability Assessment: DPSIR Approaches

The DPSIR framework, which refers to driving forces, pressures, state, impact, and response, offers a clear and policy-oriented way to examine socio-ecological vulnerability in relation to climate disasters. In coastal regions, vulnerability is shaped not only by climate hazards such as cyclones or sea-level rise, but also by long-term social, economic, and environmental processes. This framework is useful because it connects human activities, environmental change, and social outcomes within a single structure. Driving forces describe the broad social, economic, and political trends that influence coastal development and resource use. These include population growth, urban expansion, tourism, industrial activity, market integration, and climate change. In coastal areas, these forces often increase pressure on land, fisheries, and infrastructure, pushing human activities into fragile and hazard-prone environments. These trends create pressures such as mangrove loss, wetland conversion, overfishing, pollution, and shoreline modification, which reduce the natural capacity of ecosystems to protect against climate extremes (Newton & Weichselgartner, 2014; Sun et al., 2025).

The state component reflects the condition of coastal ecosystems and social systems at a given time. This includes the health of mangroves, coral reefs, and estuaries, shoreline stability, water quality, and the condition of housing, infrastructure, and livelihoods. When ecosystems are degraded and social systems are weakened, coastal environments become more fragile and less able to withstand climate hazards. Impacts refer to the consequences of changes in system conditions for ecosystems, livelihoods, and human well-being. In climate disaster settings, these impacts may include flooding, loss of fisheries, damage to homes, livelihood insecurity, displacement, and ecosystem decline. These outcomes show how environmental degradation and social inequality together increase vulnerability. Responses include the actions taken by governments, communities, and institutions to reduce risk and adapt to change. These may involve ecosystem restoration, land-use regulation, disaster preparedness, livelihood support, and climate adaptation policies. By linking long-term drivers with impacts and responses, the DPSIR framework helps identify both existing vulnerabilities and pathways toward more sustainable and climate-resilient coastal planning (Khajuria & Ravindranath, 2012; Wang et al., 2022).

2.2.3 Socio-Ecological Vulnerability Assessment: SES and PSR Approaches

The social ecological systems framework and the pressure state response (PSR) framework are widely used to study vulnerability in systems where human and environmental processes are closely linked (Fig. 2.3). In climate-sensitive coastal regions, livelihoods, ecosystems, and exposure to hazards are strongly connected, making these frameworks especially useful. The Social ecological systems framework focuses on interactions and feedbacks between social and ecological components, while the pressure state response framework offers a clear structure that links human activities with environmental change and societal responses. Together, they provide both analytical depth and practical guidance for socio-ecological vulnerability assessment. The social ecological systems framework views coastal regions as integrated systems made up of natural resources, ecosystems, users, and governance arrangements. Resources such as fish stocks and mangroves are connected to larger systems like estuaries and coastlines, and these are shaped by the actions of users such as fishers, farmers, and coastal residents. Governance systems, including institutions and rules, influence how resources are managed. Vulnerability develops through interactions among these elements, especially under climate stress. Overfishing combined with weak governance, for example, can degrade ecosystems and reduce their ability to protect coastlines from storms, increasing the vulnerability of dependent communities. This framework highlights feedbacks and gradual change, helping explain how slow processes such as sea-level rise can lead to serious social and ecological impacts over time. The pressure state response framework provides a more structured way to organize vulnerability information. It focuses on how pressures such as coastal development, pollution, and climate extremes affect the condition of ecosystems, infrastructure, and social well-being. These changes then lead to responses from governments, institutions, or communities, including policies, ecosystem restoration, and disaster preparedness efforts. In coastal vulnerability studies, this framework helps trace how pressures weaken ecological and social conditions and how responses can either reduce or increase vulnerability. Using the social ecological systems and pressure state response frameworks together strengthens socio-ecological vulnerability assessment. The first explains why vulnerability develops and persists through complex interactions, while the second helps organize this understanding for mapping, monitoring, and policy action. In coastal regions exposed to climate disasters, this combined approach supports more informed decisions and promotes resilience rooted in ecological health and social realities (Bao et al., 2025; Sun et al., 2025; Zhang et al., 2017).

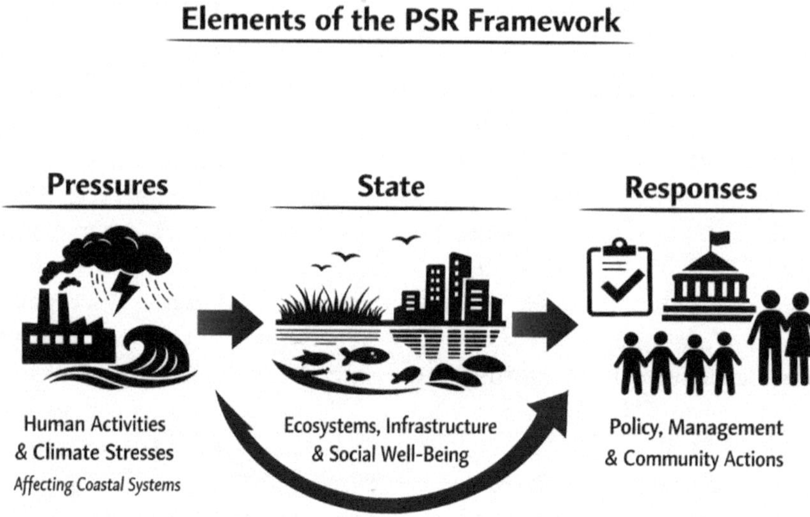

Fig. 2.3 Elements of the PSR framework (Prepared by the authors)

2.3 Indicators of Socio-Ecological Vulnerability: Ecological and Social Subsystems

Socio-ecological vulnerability in coastal regions reflects the combined condition of ecosystems and the social systems that depend on and manage them. Climate disasters such as cyclones, storm surges, flooding, and sea-level rise affect nature and society at the same time. These events act on linked systems where environmental damage increases social risk, and social disadvantage places further stress on ecosystems. Indicators of socio-ecological vulnerability must therefore capture both ecological and social conditions, along with the links between them. This combined view helps explain where vulnerability is high, why it continues, and how it can be reduced in coastal areas.

Ecological indicators describe the sensitivity and protective capacity of coastal environments (Fig. 2.4). Ecosystems such as mangroves, coral reefs, salt marshes, dunes, and estuaries reduce climate risk by stabilizing shorelines, weakening wave energy, and supporting livelihoods. When these ecosystems are degraded, coastal areas become more exposed to hazards and less able to recover. Ecological indicators help assess whether natural systems can still perform these protective and supportive functions. Common ecological indicators include mangrove and wetland cover, rates of coastal erosion, coral reef condition, habitat quality, water and salinity levels, sediment supply, and the frequency of ecosystem disturbances. These indicators show how exposure and sensitivity increase when ecosystems are weakened by pressures such as pollution, overfishing, and land-use change (Berrouet et al., 2018; Gupta et al., 2019; Xu et al., 2024).

Social vulnerability refers to the characteristics of populations, livelihoods, and institutions that shape how people experience climate stress. Coastal livelihoods often depend on fragile ecosystems, while poverty, informal settlements, and limited access to services reduce the ability to cope with hazards. Social vulnerability differs across groups according to income, gender, age, occupation, caste, ethnicity, and location, making detailed social indicators essential. Key social indicators include population density, income levels, dependence on coastal resources, housing quality, education, access to health care and basic services, social networks, governance capacity, and disaster preparedness (Fig. 2.5). These indicators reflect differences in sensitivity and adaptive capacity among households and communities (Malakar et al., 2023; Sebesvari et al., 2016).

The integration of ecological and social indicators is central to socio-ecological vulnerability assessment. In many coastal regions, environmental degradation and social disadvantage occur in the same places, creating areas of high risk. Integrated indicators make it possible to identify and map these patterns, showing how ecological fragility and social marginalization reinforce each other. Combining ecological and social indicators supports more just and sustainable climate adaptation. This approach helps guide actions that restore ecosystems while strengthening livelihoods and institutions. In coastal regions exposed to climate disasters, integrated assessment plays a key role in reducing vulnerability, improving resilience, and supporting long-term socio-ecological sustainability.

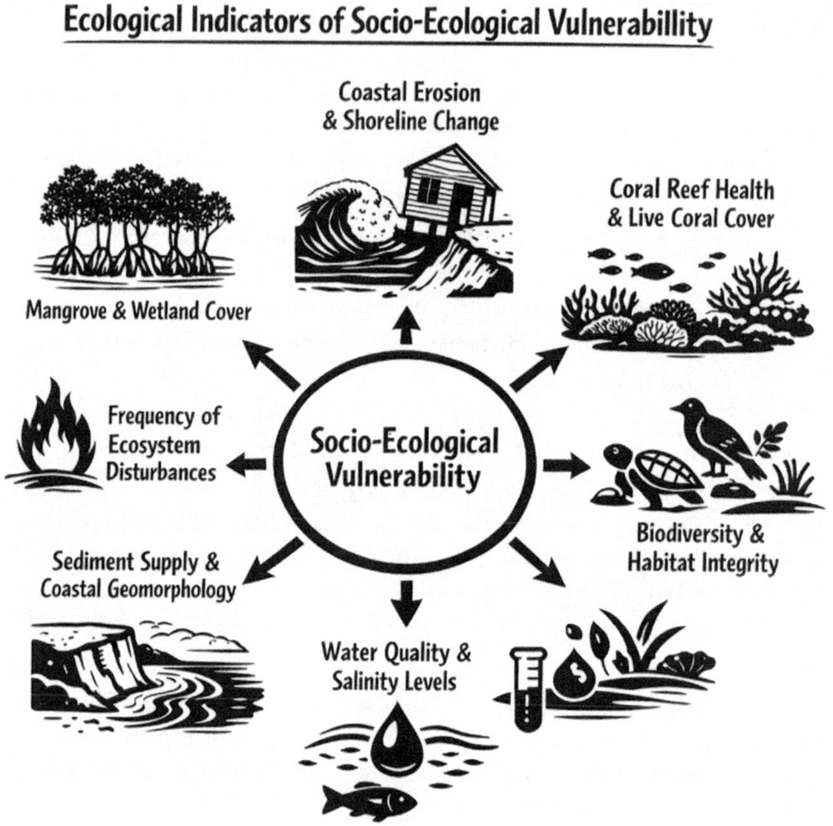

Fig. 2.4 Ecological indicators of socio-ecological vulnerability (Prepared by the authors)

2.4 Ecological, Social Vulnerability, and Integrated Socio-Ecological Vulnerability Assessment

2.4.1 Assessment of Ecological Vulnerability

Ecological vulnerability assessment examines how coastal ecosystems are exposed to climate-related hazards, how sensitive they are to disturbance, and how well they can respond to environmental change. In coastal

Fig. 2.5 Social indicators of socio-ecological vulnerability (Prepared by the authors)

regions affected by climate disasters, ecosystems such as mangroves, coral reefs, seagrass beds, dunes, wetlands, and estuaries reduce the impacts of storms, flooding, erosion, and waves. These ecosystems also support biodiversity and local livelihoods. When ecosystems are degraded or fragmented, their protective capacity weakens, increasing vulnerability for both natural systems and coastal communities. Coastal ecosystems change naturally through tides, sediment movement, freshwater flows, and biological processes. Climate change adds new pressures, including sea-level rise, warmer ocean temperatures, ocean acidification, and stronger storms. These stresses often combine with human activities such as land reclamation, pollution, overfishing, and coastal infrastructure development. Ecological vulnerability assessment therefore considers both climate exposure and the ways in which human pressures reduce ecosystem resilience and increase the risk of long-term damage. A key aim of ecological vulnerability assessment is to examine how much disturbance an ecosystem can absorb while maintaining its structure and functions. Healthy mangrove forests, for example, can recover after storm events and continue to protect shorelines. Degraded mangroves, in contrast, may fail to regenerate, leading to increased erosion and flooding. Coral reefs weakened by warming waters and pollution are more likely to experience bleaching and loss, reducing their role in coastal protection and fisheries support. Common ecological indicators used in coastal vulnerability assessment include the extent and condition of mangroves and wetlands, coral reef health, rates of shoreline erosion, habitat fragmentation, water quality, biodiversity levels, frequency of ecosystem disturbance, and the capacity for natural recovery. These indicators help distinguish resilient ecosystems from those at greater risk. Ecological vulnerability assessment also has a strong spatial focus. Different coastal habitats respond differently to climate stress. Mapping ecological indicators helps identify fragile areas and supports conservation planning, coastal zoning, and ecosystem-based adaptation (Das & Behera, 2024; He et al., 2018; Xu, et al., 2024). Ecological vulnerability assessment provides the environmental foundation for understanding climate risk in coastal regions. It draws attention to areas where natural protection is weakening and highlights the importance of ecosystem conservation and restoration for building long-term coastal resilience.

2.4.2 Assessment of Social Vulnerability

Social vulnerability assessment examines how the characteristics of individuals, households, and communities influence their exposure to climate-related hazards and their ability to cope, adapt, and recover. In coastal regions affected by climate disasters, vulnerability depends on more than physical exposure to cyclones, flooding, or sea-level rise. Social and economic conditions, population characteristics, livelihood systems, and institutional support play a major role. Communities located in similar coastal settings may face the same hazards, but experience very different outcomes based on access to resources, services, and social support. Livelihood dependence is a key factor shaping social vulnerability in coastal areas. Many households rely on climate-sensitive activities such as fishing, aquaculture, salt production, and coastal farming. These livelihoods are easily disrupted by storms, erosion, and salinity intrusion. When natural systems are damaged, income sources often disappear, increasing insecurity among already vulnerable groups. Social vulnerability assessment therefore pays close attention to livelihood structure, income diversity, and employment stability (Arunachalam et al., 2023; Savelberg et al., 2025).

Social vulnerability is also shaped by inequality and social differences within communities. Gender, age, caste, ethnicity, disability, and migration status influence access to land, credit, information, and decision-making. Women, older people, and informal settlers in coastal zones often face greater difficulties during evacuation, recovery, and access to relief after disasters. These forms of disadvantage increase the severity of climate impacts and slow recovery. Institutional and governance conditions strongly influence social vulnerability. Disaster management systems, early warning services, land-use planning, and social protection programs affect how communities prepare for and respond to hazards. Weak governance, limited institutional reach, and poor planning can turn moderate hazards into serious social disruption. Social vulnerability assessment therefore considers both household-level conditions and the wider institutional environment. Common indicators used in social vulnerability assessment include income levels, livelihood dependence on natural resources, housing quality, education, health access, social marginalization, access to basic services, social networks, disaster preparedness, and governance capacity. When combined and mapped, these indicators help identify areas where climate hazards intersect with poverty, inequality, and weak institutional support (Ashokan et al., 2024; Xiao et al., 2022). Social vulnerability assessment plays an important role in supporting fair and effective climate adaptation in coastal regions. By identifying who is most at risk and why, it guides actions

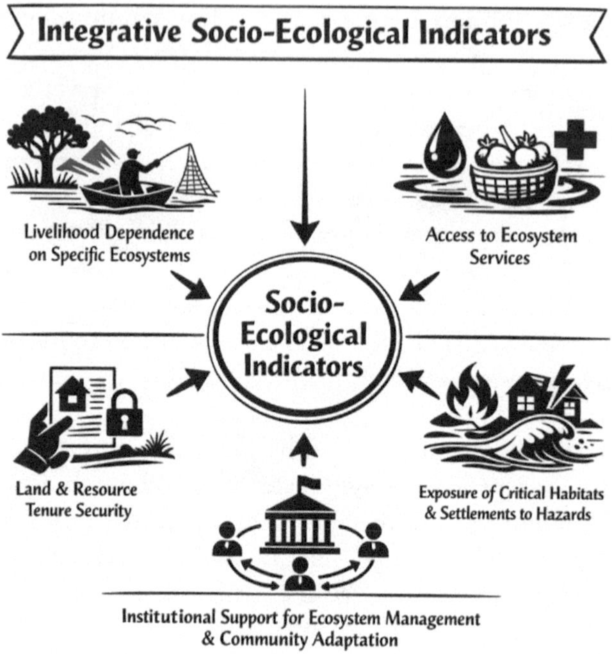

Fig. 2.6 Integrative socio-ecological indicators (Prepared by the authors)

that strengthen livelihoods, improve governance, and build long-term community resilience (Fig. 2.6).

2.4.3 Assessment of Socio-Ecological Vulnerability

Integrated socio-ecological vulnerability assessment examines how climate hazards affect connected human and environmental systems, rather than treating society and nature as separate. In coastal regions, livelihoods, settlements, and ecosystems are closely linked, and climate disasters such as cyclones, storm surges, flooding, and sea-level rise often trigger a chain of impacts that disrupt both ecological processes and social well-being. This approach helps explain vulnerability as a result of interactions between environmental fragility, social inequality, and institutional capacity. Ecological degradation and social vulnerability often reinforce each other over time. The loss of mangroves, for example, increases exposure to storm surges and flooding, which damages homes, fisheries, and infrastructure. These losses can push households into poverty and increase pressure on remaining natural resources, further weakening ecosystems. Integrated socio-ecological vulnerability assessment captures these interactions and treats vulnerability as a changing process shaped by feedbacks. This approach draws on established ideas from climate risk assessment, social and ecological systems thinking, and environmental management, while bringing social and ecological indicators into a single analytical structure. Ecological indicators describe ecosystem condition, habitat integrity, and natural protection, while social indicators reflect livelihood dependence, poverty, governance, and access to services. When combined, these indicators reveal areas where ecological weakness and social disadvantage occur together, creating zones of high vulnerability in coastal landscapes.

Spatial analysis plays a key role in integrated vulnerability assessment. Coastal regions show sharp contrasts between urban and rural areas, healthy and degraded ecosystems, and socially advantaged and disadvantaged communities. Mapping social and ecological indicators together helps identify locations where actions such as ecosystem restoration, infrastructure improvement, or social protection can most effectively reduce vulnerability. Indicators that link social and ecological systems include dependence on specific ecosystems for livelihoods, access to ecosystem services such as coastal protection and freshwater, security of land and resource tenure, exposure of settlements and habitats to hazards, and institutional capacity for ecosystem-based adaptation. These indicators show how changes in ecosystems affect social well-being and how social conditions influence environmental outcomes (Shah & Wang, 2024; Sultana et al., 2023; Tanim et al., 2022).

Integrated socio-ecological vulnerability assessment supports climate adaptation and disaster risk reduction by promoting approaches that combine ecosystem restoration, livelihood support, and institutional strengthening. It also highlights issues of equity by identifying who faces the greatest risks and who benefits from ecosystem services. In climate-disaster-prone coastal regions, this assessment approach encourages a shift from short-term response to long-term resilience. It helps clarify how development choices, environmental management, and social policies shape future climate risk and provides a practical foundation for planning and governing coastal areas in a changing climate (Fig. 2.7).

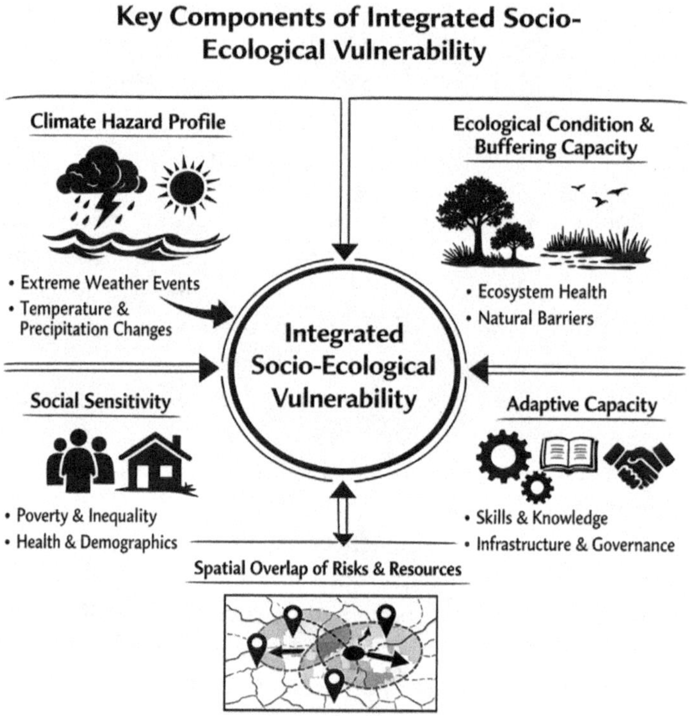

Fig. 2.7 Key components of integrated socio-ecological vulnerability (Prepared by the authors)

2.5 Climate Disasters and Socio-Ecological Vulnerability: Coastal Ecologies and Sustainability

Climate disasters are increasingly shaping the future of coastal regions. In these areas, exposure to extreme events intersects with fragile ecosystems and deep social inequalities. Cyclones, storm surges, coastal flooding, sea-level rise, saline intrusion, heatwaves, and changing rainfall patterns are no longer occasional events. They now act as continuous pressures that reshape coastal landscapes, livelihoods, and settlement patterns. These hazards affect more than infrastructure or ecosystems alone. They disrupt connected socio-ecological systems in which environmental health, economic activity, and human well-being are closely linked. An integrated perspective is therefore essential for understanding coastal vulnerability. Coastal ecosystems provide important protective and productive functions. Mangroves, coral reefs, dunes, wetlands, and seagrass meadows reduce wave energy, stabilize shorelines, regulate water movement, and support fisheries, tourism, and food security. These ecosystems act as natural protection while sustaining local economies. Rapid urban growth, port development, aquaculture, tourism, and resource extraction have weakened many coastal ecosystems. When climate disasters strike degraded coasts, the loss of natural buffering greatly increases physical damage and social disruption, often turning hazards into large-scale ecological and humanitarian crises (Jeganathan et al., 2021; Shah et al., 2025).

Socio-ecological vulnerability in coastal regions develops through the interaction of rising climate hazards, declining ecosystem resilience, and unequal social capacity. Poor and marginalized groups often live in low-lying deltas, eroding shorelines, and reclaimed wetlands where risks are highest. Their reliance on climate-sensitive livelihoods such as fishing, coastal farming, salt production, and informal tourism increases their exposure to environmental shocks. Weak governance, insecure land tenure, and limited access to finance and information further restrict their ability to adapt or recover. Vulnerability is therefore shaped not only by natural processes but also by social and political choices. Climate disasters also create feedback processes that deepen vulnerability over time. Damage to mangroves, reefs, and protective infrastructure increases exposure to future hazards. Loss of livelihoods often forces communities to rely more heavily on remaining natural resources, accelerating environmental degradation. Migration from damaged coastal areas can place new pressure on urban zones and inland ecosystems. These processes show that climate disasters are part of longer cycles of risk rather than isolated events.

Sustainability challenges become clear when these trends are viewed together. Sustainable coastal development depends on balancing economic growth, social equity, and ecological integrity. Development choices that favor short-term gains, such as wetland reclamation, mangrove removal, or construction in flood-prone areas, weaken long-term resilience. As climate hazards intensify, these choices lead to higher losses, wider inequality, and increasing recovery costs. A sustainability-focused response to climate disasters in coastal regions requires socio-ecological vulnerability assessment and integrated planning. Ecosystems must be recognized as essential assets for risk reduction and climate adaptation. Social drivers of vulnerability, including poverty, insecure tenure, gender inequality, and weak institutions, also need to be addressed. Without attention to these social conditions, ecological interventions alone cannot protect those most at risk.

Sustainable coastal futures depend on breaking the cycle of hazard exposure, environmental degradation, and social inequality. Ecosystem-based adaptation, such as mangrove restoration, wetland conservation, and dune stabilization, can rebuild natural protection while supporting livelihoods. Social protection, livelihood diversification, inclusive governance, and secure land and resource rights can strengthen the ability of communities to cope with change. When combined, these measures reduce vulnerability and support more equitable development. Climate disasters also create opportunities for change. Recovery efforts can support safer settlement patterns, restore ecosystems, and promote resilient livelihoods. Relocating communities from high-risk zones, investing in natural protection, and strengthening institutions can shift coastal regions toward more sustainable pathways. Such transformation depends on governance capacity, political commitment, and recognition of vulnerability as a socio-ecological challenge. Climate disasters are reshaping coastal regions through complex interactions between environmental change and social systems. Sustainability cannot be achieved through technical solutions alone. It requires an integrated understanding of climate hazards, ecosystems, and societies. Linking climate disasters with socio-ecological vulnerability and sustainability provides a strong foundation for assessment, mapping, and policy aimed at building resilient and just coastal futures (Kanan & Giupponi, 2024; Li & Ford, 2019).

2.6 Conclusion and Arguments

The assessment of climate disasters, socio-ecological vulnerability, and coastal ecologies shows that vulnerability is shaped by more than physical exposure to hazards. It develops through the interaction of fragile ecosystems, social inequality, and the ability of institutions to manage risk. Socio-ecological vulnerability assessment offers a strong framework for explaining why climate risks are unevenly distributed across coastal landscapes and social groups. By linking exposure, sensitivity, and adaptive capacity with ecosystem condition and governance arrangements, this approach explains how climate disasters affect both human well-being and ecological stability. Coastal ecologies play an important role in these outcomes. Ecosystems such as mangroves, coral reefs, wetlands, and dunes reduce disaster impacts while supporting livelihoods and local economies. When these systems are weakened by unsustainable development or poor governance, climate hazards become more destructive and lead to wider social and ecological damage. Social vulnerability, shaped by poverty, marginalization, insecure livelihoods, and limited access to institutions, further determines who experiences the greatest losses and who recovers more slowly. The interaction between ecological degradation and social disadvantage explains why similar hazards lead to very different outcomes across coastal regions.

Socio-ecological vulnerability assessment is important because it connects scientific understanding with practical decision-making.

Through integrated indicators and spatial analysis, it helps identify areas of high risk, clarify the drivers of vulnerability, and support more equitable and effective adaptation strategies. Sustainable coastal development depends on recognizing vulnerability as a shared social and environmental challenge. Protecting ecosystems, strengthening livelihoods, and improving governance are closely linked goals that together support climate-resilient coastal futures. The chapter discussions lead to the following arguments:

(a) Climate disasters create risk through the interaction of social conditions and ecological systems, rather than through physical hazards alone.
(b) Degradation of coastal ecosystems increases exposure to climate extremes and weakens natural protective functions.
(c) Social inequality and weak governance place a disproportionate burden of climate risk on marginalized coastal communities.
(d) Integrated socio-ecological assessment helps identify vulnerability hotspots and reveals the deeper structural causes of risk.
(e) Long-term coastal sustainability depends on adaptation approaches that combine ecosystem restoration with social inclusion and effective governance.

References

Arunachalam, M., Saravanavel, J., & Joseph Kochuparampil, A. (2023). PCA-based approach for mapping social vulnerability to hazards in the Chennai metropolitan area, east coast of India. *Annals of GIS, 29*(4), 529–552. https://doi.org/10.1080/19475683.2023.2226189

Ashokan, A. M., Rajagopal, J., Krishnaswamy, P., et al. (2024). Social vulnerability assessment in the health and disease context: Review. *Discover Social Science and Health, 4*, 55. https://doi.org/10.1007/s44155-024-00116-x

Bao, Y., Zhang, H., & Wu, C. (2025). Uncovering impacts of tourism on social–ecological vulnerability using geospatial analysis and big earth data:

A karst Ethnic Village perspective. *Land, 14*(5), 1030. https://doi.org/10.3390/land14051030

Behnassi, M., Gupta, H., Haiba, M. E., & Ramachandran, G. (Eds.). (2021). *Social-ecological systems (SES): From risks and insecurity to viability and resilience*. Springer Cham. https://doi.org/10.1007/978-3-030-76247-6

Berrouet, L. M., Machado, J., & Villegas-Palacio, C. (2018). Vulnerability of socio—Ecological systems: A conceptual framework. *Ecological Indicators, 84*, 632–647. https://doi.org/10.1016/j.ecolind.2017.07.051

Cheng, C. (2019). EcoWisdom for climate justice planning: Social-ecological vulnerability assessment in Boston's Charles River watershed. In B. Yang & R. Young (Eds.), *Ecological Wisdom. EcoWISE*. Springer. https://doi.org/10.1007/978-981-13-0571-9_13

Das, U., & Behera, B. (2024). Geospatial assessment of ecological vulnerability of fragile eastern Duars Forest integrating GIS-based AHP, CRITIC and AHP-TOPSIS models. *Geomatics, Natural Hazards and Risk, 15*(1), 2330529. https://doi.org/10.1080/19475705.2024.2330529

Das, S., Ghosh, A., Hazra, S., Ghosh, T., de Campos, R. S., & Samanta, S. (2020). Linking IPCC AR4 & AR5 frameworks for assessing vulnerability and risk to climate change in the Indian Bengal Delta. *Progress in Disaster Science, 7*, 100110. https://doi.org/10.1016/j.pdisas.2020.100110

Gupta, A. K., Negi, M., Nandy, S., Alatalo, J. M., Singh, V., & Pandey, R. (2019). Assessing the vulnerability of socio-environmental systems to climate change along an altitude gradient in the Indian Himalayas. *Ecological Indicators, 106*, 105512. https://doi.org/10.1016/j.ecolind.2019.105512

He, L., Shen, J., & Zhang, Y. (2018). Ecological vulnerability assessment for ecological conservation and environmental management. *Journal of Environmental Management, 206*, 1115–1125. https://doi.org/10.1016/j.jenvman.2017.11.059

Jeganathan, A., Andimuthu, R., & Kandasamy, P. (2021). Climate risks and socio-economic vulnerability in Tamil Nadu, India. *Theoretical and Applied Climatology, 145*, 121–135. https://doi.org/10.1007/s00704-021-03595-z

Jha, S. K., Negi, A. K., Alatalo, J. M., et al. (2021). Socio-ecological vulnerability and resilience of mountain communities residing in capital-constrained environments. *Mitigation and Adaptation Strategies for Global Change, 26*, 38. https://doi.org/10.1007/s11027-021-09974-1

Kanan, A. H., & Giupponi, C. (2024). Coastal socio-ecological systems adapting to climate change: A global overview. *Sustainability, 16*(22), 10000. https://doi.org/10.3390/su162210000

Khajuria, A., & Ravindranath, N. H. (2012). Climate change vulnerability assessment: Approaches DPSIR framework and vulnerability index. *Journal of Earth Science and Climatic Change, 3*, 109. https://doi.org/10.4172/2157-7617.1000109

Khromova, S., Méndez, G. V., Eckelman, M. J., Herreros-Cantis, P., & Langemeyer, J. (2025). A social-ecological-technological vulnerability approach for assessing urban hydrological risks. *Ecological Indicators, 173*, 113334. https://doi.org/10.1016/j.ecolind.2025.113334

Kok, M., Lüdeke, M., Lucas, P., et al. (2016). A new method for analysing socio-ecological patterns of vulnerability. *Regional Environmental Change, 16*, 229–243. https://doi.org/10.1007/s10113-014-0746-1

Kumar, P., Fürst, C., & Joshi, P. K. (2024). Differentiated socio-ecological system approach for vulnerability and adaptation assessment in the central Himalaya. *Mitigation and Adaptation Strategies for Global Change, 29*, 7. https://doi.org/10.1007/s11027-023-10097-y

Lalyer, C. R., von Gleich, A., & Giese, B. (2020). Vulnerability analysis of ecological systems. In A. von Gleich & W. Schröder (Eds.), *Gene drives at tipping points*. Springer. https://doi.org/10.1007/978-3-030-38934-5_3

Li, A., & Ford, J. (2019). Understanding socio-ecological vulnerability to climatic change through a trajectories of change approach: A case study from an indigenous community in Panama. *Weather, Climate, and Society, 11*(3), 577–593.

Lyngdoh, V., Lynrah, M. M., Wahlang, E., et al. (2025). District-level climate vulnerability assessment in Meghalaya using an IPCC AR5 framework. *Scientific Reports, 15*, 42486. https://doi.org/10.1038/s41598-025-26651-4

Maina, J., Kithiia, J., Cinner, J., et al. (2016). Integrating social–ecological vulnerability assessments with climate forecasts to improve local climate adaptation planning for coral reef fisheries in Papua New Guinea. *Regional Environmental Change, 16*, 881–891. https://doi.org/10.1007/s10113-015-0807-0

Malakar, K. D., Kumar, M., Anand, S., & Kuzur, G. (2023). *Climate change and socio-ecological transformation: Vulnerability and sustainability*. Springer. https://doi.org/10.1007/978-981-99-4390-6

Newton, A., & Weichselgartner, J. (2014). Hotspots of coastal vulnerability: A DPSIR analysis to find societal pathways and responses. *Estuarine, Coastal and Shelf Science, 140*, 123–133. https://doi.org/10.1016/j.ecss.2013.10.010

Savelberg, L., Casali, Y., van den Homberg, M., et al. (2025). Comparing hierarchical and inductive methods reveals fundamental differences in social vulnerability rankings. *Scientific Reports, 15*, 34541. https://doi.org/10.1038/s41598-025-17860-y

Sebesvari, Z., Renaud, F. G., Haas, S., et al. (2016). A review of vulnerability indicators for deltaic social–ecological systems. *Sustainability Science, 11*, 575–590. https://doi.org/10.1007/s11625-016-0366-4

Shah, M. A. R., Van Dau, Q., & Wang, X. (2025). Social-ecological vulnerability and risk to coastal flooding and erosion in major coastal cities. *International Journal of Disaster Risk Reduction, 118*, 105286. https://doi.org/10.1016/j.ijdrr.2025.105286

Shah, M. A. R., & Wang, X. (2024). Assessing social-ecological vulnerability and risk to coastal flooding: A case study for Prince Edward Island, Canada. *International Journal of Disaster Risk Reduction, 106*, 104450. https://doi.org/10.1016/j.ijdrr.2024.104450

Sultana, R., Irfanullah, H. M., Selim, S. A., & Alam, M. S. (2023). Social-ecological vulnerability to climate change and risk governance in coastal fishing communities of Bangladesh. *Frontiers in Marine Science, 10*, 1174659. https://doi.org/10.3389/fmars.2023.1174659

Sun, L., Yu, Y., He, J., Li, C., Yu, X., Zhang, L., et al. (2025). Vulnerability assessment of social-ecological systems in arid regions: A cross-efficiency modified DEA model with entropy weight aggregation. *Ecological Indicators, 179*, 114149. https://doi.org/10.1016/j.ecolind.2025.114149

Tanim, A. H., Goharian, E., & Moradkhani, H. (2022). Integrated socio-environmental vulnerability assessment of coastal hazards using data-driven and multi-criteria analysis approaches. *Scientific Reports, 12*, 11625. https://doi.org/10.1038/s41598-022-15237-z

Wang, B., Yu, F., Teng, Y., Cao, G., Zhao, D., & Zhao, M. (2022). A SEEC model based on the DPSIR framework approach for watershed ecological security risk assessment: A case study in Northwest China. *Water, 14*(1), 106. https://doi.org/10.3390/w14010106

Xiao, Y., Tang, X., Li, Y., Huang, H., & An, B. W. (2022). Social vulnerability assessment of landslide disaster based on improved TOPSIS method: Case study of eleven small towns in China. *Ecological Indicators, 143*, 109316. https://doi.org/10.1016/j.ecolind.2022.109316

Xu, M., Cao, C., Zhong, S., Yang, X., Bashir, B., Wang, K., et al. (2024). Ecological vulnerability assessment and spatial-temporal variations analysis in typical ecologically vulnerable areas of China. *Frontiers in Ecology and Evolution, 12*, 1406444. https://doi.org/10.3389/fevo.2024.1406444

Zango-Palau, A., Jolivet, A., Lurgi, M., & Claramunt-López, B. (2024). A quantitative approach to the understanding of social-ecological systems: A

case study from the Pyrenees. *Regional Environmental Change, 24*, 9. https://doi.org/10.1007/s10113-023-02177-1
Zhang, X., Wang, L., Fu, X., Li, H., & Xu, C. (2017). Ecological vulnerability assessment based on PSSR in Yellow River Delta. *Journal of Cleaner Production, 167*, 1106–1111. https://doi.org/10.1016/j.jclepro.2017.04.106

Part II

Understanding the Coastal Socio-Ecological Systems: Climate Disasters, Coastal Resilience, and Management

3

Climate Disasters and Coastal Infrastructural Resilience

Coastal regions represent some of the most dynamic and biologically productive environments on the planet, but they are also increasingly exposed to climate-related hazards. This chapter introduces the evolving challenges faced by coastal ecologies as climate change intensifies processes such as sea-level rise, coastal flooding, extreme storms, and shoreline erosion. Coastal zones, long recognized for their ecological value and economic significance, are now being reshaped by climate risks that affect both natural ecosystems and human livelihoods. The chapter serves as an entry point for understanding the complex interactions between climate disasters and coastal ecological systems. It examines the physical processes that drive coastal hazards while also emphasizing the influence of human actions, including land-use decisions, development patterns, and governance arrangements, in shaping vulnerability. In this context, coastal vulnerability is understood not only as a consequence of environmental change but also as an outcome of how societies use, manage, and transform coastal landscapes over time. By integrating ecological perspectives with social and institutional considerations, the chapter presents coastal ecologies as interconnected systems under increasing stress. This approach highlights the need to consider

K. D. Malakar et al., *Coastal Ecologies at Risk*, Palgrave Studies in Environmental Sustainability, https://doi.org/10.1007/978-3-032-26081-9_3

both natural dynamics and human interventions when assessing risk and planning responses. The discussion establishes a foundation for later chapters that focus on adaptation strategies, resilience-building, and sustainable coastal management, and it supports a deeper engagement with the challenges posed by climate change in coastal environments.

Key Points of the Chapter

- Describes how climate disasters have turned coastal infrastructure into systems that require constant adaptation.
- Explains the need to combine engineering strength, ecological protection, and adaptive planning for resilience.
- Examines how risk assessment links climate hazards, exposure, and vulnerability to guide infrastructure decisions.
- Highlights the role of nature-based and hybrid solutions in improving protection and long-term sustainability.
- Describes the importance of forward-looking planning and inclusive policy for resilient coastal development.

3.1 Climate Disasters on Coastal Infrastructure: Impacts

Climate disasters are now a major force shaping the physical, economic, and social conditions of coastal regions across the world. Sea-level rise, stronger cyclones, storm surges, coastal erosion, saltwater intrusion, and extreme rainfall are no longer occasional events but recurring pressures that steadily weaken coastal infrastructure. Ports, roads, bridges, seawalls, housing, energy facilities, water systems, and communication networks are increasingly exposed to hazards that go beyond the conditions for which they were originally designed. These events reveal long-standing weaknesses in infrastructure planning, governance systems, and development practices rather than acting as isolated natural shocks. Coastal infrastructure supports key economic activities, especially in regions dependent on maritime trade, fisheries, tourism, and coastal industries. Disruption to these systems often leads to effects that extend far beyond physical damage. Flooded ports interrupt supply chains, damaged transport routes isolate settlements, power failures slow economic activity, and unsafe water systems increase health risks. These cascading impacts show that infrastructure resilience is closely linked to economic stability, social well-being, and environmental security. Communities with limited resources are often the most affected, as they tend to live and work in areas with the highest exposure to climate hazards (Mashwama & Phesa, 2025).

Climate disasters affect coastal infrastructure through both direct and indirect processes. Direct impacts include structural damage caused by flooding, storm surges, and wave action, along with material deterioration from prolonged exposure to saline conditions. Indirect impacts arise through shoreline retreat, land subsidence, and the loss of natural protective features such as mangroves, coral reefs, and sand dunes. As these natural buffers decline due to climate change and human pressures, built infrastructure becomes more exposed to extreme events. Infrastructure vulnerability is therefore closely tied to ecological conditions and land-use patterns, not only to engineering design. Addressing these challenges calls for moving beyond narrow engineering solutions toward a broader systems-based perspective. Climate risks are becoming more intense and less predictable, making reliance on past design standards increasingly ineffective. Flexible and adaptive infrastructure approaches are needed to respond to changing conditions over time. These include elevating critical structures, improving drainage capacity, using climate-resilient materials, and integrating nature-based measures such as wetland restoration and mangrove protection into infrastructure planning (Adshead et al., 2024; Rahman & Rahman, 2015).

Governance and institutional capacity play a critical role in shaping how coastal infrastructure responds to climate risks. Weak planning coordination, short-term investment priorities, and limited regulatory enforcement often result in infrastructure being placed in high-risk zones without sufficient protection. Disaster responses frequently focus on rebuilding after events rather than reducing risks beforehand. A resilience-oriented approach requires climate risk assessment to be embedded in infrastructure planning, stronger coastal zoning regulations, and closer alignment between public and private investment decisions and long-term adaptation goals. Through such efforts, infrastructure can serve not only as a physical asset but also as a foundation for building resilience in coastal regions. Climate disasters and coastal infrastructure are deeply connected through the ways natural processes interact with human decisions. Treating disasters as predictable outcomes of a changing climate highlights the need for new approaches to coastal development. Resilient infrastructure depends on strong engineering, healthy coastal ecosystems, effective governance, and social inclusion. This integrated approach offers a pathway for coastal regions to cope with increasing climate pressures while sustaining their economic and ecological functions over the long term.

3.2 Resilient Coastal Infrastructure: Concept and Importance

The design of resilient coastal infrastructure has become one of the major challenges of the twenty-first century as climate disasters increasingly influence the future of coastal regions across the world. Coastal areas are densely populated and economically important, supporting ports, industries, urban settlements, tourism, and essential service systems. These areas are also highly exposed to climate-related hazards such as sea-level rise, storm surges, coastal flooding, cyclones, saltwater intrusion, and shoreline erosion. The combination of growing climate risks and rapid coastal development has made resilience a central concern for long-term planning. In this context, resilient coastal infrastructure refers not only to the strength of individual structures, but also to the ability of systems to absorb shocks, adjust to changing conditions, and continue to deliver essential services during periods of stress.

The idea of resilience is closely linked to preparation and response under uncertain climate conditions. Traditional infrastructure design has largely relied on past climate records and assumptions of environmental stability. These assumptions no longer provide reliable guidance. Changes in temperature, rainfall patterns, and storm behaviors mean that historical conditions offer limited insight into future risks. Designing resilient infrastructure requires flexibility, redundancy, and adaptive capacity so that systems can function across a wide range of possible scenarios. This approach shifts infrastructure planning away from static designs toward strategies that account for long-term change and uncertainty. Physical strength remains an important element of resilient coastal infrastructure. Buildings, bridges, seawalls, ports, and utility networks must be designed to withstand stronger winds, higher waves, deeper flooding, and prolonged exposure to saline conditions. Measures such as elevated foundations, reinforced materials, effective drainage, and flood-resistant construction help reduce damage during extreme events. Equally important is the ability of infrastructure systems to recover quickly after disruption. Rapid repair and restoration are critical in coastal areas where transport, energy, water supply, and emergency services support daily life and economic activity. Adaptability is another essential feature of resilient infrastructure. Most infrastructure assets are designed to last for several decades, during which climate conditions are expected to change significantly. Adaptive design allows infrastructure to be modified over time through actions such as raising structures, expanding capacity, or relocating facilities as risk zones shift. Flexible land-use planning also plays a key role in reducing long-term exposure to hazards and avoiding development pathways that increase future vulnerability. Resilient coastal infrastructure depends on system-level thinking. Infrastructure operates through interconnected networks that include transportation, energy, water, sanitation, communication, and emergency services. Failure in one part of the system can quickly affect others. Flooding of an energy facility, for example, can disrupt water supply, healthcare services, and communication systems. Identifying critical components, providing backup options, and ensuring alternative service routes help limit cascading impacts. Natural and hybrid infrastructure approaches strengthen coastal resilience by working with natural processes. Coastal ecosystems such as mangroves, coral reefs, salt marshes, and sand dunes reduce wave energy, stabilize shorelines, and limit storm surge impacts. Incorporating these ecosystems into infrastructure planning enhances protection while supporting biodiversity and local livelihoods. Combining engineered structures with natural features often provides more durable and cost-effective protection than relying on hard infrastructure alone (Chávez et al., 2021; Kantamaneni, 2016; Oktari et al., 2020).

The importance of resilient coastal infrastructure becomes clear when considering the scale of losses associated with climate disasters. Damage to ports can disrupt international trade, flooded transport networks can isolate communities, and failures in water, sanitation, and energy systems can create serious public health risks. Investments in resilience reduce these impacts and help protect both physical assets and human well-being. Resilient infrastructure also supports social equity and sustainable development. Climate disasters often affect vulnerable communities most severely, particularly those living in high-risk coastal zones with limited access to resources. Infrastructure that is designed to be resilient, accessible, and inclusive can reduce these inequalities and strengthen the ability of communities to cope with future hazards. The economic case for resilient coastal infrastructure is strong. While resilient design may involve higher initial costs, it lowers long-term expenses related to disaster damage, emergency response, and reconstruction. In regions facing increasing climate risks, these long-term savings are particularly important. Resilient infrastructure also encourages investment by reducing uncertainty and improving confidence in long-term stability. Governance and policy frameworks play a decisive role in turning resilience concepts into practice. Effective coordination across sectors, strong planning institutions, and the integration of climate risk into development decisions are essential. Building codes, zoning regulations, and investment strategies must reflect current and future climate conditions rather than outdated assumptions. Participation by local communities further improves the relevance and effectiveness of resilience measures. Designing resilient coastal infrastructure is both a practical necessity and a strategic response to climate change. It reflects changing approaches to risk, uncertainty, and development in coastal regions. When resilience guides infrastructure planning, coastal areas are better positioned to reduce disaster impacts, support sustainable growth, and protect both people and ecosystems in a changing climate (Chaudhuri, 2025; Powell et al., 2019; Singhvi et al., 2022).

3.3 Coastal Infrastructure Adaptation Strategies

Coastal infrastructure adaptation strategies have become a central response to the increasing risks created by climate change in coastal regions. Sea-level rise, storm surges, coastal flooding, heat stress, and shoreline erosion are placing growing pressure on settlements, transport networks, ports, energy facilities, and water systems. As these threats intensify, adaptation has moved from a supportive policy option to a core development priority. While mitigation focuses on reducing the causes of climate change, adaptation addresses its consequences by adjusting physical systems, institutions, and social practices. In coastal areas, where populations and economic activities are highly concentrated, effective adaptation plays a key role in safeguarding infrastructure, livelihoods, and regional development.

Coastal infrastructure adaptation involves changes in location, design, materials, management practices, and governance arrangements so that infrastructure can function under changing environmental conditions (Fig. 3.1). Earlier planning approaches relied heavily on historical climate patterns and relatively stable coastlines. These assumptions are no longer reliable as sea levels continue to rise and extreme events become more frequent. Adaptation strategies now rely on forward-looking risk assessments, scenario-based planning, and flexible design approaches that allow infrastructure to perform under a range of future conditions. This shift calls for closer coordination between climate science, engineering, urban planning, environmental management, and disaster risk reduction. Protection remains one of the most commonly applied adaptation approaches. Structural measures such as seawalls, dikes, storm surge barriers, breakwaters, and floodgates are used to reduce exposure to waves and flooding, particularly for high-value assets such as ports, industrial areas, and urban centers. Non-structural measures such as improved drainage systems, land-use zoning, early warning systems, and emergency planning support these defenses by lowering exposure and strengthening response capacity. Accommodation strategies recognize that some degree of flooding and coastal change is unavoidable and adjust infrastructure to continue functioning under such conditions. These measures include elevating buildings and roads, using flood-tolerant designs and materials, relocating critical equipment, and developing floating or amphibious structures in highly flood-prone locations. In transport and port systems, accommodation may involve raising quay levels, strengthening surfaces, and improving stormwater management. Such strategies are especially relevant in densely built coastal areas where complete protection or relocation is not feasible. Planned retreat or relocation represents a more transformative form of adaptation in areas facing long-term and severe risk from sea-level rise and erosion. In these locations, maintaining existing infrastructure can become economically, socially, and environmentally unsustainable. Relocation involves moving infrastructure and settlements to safer areas, reducing repeated losses and long-term risk. Although this approach presents social and political challenges, it can offer a lasting solution for critical facilities such as power plants, wastewater treatment plants, and healthcare infrastructure (Malakar et al., 2023; Leal Filho, 2018; Rosenzweig et al., 2011; Sauer et al., 2021).

Nature-based adaptation strategies are increasingly recognized for their role in supporting coastal resilience. Coastal ecosystems such as mangroves, coral reefs, salt marshes, seagrass beds, and sand dunes reduce wave energy, stabilize shorelines, and limit storm surge impacts. Integrating these systems into infrastructure planning enhances protection while also supporting fisheries, tourism, biodiversity, and carbon storage. Combining natural systems with engineered structures can provide flexible and cost-effective solutions that adapt over time. Effective adaptation also depends on governance, planning, and financing frameworks. Climate risks need to be incorporated into infrastructure investment decisions, design standards, and regulatory systems through the use of climate projections, hazard mapping, and vulnerability assessments. Financial mechanisms such as adaptation funds, insurance schemes, and climate-resilient investments help support these efforts. Community involvement is equally important, as local knowledge and participation improve the relevance, acceptance, and success of adaptation measures. Coastal infrastructure adaptation strategies support both risk reduction and long-term development goals. As climate risks continue to intensify, regions that invest in adaptive infrastructure are better positioned to protect lives, sustain economic activity, and preserve coastal ecosystems.

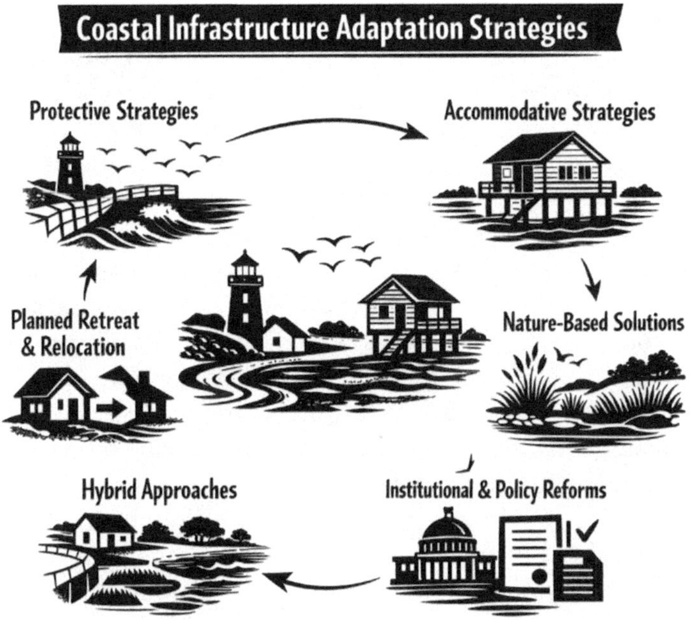

Fig. 3.1 Coastal infrastructure adaptation strategies (Prepared by the authors)

Therefore, the effective adaptation allows societies to adjust to changing coastal conditions while strengthening resilience for the future (Geukes et al., 2026; Leal Filho et al., 2025; Lucatello, 2025).

3.4 Coastal Infrastructure: Assessment and Management of Risk

Risk assessment and management play a central role in strengthening coastal infrastructural resilience in the face of growing climate-related disasters. Coastal regions are increasingly exposed to multiple hazards, including sea-level rise, storm surges, cyclones, extreme rainfall, coastal flooding, heat stress, and long-term shoreline erosion. These hazards interact with existing patterns of urban growth, industrial development, and ecological decline, creating complex risk conditions. In coastal settings, risk is best understood as the combined effect of hazards, the exposure of people and assets, and the vulnerability of infrastructure systems. Ports, transport networks, power supply systems, water and sanitation facilities, housing, and communication networks are located where these factors intersect, making careful risk assessment and active management essential for sustainable coastal development. Risk assessment for coastal infrastructure begins with identifying climate-related hazards and examining how their frequency and intensity may change over time. This work draws on climate science, ocean studies, and coastal processes to understand trends in sea levels, storm behavior, wave action, and rainfall. Climate projections, satellite observations, and historical data are commonly used to identify flood-prone areas, erosion zones, and storm surge pathways. These analyses help define where infrastructure is most likely to face future impacts. Exposure analysis focuses on identifying infrastructure assets and populations located in hazard-prone areas. Coastal zones often contain dense concentrations of ports, industries, tourism facilities, and urban settlements, resulting in high exposure levels. A single extreme event can disrupt several infrastructure systems at once, increasing damage and recovery costs. Identifying exposed assets allows planners to prioritize critical facilities such as hospitals, power plants, transport corridors, and water supply systems that are essential for public safety and economic stability (Kantamaneni, 2016; Nguyen et al., 2025; Tumawu et al., 2024).

Vulnerability assessment examines how sensitive infrastructure systems are to climate stresses. Factors such as design standards, construction materials, age, maintenance practices, and the availability of backup systems influence how infrastructure performs during extreme events. Social and institutional conditions also affect vulnerability. Limited technical capacity, weak governance, and insufficient funding can delay recovery and increase long-term impacts. Combining information on hazards, exposure, and vulnerability helps identify where risks are greatest and why certain systems are more likely to fail. Risk management builds on assessment by applying strategies that reduce losses and improve preparedness. Avoiding risk is one of the most effective long-term approaches and involves limiting new infrastructure development in high-risk coastal zones through land-use planning and zoning regulations. Reducing risk focuses on strengthening existing infrastructure through design improvements, better drainage, elevation of critical facilities, and the use of natural features such as wetlands and mangroves to reduce flood and wave impacts. Retrofitting older infrastructure is particularly important in many coastal cities where systems were designed for past climate conditions. Risk transfer supports recovery by spreading financial losses through insurance and other financial mechanisms. Although these tools do not reduce physical damage, they help ensure that funds are available for repair and reconstruction after disasters. Risk retention remains unavoidable, as no strategy can remove all climate threats. Emergency planning, early warning systems, backup services, and rapid repair arrangements help manage remaining risks and reduce disruption during extreme events. Institutional coordination and governance strongly influence how well risk assessment and management are carried out. Coastal infrastructure involves multiple sectors and agencies, making coordination essential. Integrated planning approaches and climate-informed policies help align investments, standards, and responsibilities. Involving local communities improves decision-making by incorporating local knowledge and addressing social priorities. Risk assessment and management form the foundation of broader coastal resilience efforts. By identifying where risks exist and how they develop, these processes support better planning, smarter investment, and more equitable outcomes (Fig. 3.2). As climate change continues to reshape coastal environments, the ability to assess and manage risk will play a key role in protecting lives, infrastructure, and long-term coastal sustainability (Pal et al., 2023; Tumawu et al., 2024).

3.5 Resilient Coastal Engineering and Construction

Resilient coastal engineering and construction have become essential for the long-term safety and sustainability of coastal regions as climate-related disasters continue to intensify. Rising sea levels, stronger storms,

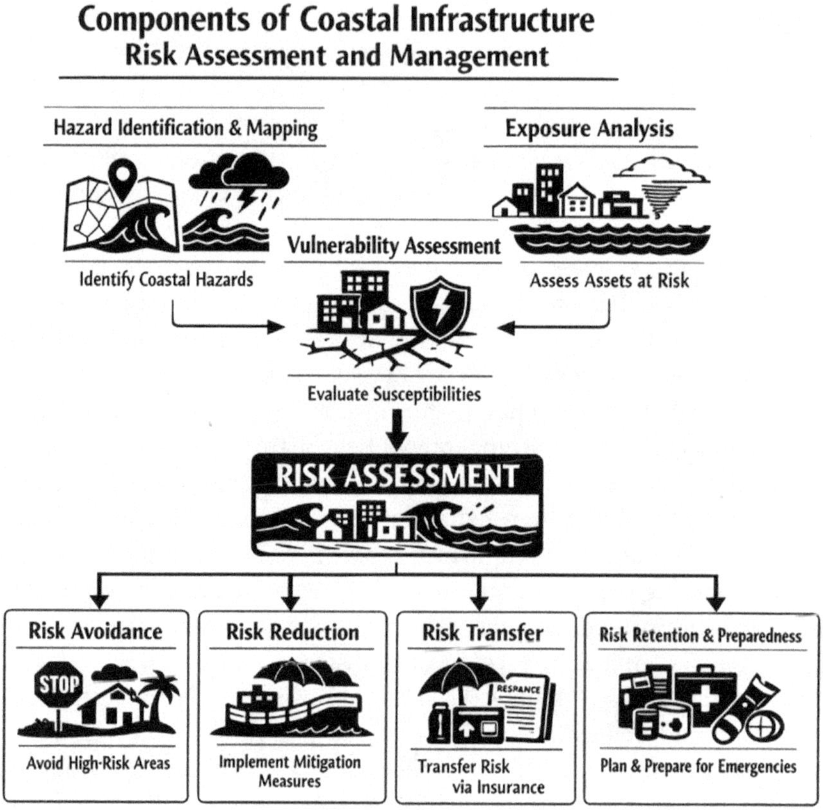

Fig. 3.2 Components of coastal infrastructure risk assessment and management (Prepared by the authors)

and increasing coastal erosion are placing growing pressure on infrastructure systems that were designed for more stable environmental conditions. Ports, bridges, roads, seawalls, housing, power plants, and water and sanitation systems in coastal areas must now operate under conditions marked by uncertainty, frequent extremes, and long-term environmental change. In response, resilient coastal engineering combines sound technical design with climate knowledge, ecological awareness, and risk-based planning to help infrastructure withstand damage, adjust to change, and recover after disruption. Resilient coastal engineering goes beyond simply building stronger structures. Resistance to waves, wind, and flooding remains important, but resilience also depends on flexibility, backup capacity, and the ability to recover quickly. Infrastructure should be able to absorb stress without complete failure, continue providing essential services during extreme events, and return to normal use within a reasonable time. This approach reflects a shift away from designs that assume damage can always be avoided, toward systems that accept the possibility of failure and allow for controlled damage and efficient repair. Such thinking is especially relevant in coastal areas where hazards are frequent and difficult to predict. Effective coastal engineering begins with a clear understanding of the physical processes that shape coastal environments. Tides, waves, currents, sediment movement, and shoreline change interact with storms and heavy rainfall to influence how water and energy move through coastal landscapes. Climate change is altering these processes by raising sea levels and increasing the intensity of extreme events. Engineers rely on scientific analysis and simulation tools that combine ocean, climate, and geological information to explore future conditions and identify weak points in infrastructure systems. These tools support better design choices by testing how structures may perform under different climate conditions (Slinger et al., 2021; Tschirky, 2024; Zhou et al., 2025).

In practice, resilient coastal construction applies design approaches and materials suited to harsh marine environments. Elevated structures reduce exposure to flooding and storm surges, while well-designed foundations help limit damage from erosion and scour. Materials such as corrosion-resistant steel, durable concrete, and reinforced composites are commonly used to withstand saltwater exposure, humidity, and temperature changes. Construction methods such as modular design and prefabrication allow damaged components to be repaired or replaced more quickly after extreme events, reducing downtime and recovery costs. The arrangement and layout of infrastructure systems also shape resilience outcomes. Linear systems such as roads, railways, and pipelines are especially vulnerable because damage at one point can disrupt the entire network. Resilient design introduces alternative routes and backup systems so that services can continue even when part of the system fails. In ports and industrial zones, critical facilities such as power supply units, fuel storage, and communication systems are often elevated or spatially separated to reduce the risk of widespread disruption during floods or storms. Natural and hybrid solutions play an increasingly important role in resilient coastal engineering. Coastal ecosystems such as mangroves, salt marshes, coral reefs, and sand dunes reduce wave energy, stabilize shorelines, and limit erosion. Integrating these natural features into engineering designs reduces stress on built structures while supporting biodiversity and local livelihoods. Combining engineered defenses with natural systems often provides more adaptable and sustainable protection than relying on hard structures alone (Slinger et al., 2021; Tschirky, 2024; Zhou et al., 2025).

Construction practices and project management also influence resilience. Coastal construction sites are exposed to weather extremes that can delay work, damage unfinished structures, and increase safety risks. Careful scheduling, including avoiding peak storm seasons and using temporary protective measures, helps reduce these challenges. Strong quality control during construction is equally important, as small defects can become serious weaknesses under extreme conditions. Governance, regulation, and institutional capacity strongly affect how resilient coastal engineering is applied. Design standards, building codes, and approval processes need to reflect present and future climate risks rather than past conditions. This includes higher flood levels, stronger wind requirements, and longer planning horizons. Education and training ensure that engineers, planners, and construction professionals are prepared to apply resilient practices effectively. The social and economic value of resilient coastal construction is significant. Infrastructure supports daily life, livelihoods, and economic activity, and its failure can have serious consequences for communities. Resilient ports support trade and employment, reliable transport systems enable emergency response, and stable water and energy services protect public health. Although resilient construction may involve higher initial costs, these investments reduce long-term losses, lower repair expenses, and increase confidence among communities, investors, and insurers. Resilient coastal engineering and construction offer a practical response to the changing nature of coastal risk. This approach recognizes the dynamic character of coastal environments and seeks to work with change rather than resist it completely. By combining scientific knowledge, technical innovation, ecological awareness, and effective governance, resilient coastal engineering supports safer, more adaptable, and more sustainable coastal regions in a future shaped by climate change.

3.6 Coastal Infrastructure Planning and Policy

Coastal infrastructure planning and policy provide the institutional and strategic base for building resilience in a period of increasing climate-related disasters (Fig. 3.3). Coastal zones support a large share of national economic activity, population, and critical assets such as ports, energy facilities, transport networks, and urban centers. These same areas are becoming more exposed to sea-level rise, storm surges, coastal flooding, erosion, and extreme weather linked to climate change. In this setting, planning and policy choices directly shape whether coastal development reduces risk or increases long-term vulnerability. Well-designed infrastructure policy brings together scientific knowledge, social needs, and economic priorities to guide how infrastructure is planned, built, maintained, and adapted over time. Coastal infrastructure planning needs to move away from short-term, growth-focused approaches and adopt a long-term perspective that recognizes climate risk. Earlier planning practices often assumed stable environmental conditions and prioritized immediate economic returns, which led to infrastructure being located in hazard-prone areas. Climate change has altered these assumptions. Rising sea levels and stronger storms mean that decisions made today will influence exposure and risk for many decades. Planning processes must therefore account for future conditions by using climate projections, hazard maps, and vulnerability assessments to guide land-use zoning, infrastructure location, and investment priorities. Aligning development with expected climate conditions helps reduce future losses and avoids unsafe patterns of growth. Integration across sectors and levels of governance is essential for resilient coastal infrastructure policy. Coastal risks affect transport, energy, water, housing, environmental management, and disaster response at the same time. When these sectors plan independently, gaps and inconsistencies often emerge, increasing overall vulnerability. Coordinated approaches such as integrated coastal management and climate-resilient development frameworks help align policies and investments across agencies and government levels. National strategies, regional plans, and local regulations need to work toward shared resilience goals (Jha et al., 2022; Johnson & Bayley, 2016; Jordan & Fröhle, 2022).

Policy also shapes the standards and rules that influence the quality and safety of coastal infrastructure. Building codes, engineering guidelines, and environmental regulations determine how infrastructure is designed, constructed, and maintained. In many coastal regions, these standards are still based on past climate conditions and do not reflect current or future risks. Updating regulations to include higher flood levels, stronger wind requirements, and longer design lifetimes is necessary to improve resilience. These changes help ensure that new infrastructure is better prepared for climate extremes and that existing assets can be strengthened through retrofitting. Financing and investment policies play an important role in coastal infrastructure planning. Resilient infrastructure may require higher initial investment, but it reduces long-term costs linked to disaster damage, repairs, and economic disruption. Public spending priorities, private sector incentives, and financial instruments need to support these long-term benefits. Adaptation funds, insurance schemes, climate bonds, and public-private partnerships can help mobilize resources for resilient infrastructure. Transparent financial management helps ensure that investments are effective and fairly distributed (Chávez et al., 2021; Kasthala et al., 2025).

Social equity and community participation are central to effective coastal infrastructure policy. Climate disasters often affect low-income and marginalized groups most severely, particularly those living in informal or poorly serviced coastal settlements. Planning that ignores these realities can deepen existing inequalities by protecting high-value areas while leaving vulnerable communities exposed. Inclusive planning processes that involve local residents, civil society groups, and indigenous communities help ensure that infrastructure decisions reflect diverse needs and local knowledge. Participation also builds trust and supports successful policy implementation. Strong institutions and access

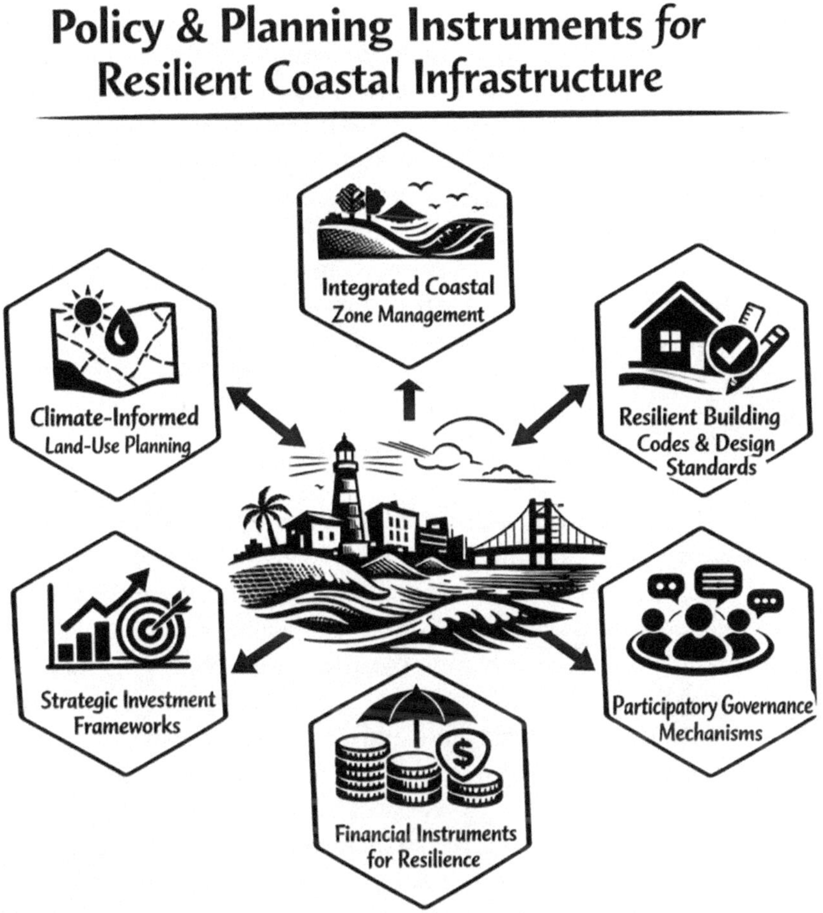

Fig. 3.3 Policy and planning instruments for resilient coastal infrastructure (Prepared by the authors)

to reliable data are essential for effective planning and policy. Climate information, coastal monitoring systems, and accurate infrastructure records support informed decision-making. Governments need to invest in technical capacity, data sharing, and coordination across agencies to translate knowledge into practice. Local authorities play a key role, as they are often responsible for zoning regulations, building approvals, and emergency preparedness. Coastal infrastructure planning and policy link scientific understanding, engineering practice, financial decision-making, and social priorities. These frameworks shape the rules and incentives that determine whether infrastructure increases risk or supports adaptation. When climate risk, social equity, and long-term sustainability are built into planning and policy, coastal regions can move from reactive responses toward proactive resilience-building. This approach protects physical infrastructure while also supporting livelihoods, ecosystems, and the long-term well-being of coastal communities in a changing climate (Hong et al., 2023; Van Assche et al., 2020).

3.7 Conclusion and Arguments

The relationship between climate disasters and coastal infrastructure makes it clear that the future of coastal regions depends on how effectively environmental risks are anticipated, addressed, and managed. Climate-related hazards such as sea-level rise, stronger storms, and accelerating coastal erosion are reshaping coastlines and challenging long-held ideas about safety and stability. Coastal zones can no longer be treated as stable boundaries between land and sea. They are dynamic spaces where infrastructure must function under growing uncertainty and frequent exposure to extreme events. In this context, resilience needs to guide every stage of coastal infrastructure development, from design and construction to operation and governance. Traditional infrastructure models, often based on historical climate conditions and short-term economic priorities, are increasingly misaligned with present and future realities. Climate disasters are recurring processes that reveal deep weaknesses in physical systems, institutional arrangements, and development choices. Coastal infrastructure systems such as ports, transport networks, housing, energy supply, and water services are closely connected. Failure in one part of the system can quickly spread to others, leading to wider social and economic disruption. Resilience therefore depends not only on the strength of individual structures but also on the ability of entire systems to adapt, respond, and continue functioning during periods of stress. Progress toward resilient coastal futures relies on the coordination of multiple approaches. Engineering innovation provides the technical base for infrastructure that can withstand damage and recover after extreme events. Adaptation strategies such as protection, accommodation, and planned retreat offer practical responses to different levels of climate risk. Risk assessment and management support informed decision-making by identifying priority areas for intervention and investment. Planning and policy frameworks connect scientific knowledge with coordinated action across sectors and levels of governance. Together, these elements help shift coastal infrastructure away from vulnerability and toward long-term stability. Building climate-resilient coastal infrastructure is also a social and political choice. Development in coastal areas must recognize environmental limits and changing climate conditions. When resilience guides planning and investment, coastal regions are better prepared to protect lives, support economic activity, and conserve fragile ecosystems. Such an approach encourages societies to adjust to coastal change in a planned and responsible way, rather than responding only after disasters occur. The chapter discussions lead to the following arguments:

(a) Climate disasters have significantly changed the nature of risk in coastal regions, making conventional infrastructure planning and design approaches insufficient for current and future conditions.
(b) Coastal infrastructure should be viewed as an interconnected system, where failure in one component can trigger wider economic, social, and environmental impacts.
(c) Resilient engineering and construction practices are necessary to ensure that coastal infrastructure can endure extreme events, adjust to changing conditions, and recover efficiently after disruption.
(d) Adaptation strategies such as protection, accommodation, and planned retreat offer flexible and location-specific responses to different levels of climate risk.
(e) Risk assessment and management provide a structured basis for identifying vulnerabilities, setting priorities, and reducing potential damage within coastal infrastructure systems.

(f) Nature-based and hybrid solutions contribute to resilience by combining physical protection with ecological benefits and social value.
(g) Planning and policy frameworks play a key role in integrating climate risk, social equity, and long-term sustainability into coastal infrastructure development.
(h) Inclusive governance and active community participation improve the effectiveness, fairness, and long-term success of resilience-building initiatives.

References

Adshead, D., Paszkowski, A., Gall, S. S., Peard, A. M., Adnan, M. S. G., Verschuur, J., & Hall, J. W. (2024). Climate threats to coastal infrastructure and sustainable development outcomes. *Nature Climate Change, 14*, 344–352. https://doi.org/10.1038/s41558-024-01950-2

Chaudhuri, S. K. (2025). Socioecological resilience among coastal communities for the sustainable blue economy in India. In W. Leal Filho, A. L. Salvia, J. P. P. Eustachio, & M. A. P. Dinis (Eds.), *Handbook of sustainable blue economy*. Springer. https://doi.org/10.1007/978-3-031-32671-4_19-1

Chávez, V., Lithgow, D., Losada, M., & Silva-Casarin, R. (2021). Coastal green infrastructure to mitigate coastal squeeze. *Journal of Infrastructure Preservation and Resilience, 2*, 7. https://doi.org/10.1186/s43065-021-00026-1

Geukes, H. H., Visser, E. S., Bondarouk, E. V., van Bodegom, P. M., & van Oudenhoven, A. P. (2026). Navigating the unknown: Nature-based solutions for coastal climate adaptation under deep uncertainty. *Journal of Environmental Management, 398*, 128483. https://doi.org/10.1016/j.jenvman.2025.128483

Hong, G., Vivian, C., Vogt, C., Haag, F., Zuo, F., & Qin, K. (2023). Enhancing the Anthropocene coastal infrastructure sustainability using the approaches developed by the London convention and protocol. *Anthropocene Coasts, 6*, 7. https://doi.org/10.1007/s44218-023-00022-w

Jha, R., Singh, V. P., Singh, V., Roy, L. B., & Thendiyath, R. (Eds.). (2022). *River and coastal engineering: Hydraulics, water resources and coastal engineering*. Springer. https://doi.org/10.1007/978-3-031-05057-2

Johnson, M., & Bayley, A. (Eds.). (2016). *Coastal change, ocean conservation and resilient communities*. Springer. https://doi.org/10.1007/978-3-319-41914-5

Jordan, P., & Fröhle, P. (2022). Bridging the gap between coastal engineering and nature conservation? *Journal of Coastal Conservation, 26*, 4. https://doi.org/10.1007/s11852-021-00848-x

Kantamaneni, K. (2016). Coastal infrastructure vulnerability: An integrated assessment model. *Natural Hazards, 84*(1), 139–154. https://doi.org/10.1007/s11069-016-2413-y

Kasthala, S., Parthasarathy, D., Narayanan, K., & Inamdar, A. B. (2025). A methodological framework for assessing anthropogenic vulnerability caused by the coastal regulation zone rules in India. *Discover Oceans, 2*, 47. https://doi.org/10.1007/s44289-025-00079-0

Leal Filho, W. (Ed.). (2018). *Climate change impacts and adaptation strategies for coastal communities*. Springer. https://doi.org/10.1007/978-3-319-70703-7

Leal Filho, W., Nagy, G. J., & Ayal, D. Y. (Eds.). (2025). *Handbook of nature-based solutions to mitigation and adaptation to climate change*. Springer. https://doi.org/10.1007/978-3-030-98067-2

Lucatello, S. (2025). Nature-based solutions to climate change adaptation and mitigation in Latin America and the Caribbean: An overview of projects and implementation. In W. Leal Filho, G. J. Nagy, & D. Y. Ayal (Eds.), *Handbook of nature-based solutions to mitigation and adaptation to climate change*. Springer. https://doi.org/10.1007/978-3-030-98067-2_87-1

Mashwama, N. X., & Phesa, M. (2025). Systematic review of multidimensional assessment of coastal infrastructure resilience to climate-induced flooding: Integrating structural vulnerability, system capacity, and organizational preparedness. *Climate, 13*(9), 192. https://doi.org/10.3390/cli13090192

Malakar, K. D., Kumar, M., Anand, S., & Kuzur, G. (2023). Climate Crisis and Socio-Ecological Sustainability. In: *Climate Change and Socio-Ecological Transformation*. Advances in Geographical and Environmental Sciences. Springer, Singapore. https://doi.org/10.1007/978-981-99-4390-6_9

Nguyen, T. H. H., Wang, G., Chen, W., Yu, J., Liu, R., Huang, X., et al. (2025). Risk assessment and management strategy of coastal erosion in the red River Delta, Vietnam. *Land, 14*(6), 1247. https://doi.org/10.3390/land14061247

Oktari, R. S., Comfort, L. K., & Dwitama, P. (2020). Measuring coastal cities' resilience toward coastal hazards: Instrument development and validation. *Progress in Disaster Science, 5*, 100057. https://doi.org/10.1016/j.pdisas.2019.100057

Pal, I., Kumar, A., & Mukhopadhyay, A. (2023). Risks to coastal critical infrastructure from climate change. *Annual Review of Environment and Resources, 48*(1), 681–712. https://doi.org/10.1146/annurev-environ-112320-101903

Powell, E. J., Tyrrell, M. C., Milliken, A., Tirpak, J. M., & Staudinger, M. D. (2019). A review of coastal management approaches to support the integration of ecological and human community planning for climate change. *Journal of Coastal Conservation, 23*, 1–18. https://doi.org/10.1007/s11852-018-0632-y

Rahman, S., & Rahman, M. A. (2015). Climate extremes and challenges to infrastructure development in coastal cities in Bangladesh. *Weather and Climate Extremes, 7*, 96–108. https://doi.org/10.1016/j.wace.2014.07.004

Rosenzweig, C., Solecki, W. D., Blake, R., Bowman, M., Faris, C., Gornitz, V., et al. (2011). Developing coastal adaptation to climate change in the new York City infrastructure-shed: Process, approach, tools, and strategies. *Climatic Change, 106*, 93–127. https://doi.org/10.1007/s10584-010-0002-8

Sauer, I. J., Roca, E., & Villares, M. (2021). Integrating climate change adaptation in coastal governance of the Barcelona metropolitan area. *Mitigation and Adaptation Strategies for Global Change, 26*, 16. https://doi.org/10.1007/s11027-021-09953-6

Singhvi, A., Luijendijk, A. P., & van Oudenhoven, A. P. (2022). The grey–green spectrum: A review of coastal protection interventions. *Journal of Environmental Management, 311*, 114824. https://doi.org/10.1016/j.jenvman.2022.114824

Slinger, J., Stive, M., & Luijendijk, A. (2021). Nature-based solutions for coastal engineering and management. *Water, 13*(7), 976. https://doi.org/10.3390/w13070976

Tschirky, P. (2024). Coastal engineering: The frontline of resilience and climate adaptation. *Coastal Engineering Proceedings, 38*, 232–232. https://doi.org/10.9753/icce.v38.management.232

Tumawu, A. F., Kafu, G. Y., Yiran, G. A. B., & Frimpong, L. K. (2024). Assessment of coastal flood risk scenarios on infrastructure in the Keta municipality in Ghana using a GIS approach. *Heliyon, 10*(21). https://doi.org/10.1016/j.heliyon.2024.e39824

Van Assche, K., Hornidge, A. K., Schlüter, A., & Vaidianu, N. (2020). Governance and the coastal condition: Towards new modes of observation, adaptation and integration. *Marine Policy, 112*, 103413. https://doi.org/10.1016/j.marpol.2019.01.002

Zhou, Y., Jiang, C., Jiang, Y., Zhu, Y., Jin, Y., Wang, X., et al. (2025). A whole process resilience management practice in coastal engineering. *Frontiers in Marine Science, 11*, 1518249. https://doi.org/10.3389/fmars.2024.1518249

4

Risk Perceptions, Communication, and Early Warning Systems

This chapter places risk communication at the center of disaster risk management by exploring how information, public perception, trust, and technology come together within early warning systems. It explains that communication during disasters is not simply about sending out warnings, but about how people receive, interpret, and act on those messages. The chapter looks at the many ways warnings are shared, including traditional media, digital platforms, and direct community interactions, and discusses how these channels affect the clarity, reach, and credibility of early warnings. The chapter also pays close attention to human behavior during emergencies. It discusses how emotions such as fear and uncertainty, along with trust in authorities and information sources, influence decision-making in high-risk situations. Issues like misinformation and misunderstanding are examined to show how they can weaken response efforts if not addressed carefully. Cultural context is another key focus of the chapter. It highlights the importance of local languages, indigenous knowledge, and community traditions in making risk communication meaningful and inclusive. By bringing together social, cultural, and technical perspectives, the chapter sets the stage for understanding how early warning systems can support timely

K. D. Malakar et al., *Coastal Ecologies at Risk*, Palgrave Studies in Environmental Sustainability, https://doi.org/10.1007/978-3-032-26081-9_4

action and help build stronger and more resilient communities in the face of growing disaster risks.

Key Points of the Chapter

- Describes how risk communication connects early warnings with lifesaving actions.
- Explains how human behavior, emotions, and trust shape responses to warnings.
- Highlights the role of multi-channel communication in reaching communities at the last mile.
- Examines how feedback and learning improve the effectiveness of warning systems.
- Describes the value of cultural sensitivity and indigenous knowledge in risk communication.

4.1 Risk Communication

Risk communication refers to the ongoing process through which information about hazards, uncertainty, and protective actions is shared between authorities, experts, the media, and the public. It goes beyond the simple transfer of technical facts and focuses on helping people understand risk in ways that support timely and practical decisions. In disaster and early warning contexts, risk communication is a central part of preparedness, because even accurate forecasts or advanced warning technologies are ineffective if people do not receive the message clearly or feel confident enough to act on it. At a conceptual level, risk communication links scientific knowledge with everyday decision-making. Specialists assess hazards using data and forecasts, but communities experience risk through daily life, past events, and local conditions. Risk communication translates technical information, such as warning levels or expected impacts, into messages that relate to people's real situations. This process helps connect expert assessments with public understanding. Risk perception plays a key role in shaping responses. People judge risk based not only on official information but also on personal experience, social memory, media influence, and cultural values. Communities that have faced disasters before may respond quickly, while others may underestimate danger due to long periods without major events. Effective communication recognizes these differences and adjusts messages to local contexts rather than relying only on technical detail. Trust is another essential element. People are more likely to follow warnings when they trust the institutions and individuals who issue them. Trust develops through clear, consistent, and transparent communication over time. When uncertainty is acknowledged and explained honestly, people are more willing to take protective action. Poor coordination or unclear messages can weaken confidence and reduce response. Clarity and accessibility are equally important. Messages should use simple language and practical guidance so they can be understood by people with different levels of education and access to technology. Combining digital tools with local communication channels helps ensure that warnings reach everyone. Risk communication also values dialogue and community involvement. Local knowledge and feedback strengthen early warning systems and encourage shared responsibility. In this way, risk communication turns early warnings into meaningful action and supports long-term resilience (Fig. 4.1) (Bourrier & Bieder, 2018; Covello et al., 1988; Shepherd & Frewer, 1997).

Risk communication is a key part of disaster risk management and early warning systems because it connects scientific knowledge about hazards with the ways people understand and respond to danger. While monitoring systems and risk assessments help identify possible threats, communication decides whether this information leads to timely and appropriate action. In regions exposed to cyclones, floods, earthquakes, tsunamis, heat waves, and industrial hazards, the effectiveness of early warnings depends less on advanced tools and more on how clearly, credibly, and inclusively risk information reaches those at risk. Risk communication involves an ongoing exchange of information among scientists, government agencies, emergency managers, the media, and communities. It goes beyond sharing technical details and focuses on explaining uncertainty, likely impacts, and protective actions in ways that support everyday decision-making. Within early warning systems, risk

Fig. 4.1 Elements of risk communication (prepared by the authors)

communication turns forecasts into guidance that helps people understand what a threat means for their safety, livelihoods, and immediate choices. In this sense, it plays a central role in building disaster-resilient societies. Several elements influence effective risk communication. The source of information strongly shapes public response. Messages delivered by trusted institutions or respected community leaders are more likely to be believed and acted upon. Trust develops over time through accuracy, consistency, and openness. When warnings are seen as reliable and honest, people are more willing to respond during emergencies (Ali & Kamraju, 2026; Trogrlić et al., 2022).

The content of the message is equally important. Risk messages should explain the nature of the hazard, the level of risk, who may be affected, and what actions should be taken. Technical language and complex terms can limit understanding, especially in stressful situations. Clear and practical instructions help people respond quickly and with confidence.

Communication channels also play a critical role. Risk information may be shared through television, radio, mobile phones, social media, sirens, and local networks. Each channel reaches different groups. Using several channels together helps ensure that warnings reach both urban and remote populations in time. Audience characteristics must be considered. Differences in language, age, culture, disability, and livelihood affect how people receive and interpret warnings. Communication that reflects these differences is more likely to reach vulnerable groups and reduce confusion. Timing, feedback, and interaction further strengthen communication. Warnings should be issued early enough to allow action and updated as conditions change. Opportunities for dialogue allow communities to share local knowledge and build shared responsibility. Risk communication is important because it turns knowledge into action. Clear and trusted warnings reduce loss of life and property, improve risk awareness, build trust in institutions, limit misinformation, and support inclusive disaster management. Within early warning systems, risk communication links forecasts with protective action and long-term resilience (Fig. 4.2) (Graf et al., 2026; Gündüz & Dinçer, 2025; Kalligeris et al., 2026).

4.1.1 Role of Risk Perception in Communication

Risk perception plays an important role in shaping how coastal communities receive, interpret, and respond to risk communication (Fig. 4.3). People living in coastal systems are often exposed to hazards such as cyclones, storm surges, flooding, erosion, and sea-level rise. The way these hazards are perceived largely determines whether warnings are accepted and acted upon or ignored. Risk perception is shaped not only by scientific information but also by past experiences, local knowledge, shared memories of previous events, and daily interaction with the coastal environment. Communities that have faced serious coastal disasters in the past often respond more quickly to warnings, while those without recent experience may underestimate danger even when official alerts are issued. Risk perception strongly influences trust, attention, and decision-making during emergencies. Coastal residents commonly

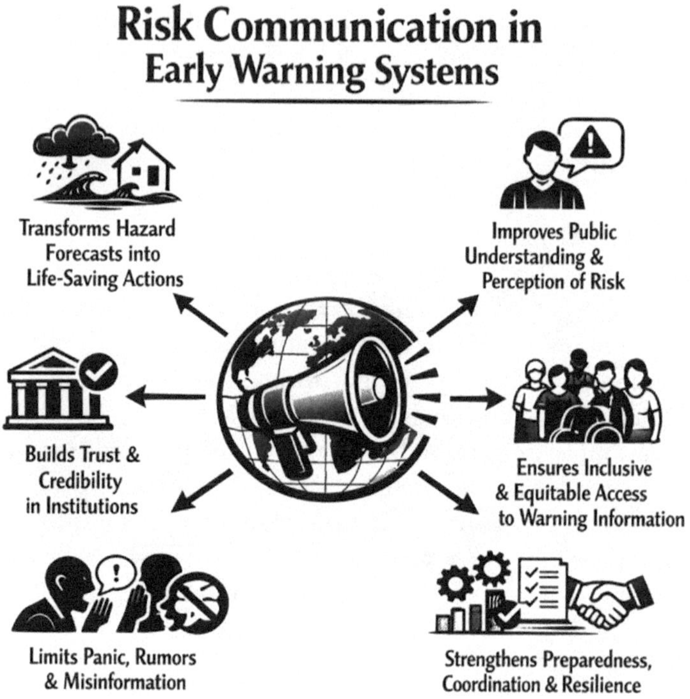

Fig. 4.2 Risk communication in early warning systems (prepared by the authors)

judge the reliability of warnings based on what happened in the past. When earlier alerts were unclear or did not lead to visible impacts, people may lose confidence and delay action during later events. At the same time, repeated exposure to hazards can create a sense of familiarity that reduces perceived risk and encourages unsafe behavior. Risk communication needs to address both fear and overconfidence in a careful and balanced way (Lechowska, 2018; Moreschini et al., 2025).

Social and cultural factors further shape how risk is perceived in coastal areas. Livelihoods such as fishing, tourism, and coastal farming depend directly on the sea, and economic pressures may lead people to downplay risk to protect daily income. Cultural beliefs, traditions, and community norms also influence how danger is understood and discussed. Messages that reflect local language, values, and experiences are more likely to be trusted and followed. Risk perception also affects how uncertainty

Fig. 4.3 Role of coastal risk perception in communication (prepared by the authors)

is understood. Coastal hazards often change quickly, and forecasts may be updated as conditions evolve. When uncertainty is explained in clear and simple terms, people are better able to prepare and respond. Over time, regular communication, education, and community involvement help align public perception with real levels of risk, supporting preparedness and resilience in coastal systems (Aude et al., 2025; Matyas et al., 2011; Monaco, 2025).

4.1.2 Challenges in Risk Communication

Risk communication in coastal systems faces many challenges because coastal hazards are complex and coastal communities are highly diverse. Hazards such as storm surges, coastal erosion, flooding, and sea-level rise are often explained using technical language that is difficult for many people to understand, especially during emergencies. Even when information is accurate, unclear or complicated messages can lead to misunderstanding or inaction. If people do not clearly understand what the risk is and what they should do, warnings may fail to achieve their purpose. Trust and credibility present another major challenge. In many coastal areas, earlier experiences with false alarms, delayed warnings, or inconsistent messages from different authorities have reduced public confidence. When trust is weak, people may question new warnings or wait too long before responding. Language differences, low literacy levels, and limited access to communication technology further affect the delivery of timely information, particularly in remote fishing villages and informal coastal settlements. Social and economic conditions also influence how risk messages are received. Many coastal residents depend on fishing, tourism, or marine activities for their income, and concerns about losing daily earnings may lead them to downplay risk. Cultural beliefs, traditions, and local norms shape how hazards are understood and discussed within communities. At the same time, misinformation can spread quickly through informal networks and digital media, creating confusion and fear. These challenges show the need for risk communication that is clear, trusted, and adapted to local coastal contexts (Monaco, 2025; Rollason et al., 2018; Saengsupavanich et al., 2023).

4.2 Risk Communication Channels and Tools

4.2.1 Risk Communication Channels and Tools: Traditional Media

Traditional media continues to play an important role in risk communication within early warning systems, particularly in disaster-prone and resource-limited regions (Fig. 4.4). Even with the rapid growth of digital platforms, newspapers, radio, television, and public broadcasting remain widely used and trusted sources of information. Their strength lies in their wide reach, long-standing credibility, and familiarity among communities. During disasters, traditional media helps link scientific agencies, government authorities, and the public by sharing warnings and preparedness messages in a timely and reliable way. Radio has long been one of the most effective tools for disaster-related communication, especially in rural, coastal, and remote areas. Battery-operated radios and local community stations can continue broadcasting even when electricity and internet services are disrupted. This makes radio especially valuable during cyclones, floods, and earthquakes. Because radio relies on spoken messages, it can overcome literacy barriers and communicate in local languages and dialects. Regular weather updates, emergency announcements, and interviews with officials help turn technical forecasts into clear guidance that people can understand and act upon. Television supports risk communication by combining sound and visual information. Live broadcasts, maps, graphics, and on-screen alerts help people understand where hazards are occurring and how serious they may be. Continuous news coverage and official briefings during emergencies keep the public informed as situations change. Visual information, such as images of flood levels or storm paths, can improve risk awareness and encourage protective action. Print media, including newspapers and magazines, plays a stronger role before and after disasters. While it is less effective for urgent warnings, it helps build long-term awareness through detailed reporting and explanation. Public broadcasting and government information services also play a key role by sharing verified information quickly and reducing confusion. Together, traditional media remains a vital part of early warning systems by ensuring that risk information

Fig. 4.4 Risk communication channels and tools: traditional media (prepared by the authors)

reaches a wide and diverse audience (Nishizawa, 2018; Onyeaka, 2025; Vicari et al., 2025).

4.2.2 Risk Communication Channels and Tools: Social Media

Social media has become an important channel for risk communication within modern early warning systems (Fig. 4.5). Platforms such as X, Facebook, WhatsApp, Telegram, and YouTube allow hazard-related information to be shared quickly with large and diverse audiences. This speed is especially valuable during sudden events such as cyclones, floods, earthquakes, and heat waves. Social media allows both authorities and community members to share information at the same time, creating an interactive space that strongly influences how risk is understood and how people respond. One key strength of social media is its ability to spread information rapidly. Government agencies, weather services, and disaster management authorities can issue warnings and updates within seconds. These messages are often shared widely by users, allowing critical information to travel through personal networks. This process is particularly effective among younger people, urban populations, and migrant communities who rely heavily on mobile phones and digital platforms for updates. Social media also supports two-way communication, which is essential for effective risk communication. Community members can ask questions, report local conditions, and share photos or videos during emergencies. Such information helps authorities understand conditions on the ground, including damaged infrastructure, flooded areas, or people needing assistance. This exchange strengthens emergency response and makes communication more responsive to local needs. Visual content and personal stories shared on social media also shape risk perception. Images and videos can show the seriousness of a hazard more clearly than written warnings. Messages shared by friends, family members, or local community groups are often seen as more relatable and trustworthy, which can encourage quicker action. At the same time, social media presents challenges. False information and rumors can spread quickly and create confusion or fear. Clear and regular communication from trusted official sources is therefore essential. When used carefully and alongside traditional media, social media can improve the reach, clarity, and inclusiveness of risk communication and support stronger community response during disasters (Chen et al., 2024; Erokhin & Komendantova, 2024; Renn, 1991).

Fig. 4.5 Risk communication channels and tools: social media (prepared by the authors)

4.2.3 Risk Communication Channels and Tools: Digital Platforms

Digital platforms have become an important part of risk communication within modern early warning systems (Fig. 4.6). These platforms include official websites, mobile applications, SMS alert services, mapping dashboards, and disaster management portals managed by governments and emergency agencies. They are designed specifically to share risk-related information and support emergency management. This allows them to provide clear, reliable, and regularly updated information to both institutions and the public. A key strength of digital platforms is their ability to deliver real-time and location-specific information. Mobile apps and online portals linked to monitoring systems can show live maps of storms, flood levels, earthquakes, and evacuation zones. This helps people understand not only that a hazard exists, but whether they are personally at risk. When people can see how a hazard affects their own location, they are better able to make informed decisions such as choosing safer routes, evacuating early, or taking protective measures. Digital platforms also improve the delivery of early warnings through direct and automated communication. SMS alerts, push notifications, and cell broadcast messages can reach large numbers of people at the same time, even when they are not actively seeking information. These alerts often include short and clear instructions, such as moving to higher ground or avoiding coastal areas, which encourage quick action. Because these messages come directly from official systems, they reduce delays and limit confusion. Another important role of digital platforms is coordination. Disaster management portals can bring together information from weather agencies, water authorities, emergency services, and local governments in one place. This helps ensure that institutions work with the same information and that the public receives consistent guidance. Evacuation notices, shelter locations, and relief updates can be shared through a single platform, reducing uncertainty during emergencies. Digital platforms also support preparedness and learning. Many provide hazard maps, safety guidelines, and training materials that help people understand risks before disasters occur. However, access to technology and digital skills is not equal for everyone. For this reason, digital platforms should be used alongside traditional and social media, with simple design and local languages to ensure inclusive risk communication (Dinçer, 2024; Höppner et al., 2012; Strauss et al., 2025).

4.2.4 Risk Communication Channels and Tools: Face-to-Face Communication

Face-to-face communication remains one of the most effective and trusted forms of risk communication within early warning systems, especially in communities that are socially close, culturally diverse, or have limited access to digital technology (Fig. 4.7). Direct interaction allows

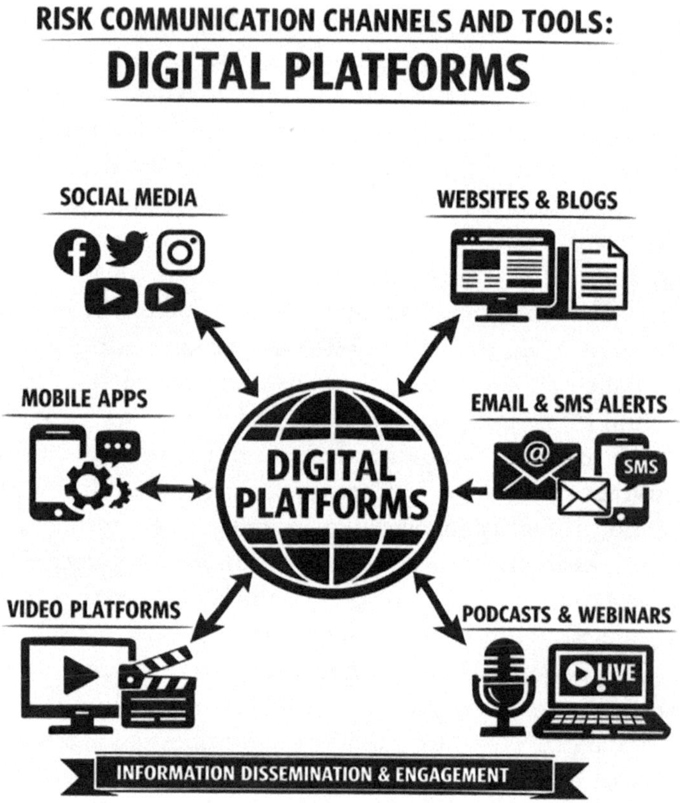

Fig. 4.6 Risk communication channels and tools: digital platforms (prepared by the authors)

risk messages to be shared with clarity, empathy, and immediate feedback. In disaster-prone settings such as coastal villages, informal urban settlements, and remote rural areas, this form of communication often decides whether warnings are clearly understood, trusted, and acted upon. A major strength of face-to-face communication is its ability to build trust. When warnings and safety messages are delivered by familiar and respected individuals such as community leaders, teachers, health workers, disaster volunteers, or elders, people are more likely to believe the information and follow advice. This is especially important in areas where confidence in distant authorities may be low due to past experiences with delayed or inaccurate warnings. Personal interaction makes risk messages feel relevant and credible. Face-to-face communication also allows two-way dialogue. Community members can ask questions, raise concerns, and clarify doubts about evacuation plans, shelters, or possible impacts on livelihoods. This exchange helps identify misunderstandings, cultural sensitivities, and practical barriers that could limit effective response. Local knowledge shared during these discussions, such as information on vulnerable households or safe routes, can improve early warning planning and delivery. Another important role of face-to-face communication is reaching vulnerable groups. Elderly people, persons with disabilities, women, children, and language minorities may not always receive information through mass or digital media. Door-to-door visits, community meetings, and local networks help ensure that no one is excluded. Messages can be adapted to local languages, customs, and specific needs. Face-to-face communication also supports preparedness through training, drills, and community discussions. These activities build shared understanding and practical skills, so people are better prepared when real warnings are issued. In early warning systems, personal outreach often provides the final link that turns alerts into action, helping reduce loss of life and strengthen community resilience (Berger-Sabbatel & Journé, 2018; Bica et al., 2020; Rafliana et al., 2022).

4.3 Early Warning Systems

Early warning systems are organized arrangements of people, institutions, technologies, and procedures that work together to detect hazards, assess risk, and share timely warnings with people who may be affected. They are not simply technical tools for predicting hazards but systems that link scientific monitoring with human decision-making. Their main purpose is to provide enough time and clear guidance so that individuals, communities, and authorities can take action to reduce harm before a hazard becomes a disaster. An effective early warning system depends on the integration of four closely connected components. Risk knowledge helps identify hazards, exposed populations, vulnerable infrastructure,

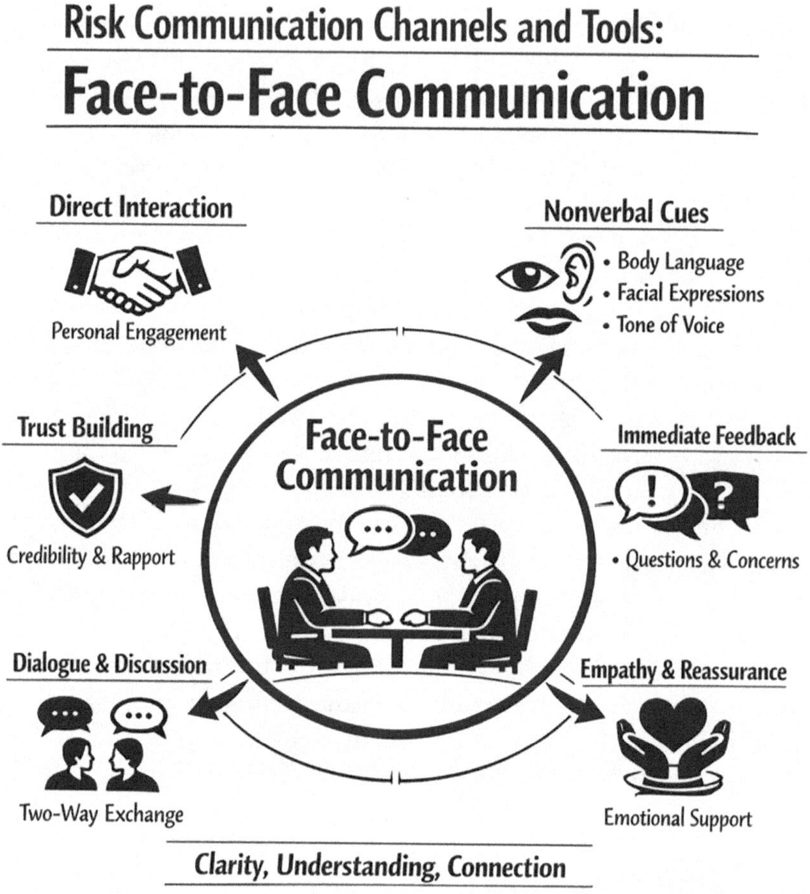

Fig. 4.7 Risk communication channels and tools: face-to-face communication (prepared by the authors)

and possible impacts. Hazard monitoring and forecasting track changing threats and estimate their timing and intensity. Communication and dissemination ensure that warnings reach people in ways they can understand and trust. Response capability refers to the readiness of individuals, communities, and institutions to act on warnings through evacuation, sheltering, or other protective measures. These components need to function together rather than separately. Strong coordination among institutions, clear rules and procedures, and active community involvement are essential to ensure that warnings are credible and followed. Within the wider context of risk perception and communication, the success of an early warning system depends on its ability to turn technical alerts into messages that people find meaningful and actionable. When this connection is achieved, early warning systems become a key tool for building disaster resilience and supporting proactive risk management (Islam et al., 2025; Lamsal & Vijay Kumar, 2020; Yadav et al., 2026).

Early warning systems are built on a set of closely linked components that work together to turn hazard detection into protective action. Risk knowledge is the first component and involves identifying hazards, exposed populations, vulnerable infrastructure, and possible impacts. This understanding helps ensure that warnings are focused on the people and places most at risk and are therefore more relevant and useful. Hazard monitoring and forecasting form the second component. Scientific instruments and observation networks track changing threats such as storms, floods, earthquakes, and heat waves. Forecasting processes use this information to estimate how severe an event might be and when it is likely to occur. Reliable monitoring and forecasting provide the foundation for timely warnings. Communication and dissemination make up the third and most critical component. This includes the systems, messages, and institutions that convert technical forecasts into clear and practical warnings for the public and decision-makers. When communication is weak or unclear, even accurate scientific information cannot prevent loss of life. Effective communication helps people understand the risk and take appropriate action. Response capability is the fourth component. It refers to the readiness of individuals, communities, and authorities to act on warnings through evacuation, sheltering, or other protective measures. Preparedness, available resources, and coordination all influence how well responses are carried out. These components function as a connected system rather than separate parts. When risk knowledge, monitoring, communication, and response work well together, early warnings become not only scientifically reliable but also socially meaningful and effective in practice (Chovanec et al., 2025; Kelman & Glantz, 2014; Tora et al., 2025).

Data collection, analysis, and interpretation form the scientific foundation of early warning systems and strongly influence how well hazards are detected and risks are anticipated. An early warning system is only as effective as the data it relies on. For hazards such as cyclones, floods, droughts, heat waves, landslides, and earthquakes, timely and accurate data allows authorities to move from uncertainty to informed action. These processes turn environmental and social information into forecasts and alerts that help protect lives, livelihoods, and infrastructure. Data collection begins with regular observation of natural and human-related processes that create risk. This includes weather data such as rainfall, wind speed, temperature, and air pressure; water-related data such as river levels, soil moisture, and groundwater; geological data linked to earthquakes and slope stability; and ocean data such as tides, wave height, and sea surface temperature. Socioeconomic information, including population distribution, land use, housing conditions, and critical infrastructure, is also important for understanding exposure and vulnerability. Early warning systems increasingly use a mix of ground instruments, satellite observations, remote sensing, and community-based monitoring to maintain continuous and updated data. Data analysis is the next key stage. Collected information is processed to identify patterns, warning thresholds, and signs of emerging danger. Forecasting methods are used to estimate the likely timing, intensity, and spread of hazards. Comparisons with historical records help distinguish normal conditions from situations that indicate risk. Through analysis, large volumes of data are converted into indicators that support early decision-making. Interpretation is equally important. Technical results must be translated into clear explanations of what may happen and what it could mean for people and places. This requires expert judgment, local knowledge, and attention to uncertainty. The same hazard can have very different impacts depending on local conditions. Interpreting data in relation to these conditions ensures that warnings are both accurate and meaningful. Modern early warning systems rely on combining many sources of data to build a fuller picture of risk. Mapping tools help identify high-risk areas and possible connected impacts. Data quality, timely sharing, and coordination among institutions are essential for maintaining trust in warnings. Community involvement in observation and reporting further strengthens early warning systems by adding local insight and improving response (Chavula et al., 2025; Islam et al., 2025).

4.3.1 Early Warning Systems and Science and Technology

The integration of technology and automated alert systems has greatly improved how modern early warning systems function. As climate conditions become more unpredictable and disasters occur with little notice, the ability to detect hazards, process information quickly, and share warnings without delay has become essential. Technology allows early warning systems to move away from slow and manual methods toward continuous, data-based processes that provide timely and useful information to people at risk. Monitoring technologies form the foundation of this integration. Sensors, satellites, weather radars, seismic instruments, ocean buoys, and river gauges collect information on environmental conditions on a continuous basis. These tools help track hazards such as cyclones, floods, earthquakes, tsunamis, and heat waves with greater accuracy. Automated data collection reduces reliance on human observation and allows monitoring even in remote or dangerous locations, improving early detection and forecasting (Wenzel & Zschau, 2014).

The collected data is processed through advanced computing systems and analytical models that can manage large volumes of information. These systems help identify warning signs, estimate hazard strength, follow movement patterns, and assess possible impacts. Automation makes it possible to generate warnings much faster than manual processes, which is critical for fast-moving hazards such as flash floods or earthquakes. Automated alert systems connect scientific analysis with public action. When predefined warning levels are reached, alerts can be issued automatically. These messages are shared through multiple channels such as SMS, mobile applications, sirens, public address systems, and digital displays. Using several channels at the same time increases the likelihood that warnings reach people quickly and support timely action. Technology also improves how warnings are targeted. Location-based systems allow alerts to be sent only to people in affected areas.

Digital maps help identify evacuation zones, safe routes, and shelter locations with greater accuracy. This makes warnings more relevant and easier to follow. Technology also supports coordination among agencies. Shared platforms allow institutions to access the same information in real time, reducing confusion and improving decision-making. For these systems to work well, they must be reliable and trusted. Backup arrangements, regular testing, and clear messages are essential. When combined with public awareness and preparedness, technology and automation greatly strengthen early warning systems and support safer and more resilient communities (Hermans et al., 2022; Marchezini et al., 2017).

4.4 Stakeholder Engagement in Early Warning Systems: Roles and Responsibilities

Early warning systems are not only technical arrangements; they are shared systems that depend on the coordinated efforts of many actors. From monitoring hazards to public response, each stage of an early warning system relies on the contribution of different stakeholders, each bringing specific skills, authority, and local understanding. Active stakeholder engagement is essential to ensure that warnings are accurate, trusted, and acted upon. When roles and responsibilities are clear and coordination is strong, early warning systems function more effectively and remain reliable over time. Scientific and technical institutions form the foundation of early warning systems. Agencies responsible for weather, water, geology, and environmental monitoring collect data, track hazards, and prepare forecasts and risk assessments. Their work provides the information needed to identify emerging threats and estimate possible impacts. Accuracy, openness, and regular improvement in monitoring practices are critical, as the credibility of warnings depends on the quality of this scientific input. Government and disaster management authorities play a central coordinating role. They convert scientific information into official warnings, activate emergency procedures, and ensure that alerts are shared through suitable channels. These authorities also establish policies, legal arrangements, and operational guidelines that define warning levels, evacuation procedures, and coordination among agencies (Fig. 4.8). Their leadership helps ensure that warnings are timely, clear, and authoritative (Jayasekara et al., 2022; Trogrlić et al., 2022; Tayebi et al., 2024).

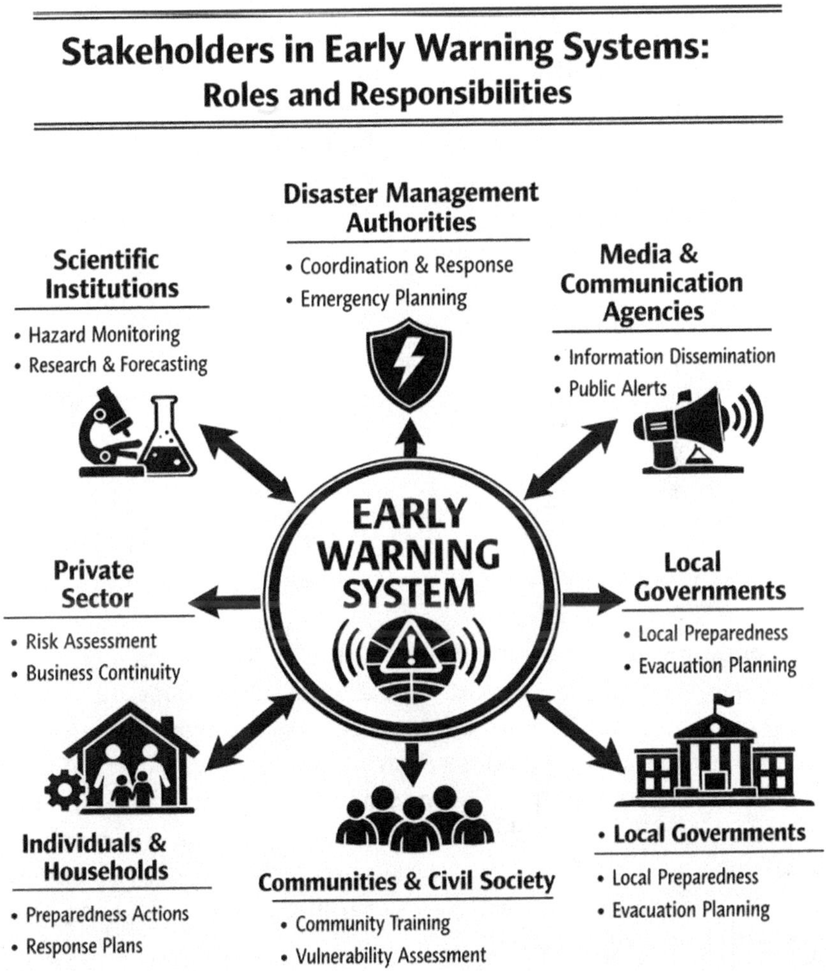

Fig. 4.8 Stakeholders in early warning systems: roles and responsibilities (prepared by the authors)

Media and communication agencies act as important links between institutions and the public. Through television, radio, print, and digital platforms, they help spread warnings and preparedness messages to large audiences. Responsible reporting is essential to avoid confusion, panic, or the spread of false information. Close cooperation between media organizations and disaster authorities improves clarity and consistency in public messaging. Local governments play a key role in turning wider warnings into local action. They manage shelters, evacuation routes, and emergency services and understand local risks and capacities. Communities and civil society groups also take an active role by sharing information, supporting vulnerable groups, and providing feedback. The private sector supports early warning systems through communication networks, technology, and essential services. Individuals and households complete the system by understanding warnings and taking protective action. Strong early warning systems depend on ongoing communication, trust, and cooperation among all stakeholders. When these relationships are in place, warnings are more likely to lead to timely and lifesaving action and help build resilience to disasters (Agbehadji et al., 2023; Pham et al., 2024).

4.5 Risk Communication in Different Phases of Disasters

Risk communication is not limited to the moment when a disaster occurs. It is a continuous process that extends across the entire disaster cycle, from the period before a hazard occurs, through the emergency itself, and into recovery and reconstruction. At each stage, communication influences how people understand risk, how they respond to warnings, and how they rebuild their lives. Examining risk communication across the pre-disaster, during-disaster, and post-disaster phases is therefore essential for improving early warning systems, reducing losses, and supporting long-term resilience.

In the pre-disaster phase, risk communication focuses on awareness, preparedness, and the shaping of risk perception. This phase includes public education campaigns, preparedness training, sharing of hazard maps, and regular weather or hazard updates. The aim is to help communities understand the risks they face and prepare for possible impacts. Clear communication during this stage helps people see disasters as events that can be anticipated and managed rather than sudden and unavoidable shocks. By explaining hazard patterns, seasonal risks, and likely impacts, communication builds a shared understanding between experts and the public, which is critical for trust in early warning systems. Pre-disaster communication also helps create a culture of readiness. Information about evacuation routes, emergency supplies, shelter locations, and family communication plans prepares people to act quickly when warnings are issued. Drills, training sessions, and community meetings reinforce this knowledge and turn abstract risk into practical action. This phase also allows space for dialogue, where communities can share local knowledge, raise concerns, and point out barriers to preparedness. Such interaction makes communication more grounded in local realities and increases the likelihood that warnings will be taken seriously later (Babič et al., 2023; Bradley et al., 2016).

When a hazard is imminent or already unfolding, communication moves into the during-disaster phase. At this point, speed, clarity, and authority are crucial. Messages shift from general awareness to urgent instructions. Early warnings, evacuation orders, and safety guidance must be delivered quickly and repeatedly through multiple channels so that everyone at risk receives them. The main goal during this phase is to protect lives and reduce immediate harm by guiding people toward safe actions. During-disaster communication needs to be simple and direct. People under stress have limited time and attention, so messages should clearly state what is happening, what actions are required, where to go, and how much time is available. Consistency across communication channels such as radio, television, mobile alerts, social media, and local announcements is essential to avoid confusion. Trust built during the pre-disaster phase strongly influences whether people follow these urgent instructions. This phase also requires careful handling of uncertainty and misinformation. Conditions can change rapidly, and incomplete information is common during emergencies. Regular updates, honest explanations of what is known and unknown, and quick correction of false information help maintain public confidence. Transparent communication reduces panic and supports informed decision-making in difficult circumstances (Bradley et al., 2016; Hansson et al., 2020; Rahman & Munadi, 2019).

After the immediate danger has passed, communication enters the post-disaster phase. This stage focuses on recovery, reconstruction, and learning. It often lasts much longer than the emergency itself and involves helping communities understand what happened, what support is available, and how recovery will take place. Communication during this phase supports emotional recovery by acknowledging loss and trauma, while also providing practical information about relief, compensation, housing, and livelihoods. Post-disaster communication is also important for restoring trust and social cohesion. Communities need clear and honest information about the causes of the disaster, how warning systems performed, and what steps will be taken to reduce future risk. Open communication about successes and failures creates opportunities for learning and improvement. When people feel informed and included, confidence in institutions is more likely to be maintained. Another key role of post-disaster communication is supporting long-term resilience. Recovery is not only about rebuilding what was lost but about reducing future vulnerability. Communication can promote safer construction practices, better land-use planning, and stronger community preparedness. By linking recovery efforts to lessons learned, communication helps turn experience into knowledge and knowledge into safer development (Dehghani et al., 2022; Lin et al., 2020).

Across all phases, risk communication interacts closely with risk perception. Before disasters, it shapes how people understand and prioritize risk. During emergencies, it influences how seriously warnings are taken. After disasters, it affects how people interpret events and whether they support changes that reduce future risk. This continuity shows that risk communication is cumulative, with each phase building on trust and understanding developed earlier. Risk communication across the disaster cycle also strengthens early warning systems. Pre-disaster education helps people recognize warnings and respond appropriately. During-disaster communication turns forecasts into action. Post-disaster communication provides feedback that improves monitoring, messaging, and coordination. Together, these phases create a learning system that improves over time. Risk communication across the pre-disaster, during-disaster, and post-disaster phases forms a continuous link between preparedness, response, and recovery. By building awareness and trust before disasters, guiding action during emergencies, and supporting recovery afterward, communication becomes a central element of disaster risk reduction. This phase-based approach shows how timely, clear, and inclusive communication can reduce vulnerability and support safer and more resilient communities.

4.6 Behavioral and Psychological Aspects of Risk Communication

4.6.1 Human Behavior and Decision-Making During Disasters

Human behavior and psychological responses play a central role in the success of risk communication and early warning systems. Hazards such as storms, floods, earthquakes, and heat waves are physical events, but their impacts depend largely on how people understand risk information and how they choose to act. Disasters often occur in situations marked by uncertainty, emotional stress, and limited time, where decisions are shaped less by careful analysis and more by habits, feelings, and social cues. For this reason, attention to behavioral and psychological factors is essential for designing communication that leads to timely and protective action.

Risk perception strongly influences how people respond to warnings. Individuals do not judge risk only on scientific estimates or probabilities. Their views are shaped by past experiences, cultural beliefs, media coverage, and personal circumstances. Communities that have faced serious disasters before often react more quickly to warnings, while those without recent experience may underestimate danger. Many people also believe that a disaster will not affect them personally, which can delay evacuation or other safety measures even when warnings are clear.

Emotions have a powerful effect on decision-making during emergencies. Fear, anxiety, and stress can encourage action, but they can also cause confusion or inaction if messages are poorly framed. Very alarming messages may create panic or denial, while messages that sound too reassuring may reduce urgency. Effective risk communication balances these effects by explaining the risk clearly and providing simple guidance on what to do. Clear instructions help turn fear into purposeful behavior and give people a sense of control. Trust is another key factor in whether warnings are followed. People are more likely to act when they trust the institutions and individuals issuing warnings. Trust grows through clear, consistent, and honest communication over time. When authorities explain what is known and what is uncertain, confidence increases. False alarms or mixed messages, however, can weaken trust and lead people to ignore future alerts (Pan, 2020; Robinson & Botzen, 2025).

Social influence also shapes behavior during disasters. People often look to family members, neighbors, and community leaders for guidance. When others take action, individuals are more likely to follow. If those around them remain inactive, the risk may seem low. Communication that involves trusted local figures and encourages collective action is often more effective. Stress limits people's ability to process complex information. Simple, repeated, and action-focused messages work better than detailed explanations. Clear instructions that state what to do and where to go are easier to follow in emergency situations. Biases such as the belief that conditions will remain normal can delay response, especially during slow-developing hazards like floods or heat waves. Regular communication and education before disasters help people recognize early warning signs and understand how quickly situations can change. Experience, culture, and social values also influence behavior. Past disasters can increase awareness, while repeated mild events may reduce concern. Strong ties to land, property, or community can affect willingness to evacuate. Communication that respects these realities is more likely to succeed. Human behavior ultimately determines whether risk communication achieves its purpose. Early warning systems are most effective when communication reflects how people think, feel, and act under pressure, helping them make safer decisions when it matters most (Bakhshian & Martinez-Pastor, 2023).

4.6.2 Emotional Intelligence and Empathy

Emotional intelligence and empathy play an important role in effective risk communication within disaster risk management and early warning systems. Accurate information and timely warnings are essential, but they are often not enough to motivate people to act. Disasters occur in situations filled with fear, uncertainty, stress, and loss, and the emotional state of affected communities strongly influences how messages are received. Emotional intelligence, understood as the ability to recognize and manage emotions in oneself and others, helps communicators deliver messages in ways that build trust, understanding, and cooperation. Empathy in risk communication means recognizing and acknowledging the feelings, concerns, and everyday realities of people facing danger. When authorities communicate with empathy, they show that communities are seen as people, not just as numbers or risk categories. This reduces emotional distance and makes warnings feel more genuine and relevant. Acknowledging the difficulty of leaving homes, livelihoods, or family members during evacuation can make safety messages easier to accept.

Emotional intelligence also shapes how messages are framed and delivered. Disasters can trigger strong emotional reactions that either support or block protective action. Messages that are too alarming may cause panic or denial, while messages that sound distant or overly formal may fail to convey urgency. Communicators who are emotionally aware can balance seriousness with reassurance. Clear instructions combined with expressions of care help people move from fear to purposeful action. Trust is central to early warning systems and is closely linked to emotional factors. People are more likely to follow guidance from sources they see as honest, respectful, and caring. Empathetic communication strengthens trust by showing concern for community well-being. This is especially important in places where past experiences of neglect or unclear communication have reduced confidence in institutions. Emotional intelligence is also important for responding to the needs of different groups. Children, older adults, persons with disabilities, and people affected by past trauma may react differently to warnings. Some may need reassurance and simple explanations, while others may need more detailed information. Adapting tone, language, and delivery makes communication more inclusive and effective. After a disaster, empathy becomes even more important. Communities dealing with loss and uncertainty need communication that recognizes their experiences and supports emotional recovery. Clear and compassionate information about what happened, what help is available, and what steps will follow helps restore trust and support recovery. When emotional intelligence and empathy are part of risk communication, early warning systems become more responsive and effective. Communication that respects emotions and lived experience helps people make safer decisions and strengthens resilience during crises (Kuday & Erdoğan, 2023; Suhaimi et al., 2014).

4.6.3 Misinformation and Rumors Through Communication Strategies

Misinformation and rumors pose serious challenges to effective risk communication during disasters. In situations marked by fear, uncertainty, and rapidly changing conditions, people naturally seek information to understand what is happening and decide how to respond. When official information is delayed, unclear, or not trusted, informal channels such as social networks, messaging apps, and word of mouth often take its place. These channels can spread false or misleading information quickly, sometimes faster than verified warnings, shaping how people perceive risk and influencing their actions. Addressing misinformation therefore requires attention to the psychological and social factors that allow it to spread. Rumors tend to grow in environments of anxiety and uncertainty. During disasters, people look for certainty and reassurance, and unverified information may appear helpful if it offers simple answers. Cognitive tendencies also influence this process. Individuals are more likely to accept messages that fit their existing beliefs or emotions. Those who already distrust authorities may believe rumors that challenge official warnings, while others may prefer information that reduces the perceived severity of a threat. Social relationships strengthen this effect, as people often trust information shared by family members, friends, or local figures more than messages from distant institutions (Cuadra & Cotoron, 2025; Rizal et al., 2025).

The consequences of misinformation during disasters can be severe. False claims about evacuation routes, damage levels, or the authenticity of warnings can cause panic, unsafe behavior, or refusal to follow instructions. Over time, repeated exposure to rumors can weaken trust in early warning systems, leading people to ignore future alerts even when they are accurate. Managing misinformation is therefore not only about correcting false details but also about protecting the credibility of risk communication as a whole.

Clear and transparent communication is one of the most effective ways to limit the spread of rumors. Regular updates, simple explanations, and honest acknowledgment of uncertainty help reduce speculation. Delays or silence can create information gaps that rumors quickly fill. Trusted messengers also matter. Local leaders, health workers, community groups, and local media can help carry accurate information through informal networks where rumors often circulate. Message clarity and tone are equally important. Short, consistent, and action-focused messages repeated across multiple channels are easier to understand and harder to replace with false information. Respectful and empathetic responses to misinformation are more effective than confrontational ones, as they acknowledge fear and encourage trust rather than resistance. Education before disasters also plays a key role. Helping people recognize reliable sources and think critically about information reduces the spread of rumors during emergencies. By combining clarity, trust, empathy, and timely engagement, risk communication can reduce the harmful effects of misinformation and support safer public response during disasters (Malakar et al., 2023a, 2023b; Cuadra & Cotoron, 2025).

4.7 Risk Communication

4.7.1 Monitoring, Evaluation, and Feedback Mechanisms

Monitoring and evaluation are essential for ensuring that risk communication within early warning systems is effective, not just active. Sending warnings and preparedness messages is only meaningful if people understand them, trust them, and act on them. Monitoring and evaluation provide structured ways to review communication practices, identify gaps, and improve future efforts. In disaster contexts, this learning process is vital for reducing losses and strengthening long-term resilience. Monitoring focuses on the ongoing observation of how risk communication activities are carried out. This includes checking whether warnings are issued on time, which channels are used, and how widely messages reach people. Authorities may track the delivery of SMS alerts, radio broadcasts, or information shared through digital platforms. Monitoring helps identify practical problems, such as communities that are not being reached or communication channels that do not function well during emergencies. Evaluation examines outcomes and impact. It asks whether people received the message, understood it, and took action. Post-disaster surveys, interviews, and community discussions help reveal how warnings were interpreted and what influenced people's decisions. This helps distinguish between technical success, such as sending a warning, and real effectiveness, which involves protecting lives and reducing harm. Trust and risk perception are central to evaluation. People may receive warnings but choose not to act if they do not trust the source or believe the risk is low. Evaluation therefore needs to examine public confidence in authorities, clarity of messages, and credibility of communication channels. These insights help improve message design and delivery.

Monitoring and evaluation also support institutional learning and accountability. Reviewing past experiences helps organizations improve coordination, update procedures, and strengthen communication capacity. Transparent evaluation shows a commitment to learning, which can build public confidence. Community feedback is an important part of this process. Local people, as the main users of warnings, provide valuable insights into what works and what does not. Community meetings and feedback tools allow concerns to be shared and improvements suggested. Through regular monitoring, evaluation, and community engagement, risk communication becomes more responsive and better able to support timely and lifesaving action during disasters. Feedback mechanisms are an essential part of effective risk communication within early warning systems because they support learning, adjustment, and trust between institutions and communities. Risk communication is not a one-way flow of information where authorities send warnings and expect people to respond automatically. It is an ongoing interaction in which people interpret messages, act on them, and form opinions about their usefulness. Feedback mechanisms allow these experiences and responses to reach decision-makers, helping improve how warnings are designed and delivered (Melet et al., 2020; Turki et al., 2025).

During disasters, feedback helps authorities understand how warnings were received and used by communities. After an event, people can share whether alerts arrived on time, whether messages were clear, and whether they were able to take protective action. This information can be gathered through surveys, community meetings, helplines, digital platforms, and discussions with local organizations. Such feedback often highlights practical problems, such as unclear instructions, language barriers, or lack of access to communication channels, that may not be visible to central agencies but are important for improving future communication. Feedback mechanisms also strengthen trust and accountability. When communities see that their concerns are heard and lead to real changes, they are more likely to trust warning authorities. This trust is especially important in areas that face repeated hazards and rely regularly on early warning systems. Sharing how feedback has led to improvements, such as clearer messages or better delivery methods, helps reinforce this relationship and encourages continued participation. Feedback during emergencies is equally valuable. Information shared through local networks, volunteers, and digital platforms can show whether warnings have reached people, which areas are most affected, and where help is needed. This real-time input allows authorities to adjust messages, focus on specific locations, and correct misunderstandings while the situation is still unfolding. Feedback mechanisms help link technical warning systems with real human experience. They reduce the gap between what institutions aim to communicate and what people actually understand and do. When feedback is built into risk communication, early warning systems become more responsive, trusted, and effective. By turning communication into a process of dialogue and learning, feedback helps early warning systems adapt to changing risks and community needs. Listening to people and responding to their experiences strengthens risk communication and supports timely and lifesaving action during disasters (Jacobson et al., 2014; Malakar et al., 2023b).

4.7.2 Adaptive Communication Strategies

Adaptive communication strategies are essential for keeping risk communication effective as hazards change, technologies develop, and social conditions vary across communities. Disasters do not follow a fixed path, and people's understanding of risk can shift quickly as situations evolve. Adaptive approaches allow early warning systems to move away from rigid messaging and adopt flexible methods that can be adjusted during an event and improved over time. In this way, communication becomes a process of learning that strengthens through experience and reflection. Adaptive communication is based on the idea that risk messages must suit different audiences and situations. Communities differ in language, culture, education, and access to communication tools, all of which affect how messages are received. An adaptive approach therefore uses a range of communication channels, including radio, mobile alerts, social media, and direct community interaction, to reach everyone. Messages also need to change as a disaster progresses, focusing on preparedness before an event, clear instructions during an emergency, and recovery information afterward. Real-time monitoring and feedback play a central role in adaptive communication. By observing how messages are shared and how people respond, authorities can quickly identify gaps or confusion. If certain groups or locations are not receiving warnings, communication can be redirected to reach them. If instructions are misunderstood, messages can be clarified and repeated. This ability to adjust communication while a crisis is unfolding greatly improves the performance of early warning systems. Learning from past events is another important element. Reviews after disasters and feedback from communities help identify which messages and channels worked well and where problems occurred. These lessons can then be used to revise communication plans, improve training, and strengthen delivery methods. Over time, this process builds stronger institutional capacity and reduces the risk of repeating earlier mistakes. Adaptive communication strategies help connect technical warning systems with real human behavior and social diversity. They recognize that risk perception changes over time and across situations. By remaining flexible, responsive, and guided by experience, adaptive communication builds trust and encourages people to act on warnings. This adaptability helps ensure that risk communication remains relevant, credible, and effective in supporting safer responses to disasters (Attems et al., 2020a, 2020b).

4.8 Cultural and Contextual Considerations in Risk Communication

4.8.1 Cultural Sensitivity and Language Barriers

Cultural sensitivity and careful attention to language barriers are essential for effective risk communication within early warning systems. Disasters affect communities that are socially and culturally diverse, with differences in language, beliefs, traditions, and social structures. These factors influence how people understand danger, whom they trust, and how they respond to warnings. When communication does not reflect this diversity, even timely and accurate alerts may fail to lead to action, placing vulnerable groups at greater risk. Cultural beliefs and social norms strongly shape how hazards are understood and how protective actions are taken. In some communities, strong emotional ties to land, homes, or livestock make evacuation difficult, even when risks are clear. In others, decisions are guided by collective discussion or advice from respected local leaders rather than by individual judgment. Risk communication that is culturally sensitive recognizes these realities and works within them. Messages that connect with local values, experiences, and priorities are more likely to be understood and accepted (Mokhele & Mvanyashe, 2025; Ogie et al., 2018).

The involvement of trusted local figures can greatly improve the effectiveness of communication. Religious leaders, elders, teachers, and community representatives often hold influence and trust within their communities. When they help share warnings and safety advice, messages gain credibility and relevance. Their participation also helps ensure that information is shared in ways that respect local customs and social norms. Language barriers remain a major challenge, especially in multilingual societies and areas with migrant, indigenous, or marginalized populations. When warnings are delivered only in official or dominant languages, many people may not fully understand the threat or the actions required. This can result in confusion, delayed response, or exclusion from early warning systems. Providing information in multiple local languages, using clear and simple wording, and supporting messages with symbols or visuals can greatly improve understanding and inclusion. Effective communication involves more than translation alone. The way information is expressed also matters. Tone, examples, and explanations should reflect local ways of speaking and thinking. Referring to familiar events, environmental signs, or shared experiences can make risks easier to grasp. Avoiding insensitive or unfamiliar expressions helps reduce misunderstanding and resistance (Weitkamp et al., 2020; Yin et al., 2025).

Therefore, the cultural sensitivity and language accessibility ensure that early warnings are not only delivered but also understood and trusted. By recognizing diversity and adapting communication to local contexts, early warning systems become more inclusive and effective. This approach strengthens trust, supports timely action, and contributes to building resilient communities in the face of increasing disaster risks.

4.8.2 Indigenous Knowledge and Communication Approaches

Indigenous knowledge and traditional communication approaches play an important role in effective risk communication, especially in regions where communities have lived with environmental hazards for many generations. This knowledge is built through long-term observation of weather, land, seas, and ecosystems. Over time, communities have learned to recognize early signs of hazards such as cyclones, floods, droughts, and landslides. When included within early warning systems, indigenous knowledge adds local meaning and strengthens the credibility of risk communication alongside scientific forecasts.

Many indigenous communities rely on natural signs to anticipate environmental change. Shifts in animal behavior, cloud patterns, wind direction, sea conditions, or plant cycles often signal coming hazards. These observations are shared through oral traditions and daily practices and function as local early warning methods. When formal risk communication recognizes these signs, it connects scientific information with lived experience. People are more likely to trust official warnings when they see their own knowledge reflected in them. Traditional communication within indigenous societies is often based on relationships and collective interaction. Knowledge is shared through storytelling, community gatherings, rituals, and guidance from elders or traditional leaders. Decisions are usually made together rather than individually. Using these familiar channels for risk communication allows warnings and preparedness messages to fit naturally into community life. This approach helps messages remain clear, memorable, and more likely to lead to action. Respecting indigenous communication practices is also important for inclusion and fairness. Many indigenous groups have faced long periods of exclusion from formal decision-making, which has sometimes created mistrust toward external authorities. When disaster management agencies work closely with indigenous leaders and adapt communication to local customs, they show respect and partnership. This helps build cooperation and improves response during emergencies (Kibria et al., 2025; Mikulecký et al., 2023).

Indigenous knowledge systems are not fixed or isolated. They continue to evolve as communities interact with modern tools and institutions. Combining traditional practices with mobile phones, local radio, or community monitoring groups can create systems that are both culturally rooted and practical. For example, traditional weather signs can be discussed alongside official forecasts in village meetings or local broadcasts, strengthening awareness and preparedness. Indigenous knowledge and traditional communication approaches make risk communication more locally grounded, trusted, and culturally appropriate. Valuing these systems supports more inclusive early warning practices and strengthens resilience. Effective risk communication is not only about sharing information but about engaging with how people understand their environment and respond to risk in their daily lives.

4.8.3 Cultural Perspectives on Risk and Vulnerability

Risk and vulnerability are not understood in the same way across all cultures. They are shaped by history, livelihoods, belief systems, social relationships, and everyday interaction with the environment. In early warning systems, these cultural differences strongly influence how people perceive hazards, how seriously they take warnings, and how they decide to respond. Risk communication therefore needs to move beyond technical explanations of danger and engage with the cultural meanings that communities attach to safety, loss, and survival.

In many societies, risk is understood through more than physical impacts alone. Spiritual beliefs, moral values, and shared worldviews often shape how hazards are interpreted. Some communities may view floods, storms, or earthquakes as part of natural cycles, fate, or divine will, which can affect their willingness to take preventive action. Others understand risk mainly through livelihood needs. Fishing, farming, or coastal communities may continue working in dangerous conditions because daily survival depends on it. These responses reflect lived realities rather than a lack of awareness. Risk communication that ignores these perspectives may seem unrealistic or disconnected from everyday life, reducing its influence (Cunha Ferreira et al., 2023; Rodríguez-Rosales et al., 2021).

Cultural views of vulnerability also vary widely. In some contexts, vulnerability is closely linked to age, gender, disability, or social position, shaping who receives information first and who makes decisions during emergencies. In other settings, strong family ties or community networks provide informal support systems that influence response. Recognizing these social patterns helps communicators identify those most at risk and design messages that encourage shared responsibility rather than individual action alone. Language, symbols, and stories are central to how cultures understand risk. People make sense of danger through shared memories, local expressions, and past experiences. Referring to earlier disasters, familiar places, or everyday signs of change can make warnings more meaningful. In contrast, technical terms or unfamiliar language can distance people from the message. Communication that reflects local ways of speaking and thinking helps people connect warnings to their own lives. Respecting cultural perspectives strengthens both the effectiveness and fairness of early warning systems. Communication grounded in local realities builds trust, improves understanding, and supports timely action. Cultural diversity should not be seen as a challenge to risk communication but as an essential context. When early warning systems recognize and include different cultural views of risk and vulnerability, they become more inclusive, credible, and better able to protect all sections of society from disaster risk (Martinez, 2021; Mirenda & Lazos Chavero, 2022).

4.9 Conclusion and Arguments

Risk communication is central to effective disaster risk management because it links scientific knowledge, early warning systems, and human behavior into a shared process of preparedness and response. As discussed across this chapter, risk communication is not limited to sending warnings. It shapes how risks are understood, whether information is trusted, and how people decide to act. When communication is done well, complex hazard information is turned into clear and meaningful messages that people can relate to and use to protect themselves and their communities. Using multiple communication channels strengthens this process. Traditional media, social media, digital platforms, and face-to-face interaction together create a layered communication system that improves reach and inclusion. Each channel serves different groups and situations, helping ensure that warnings are delivered widely and in time. At the same time, communication must consider how people think and feel. Emotions, trust, past experience, social influence, and common biases all shape how warnings are received. When these human factors are ignored, even advanced warning technologies may fail to prompt action.

Risk communication also depends on continuous learning and adaptation. Monitoring, evaluation, and feedback help identify strengths and weaknesses in communication practice. Adaptive strategies allow messages to change as hazards, technologies, and social conditions evolve. Cultural sensitivity, respect for indigenous knowledge, and attention to language differences further strengthen communication by connecting messages to everyday life and local experience. These approaches help build trust and encourage people to follow guidance. Risk communication is therefore a dynamic and participatory process. It involves responsibility, openness, and respect for diversity. When communication strategies reflect technology, human behavior, and cultural context, early warning systems become more effective. They support timely response, learning after disasters, and progress toward safer and more resilient societies. The chapter discussions lead to the following arguments:

(a) Risk communication links scientific early warnings with timely human action.
(b) Human behavior, emotions, and trust are as critical as technology in disaster response.
(c) Multi-channel communication ensures inclusiveness and effective last-mile reach.
(d) Feedback, monitoring, and adaptive approaches support continuous improvement.
(e) Cultural sensitivity and indigenous knowledge enhance credibility and local relevance.

References

Agbehadji, I. E., Mabhaudhi, T., Botai, J., & Masinde, M. (2023). A systematic review of existing early warning systems' challenges and opportunities in cloud computing early warning systems. *Climate, 11*(9), 188. https://doi.org/10.3390/cli11090188

Ali, M. A., & Kamraju, M. (2026). Harnessing community strengths: Disaster management and mitigation in coastal areas. In K. K. B. Ahammed, A. C. Pandey, B. R. Parida, & I. W. G. A. Karang (Eds.), *Monitoring coastal marine ecosystems under climate change*. Springer. https://doi.org/10.1007/978-3-032-09137-6_20

Attems, M. S., Schlögl, M., Thaler, T., Rauter, M., & Fuchs, S. (2020a). Risk communication and adaptive behaviour in flood-prone areas of Austria: A Q-methodology study on opinions of affected homeowners. *PLoS ONE, 15*(5), Article e0233551. https://doi.org/10.1371/journal.pone.0233551

Attems, M. S., Thaler, T., Snel, K. A., Davids, P., Hartmann, T., & Fuchs, S. (2020b). The influence of tailored risk communication on individual adaptive behaviour. *International Journal of Disaster Risk Reduction, 49*, Article 101618. https://doi.org/10.1016/j.ijdrr.2020.101618

Aude, N., Oscar, N., Manon, C., Martin, J., Marc, R., Céline, C., & Ghozlane, F. B. (2025). Communicating the risk of erosion: the effects of map-based communication on risk perception and affect. *Climate Risk Management*, 100738. https://doi.org/10.1016/j.crm.2025.100738

Babič, A., Sinković, N. L., & Dolšek, M. (2023). A model for communication and management support of natural hazards risk. *International Journal of Disaster Risk Reduction, 90*, Article 103672. https://doi.org/10.1016/j.ijdrr.2023.103672

Bakhshian, E., & Martinez-Pastor, B. (2023). Evaluating human behaviour during a disaster evacuation process: A literature review. *Journal of Traffic and Transportation Engineering, 10*(4), 485–507. https://doi.org/10.1016/j.jtte.2023.04.002

Berger-Sabbatel, A., & Journé, B. (2018). Organizing risk communication for effective preparedness: Using plans as a catalyst for risk communication. In M. Bourrier & C. Bieder (Eds.), *Risk communication for the future. SpringerBriefs in applied sciences and technology*. Springer. https://doi.org/10.1007/978-3-319-74098-0_3

Bica, M., Weinberg, J., & Palen, L. (2020). Achieving accuracy through ambiguity: The interactivity of risk communication in severe weather events. *Computer Supported Cooperative Work, 29*, 587–623. https://doi.org/10.1007/s10606-020-09380-2

Bourrier, M., & Bieder, C. (Eds.). (2018). *Risk communication for the future: Towards smart risk governance and safety management*. Springer. ISBN: 978-3-319-74098-0. https://doi.org/10.1007/978-3-319-74098-0

Bradley, D. T., McFarland, M., & Clarke, M. (2016). The effectiveness of disaster risk communication: A systematic review of intervention studies. *Effective Communication During Disasters*, 81–120. https://doi.org/10.1371/currents.dis.349062e0db1048bb9fc3a3fa67d8a4f8

Chavula, P., Kayusi, F., Lungu, G., & Uwimbabazi, A. (2025). The current landscape of early warning systems and traditional approaches to disaster detection. *LatIA, 3*, 77. https://doi.org/10.62486/latia202577

Chen, Y., Chen, Y., Yu, S., et al. (2024). Utilizing social media for community risk communication in megacities: Analysing the impact of WeChat group information interaction and perception on communication satisfaction during the COVID-19 pandemic in Shanghai. *BMC Public Health, 24*, 1889. https://doi.org/10.1186/s12889-024-19276-1

Chovanec, D., Kollár, B., Halúsková, B., Kubás, J., Pawęska, M., & Ristvej, J. (2025). A component-based approach to early warning systems: A theoretical model. *Applied Sciences, 15*(6), 3218. https://doi.org/10.3390/app15063218

Covello, V. T., von Winterfeldt, D., & Slovic, P. (1988). Risk communication. In C. C. Travis (Ed.), *Carcinogen risk assessment. Contemporary issues in risk analysis* (Vol. 3). Springer. https://doi.org/10.1007/978-1-4684-5484-0_15

Cuadra, J. M., & Cotoron, V. N. (2025). Addressing false information through local capacity building in community-based disaster risk management. *Jàmbá: Journal of Disaster Risk Studies, 17*(1), 1–7. https://doi.org/10.4102/jamba.v17i1.1836

Cunha Ferreira, T., Romão, X., Freitas, P. M., & Mendonça, H. (2023). Risk assessment and vulnerability analysis of a coastal concrete heritage structure. *Heritage, 6*(9), 6153–6171. https://doi.org/10.3390/heritage6090323

Dehghani, A., Ghomian, Z., Rakhshanderou, S., Khankeh, H., & Kavousi, A. (2022). Process and components of disaster risk communication in health systems: A thematic analysis. *Jàmbá-Journal of Disaster Risk Studies, 14*(1), 1367. https://doi.org/10.4102/jamba.v14i1.1367

Dinçer, E. (2024). Using digital tools for an enhanced risk communication in health professionals-patient interactions. *Applied Scientific Research, 3*(1), 156–167.

Erokhin, D., & Komendantova, N. (2024). Social media data for disaster risk management and research. *International Journal of Disaster Risk Reduction, 114*, Article 104980. https://doi.org/10.1016/j.ijdrr.2024.104980

Graf, J., Renner, R., & Klebel, T. (2026). Bridging warning and adaptation addressing risk communication strategies for short-term natural hazard warnings and long-term risk adaptation—A scoping review. *Progress in Disaster Science*, 100515. https://doi.org/10.1016/j.pdisas.2026.100515

Gündüz, F., & Dinçer, S. (2025). Communication of flood early warnings and community response: The case of Sinop-Kastamonu, Türkiye. *Natural Hazards, 121*, 12575–12588. https://doi.org/10.1007/s11069-025-07286-0

Hansson, S., Orru, K., Siibak, A., Bäck, A., Krüger, M., Gabel, F., & Morsut, C. (2020). Communication-related vulnerability to disasters: A heuristic framework. *International Journal of Disaster Risk Reduction, 51*, Article 101931. https://doi.org/10.1016/j.ijdrr.2020.101931

Hermans, T. D. G., Trogrlić, R. S, van den Homberg, M. J. C., et al. (2022). Exploring the integration of local and scientific knowledge in early warning systems for disaster risk reduction: A review. *Natural Hazards, 114*, 1125–1152. https://doi.org/10.1007/s11069-022-05468-8

Höppner, C., Whittle, R., Bründl, M., et al. (2012). Linking social capacities and risk communication in Europe: A gap between theory and practice? *Natural Hazards, 64*, 1753–1778. https://doi.org/10.1007/s11069-012-0356-5

Islam, M. M., Hasan, M., Mia, M. S., Al Masud, A., & Islam, A. R. M. T. (2025). Early warning systems in climate risk management: Roles and implementations in eradicating barriers and overcoming challenges. *Natural Hazards Research*. https://doi.org/10.1016/j.nhres.2025.01.007

Jacobson, C., Carter, R. W., Thomsen, D. C., & Smith, T. F. (2014). Monitoring and evaluation for adaptive coastal management. *Ocean & Coastal Management, 89*, 51–57. https://doi.org/10.1016/j.ocecoaman.2013.12.008

Jayasekara, R., Siriwardana, C., Amaratunga, D., & Haigh, R. (2022). Evaluating the network of stakeholders in Multi-Hazard Early Warning Systems for multiple hazards amidst biological outbreaks: Sri Lanka as a case in point. *Progress in Disaster Science, 14*, Article 100228. https://doi.org/10.1016/j.pdisas.2022.100228

Kalligeris, N., et al. (2026). Communicating Tsunami risk and uncertainties. In M. B. Sørensen, et al. (Eds.), *Probabilistic Tsunami hazard and risk analysis. Mathematics of planet earth* (Vol. 14). Springer. https://doi.org/10.1007/978-3-031-98115-9_8

Kelman, I., & Glantz, M. H. (2014). Early warning systems defined. In A. Singh & Z. Zommers (Eds.), *Reducing disaster: Early warning systems for climate change*. Springer. https://doi.org/10.1007/978-94-017-8598-3_5

Kibria, G., Sharmin, N., Islam, M. R., et al. (2025). Assessment of indigenous knowledge and technology used for climate adaptation and resilience in coastal area of Bangladesh. *Discov Environ, 3*, 178. https://doi.org/10.1007/s44274-025-00381-w

Kuday, A. D., & Erdoğan, Ö. (2023). Relationship between emotional intelligence and disaster response self-efficacy: A comparative study in nurses. *International Emergency Nursing, 70*, Article 101319. https://doi.org/10.1016/j.ienj.2023.101319

Lamsal, R., & Kumar, T. V. V. (2020). Artificial Intelligence based early warning system for coastal disasters. In A. Singh, R. L. S. Fernando, & N. P. Haran (Eds.), *Development in coastal zones and disaster management. Disaster research and management series on the global south*. Palgrave Macmillan. https://doi.org/10.1007/978-981-15-4294-7_21

Lechowska, E. (2018). What determines flood risk perception? A review of factors of flood risk perception and relations between its basic elements. *Natural Hazards, 94*, 1341–1366. https://doi.org/10.1007/s11069-018-3480-z

Lin, K. H. E., Khan, S., Acosta, L. A., Alaniz, R., & Olanya, D. R. (2020). The dynamism of post disaster risk communication: A cross-country synthesis. *International Journal of Disaster Risk Reduction, 48*, Article 101556. https://doi.org/10.1016/j.ijdrr.2020.101556

Malakar, K. D., Kumar, M., Anand, S., & Kuzur, G. (2023a). Climate, climate change ecology, and socio-ecological environments. In *Climate change and socio-ecological transformation. Advances in geographical and environmental sciences*. Springer. https://doi.org/10.1007/978-981-99-4390-6_2

Malakar, K. D., Kumar, M., Anand, S., & Kuzur, G. (2023b). Disaster, policy instruments, and sustainability. In *Climate change and socio-ecological transformation. Advances in geographical and environmental sciences*. Springer. https://doi.org/10.1007/978-981-99-4390-6_6

Marchezini, V., Trajber, R., Olivato, D., et al. (2017). Participatory early warning systems: Youth, citizen science, and intergenerational dialogues on disaster risk reduction in Brazil. *International Journal of Disaster Risk Science, 8*, 390–401. https://doi.org/10.1007/s13753-017-0150-9

Martinez, G. (2021). Coastal risk cultures: Local and regional formation of knowledge and action. *Frontiers in Environmental Science, 9*, Article 578238. https://doi.org/10.3389/fenvs.2021.578238

Matyas, C., Srinivasan, S., Cahyanto, I., et al. (2011). Risk perception and evacuation decisions of Florida tourists under hurricane threats: A stated preference analysis. *Natural Hazards, 59*, 871–890. https://doi.org/10.1007/s11069-011-9801-0

Melet, A., Teatini, P., Le Cozannet, G., et al. (2020). Earth observations for monitoring marine coastal hazards and their drivers. *Surveys in Geophysics, 41*, 1489–1534. https://doi.org/10.1007/s10712-020-09594-5

Mikulecký, P., Punčochářová, A., Babič, F., Bureš, V., Čech, P., Husáková, M., Mls, K., Nachazel, T., Pance, D., Stekerova, K., Triantafyllou, I., Tucnik, P., Sunanda, V., & Zanker, M. (2023). Dealing with risks associated with tsunamis using indigenous knowledge approaches. *International Journal of Disaster Risk Reduction, 86*, Article 103534. https://doi.org/10.1016/j.ijdrr.2023.103534

Mirenda, C., & Lazos Chavero, E. (2022). Cultural vulnerability, risk reduction and gender equity: Two Mexican coastal communities. *Environmental Hazards, 21*(3), 235–253. https://doi.org/10.1080/17477891.2021.1945996

Mokhele, M. O., & Mvanyashe, A. (2025). The role of language as a catalyst in communicating early warning messages: a case study in Walmer Airport Valley Informal Settlement, Gqeberha, South Africa. *International Journal of Disaster Risk Reduction*, 105573. https://doi.org/10.1016/j.ijdrr.2025.105573

Monaco, S. (2025). Risk communication in coastal cities: The case of Naples, Italy. *Land, 14*(6), 1288. https://doi.org/10.3390/land14061288

Moreschini, I., Cugliari, L., Cerbara, L., et al. (2025). Tsunami risk perception of the touristic population of Stromboli Island: Towards effective risk communication strategies. *Natural Hazards, 121*, 519–542. https://doi.org/10.1007/s11069-024-06845-1

Nishizawa, M. (2018). How risk communication can contribute to sharing accurate health information for individual decision-making. In M. Bourrier & C. Bieder (Eds.), *Risk communication for the future*. SpringerBriefs in Applied Sciences and Technology. Springer. https://doi.org/10.1007/978-3-319-74098-0_6

Ogie, R., Castilla Rho, J., Clarke, R. J., & Moore, A. (2018). Disaster risk communication in culturally and linguistically diverse communities: The role of technology. In *Proceedings* (Vol. 2, No. 19, p. 1256). MDPI. https://doi.org/10.3390/proceedings2191256

Onyeaka, H. (2025). Risk communication in food safety: Addressing the gap between scientific risk assessments and public perception, and strategies to improve transparency and trust in food safety regulations. *Food Safety and Risk, 12*, 7. https://doi.org/10.1186/s40550-025-00116-y

Pan, A. (2020). Study on the decision-making behavior of evacuation for coastal residents under typhoon storm surge disaster. *International Journal of Disaster Risk Reduction, 45*, Article 101522. https://doi.org/10.1016/j.ijdrr.2020.101522

Pham, T. D. M., Thieken, A. H., & Bubeck, P. (2024). Community-based early warning systems in a changing climate: An empirical evaluation from coastal central Vietnam. *Climate and Development, 16*(8), 673–684. https://doi.org/10.1080/17565529.2024.2307398

Rafliana, I., Jalayer, F., Cerase, A., Cugliari, L., Baiguera, M., Salmanidou, D., Hancilar, U., et al. (2022). Tsunami risk communication and management: Contemporary gaps and challenges. *International Journal of Disaster Risk Reduction, 70*, Article 102771. https://doi.org/10.1016/j.ijdrr.2021.102771

Rahman, A., & Munadi, K. (2019). Communicating risk in enhancing disaster preparedness: A pragmatic example of disaster risk communication approach from the case of smong story. In *IOP conference series: Earth and environmental science* (Vol. 273, No. 1, p. 012040). IOP Publishing.

Renn, O. (1991). Risk communication and the social amplification of risk. In R. E. Kasperson & P. J. M. Stallen (Eds.), *Communicating risks to the public*. Technology, Risk, and Society (Vol. 4). Springer. https://doi.org/10.1007/978-94-009-1952-5_14

Rizal, E., Winoto, Y., Sugito, T., Nugroho, C., & Septian, F. I. (2025). Disaster communication in the digital age: A community-based case study of media, education, and local knowledge in Pangandaran, Indonesia. *Frontiers in Communication, 10*, 1632436. https://doi.org/10.3389/fcomm.2025.1632436

Robinson, P. J., & Botzen, W. J. W. (2025). Behavioural public policy for natural disaster preparedness and the role of economic experiments. *Humanities Social Sciences Communications, 12*, 719. https://doi.org/10.1057/s41599-025-05086-2

Rodríguez-Rosales, B., Abreu, D., Ortiz, R., Becerra, J., Cepero-Acán, A. E., Vázquez, M. A., & Ortiz, P. (2021). Risk and vulnerability assessment in coastal environments applied to heritage buildings in Havana (Cuba) and Cadiz (Spain). *Science of the Total Environment, 750*, Article 141617. https://doi.org/10.1016/j.scitotenv.2020.141617

Rollason, E., Bracken, L. J., Hardy, R. J., et al. (2018). Rethinking flood risk communication. *Natural Hazards, 92*, 1665–1686. https://doi.org/10.1007/s11069-018-3273-4

Saengsupavanich, C., Ratnayake, A. S., Yun, L. S., et al. (2023). Current challenges in coastal erosion management for southern Asian regions: Examples from Thailand, Malaysia, and Sri Lanka. *Anthropocene Coasts, 6*, 15. https://doi.org/10.1007/s44218-023-00030-w

Shepherd, R., & Frewer, L. J. (1997). Risk communication. In D. R. Tennant (Ed.), *Food chemical risk analysis*. Springer. https://doi.org/10.1007/978-1-4613-1111-9_18

Strauss, C., Harr, M. D., & Pieper, T. M. (2025). Analyzing digital communication: A comprehensive literature review. *Management Review Quartely, 75*, 3119–3157. https://doi.org/10.1007/s11301-024-00455-8

Suhaimi, A. W., Marzuki, N. A., & Mustaffa, C. S. (2014). The relationship between emotional intelligence and interpersonal communication skills in disaster management context: A proposed framework. *Procedia-Social and Behavioral Sciences, 155*, 110–114. https://doi.org/10.1016/j.sbspro.2014.10.265

Tayebi, S., Jabed, M. A., Ruano, A. L., et al. (2024). Stakeholder perspectives on landslide triggers and impacts in five countries. *Landslides, 21*, 2033–2043. https://doi.org/10.1007/s10346-024-02270-4

Tora, T. T., Andarge, L. T., Abebe, A. T., et al. (2025). An expert-based assessment of early warning systems effectiveness in South Ethiopia Regional State. *Discover Sustainability, 6*, 188. https://doi.org/10.1007/s43621-025-00943-1

Trogrlić, R. S., van den Homberg, M., Budimir, M., McQuistan, C., Sneddon, A., & Golding, B. (2022). Early warning systems and their role in disaster risk reduction. In B. Golding (Ed.), *Towards the "perfect" weather warning*. Springer. https://doi.org/10.1007/978-3-030-98989-7_2

Turki, E. I., Ojeda, E., Mendoza, E. T., Medellín, G., Price, T. D., Salameh, E., Wang, X. H., Li, L., Franklin, G. L., & Torres-Freyermuth, A. (2025). Monitoring resilience in coastal systems: A comprehensive assessment. *Journal of Marine Science and Engineering, 13*(11), 2113. https://doi.org/10.3390/jmse13112113

Vicari, R., Komendatova, N., Elroy, O., et al. (2025). A toolbox to deal with misinformation in disaster risk management. *AI & SOCIETY*. https://doi.org/10.1007/s00146-025-02680-5

Weitkamp, E., McEwen, L., & Ramirez, P. (2020). Communicating the hidden: Toward a framework for drought risk communication in maritime climates. *Climatic Change, 163*, 831–850. https://doi.org/10.1007/s10584-020-02906-z

Wenzel, F., & Zschau, J. (Eds.). (2014). *Early warning for geological disasters: Scientific methods and current practice*. Springer. ISBN: 978-3-642-12232-3. https://doi.org/10.1007/978-3-642-12233-0

Yadav, N., Thalakiya, R., Srivastava, S., et al. (2026). Scoping review of early warning systems in improving disaster risk outcomes: Perspective from evolution to usage in Southeast Asian countries. *BMC Public Health*. https://doi.org/10.1186/s12889-026-26292-w

Yin, J., Khan, R. U., Afzaal, M., Almalki, H. M., Khasawneh, M. A. S., & Al Sulaie, S. (2025). Quantitative risk assessment of speech acts and lexical factors in maritime communication failures and accidents. *Safety Science, 191*, Article 106968. https://doi.org/10.1016/j.ssci.2025.106968

5

Climate Disasters and Enhancing Coastal Resilience: Applications of Ecosystem-Based Adaptation

Coastal regions are among the most vulnerable landscapes to climate-related disasters. Rising sea levels, stronger cyclones, storm surges, flooding, and the gradual loss of natural habitats are placing increasing pressure on both coastal ecosystems and the communities that depend on them. These growing risks call for adaptation approaches that go beyond short-term or purely engineering-based solutions and instead work with natural systems. This chapter introduces Ecosystem-based Adaptation (EbA) as a useful way to understand and respond to climate disasters in coastal areas. It explains how coastal ecosystems such as mangroves, coral reefs, wetlands, and seagrass meadows act as natural protective barriers, reducing the impacts of extreme events while supporting livelihoods, biodiversity, and carbon storage. The chapter outlines the basic concepts of EbA, the role of ecosystem services, and how vulnerability in coastal regions is shaped by both environmental and human factors. Attention is also given to the social, economic, and governance dimensions of ecosystem-based adaptation. The chapter highlights the importance of community participation, supportive policies, and flexible management practices for successful implementation. It also acknowledges the practical challenges, uncertainties, and limitations involved in applying EbA

K. D. Malakar et al., *Coastal Ecologies at Risk*, Palgrave Studies in Environmental Sustainability, https://doi.org/10.1007/978-3-032-26081-9_5

in complex and changing coastal environments. Together, these discussions set the foundation for understanding climate disasters and coastal ecologies through a resilience-oriented and sustainable perspective.

Key Points of the Chapter

- Describes coastal ecosystems as living infrastructure that reduce disaster risks and build climate resilience.
- Explains how healthy ecosystems lower hazards while supporting livelihoods and biodiversity.
- Examines nature-based solutions as more adaptive and cost-effective than hard engineering alone.
- Highlights the need to integrate ecosystem-based adaptation into policy, planning, and community governance.
- Explains the co-benefits of ecosystem-based adaptation, including carbon storage, food security, and sustainable development.

5.1 Ecosystem-Based Adaptation (EbA)

Ecosystem-based Adaptation (EbA) refers to the planned use of biodiversity and ecosystem services to help societies adjust to the impacts of climate change while maintaining the health and functioning of natural systems. As climate-related disasters continue to intensify, especially in fragile coastal regions affected by sea-level rise, cyclones, storm surges, erosion, and salinization, the need for effective and sustainable adaptation strategies has become more urgent. EbA offers a practical and inclusive pathway for building resilience by working with natural processes rather than relying mainly on hard engineering structures. Coastal ecosystems such as mangroves, coral reefs, salt marshes, dunes, and seagrass meadows function as living forms of protection that reduce exposure to hazards, lessen the impacts of extreme events, and support local livelihoods. This approach reflects a shift from short-term responses toward long-term strategies that reduce risk while strengthening both human and ecological resilience.

EbA is grounded in the idea that healthy, diverse, and well-connected ecosystems are better able to absorb shocks and continue providing essential services under changing climate conditions. Coastal ecosystems play a critical role in reducing vulnerability by weakening wave energy, stabilizing shorelines, regulating sediment movement, and supporting fisheries and other coastal resources. Protecting and restoring these ecosystems enhance natural adaptive capacity and help avoid many of the environmental and social costs linked to conventional gray infrastructure. EbA does not replace engineered solutions but encourages approaches in which natural systems and built structures complement each other, improving protection while maintaining ecological functions. The planning and implementation of Ecosystem-based Adaptation are guided by several key principles. One important principle is ecosystem integrity, which stresses that adaptation actions should conserve, restore, or enhance ecosystem structure, functions, and biodiversity. Measures such as mangrove restoration, wetland rehabilitation, and coral reef conservation are valued not only for their role in reducing disaster risks but also for maintaining long-term ecosystem health. Climate responsiveness is another guiding principle, requiring EbA measures to consider present and future climate risks, including rising sea levels, increasing temperatures, and changing rainfall and storm patterns, so that ecosystems remain effective over time (Hill et al., 2025; Mukherjee & Shaw, 2021).

Social inclusiveness and equity are also central to Ecosystem-based Adaptation. Coastal communities, particularly small-scale fishers, indigenous groups, and low-income households, often face the highest exposure to climate hazards. Effective EbA therefore depends on meaningful community participation, respect for local and indigenous knowledge, secure access to ecosystem resources, and fair sharing of benefits. EbA is also designed to deliver multiple benefits by linking disaster risk reduction with food security, carbon storage, biodiversity conservation, and livelihood opportunities. Adaptive management further shapes Ecosystem-based Adaptation by recognizing that climate risks and ecosystem conditions are constantly changing. Continuous monitoring, learning, and flexible decision-making allow adaptation measures to be adjusted as environmental conditions and social needs evolve. Through these principles and practices, Ecosystem-based Adaptation offers a clear and balanced approach for strengthening coastal resilience and responding to the growing challenges posed by climate-related disasters (Spencer et al., 2024; Vasseur, 2021).

5.1.1 Concept and Principles of Ecosystem-Based Adaptation (EbA)

Ecosystem-based Adaptation (EbA) refers to the deliberate use of biodiversity and ecosystem services to help societies cope with the impacts of climate change while maintaining the health of natural systems (Fig. 5.1). Climate-related disasters are increasing in frequency and intensity, particularly in fragile coastal regions exposed to sea-level rise, storm surges, cyclones, erosion, and salinization. These growing risks highlight the need for adaptation strategies that are effective, sustainable, and socially inclusive. EbA responds to this need by working with natural processes rather than relying mainly on hard engineering structures. Coastal ecosystems such as mangroves, coral reefs, salt marshes, dunes, and seagrass meadows act as living protective systems that reduce exposure to hazards, limit damage from extreme events, and support livelihoods closely linked to coastal environments. Within climate adaptation practice, EbA marks a move away from short-term responses that shift risk elsewhere toward long-term approaches that actively reduce risk and strengthen both human and ecological resilience.

The idea behind EbA is based on the simple understanding that healthy, diverse, and connected ecosystems are better able to absorb shocks and continue providing essential services as climate conditions change. Coastal ecosystems reduce vulnerability by weakening wave energy, stabilizing shorelines, regulating sediment movement, and supporting fisheries and other coastal resources. Conservation and restoration of these systems strengthen natural adaptive capacity and help avoid many of the environmental and social impacts associated with conventional gray infrastructure. EbA does not dismiss engineered solutions but supports approaches where natural systems and built structures are combined to improve protection while maintaining ecological

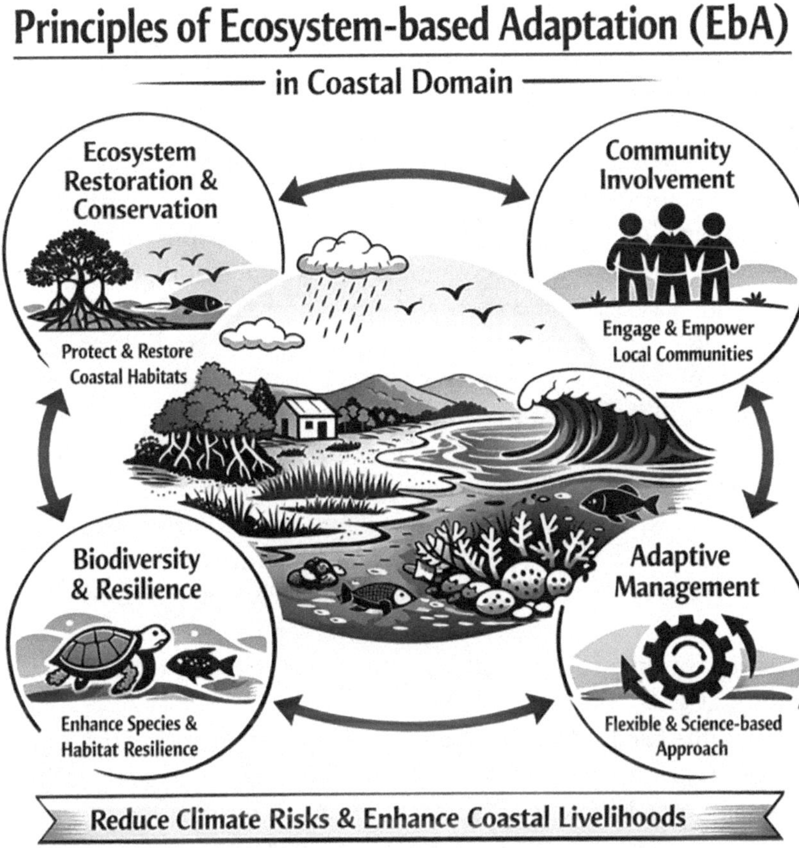

Fig. 5.1 Principles of Ecosystem-based Adaptation (EbA) (prepared by the authors)

functions. Several key principles guide the planning and application of Ecosystem-based Adaptation. Ecosystem integrity is a central principle, stressing that adaptation measures should conserve, restore, or improve ecosystem structure, functions, and biodiversity. Activities such as mangrove restoration, wetland rehabilitation, and coral reef conservation are selected not only for their role in reducing disaster risks but also for their contribution to long-term ecosystem health. Climate responsiveness is another guiding principle, requiring EbA measures to consider both present and future climate risks, including rising sea levels, increasing temperatures, and changes in rainfall and storm patterns, so that ecosystems remain effective over time (Scarano, 2017; Yáñez-Arancibia et al., 2013).

Social inclusiveness and equity are also essential to Ecosystem-based Adaptation. Coastal communities, particularly small-scale fishers, indigenous groups, and low-income households, often face the highest exposure to climate hazards. Successful EbA therefore depends on meaningful community participation, recognition of local and indigenous knowledge, secure access to ecosystem resources, and fair sharing of benefits. EbA is also designed to deliver multiple benefits by linking disaster risk reduction with food security, carbon storage, biodiversity conservation, and livelihood diversification. Adaptive management further shapes Ecosystem-based Adaptation by recognizing that ecosystems and climate risks are constantly changing. Ongoing monitoring, learning from experience, and flexible decision-making allow adaptation measures to be adjusted as environmental conditions and social needs evolve. Through these principles and practices, Ecosystem-based Adaptation offers a practical and balanced approach for strengthening coastal resilience and addressing the increasing challenges posed by climate-related disasters (Scarano, 2017).

Ecosystem-based Adaptation (EbA) has become an important response to the growing impacts of climate-related disasters, particularly in highly vulnerable coastal regions. Rising sea levels, frequent coastal flooding, stronger cyclones, heat stress, and the steady decline of coastal ecosystems are increasing risks for both people and natural environments. In many settings, conventional engineering-based adaptation measures alone have proven to be costly, limited in reach, and harmful to ecological systems. EbA addresses these challenges by using the natural capacity of ecosystems to reduce climate risks while sustaining the environmental foundations of human well-being. In coastal areas, ecosystems such as mangroves, coral reefs, seagrass beds, dunes, and tidal wetlands function as natural protective systems. They reduce wave energy, limit shoreline erosion, and lessen the impacts of storm surges and flooding. Protecting and restoring these ecosystems therefore offer a practical and climate-resilient way to safeguard coastal settlements and productive landscapes.

EbA is particularly valuable because it responds to multiple forms of vulnerability at the same time. Coastal communities often depend on ecosystems for livelihoods, food security, and cultural practices. Damage to ecosystems therefore increases social and economic risks alongside physical exposure to climate hazards. Strengthening ecosystem health through EbA supports fisheries, sustains tourism, improves water quality, and helps maintain soil and sediment stability. These outcomes increase the ability of local populations to cope with climate stress. Unlike single-purpose structural defenses, ecosystem-based measures provide long-term benefits that grow as ecosystems recover and mature. This makes EbA especially relevant in low-income and climate-exposed regions where financial and technical resources are limited. Ecosystem-based Adaptation is closely connected to disaster risk reduction and climate change adaptation efforts. Healthy coastal ecosystems reduce hazard exposure and sensitivity, which are key components of disaster vulnerability. Mangrove belts can reduce wind and wave damage during cyclones, while coral reefs and seagrass meadows help reduce erosion and coastal flooding. Through these functions, EbA links climate adaptation with disaster management by embedding ecological processes into strategies aimed at reducing loss and damage. EbA also contributes to climate mitigation, as many coastal ecosystems, particularly mangroves, salt marshes, and seagrass meadows, store large amounts of carbon and support nature-based solutions that address both climate impacts and emissions.

The relevance of EbA extends to sustainable development and biodiversity conservation. Because EbA focuses on maintaining ecosystem integrity, it aligns with sustainable coastal management, marine spatial planning, and biodiversity protection initiatives. Reducing environmental degradation and strengthening ecosystem services support poverty reduction, livelihood security, and resilient food systems. These contributions are closely connected to global development goals related to climate action, healthy ecosystems, and sustainable communities. EbA should therefore be viewed as an integrated approach that brings together environmental, social, and economic objectives within a single adaptation framework. Efforts to enhance coastal resilience benefit from the ability of Ecosystem-based Adaptation to connect climate science, ecosystem management, and community-based adaptation. Its importance lies not only in reducing immediate climate risks but also in shaping long-term development pathways that are more resilient, equitable, and environmentally sustainable. By grounding adaptation strategies in natural processes, EbA offers a practical response to climate disasters in coastal regions where human and ecological futures are closely linked.

5.1.2 Role of EbA in Reducing Vulnerability to Climate Disasters

Ecosystem-based Adaptation (EbA) plays an important role in reducing vulnerability to climate disasters by strengthening the natural systems that protect human societies from climatic extremes (Fig. 5.2). Coastal regions are facing increasing exposure to cyclones, storm surges, sea-level rise, saline intrusion, flooding, and shoreline erosion. Vulnerability in these areas depends not only on the strength of physical hazards but also on the condition of the ecosystems that influence how these hazards affect people and places. When mangroves, coral reefs, wetlands, and dunes are degraded, coastal populations are left directly exposed to climate forces. Healthy and well-managed ecosystems, on the other hand, function as living protective barriers. EbA reduces vulnerability by addressing these ecological foundations and by lowering the intensity and impacts of climate risks over time. Climate disaster vulnerability is commonly described through three elements: exposure, sensitivity, and adaptive capacity. EbA influences all of these elements at the same time. Conservation and restoration of coastal ecosystems reduce exposure to hazards such as storm surges, high waves, and coastal flooding. Mangrove forests dissipate wave energy and slow wind speeds during cyclones, while coral reefs and seagrass meadows weaken incoming waves before they reach the shoreline. Dunes and salt marshes also act as natural buffers that protect settlements and infrastructure. Through these processes, EbA reduces the physical force of climate hazards experienced by coastal communities.

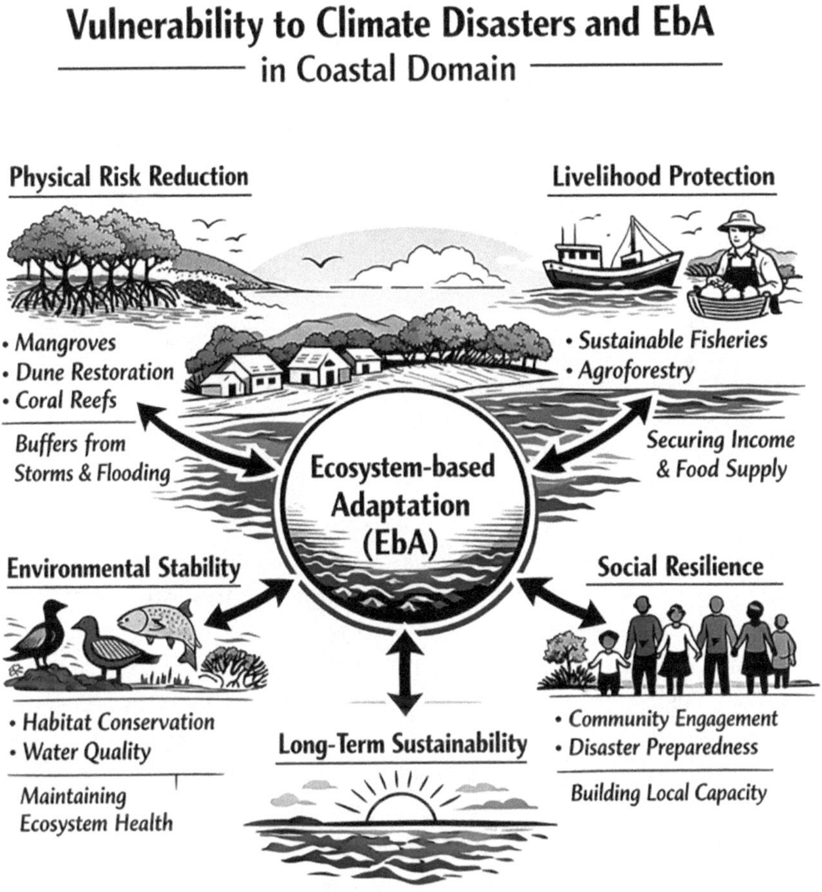

Fig. 5.2 Vulnerability to climate disasters and EbA (prepared by the authors)

EbA also reduces sensitivity by stabilizing the environmental and livelihood systems on which coastal populations depend. Climate disasters often damage not only houses and infrastructure but also fisheries, agriculture, and freshwater resources. Healthy coastal ecosystems support fish breeding, regulate salinity levels, and maintain soil and sediment balance, making livelihoods less vulnerable to climate shocks. Where ecosystems are degraded, even moderate events can cause severe social and economic losses. Restoring ecosystem functions through EbA lowers the risk that climate hazards will lead to wider livelihood disruption and food insecurity. Adaptive capacity is further strengthened through Ecosystem-based Adaptation by supporting both ecological and social resilience. Diverse and well-connected ecosystems recover more quickly after disturbances and continue to provide protective and productive services under changing climate conditions. EbA is often applied through participatory and community-based approaches that value local knowledge, improve governance, and support livelihood diversification. These processes help communities prepare for climate risks, respond more effectively to disasters, and recover more quickly after impacts occur.

EbA has also shown strong potential in managing slow-onset and combined climate risks in coastal regions. Sea-level rise, coastal erosion, and salinization gradually reduce land quality, freshwater availability, and livelihood options, increasing long-term vulnerability even in the absence of extreme events. Restoration of mangroves, tidal wetlands, and coastal vegetation helps trap sediments, stabilize shorelines, and regulate saltwater intrusion. Unlike rigid structural defenses, these ecosystems can adjust to changing sea levels, which makes them more suitable for long-term risk reduction. Additional benefits of EbA further contribute to reducing vulnerability and strengthening resilience. Coastal ecosystems support fisheries, aquaculture, tourism, and cultural practices, all of which contribute to economic stability and social cohesion. These benefits are particularly important in disaster-prone areas, where ecosystem-based livelihoods provide safety nets during climate shocks. Many coastal ecosystems also store large amounts of carbon, supporting climate mitigation and helping limit the severity of future climate change impacts. Several key pathways explain how Ecosystem-based Adaptation reduces vulnerability to climate disasters. Natural barriers such as mangroves, coral reefs, dunes, and wetlands lower wave energy, storm surge height, and erosion, reducing direct exposure to hazards. Healthy ecosystems sustain fisheries, coastal agriculture, and tourism, reducing economic sensitivity to climate shocks. Ecosystem restoration improves water regulation, soil stability, and salinity control, increasing resistance to extreme weather and gradual climate change. Community-based EbA strengthens local institutions, encourages participation, and builds skills that improve preparedness and recovery. Ecosystems also grow, regenerate, and adapt over time, allowing protective functions to continue under changing climate conditions (Khan et al., 2025; Peng et al., 2025).

Within the broader context of climate disasters and coastal resilience, Ecosystem-based Adaptation provides a strong alternative to approaches that focus only on structural protection. By working with natural systems, EbA addresses the underlying drivers of vulnerability, including ecosystem degradation, livelihood insecurity, and limited adaptive capacity. As climate risks continue to intensify, integrating Ecosystem-based Adaptation into coastal planning, disaster risk reduction, and climate policy is both environmentally responsible and socially and economically necessary for building resilient coastal futures.

5.2 Ecosystem Services and Coastal Resilience

Coastal regions are among the most productive landscapes on Earth, but they are also highly vulnerable to climate-related hazards. These areas are increasingly affected by cyclones, storm surges, coastal flooding, sea-level rise, salinity intrusion, heat stress, and shoreline erosion. Such pressures threaten human settlements, infrastructure, and livelihoods that depend closely on coastal and marine resources. In this context, ecosystem services provide a useful way to understand how natural systems contribute to coastal resilience. Ecosystem services refer to the direct and indirect benefits that people receive from ecosystems, including material resources, life-supporting processes, and protective functions. In coastal environments, ecosystems such as mangroves, coral reefs, seagrass meadows, salt marshes, dunes, and estuaries serve as vital natural assets that reduce climate risks while supporting economic and social well-being. Ecosystem services link ecological processes with human security and development. Coastal ecosystems regulate the movement of water, energy, sediments, and nutrients, shaping the stability and productivity of shorelines. Mangrove forests and salt marshes trap sediments and support land formation, helping to reduce erosion and limit the impacts of sea-level rise. Coral reefs and seagrass beds weaken wave energy and protect coastlines from storm damage. These regulating functions play a central role in coastal resilience because they reduce both exposure and sensitivity to climate hazards. When coastal ecosystems are degraded, their protective capacity declines, and even moderate climate events can cause serious damage (Munang et al., 2013; Prato et al., 2025).

The importance of ecosystem services in coastal regions extends beyond environmental protection. They support livelihoods, food security, and economic activities for millions of people. Fisheries, aquaculture, tourism, and coastal agriculture all rely on healthy ecosystems. Mangroves and estuaries provide breeding and nursery grounds for fish and shellfish, coral reefs support diverse marine life, and seagrass meadows contribute to fisheries productivity and carbon storage. In areas frequently affected by climate hazards, these services act as economic safety nets that help communities recover more quickly after disasters. Ecosystem services therefore represent not only ecological value but also natural capital that supports long-term resilience and development. As climate change intensifies, the role of ecosystem services becomes increasingly important. Stronger storms and rising sea levels place growing pressure on coastal defenses. Natural ecosystems offer flexible and self-renewing forms of protection that engineered structures often cannot provide on their own. Unlike seawalls or embankments, ecosystems can grow, adapt, and regenerate, allowing them to maintain protective functions under changing environmental conditions. They also avoid many of the negative impacts associated with hard infrastructure, such as increased erosion or habitat loss. Conservation and restoration of ecosystems therefore support long-term and cost-effective approaches to coastal resilience rather than short-term defensive solutions (Almutairi et al., 2020).

Ecosystem services also contribute to social and cultural resilience in coastal areas. Many coastal communities have strong cultural, spiritual, and historical connections to marine and coastal ecosystems. These relationships shape local knowledge systems, traditional resource management practices, and community identities. Such social and cultural ties often support collective action during and after climate disasters. Protecting ecosystems therefore helps sustain social cohesion and cultural continuity, which are important but often overlooked aspects of resilience. Several categories of ecosystem services are particularly relevant to coastal resilience. Provisioning services include food such as fish, shellfish, and seaweed, as well as freshwater, timber, fuelwood, and medicinal resources. These services support livelihoods and food security and reduce economic vulnerability to climate shocks. Regulating services include climate regulation, control of water flows, erosion reduction, and improvement of water quality. Mangroves, reefs, and wetlands reduce storm surges, flooding, and shoreline erosion, directly lowering disaster risk. Supporting services such as nutrient cycling, soil formation, and primary production maintain the ecological processes that sustain all other services and ensure long-term ecosystem functioning. Cultural services include recreational, spiritual, educational, and aesthetic benefits that strengthen community well-being and identity.

Ecosystem services contribute to coastal resilience in multiple ways. Physical protection reduces exposure to hazards and limits damage during extreme events. Economic benefits support livelihoods and reduce recovery costs after disasters. Ecological benefits strengthen biodiversity and ecosystem stability, allowing natural systems to better cope with climatic stress. Social benefits reinforce local institutions, cultural values, and cooperation within communities, all of which are essential for effective adaptation. Within the framework of Ecosystem-based Adaptation, ecosystem services form the main pathway through which nature contributes to climate risk reduction. Investment in the conservation and restoration of coastal ecosystems strengthens the flow of these services and builds resilience in a balanced and sustainable manner. This approach recognizes that ecosystem health and community safety are closely linked (Fig. 5.3). As climate disasters become more frequent and severe, safeguarding ecosystem services is not only an environmental concern but also a key strategy for securing the future of coastal regions.

5.2.1 Ecosystem Services in Climate Resilience and Coastal Adaptation

Coastal regions are among the most dynamic and productive environments on the planet, but they are also among the most exposed to

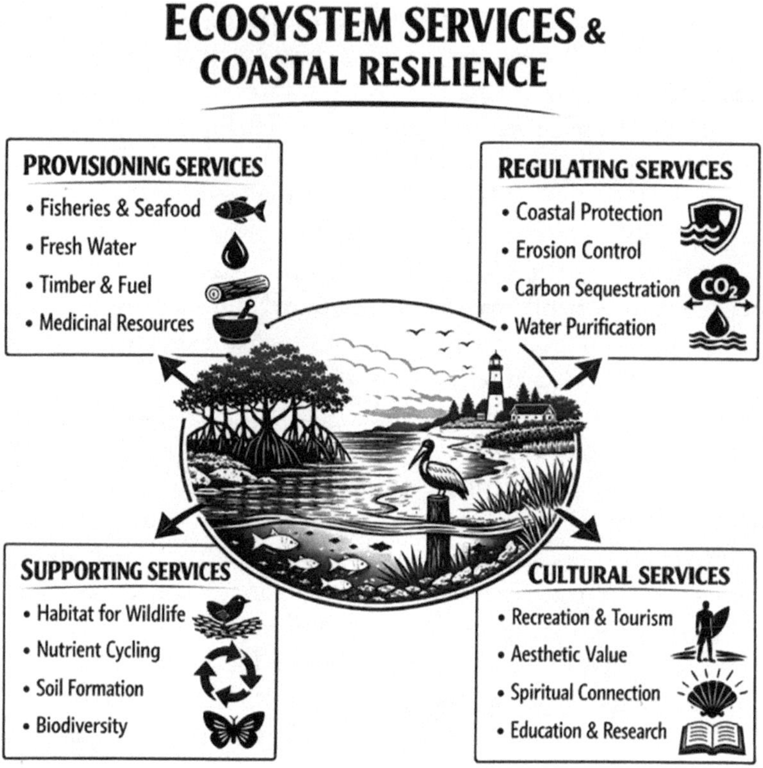

Fig. 5.3 Ecosystem services and coastal resilience (prepared by the authors)

climate-related hazards. Cyclones, storm surges, coastal flooding, sea-level rise, salinity intrusion, heat stress, and shoreline erosion increasingly affect coastal landscapes and the communities that depend on them. The capacity of coastal societies to withstand these pressures, recover from damage, and adapt to ongoing change is closely linked to the condition and functioning of coastal and marine ecosystems. Ecosystem services, understood as the benefits people derive from ecosystems, form a central foundation of coastal resilience. These services support livelihoods, reduce exposure to hazards, and help stabilize environmental conditions, making them essential to climate adaptation in coastal areas.

Ecosystem services contribute to climate resilience through their influence on exposure, sensitivity, and adaptive capacity. Regulating services such as wave reduction, flood control, erosion regulation, and water purification directly reduce the physical impacts of climate hazards. Mangrove forests can lower the height and force of storm surges, while coral reefs and seagrass meadows weaken wave energy before it reaches the shoreline. Salt marshes and tidal wetlands store excess floodwater and slow its movement into adjacent settlements. Through these processes, coastal ecosystems reduce the intensity of hazards experienced by communities and help limit damage during extreme events. Provisioning services play an equally important role by sustaining livelihoods and food security. Coastal fisheries, aquaculture, mangrove-based resources, and coastal agriculture rely on healthy ecosystems for their productivity and stability. After extreme weather events, these ecosystem-based resources often become critical for recovery and survival. When ecosystems are degraded, climate shocks are more likely to result in livelihood loss, food insecurity, and long-term poverty. In this way, ecosystem services act as economic buffers that reduce the social and economic sensitivity of coastal populations to climate stress. Supporting cultural services further strengthen resilience by maintaining ecological processes and social cohesion. Nutrient cycling, habitat provision, and primary production help ecosystems recover after disturbances and continue to function over time. Cultural services, including spiritual values, traditional ecological knowledge, and opportunities for recreation, reinforce community identity and cooperation. These social dimensions are especially important in times of crisis, as strong community ties often shape collective responses to disasters and support long-term adaptation. Together, these services create a layered form of resilience that connects ecological health with human well-being. Despite their importance, many coastal ecosystems have been degraded due to pollution, overexploitation, poorly planned development, and unsustainable infrastructure. This degradation weakens ecosystem services and increases vulnerability to climate hazards. Strengthening and restoring ecosystem services have therefore become a core objective of Ecosystem-based Adaptation in coastal regions. Conservation and restoration efforts aim not only to protect biodiversity but also to rebuild the natural functions that reduce risk and support adaptation (Fig. 5.4).

Fig. 5.4 Ecosystem services and coastal climate resilience (prepared by the authors)

Protecting existing healthy ecosystems is a critical starting point for enhancing ecosystem services. Conservation of mangroves, coral reefs, seagrass meadows, dunes, and wetlands helps maintain their protective and productive roles. Preventing deforestation, destructive fishing practices, sand mining, and unregulated coastal development is often more effective and less costly than attempting restoration after ecosystems have been lost. In areas where degradation has already occurred, active restoration becomes necessary to recover ecosystem structure and function. Restoration efforts in coastal areas include mangrove replanting, coral reef rehabilitation, re-establishing tidal flows in wetlands, stabilizing dunes with native vegetation, and improving water quality in estuaries and lagoons. These actions rebuild natural barriers, improve habitat quality, and restore ecological processes such as sediment trapping and nutrient cycling. Over time, restored ecosystems become more effective in reducing climate risks and supporting local livelihoods. Enhancing ecosystem services also depends on how coastal development and infrastructure are planned. Structures such as seawalls, ports, and embankments can disrupt sediment movement and damage ecosystems when poorly designed. Ecosystem-based and hybrid approaches, including living shorelines, mangrove buffers, and reef restoration combined with engineered structures, offer ways to provide protection while maintaining ecological functions. Integrating ecosystem considerations into planning helps ensure that development supports rather than undermines long-term resilience.

Community involvement remains essential in efforts to enhance ecosystem services for adaptation. Local communities often possess detailed knowledge of coastal ecosystems and rely on them directly for survival. Participatory restoration, community-based resource management, and recognition of customary rights help align adaptation measures with local needs and priorities. When communities have a clear stake in ecosystem health, long-term protection and sustainable management are more likely to succeed. Ecosystem services contribute to coastal climate resilience in several important ways. Natural systems such as mangroves, reefs, dunes, and wetlands reduce storm surges, waves, flooding, and erosion. Fisheries, aquaculture, tourism, and mangrove-based resources provide income and food, reducing vulnerability to climate shocks. Ecosystems regulate salinity, water quality, and sediment flows, supporting agriculture and freshwater availability. Coastal ecosystems such as mangroves and seagrasses store large amounts of carbon, contributing to climate mitigation. Cultural values, traditional knowledge, and community cooperation further strengthen social resilience. Strategies for enhancing and restoring ecosystem services for coastal adaptation include conservation of intact ecosystems through protected areas and sustainable use regulations, ecological restoration using locally appropriate and science-based methods, improved resource management to reduce pollution and overuse, integration of Ecosystem-based Adaptation into coastal planning, and inclusive governance that supports long-term stewardship and fair sharing of benefits (Fig. 5.5) (Jiménez-Navarro et al., 2025; Prato et al., 2025).

In the context of climate disasters and coastal resilience, ecosystem services should be seen as essential elements of survival and adaptation rather than optional environmental benefits. Maintaining and restoring the natural systems that protect shorelines, support livelihoods, and stabilize coastal environments offer a clear path toward resilient, sustainable, and equitable coastal futures. As climate change intensifies, continued investment in ecosystem services remains one of the most

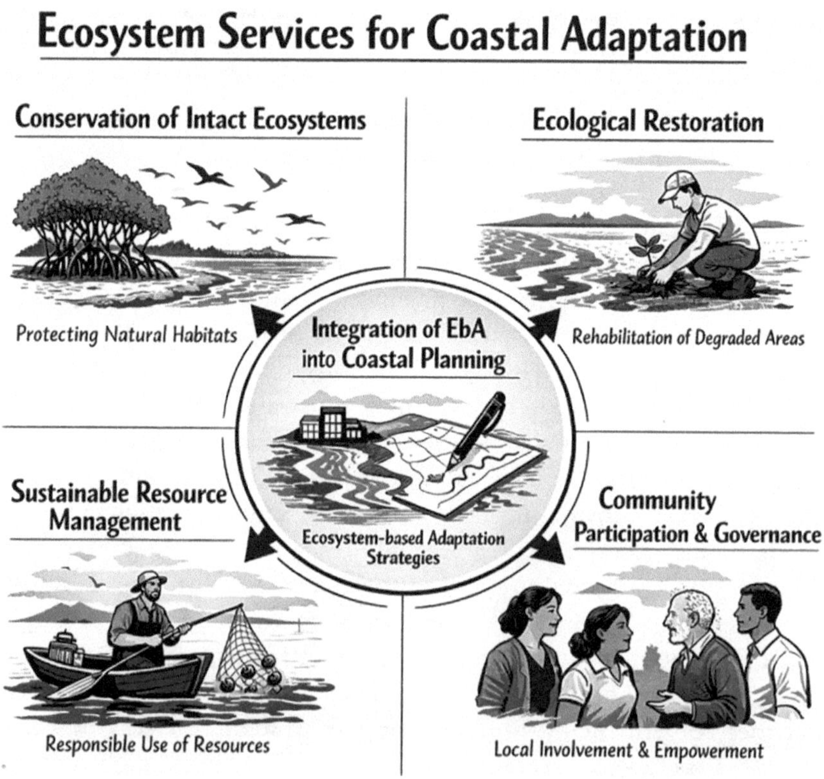

Fig. 5.5 Ecosystem services for coastal adaptation (prepared by the authors)

effective ways to reduce risk, strengthen adaptive capacity, and safeguard both people and nature in coastal regions.

5.3 Climate Disasters and Ecosystem Vulnerability

5.3.1 Climate Disasters on Coastal Ecosystems

Coastal ecosystems are among the most productive and biologically diverse environments on Earth, but they are also highly exposed to climate-related disasters. Ongoing climate change has increased both the frequency and intensity of extreme events such as cyclones, storm surges, coastal flooding, marine heatwaves, droughts, and episodes of heavy rainfall. These pressures place growing strain on coastal and marine ecosystems that already exist under naturally dynamic conditions. Mangroves, coral reefs, seagrass meadows, salt marshes, dunes, and estuaries play an essential role in protecting coastlines and supporting livelihoods. However, their ability to perform these functions is increasingly weakened by climate-driven disturbances. In many coastal regions, ecosystem degradation and climate risk now reinforce one another, leading to rising levels of vulnerability.

Extreme weather events represent one of the most immediate sources of stress for coastal ecosystems. Cyclones and storm surges can uproot mangrove forests, erode wetlands, damage coral reefs, and disrupt sediment movement along the shoreline. High-energy waves often break coral structures, while flooding and strong currents disturb seagrass beds and coastal vegetation. These impacts reduce the physical structure of ecosystems, making them less effective at reducing wave energy and stabilizing coastlines. As ecosystem structure declines, coastal areas become more exposed to future hazards, increasing the likelihood of repeated damage during subsequent events. Slow-onset climate processes further intensify ecosystem vulnerability. Sea-level rise gradually submerges low-lying wetlands and mangrove areas, while coastal erosion and saltwater intrusion degrade soils and freshwater habitats. Rising ocean temperatures cause coral bleaching and mortality, leading to declines in reef diversity and productivity. Shifts in rainfall patterns and extended dry periods affect freshwater inflows to estuaries and lagoons, altering nutrient balances and species composition. These gradual changes weaken ecosystem health over time and reduce the capacity of ecosystems to recover after disturbance. Human pressures often interact with climate-related stresses, amplifying ecosystem degradation. Overfishing, pollution, poorly planned coastal development, and habitat fragmentation reduce ecosystem resilience and increase sensitivity to climate shocks. Coral reefs that are already degraded recover more slowly after bleaching events, and mangrove areas cleared for development are more vulnerable to erosion during storms. These interactions create a cycle in which environmental degradation increases exposure to climate hazards, while climate disasters accelerate further ecosystem loss (Nagaraju & Bala, 2025; Singh et al., 2020).

The vulnerability of coastal ecosystems has serious implications for human societies. As ecosystems degrade, the services they provide such as coastal protection, fisheries support, water regulation, and carbon storage decline. Reduced ecosystem services increase physical exposure to climate hazards and weaken livelihoods, food security, and social stability in coastal communities. In this way, ecosystem vulnerability becomes a central driver of climate risk in coastal regions. Building resilience to climate disasters therefore requires close attention to the condition of coastal ecosystems. Protecting, restoring, and sustainably managing these systems through Ecosystem-based Adaptation is not only an environmental concern but also a practical strategy for reducing disaster risk. Strengthening ecosystem health helps restore natural protective functions and supports the long-term safety and well-being of coastal societies whose futures remain closely tied to the ecosystems they depend upon.

5.3.2 Vulnerability, Feedback Loops, and Ecosystem Thresholds

Coastal ecosystems operate within a delicate balance shaped by environmental forces and biological processes. This balance is increasingly disturbed by climate-related disasters. Ecosystem vulnerability refers to the extent to which natural systems are exposed to climate stress and their limited ability to cope with or recover from such pressures. Climate variability and extreme events such as cyclones, storm surges, marine heatwaves, droughts, and sea-level rise place growing strain on coastal ecosystems. Mangroves, coral reefs, salt marshes, seagrass meadows, and estuaries function within narrow limits of temperature, salinity, sediment supply, and water depth. When climate pressures push conditions beyond these limits, ecosystems begin to lose their structure, key functions, and capacity to recover, leading to long-term degradation. Feedback processes play an important role in shaping how ecosystems respond to stress. Healthy coastal ecosystems are supported by stabilizing feedbacks that help maintain ecological balance. Dense mangrove root systems trap sediments, increase surface elevation, and support the ability of mangroves to persist under rising sea levels. Coral reefs with high coral cover support fish populations that limit algae growth, allowing corals to recover after disturbance. These reinforcing processes enable ecosystems to absorb shocks and reorganize while continuing to perform their essential functions. Climate disasters can disrupt these stabilizing processes and trigger feedbacks that accelerate ecosystem decline. Storm damage, coral bleaching, and prolonged flooding often reduce vegetation cover or coral density. Reduced cover increases erosion, water turbidity, and algal growth, placing additional stress on remaining organisms. In mangrove systems, cyclone damage exposes soils to erosion, making it difficult for seedlings to establish and increasing vulnerability to future storms. In coral reef systems, bleaching reduces coral cover and allows algae to dominate, which suppresses coral regrowth and pushes the system toward a degraded condition. These reinforcing cycles make recovery increasingly difficult and deepen long-term vulnerability.

Critical ecological limits, commonly described as thresholds or tipping points, mark moments when gradual environmental stress leads to sudden and lasting changes in ecosystem structure and function. Coastal ecosystems may cross such limits due to rising sea temperatures, increasing salinity, excessive sediment loads, or repeated extreme events. Coral reefs can shift from coral-dominated systems to algae-dominated ones after repeated bleaching episodes. Mangrove forests may decline when sea-level rise exceeds sediment accumulation or when freshwater flows are altered. Once these limits are crossed, ecosystems often fail to return to their previous state even if conditions improve. Such shifts have serious implications for coastal resilience. Ecosystems that have moved into degraded states lose much of their ability to provide services such as wave reduction, erosion control, and fisheries support. This loss increases the exposure of coastal communities to climate hazards and can lead to wider social and economic instability. Degraded ecosystems are also more sensitive to future climate shocks, making subsequent disasters more damaging and recovery more uncertain (Luo et al., 2025; Setiawan et al., 2025).

Ecosystem-based Adaptation seeks to reduce these risks by strengthening stabilizing feedbacks and keeping ecosystems within conditions that support recovery and long-term functioning. Protection of biodiversity, restoration of degraded habitats, and reduction of pressures such as pollution, overfishing, and unsustainable development help maintain ecosystem health. These actions lower the risk of ecosystems crossing critical limits and preserve their ability to buffer climate hazards. Nonlinear patterns of ecosystem change are a defining feature of coastal environments under climate stress. Coastal ecosystems do not always decline in slow or predictable ways. Sudden collapse can occur once critical limits are exceeded. Effective adaptation therefore depends on early action, careful monitoring, and flexible management approaches that support ecosystem integrity and resilience in the face of increasing climate uncertainty.

5.4 Ecosystem-Based Adaptation Strategies: Coastal Habitats

Protecting and restoring coastal habitats are a central element of Ecosystem-based Adaptation for reducing vulnerability to climate disasters. Coastal ecosystems such as mangroves, coral reefs, seagrass meadows, salt marshes, dunes, and estuaries function as natural defense systems along the shoreline. They reduce exposure to storm surges, waves, flooding, and erosion while also supporting biodiversity and coastal livelihoods. Despite their importance, these ecosystems are increasingly under pressure from climate change, pollution, overexploitation, and poorly planned coastal development. Ecosystem-based adaptation is built on the idea that safeguarding and rebuilding these natural systems offers one of the most effective and lasting approaches to reducing climate risk in coastal regions. Protection of existing coastal habitats is the most immediate and cost-effective step within this approach. Intact ecosystems already provide strong protective functions and a wide range of ecosystem services. Preventing further degradation avoids the high financial and ecological costs that restoration often requires. Measures such as establishing and effectively managing coastal and marine protected areas, regulating land use, controlling destructive fishing practices, and reducing pollution are essential for maintaining ecosystem integrity. Protecting mangrove belts, coral reefs, and wetlands ensures that their ability to reduce wave energy, trap sediments, and regulate water flows is preserved, which lowers vulnerability to extreme weather events and sea-level rise.

Active restoration becomes necessary in areas where coastal habitats have already been degraded. Mangrove restoration often involves restoring natural tidal flows, planting native species, and stabilizing sediments so that forests can regenerate naturally. Coral reef restoration may include reef rehabilitation, coral transplantation, or the use of artificial structures to improve habitat complexity and wave buffering capacity. Restoration of seagrass meadows and salt marshes helps stabilize sediments, improve water quality, and enhance carbon storage. Over time, these efforts rebuild the physical structure and ecological processes that support coastal resilience. Nature-based and hybrid solutions are an important feature of restoration under Ecosystem-based Adaptation. Instead of relying only on hard structures such as seawalls and embankments, this approach promotes solutions that combine natural systems with engineered elements. Living shorelines, vegetated dunes, mangrove buffers, and reef-based defenses provide protection while maintaining ecological functions. Restoring mangroves in front of embankments, for example, can reduce wave pressure and maintenance costs while also supporting fisheries and biodiversity. These integrated approaches are more flexible and better able to respond to changing climate conditions than purely engineered structures (Gauthier et al., 2021; Peng et al., 2025).

Community involvement and local governance play a crucial role in the success of habitat protection and restoration. Coastal communities are often the main users and caretakers of local ecosystems, and their knowledge of tides, species, and environmental change is valuable for shaping effective interventions. Participatory planning, community-led restoration, and recognition of customary resource rights strengthen local ownership and long-term commitment to Ecosystem-based Adaptation initiatives. When communities experience direct benefits from healthier ecosystems, such as improved fisheries, tourism opportunities, and reduced disaster impacts, they are more likely to protect and manage these habitats sustainably. Protecting and restoring coastal habitats through Ecosystem-based Adaptation offer a resilient and sustainable pathway for reducing climate risk. Working with natural processes rather than attempting to control them strengthens the ecological foundations of coastal resilience and supports the ability of both ecosystems and communities to adjust to a changing climate.

Sustainable coastal land management and planning are also essential components of Ecosystem-based Adaptation for reducing disaster risk and strengthening resilience. Coastal regions face multiple and interacting pressures, including sea-level rise, erosion, storm surges, flooding, and rapid urban growth. Poorly planned development often disrupts natural processes, degrades ecosystems, and increases exposure to hazards. Ecosystem-based Adaptation addresses these challenges by integrating ecosystem conservation and restoration into land-use planning, infrastructure development, and disaster risk reduction strategies, aligning human activities with the natural dynamics of coastal landscapes.

A key principle of ecosystem-based coastal land management is working with natural coastal processes. Ecosystems such as mangroves, dunes, wetlands, and tidal flats regulate sediment movement, water flow, and shoreline stability. When development blocks these processes through land reclamation, embankments, or poorly located infrastructure, erosion and flood risks often increase. Sustainable land management emphasizes protecting natural buffers, limiting construction in high-risk zones, and allowing space for ecosystems to move inland as sea levels rise. Setback zones, green belts, and coastal zoning based on hazard and ecosystem assessments are important tools for reducing disaster risk. Integrating Ecosystem-based Adaptation into coastal planning also involves adopting nature-based and hybrid solutions for infrastructure and risk management. Living shorelines, vegetated embankments, mangrove buffers, and restored dunes can replace or complement hard structures such as seawalls and groynes. These approaches provide protection while maintaining biodiversity and ecosystem services, and they can adjust more easily to changing climate conditions. Planning informed by ecosystem processes also encourages the use of permeable surfaces, natural drainage systems, and wetland conservation to reduce urban flooding and manage stormwater more effectively.

Disaster risk reduction is closely linked to sustainable land-use planning within an ecosystem-based framework. Healthy ecosystems reduce hazard intensity by absorbing wave energy, storing floodwaters, and stabilizing soils. When land-use decisions take these protective functions into account, disaster risk is reduced before extreme events occur. Ecosystem-based risk assessments help identify critical habitats that should be conserved or restored to protect vulnerable communities. Emergency planning, early warning systems, and evacuation routes can also be designed in ways that reflect natural landscape features and improve safety.

Community participation and institutional coordination are essential for applying Ecosystem-based Adaptation in coastal land management and disaster risk reduction. Local knowledge of flood patterns, ecosystem change, and coastal dynamics can strengthen planning and adaptation strategies. Coordination among planning agencies, environmental authorities, and disaster management institutions helps ensure that ecosystem protection, development objectives, and risk reduction efforts support one another. Embedding ecosystems into coastal land management, spatial planning, and disaster risk reduction provides a holistic approach to climate resilience. Ecosystem-based Adaptation reduces exposure to hazards, limits environmental degradation, and supports sustainable development, making it a strong framework for addressing the growing climate risks faced by coastal regions.

5.4.1 Ecosystem-Based Adaptation Strategies: Coastal Land Management, Planning, and DRR

Sustainable coastal land management and planning are fundamental to Ecosystem-based Adaptation because they shape how human settlements, infrastructure, and economic activities interact with climate-sensitive coastal ecosystems. Coastal regions are increasingly exposed to cyclones, storm surges, sea-level rise, flooding, erosion, and salinity intrusion. These risks are often intensified by unplanned urban expansion, land reclamation, and the loss of natural buffers such as mangroves, wetlands, and dunes. Ecosystem-based Adaptation approaches coastal land management by treating ecosystems as essential protective systems whose conservation and restoration must be fully integrated into spatial planning and disaster risk reduction.

A central idea in ecosystem-based land management is working with natural coastal processes rather than attempting to control or replace them. Coastal ecosystems regulate sediment movement, stabilize shorelines, and manage water flows in ways that engineered structures cannot fully replicate. Mangroves trap sediments and reduce wave energy, salt marshes absorb floodwaters, and dunes act as flexible barriers during storms. When land-use planning ignores these functions by allowing construction in flood-prone areas, blocking tidal flows, or removing coastal vegetation, exposure to hazards increases. Sustainable coastal planning therefore gives priority to protecting these natural systems and allowing space for them to function and adjust as climate conditions change (Hill et al., 2025).

Risk-informed coastal zoning is an important tool for applying Ecosystem-based Adaptation in practice. This approach uses information on hazard exposure, ecosystem distribution, and projected climate risks to guide development decisions. Areas such as low-lying wetlands, erosion-prone shorelines, and storm surge pathways can be reserved for conservation, buffering, or low-intensity use rather than dense settlement or critical infrastructure. Shoreline setback zones, green belts, and ecological corridors allow ecosystems to move inland as sea levels rise, reducing long-term risk while preserving natural protective functions. Ecosystem-based Adaptation also encourages the use of nature-based and hybrid infrastructure in coastal development. Instead of relying only on seawalls, groynes, and embankments, planning approaches integrate mangrove belts, vegetated dunes, living shorelines, and restored reefs into coastal defense systems. These measures reduce wave energy, limit erosion, and support biodiversity while remaining flexible under changing climate conditions. Hybrid approaches that combine natural buffers with engineered structures often provide more durable and cost-effective protection than purely hard infrastructure. Urban and peri-urban coastal areas can benefit significantly from ecosystem-based planning. Rapid coastal urbanization often increases flood risk by replacing permeable surfaces with concrete, disrupting drainage systems, and encroaching on wetlands. Ecosystem-based land management promotes the protection and restoration of urban wetlands, mangroves, and floodplains, which act as natural drainage and water storage areas during heavy rainfall and storm surges. Green spaces such as parks, retention ponds, and vegetated corridors improve stormwater management while also enhancing environmental quality in coastal cities.

Disaster risk reduction is a key objective of sustainable coastal land management within an ecosystem-based framework. Healthy ecosystems reduce hazard intensity by absorbing wave energy, storing floodwaters, and stabilizing soils. When land-use decisions protect these functions and restrict settlement in high-risk zones, disaster losses are reduced before extreme events occur. Ecosystem-based risk assessments help identify priority areas for conservation and restoration that provide the greatest protective benefits. Land-use planning that aligns with these assessments strengthens long-term resilience. Community participation is essential for effective ecosystem-based land management. Coastal communities often hold detailed knowledge of local ecosystems, tidal patterns, and past hazard events. This knowledge can improve risk mapping and land-use decisions. Participatory planning helps ensure that zoning regulations, conservation measures, and relocation strategies reflect local needs and livelihood concerns. Secure land and resource rights also encourage communities to invest in long-term ecosystem stewardship, supporting both ecological health and social resilience.

Institutional coordination plays a crucial role in applying Ecosystem-based Adaptation across coastal land management and disaster risk reduction. Planning, environmental protection, urban development, and disaster management are often managed by separate agencies, leading to fragmented decision-making. Ecosystem-based approaches require integrated governance that aligns development goals with ecosystem conservation and climate risk reduction. Coordinated policies, shared data, and aligned planning processes help ensure that ecosystem services are fully considered in decision-making. Sustainable coastal land management and planning grounded in Ecosystem-based Adaptation offer a resilient pathway for development in climate-exposed coastal regions. Integrating ecosystems into spatial planning, infrastructure design, and disaster risk strategies reduces vulnerability, supports livelihoods, and maintains the natural foundations of coastal protection. As climate change continues to intensify risks along the world's coasts, this ecosystem-centered approach remains essential for safeguarding both human safety and ecological sustainability (Hill et al., 2025; Parvin et al., 2025).

5.5 Benefits and Co-benefits of Ecosystem-Based Adaptation

5.5.1 Conservation and Habitat Restoration

Ecosystem-based Adaptation places conservation and habitat restoration at the center of efforts to reduce climate risk and strengthen coastal resilience. Coastal regions are increasingly affected by cyclones, storm surges, sea-level rise, flooding, marine heatwaves, and erosion. In these settings, healthy ecosystems function as natural defense systems that protect shorelines, support livelihoods, and provide a foundation for long-term adaptation. Mangroves, coral reefs, seagrass meadows, salt marshes, dunes, and estuaries deliver a wide range of ecosystem services that reduce vulnerability while also supporting social and economic well-being. Conservation and restoration of these habitats therefore represent practical and sustainable investments for climate-resilient coastal development (Funk et al., 2020).

Maintaining natural protective functions is one of the most important outcomes of ecosystem conservation. Intact mangrove forests and salt marshes reduce the height and force of storm surges, stabilize sediments, and limit coastal erosion. Coral reefs and seagrass meadows weaken incoming waves and protect shorelines from flooding and physical damage. By conserving these habitats, Ecosystem-based Adaptation reduces exposure to climate hazards before disasters occur, lowering risks to coastal settlements and infrastructure. Unlike hard engineering structures, ecosystems have the capacity to repair themselves and adjust to changing conditions, which allows their protective roles to persist over time. Habitat restoration further strengthens coastal resilience by rebuilding degraded ecosystems and restoring lost functions. Activities such as mangrove replanting, coral reef rehabilitation, and restoration of wetlands and dunes enhance natural barriers against extreme weather while increasing biodiversity and ecosystem productivity. As restored habitats mature, they become more effective at reducing climate impacts. This form of protection improves with time rather than declining, offering a flexible response to sea-level rise and temperature change. Restoration also supports ecological recovery and reduces the risk of long-term ecosystem collapse. Conservation and restoration provide strong benefits for livelihoods and food security. Coastal ecosystems support fisheries, aquaculture, tourism, and the sustainable use of mangrove and wetland resources. Healthy habitats serve as breeding and nursery areas for fish and shellfish, which are vital sources of income and nutrition for coastal communities. Protecting and restoring these ecosystems strengthens local economies and reduces sensitivity to climate shocks. After extreme events, communities with intact ecosystems often recover more quickly because their natural resource base remains functional.

Ecosystem-based Adaptation also supports climate mitigation through the protection of blue carbon ecosystems. Mangroves, seagrass meadows, and salt marshes store large amounts of carbon in their vegetation and sediments. Conservation prevents the release of this stored carbon, while restoration enhances carbon uptake over time. This contribution to both adaptation and mitigation increase the importance of habitat conservation within climate planning and policy. Social and cultural benefits form another important dimension of conservation and restoration. Many coastal communities have strong cultural, spiritual, and historical ties to surrounding ecosystems. Mangroves, coral reefs, and wetlands are closely linked to local knowledge, traditions, and livelihoods. Conservation efforts help preserve these relationships and reinforce community identity. Restoration initiatives that involve local participation also strengthen social bonds, build skills, and encourage shared responsibility for managing natural resources. From a governance and development perspective, Ecosystem-based Adaptation offers cost-effective and flexible alternatives to conventional hard infrastructure. Nature-based measures often require lower investment and maintenance costs while delivering a wider range of benefits. They also reduce the risk of unintended impacts such as increased erosion or habitat loss. Integrating conservation and restoration into coastal planning allows governments to address disaster risk reduction, environmental protection, and sustainable development together. Key benefits and co-benefits of conservation and habitat restoration under Ecosystem-based Adaptation include disaster risk reduction through lower storm surge, wave energy, flooding, and erosion; biodiversity conservation through the protection of species and ecosystem stability; livelihood and food security through support for fisheries, tourism, and resource-based activities; climate change mitigation through carbon storage in coastal ecosystems; economic efficiency through reduced long-term costs; and social and cultural resilience through stronger community identity, knowledge, and cooperation. Within the context of climate disasters and coastal resilience, conservation and habitat restoration should be viewed as core elements of Ecosystem-based Adaptation rather than optional environmental measures. These actions provide immediate protection while generating long-term ecological, economic, and social benefits. Investment in healthy coastal ecosystems reduces vulnerability, strengthens adaptive capacity, and supports the development of resilient coastal futures under changing climate conditions (Ferreira, 2022; Funk et al., 2020).

5.5.2 Coastal Protection and Reducing Risk

Ecosystem-based Adaptation has emerged as an important approach for protecting coastlines and reducing disaster risk under accelerating climate change. Coastal regions are exposed to a wide range of climate-related hazards, including cyclones, storm surges, sea-level rise, flooding, coastal erosion, and saltwater intrusion. For decades, responses to these risks have relied largely on hard engineering measures such as seawalls and embankments. While these structures can provide localized protection, they are often expensive, difficult to maintain, and harmful to coastal ecosystems. Ecosystem-based Adaptation offers an alternative by working with natural systems such as mangroves, coral reefs, seagrass meadows, salt marshes, and dunes. These ecosystems reduce hazard impacts while supporting the ecological and economic systems on which coastal communities depend. Their protective value tends to increase over time as ecosystems grow and recover. A key contribution of Ecosystem-based Adaptation lies in its ability to reduce the physical force of climate hazards before they reach populated areas. Mangrove forests and salt marshes absorb and slow wave energy, lowering the height and speed of storm surges. Coral reefs and seagrass beds weaken offshore waves, reducing the intensity of water movement toward the coast. Dunes and coastal vegetation act as flexible barriers that protect inland areas from flooding and erosion. Maintaining and restoring these ecosystems reduces exposure to climate hazards and lowers the risk of damage to homes, infrastructure, and livelihoods (Giardino et al., 2018; Setiawati et al., 2024).

Ecosystem-based Adaptation also plays a vital role in stabilizing coastlines and managing sediment movement. Coastal ecosystems trap sediments, build land, and limit shoreline retreat, which is particularly important under conditions of rising sea levels. Mangroves and wetlands help maintain coastal elevation by capturing sediments carried by rivers and tides. This process allows shorelines to adjust gradually to sea-level rise and reduces long-term risks of inundation. Unlike rigid engineering structures, living ecosystems can respond to environmental change, making them more sustainable forms of coastal protection. Beyond physical protection, Ecosystem-based Adaptation provides important economic and social benefits that strengthen overall resilience. Healthy coastal ecosystems support fisheries, aquaculture, tourism, and other nature-based livelihoods that are central to many coastal economies. These activities provide income and food security, enabling households to recover more quickly after climate-related disasters. By sustaining ecosystem services, Ecosystem-based Adaptation reduces the sensitivity of coastal communities to climate shocks and helps prevent short-term disruptions from developing into long-term poverty or displacement. Cost-effectiveness and long-term sustainability are additional strengths of Ecosystem-based Adaptation. Nature-based approaches often require lower construction and maintenance costs than hard infrastructure, while delivering a broader range of benefits. Ecosystems grow, regenerate, and repair themselves, allowing their protective functions to improve rather than decline over time. This makes Ecosystem-based Adaptation particularly suitable for regions with limited financial and technical resources that need durable and affordable strategies for reducing climate risk (Malakar, 2026; Thupalli & Deen, 2018).

Ecosystem-based Adaptation supports coastal protection and risk reduction through several key pathways. Mangroves, reefs, and wetlands reduce wave energy and storm surges, lowering flood and erosion risk. Coastal vegetation and sediment-trapping ecosystems stabilize shorelines and reduce land loss. Living ecosystems can grow and shift landward, helping maintain protection as sea levels rise. Fisheries, tourism, and other ecosystem-based livelihoods support income and food security. Nature-based solutions often cost less over the long term than engineered structures and provide additional benefits such as improved biodiversity, water quality, and ecosystem health. Therefore, the ecosystem-based adaptation offers a comprehensive approach to coastal protection in regions facing increasing climate disasters (Fig. 5.6). Investment in healthy and resilient ecosystems reduces immediate risks from storms and flooding while building long-term capacity to cope with climate uncertainty. By linking environmental stewardship with disaster risk reduction, Ecosystem-based Adaptation provides lasting benefits for both coastal communities and the ecosystems that sustain them.

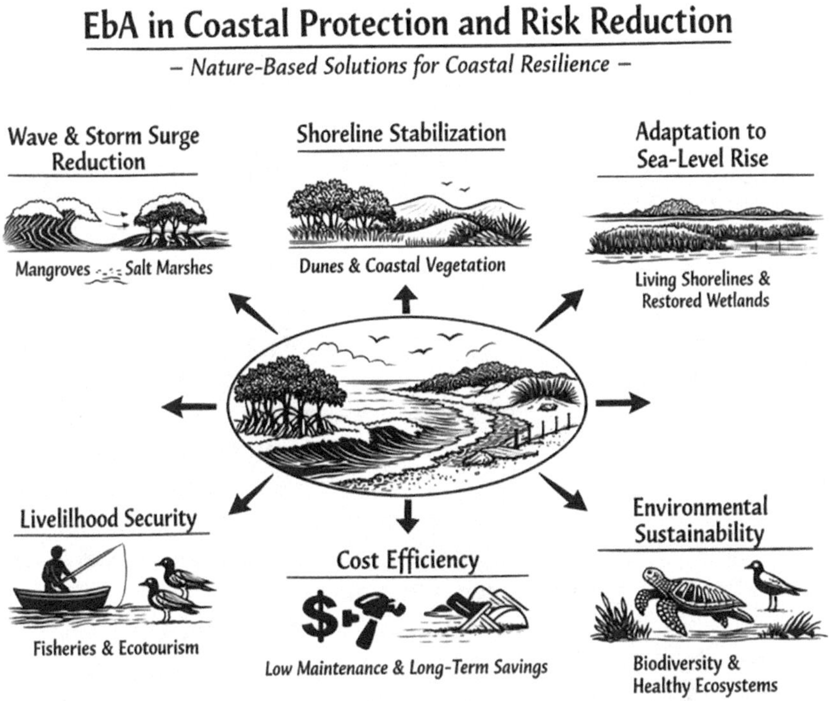

Fig. 5.6 EbA in coastal protection and risk reduction (prepared by the authors)

5.5.3 Carbon Sequestration and Climate Mitigation

Ecosystem-based Adaptation plays a clear role in reducing vulnerability to climate disasters, and it also contributes strongly to climate mitigation through carbon sequestration. Coastal and marine ecosystems such as mangroves, seagrass meadows, salt marshes, tidal wetlands, and coastal forests are among the most effective natural carbon sinks. These ecosystems absorb carbon dioxide from the atmosphere and store it in plant biomass and in deep, waterlogged soils where decomposition occurs very slowly. Carbon stored in these sediments can remain locked away for hundreds or even thousands of years. Through conservation and restoration, Ecosystem-based Adaptation addresses both the impacts of climate change and some of its underlying causes, linking adaptation and mitigation in a single approach. Degradation of coastal ecosystems increases climate risk in several ways. Damage to mangroves, wetlands, and seagrass beds reduces natural protection against storms, flooding, and erosion. At the same time, disturbance of these ecosystems releases stored carbon into the atmosphere. When mangroves are cleared, wetlands drained, or seagrass beds damaged, carbon held in vegetation and soils is exposed and released as carbon dioxide. Ecosystem-based Adaptation limits these emissions by protecting intact ecosystems and restoring degraded areas, allowing them to continue functioning as long-term carbon stores. This mitigation effect contributes to slowing global warming and helps reduce the future intensity of climate-related hazards such as heatwaves, sea-level rise, and extreme storms. The mitigation value of Ecosystem-based Adaptation is particularly high because coastal blue carbon ecosystems store carbon more densely and for longer periods than most terrestrial systems. Mangrove soils often contain several times more carbon per unit area than upland forest soils. Seagrass meadows and salt marshes continuously trap organic material and bury it in oxygen-poor sediments, where breakdown is slow. Protecting and expanding these ecosystems therefore strengthens natural carbon sinks while also supporting coastal protection and other ecosystem services (Kamal et al., 2024; Ussiri & Lal, 2017).

Carbon sequestration through Ecosystem-based Adaptation also brings benefits at local and national levels. Climate finance mechanisms and carbon markets can support conservation and restoration activities, providing new sources of funding for coastal management. Blue carbon initiatives can create economic incentives for protecting mangroves, wetlands, and seagrasses, linking climate mitigation with livelihood improvement and disaster risk reduction. In this way, ecosystem stewardship becomes a source of both environmental stability and financial resilience. The mitigation role of Ecosystem-based Adaptation strengthens long-term adaptation outcomes. Slower rates of global warming and sea-level rise reduce the scale of future climate impacts that coastal regions must manage. Healthier ecosystems are better able to store carbon and are less exposed to severe climate stress, creating positive feedbacks that benefit both natural systems and human communities. Through these interactions, Ecosystem-based Adaptation helps keep ecological and social systems within safer limits under changing climate conditions. Carbon sequestration through Ecosystem-based Adaptation provides several important benefits. Coastal ecosystems store carbon for long periods in vegetation and sediments. Protection of existing habitats prevents emissions linked to ecosystem degradation and land-use change. Increased carbon storage contributes to climate mitigation and reduces future climate risk. Blue carbon initiatives attract climate finance and support conservation and local livelihoods. Carbon-rich ecosystems also provide adaptation benefits such as coastal protection, biodiversity conservation, and livelihood support (Abdalqadir et al., 2024; Nair et al., 2021).

Carbon sequestration should therefore be treated as a core co-benefit of Ecosystem-based Adaptation rather than a secondary outcome. Integrating adaptation and mitigation within a single nature-based approach allows Ecosystem-based Adaptation to respond to climate change in a more complete and effective manner. Protecting and restoring carbon-rich coastal ecosystems reduce vulnerability to storms, flooding, and sea-level rise while also contributing to global efforts to stabilize the climate system. This combined role makes Ecosystem-based Adaptation a vital element of sustainable and resilient coastal development.

5.5.4 Socioeconomic Benefits and Livelihoods

Ecosystem-based Adaptation provides a strong framework for improving socioeconomic resilience in coastal regions that face increasing pressure from climate disasters. Coastal communities often depend heavily on natural resources for food, income, and cultural identity. This dependence increases their exposure to climate shocks such as cyclones, flooding, storm surges, and sea-level rise. Conservation and restoration of coastal ecosystems such as mangroves, coral reefs, seagrass meadows, wetlands, and dunes reduce physical exposure to hazards while also strengthening the economic and social systems that support recovery. Through these processes, Ecosystem-based Adaptation supports both immediate risk reduction and long-term livelihood sustainability.

Support for natural resource-based livelihoods is one of the most important socioeconomic outcomes of Ecosystem-based Adaptation. Healthy coastal ecosystems sustain fisheries, aquaculture, tourism, and the responsible use of forest and wetland products. Mangroves and estuaries provide breeding and nursery habitats for fish and shellfish. Coral reefs support diverse marine life that underpins both small-scale and commercial fishing. Coastal landscapes with high environmental quality also attract tourism. Protection and restoration of these ecosystems improve the stability and productivity of livelihood systems, reducing household vulnerability to climate-related disruptions. After extreme events, communities with access to healthy ecosystems often recover more quickly because their resource base remains available (Hoque et al., 2018; Riera-Spiegelhalder et al., 2023).

Ecosystem-based Adaptation also supports livelihood diversification, which is essential for reducing climate vulnerability. Activities such as ecosystem restoration, eco-tourism, sustainable aquaculture, mangrove-based products, and blue carbon initiatives create new income opportunities that are often less sensitive to climate variability than traditional livelihoods alone. These options provide employment while encouraging long-term care of natural resources. For many coastal communities with limited access to capital or markets, ecosystem-based livelihoods offer practical and locally suitable pathways for adaptation. Food and nutritional security is another important benefit of Ecosystem-based Adaptation. Coastal ecosystems supply fish, shellfish, seaweed, and other aquatic foods that are rich in protein and essential micronutrients. During and after climate disasters, these resources often act as safety nets that reduce hunger and malnutrition. Maintaining healthy ecosystems helps ensure that food supplies remain available even under climate stress, strengthening resilience at both household and community levels.

Social benefits form a central part of ecosystem-based adaptation. Community-based conservation and restoration activities encourage participation, cooperation, and local decision-making. These processes build trust, shared knowledge, and collective capacity to respond to future climate risks. Many Ecosystem-based Adaptation initiatives involve women, indigenous communities, and marginalized groups, helping to improve equity and strengthen the voices of those most affected by climate disasters. Ecosystem-based Adaptation also reduces the economic costs linked to climate disasters. Healthy ecosystems lower damage to housing, infrastructure, and productive assets by reducing the intensity of hazards. This reduces the financial burden of recovery and reconstruction for households and governments. Resources that would otherwise be spent on repeated disaster response can instead support development and long-term adaptation. Over time, these benefits contribute to poverty reduction and more stable economic growth in coastal regions. Key socioeconomic benefits of Ecosystem-based Adaptation include improved livelihood security through fisheries, aquaculture, tourism, and sustainable resource use; diversified income opportunities through restoration and nature-based enterprises; stronger food and nutritional security; reduced disaster-related losses; enhanced social cohesion through community participation; and long-term poverty reduction supported by stable ecosystem services. Socioeconomic outcomes are therefore as important as environmental gains within Ecosystem-based Adaptation. By linking ecosystem health with livelihoods, food security, and social well-being, Ecosystem-based Adaptation offers an integrated and practical pathway for building resilient and sustainable coastal societies under increasing climate risk (Kriegl et al., 2022; Sales, 2026).

5.5.5 EbA Planning and Implementation

Strategic planning and implementation are essential for Ecosystem-based Adaptation to deliver real benefits for climate-resilient coastal development. Unlike conventional adaptation measures that often focus on single sectors or isolated risks, Ecosystem-based Adaptation brings together ecosystem conservation, livelihood security, and disaster risk reduction within a unified planning approach. Placing natural systems at the center of adaptation strategies allows ecosystems such as mangroves, wetlands, coral reefs, dunes, and seagrass meadows to be treated as essential forms of infrastructure that protect coastlines and support human well-being. Risk-informed and ecosystem-focused decision-making forms a core part of Ecosystem-based Adaptation planning. Climate vulnerability assessments, ecosystem mapping, and future climate projections help identify areas where protecting or restoring ecosystems can reduce disaster risk most effectively. This process supports targeted and efficient use of resources, often at lower cost than hard engineering solutions. Aligning adaptation actions with ecological processes also helps avoid maladaptive interventions that may increase vulnerability in the long term. Implementation of Ecosystem-based Adaptation strengthens coordination across institutions and policy sectors. Responsibilities for climate adaptation, environmental management, land-use planning, and disaster risk reduction are often divided among multiple agencies, leading to fragmented responses. Ecosystem-based Adaptation provides a shared framework that links these sectors through common goals related to ecosystem protection and risk reduction. Improved coordination reduces policy conflicts, improves governance efficiency, and supports more coherent approaches to coastal development (Donatti et al., 2020; Lawlor & Jackson, 2022).

Community participation is central to the success of Ecosystem-based Adaptation planning and implementation. Adaptation measures rely on local ecosystems and long-term stewardship, making the involvement of local communities essential. Fishers, farmers, indigenous groups, and local authorities often hold detailed knowledge of coastal environments and environmental change. Participatory planning helps ensure that adaptation actions reflect local needs and priorities, building trust and encouraging shared responsibility for managing natural resources. Ecosystem-based Adaptation also delivers a range of economic and environmental benefits. Nature-based solutions often reduce the need for expensive engineered structures, lowering construction and maintenance costs. At the same time, restored and protected ecosystems support livelihoods, improve food security, conserve biodiversity, and store carbon. These multiple benefits increase the overall value of adaptation investments and strengthen the case for support from climate finance and development programs. Planning and implementation of Ecosystem-based Adaptation provide several clear benefits. Decision-making is guided by ecosystem and climate risk assessments. Adaptation costs are reduced through the use of nature-based solutions. Governance is strengthened through coordination across institutions. Community engagement is improved through participatory planning and local stewardship. Development benefits are enhanced through livelihood support, biodiversity conservation, and carbon storage (Atmaja et al., 2025; Sierra-Correa & Kintz, 2015). In coastal regions facing increasing climate disasters, careful planning and effective implementation of Ecosystem-based Adaptation help ensure that adaptation efforts reduce risk while remaining socially fair and environmentally sound. Integrating ecosystems into the core of adaptation strategies provides a strong foundation for building coastal societies that can respond and adjust to climate change over the long term.

5.6 Policy and Governance for Ecosystem-Based Adaptation

Policy and governance frameworks play a central role in turning Ecosystem-based Adaptation into meaningful outcomes for coastal disaster risk reduction. Coastal regions are exposed to multiple and interacting threats from cyclones, storm surges, sea-level rise, flooding, and ongoing ecosystem degradation. These risks are closely linked and cannot be managed through isolated or sector-specific actions. Ecosystem-based Adaptation depends on policies that treat ecosystems as essential protective systems and on governance arrangements that connect environmental management, climate adaptation, and disaster risk reduction in a coordinated manner. Integration of Ecosystem-based Adaptation into national and local development planning is a necessary foundation for effective action. Climate adaptation strategies, coastal zone management plans, biodiversity policies, and disaster risk reduction frameworks need to clearly reflect the role of ecosystems in reducing climate risk. When mangroves, wetlands, coral reefs, and dunes are formally acknowledged as assets for protection and adaptation, they are more likely to be safeguarded from unsustainable development and supported through long-term investment. Incorporating Ecosystem-based Adaptation into spatial planning and environmental regulation also helps limit activities such as coastal reclamation and deforestation that increase exposure to hazards. Governance arrangements for

Ecosystem-based Adaptation must operate across sectors and administrative levels. Coastal ecosystems often extend beyond political boundaries and involve a wide range of sectors, including fisheries, forestry, urban development, tourism, and disaster management. Coordination among ministries, agencies, and local authorities is therefore essential to align ecosystem protection, land-use planning, and risk reduction objectives. Institutional arrangements such as joint planning platforms, inter-agency coordination bodies, and integrated coastal management authorities can help reduce fragmented decision-making and support more consistent adaptation strategies (Claudet et al., 2024; Gonçalves & Pinho, 2025).

Local governance and community participation are equally important components of Ecosystem-based Adaptation. Coastal communities are often the first to experience climate disasters and are closely connected to local ecosystems through livelihoods and daily resource use. Policies that recognize customary rights, support community-based resource management, and promote participatory planning improve both the fairness and effectiveness of adaptation measures. When communities benefit directly from healthier ecosystems and have a role in decision-making, long-term stewardship and sustained adaptation become more likely. Financing and incentives also shape the success of Ecosystem-based Adaptation. Conservation and restoration of coastal ecosystems require stable and long-term investment, but funding is often fragmented or short-lived. Linking Ecosystem-based Adaptation to climate finance, disaster risk reduction budgets, and development programs helps secure ongoing support for ecosystem-based solutions. Financial mechanisms such as payments for ecosystem services, blue carbon initiatives, and biodiversity offsets can create incentives for local governments and communities to protect and restore coastal habitats. Monitoring, evaluation, and adaptive governance are necessary to keep Ecosystem-based Adaptation relevant under changing climate conditions. Coastal ecosystems and climate risks evolve over time, and governance systems must be able to respond to new information and emerging challenges. Regular assessment of ecosystem health, disaster impacts, and social outcomes provides evidence that can be used to refine policies and improve implementation. Adaptive governance allows institutions to adjust strategies as conditions change and as community needs evolve. Key elements of policy and governance for Ecosystem-based Adaptation in coastal disaster risk reduction include integration of ecosystem-based approaches into climate, development, biodiversity, and disaster policies; cross-sectoral coordination that aligns ecosystem management with land-use planning and risk reduction; participatory governance that values local knowledge and rights; sustainable financing for conservation and long-term management; and monitoring systems that support learning and policy adjustment (Fig. 5.7) (Siddiqui & Ishtiaque, 2025; Turner-Walker, 2023).

Fig. 5.7 Policy and governance elements for EbA in coastal disaster risk reduction (prepared by the authors)

In coastal regions facing increasing climate disasters, policy and governance frameworks determine whether Ecosystem-based Adaptation remains a concept or becomes a practical tool for resilience. Embedding ecosystems into decision-making processes and strengthening the capacity of institutions and communities to manage them responsibly provides a strong foundation for reducing disaster risk, conserving biodiversity, and supporting resilient coastal development in a changing climate.

5.7 Challenges and Opportunities in Ecosystem-Based Adaptation

Ecosystem-based Adaptation offers a promising approach for strengthening coastal resilience to climate disasters, but its application in practice continues to face several challenges. Coastal zones function as complex socio-ecological systems where climate risks interact with rapid development, competing land uses, and fragile ecosystems. Although the value of nature-based solutions is increasingly acknowledged, Ecosystem-based Adaptation often remains secondary in planning and investment decisions. Many adaptation strategies still favor short-term engineered responses to coastal hazards, even when these approaches create long-term environmental and social costs. Weak policy integration and fragmented governance present major obstacles to Ecosystem-based Adaptation in coastal regions. Responsibilities for climate adaptation, environmental conservation, coastal development, and disaster risk reduction are frequently divided among separate institutions with limited coordination. This separation can result in conflicting objectives, such as encouraging infrastructure development in ecologically sensitive or hazard-prone areas. In the absence of clear policy mandates that treat ecosystems as essential assets for adaptation, Ecosystem-based Adaptation struggles to compete with conventional engineering solutions in decision-making processes. Financial and technical limitations also constrain the wider adoption of Ecosystem-based Adaptation. Many coastal governments and local authorities lack sufficient funding, technical expertise, and reliable data to design and implement ecosystem-based measures.

Although Ecosystem-based Adaptation is often more cost-effective over the long term, its benefits may take time to become visible. This makes investment decisions more challenging when compared to infrastructure projects that produce immediate and visible outcomes. Restoration of mangroves, coral reefs, and wetlands also requires long-term commitment, monitoring, and maintenance, which can be difficult to sustain under short project cycles and uncertain funding. Social and land-use conflicts further complicate implementation. Coastal ecosystems are under constant pressure from aquaculture, tourism development, urban expansion, and resource extraction. Protecting or restoring these areas may require restrictions on certain activities or changes in land use, which can generate resistance from affected groups. When planning processes do not adequately involve local communities or provide fair compensation and alternative livelihood options, Ecosystem-based Adaptation may be perceived as a threat to existing livelihoods rather than a source of long-term security (O'Higgins et al., 2019; Saroar et al., 2019).

Uncertainty related to climate change and ecosystem responses adds another layer of complexity. The success of ecosystem-based measures depends on how ecosystems respond to rising temperatures, sea-level rise, and extreme events, outcomes that are difficult to predict with precision. This uncertainty can discourage policymakers from relying on nature-based approaches, even though engineered alternatives are also exposed to climate risks and may fail under extreme conditions. At the same time, the increasing impacts of climate disasters are creating new opportunities for Ecosystem-based Adaptation. Growing awareness of the limits of hard infrastructure, combined with advances in ecological science, climate finance, and community-based adaptation, is opening space for more integrated and ecosystem-focused approaches. Addressing institutional, financial, and social barriers is therefore a necessary step toward enabling Ecosystem-based Adaptation to play a larger role in coastal risk reduction. Strengthening these foundations can allow nature-based strategies to contribute more effectively to resilient and sustainable coastal futures under escalating climate pressures. Ecosystem-based Adaptation has gained attention as an important approach for strengthening coastal resilience to climate disasters, but its application in practice is shaped by several uncertainties and limitations. Coastal ecosystems such as mangroves, coral reefs, seagrass meadows, and wetlands are dynamic systems influenced by both climatic forces and human activities. Their responses to rising temperatures, sea-level rise, ocean acidification, and extreme weather events are not always predictable. This variability makes it difficult to estimate with confidence how ecosystem-based interventions will perform over long periods. Such uncertainty often leads planners and decision-makers to remain cautious, especially when ecosystem-based measures are compared with engineered structures that deliver faster and more visible outcomes. Dependence on ecological recovery processes presents a major limitation for Ecosystem-based Adaptation. Ecosystems that have been severely degraded or pushed beyond critical ecological limits may not recover sufficiently to provide strong protective or productive functions. Coral reefs exposed to repeated bleaching events or mangrove systems disrupted by altered tidal flows often require extensive restoration, and recovery is not always assured even with careful management. In these contexts, ecosystem-based measures alone may be insufficient, and additional adaptation approaches, including hybrid or engineered options, may be necessary to achieve acceptable levels of risk reduction (Hoffmann, 2022; Huynh et al., 2025; Parvin et al., 2025).

Spatial and temporal constraints also influence the performance of Ecosystem-based Adaptation in coastal regions. Ecosystem restoration and regeneration require time, often several years or decades, before full benefits are realized. This creates challenges in areas facing immediate and severe climate threats. Many densely populated or heavily modified coastlines also lack the physical space needed for ecosystems to expand or migrate inland as sea levels rise. Where such space is limited, the capacity of ecosystems to function as natural buffers is reduced, making careful planning and realistic expectations essential. Socioeconomic and governance conditions further shape the outcomes of Ecosystem-based Adaptation. Unclear land tenure arrangements, competing resource claims, and conflicting development priorities can weaken conservation and restoration efforts. Long-term success also depends on community support and stewardship. Local acceptance is closely linked to whether people experience clear benefits and perceive decision-making processes as fair. In the absence of inclusive governance structures, secure resource rights, and sustained financial support, ecosystem-based projects may struggle to deliver lasting results. At the same time, Ecosystem-based Adaptation presents important opportunities despite these challenges. Its capacity to deliver multiple benefits, including livelihood support, biodiversity conservation, and carbon storage, makes it well-suited to integrated adaptation strategies. Advances in climate science, ecosystem monitoring, and participatory planning have improved the design of more context-specific and realistic ecosystem-based interventions. As climate risks continue to intensify and the limitations of purely engineered solutions become more apparent, the role of ecosystems in coastal adaptation is gaining wider acceptance. Uncertainty and limitation should therefore be treated as factors that guide how Ecosystem-based Adaptation is applied, rather than reasons to dismiss it. Careful design, adaptive management, and combination with other adaptation measures allow ecosystem-based approaches to respond to changing conditions. With sound scientific understanding, inclusive governance, and long-term commitment, Ecosystem-based Adaptation remains one of the most promising pathways for building resilient and sustainable coastal systems in an increasingly uncertain climate future.

5.8 Conclusion and Arguments

Ecosystem-based Adaptation has developed into a scientifically grounded and socially relevant approach for responding to growing climate risks in coastal regions. The approach is based on the principle of working with natural processes rather than relying only on engineered solutions. Coastal ecosystems such as mangroves, coral reefs, seagrass meadows, wetlands, and dunes serve as essential elements of coastal protection and human security. This chapter has shown that Ecosystem-based Adaptation operates at the intersection of climate resilience, disaster risk reduction, biodiversity conservation, and sustainable development, offering an integrated framework for addressing complex coastal challenges. Healthy and well-managed ecosystems play a central role in reducing vulnerability to climate hazards. They reduce exposure to storms and flooding, stabilize shorelines, regulate water and sediment movement, and support livelihoods that coastal communities depend on. These functions help limit damage from sudden disasters such as cyclones and storm surges, while also addressing slower processes such as sea-level rise, erosion, and salinity intrusion. Through these mechanisms, Ecosystem-based Adaptation strengthens resilience across both short and long-time scales.

Ecosystem services also generate a wide range of additional benefits that extend beyond physical protection. Coastal ecosystems contribute to food security, store carbon, support cultural practices, and create economic opportunities through fisheries, tourism, and other nature-based livelihoods. These co-benefits increase the overall value of Ecosystem-based Adaptation and make it a cost-efficient and multi-functional strategy, especially in regions where financial and technical resources are limited. Constraints related to ecological thresholds, governance capacity, and uncertainty in ecosystem responses can limit the performance of Ecosystem-based Adaptation in some settings. These challenges point to the need for careful design, continuous learning, and inclusive decision-making. Adaptive management, strong institutional coordination, and meaningful community participation are essential for ensuring that ecosystem-based measures remain effective as climate conditions continue to change. Climate change is expected to intensify existing pressures on coastal regions, making long-term resilience increasingly dependent on the condition of natural systems. Protection and restoration of coastal ecosystems therefore represent both an environmental responsibility and a development priority. Ecosystem-based Adaptation offers a strategic pathway toward coastal futures that are resilient, socially equitable, and environmentally sustainable, grounded in the recognition that human well-being and ecosystem health are deeply interconnected. The chapter discussions lead to the following arguments:

(a) Ecosystems act as natural infrastructure that reduce exposure to climate hazards and disaster risk in coastal areas.
(b) Ecosystem-based Adaptation reduces vulnerability by strengthening ecological, social, and economic resilience together.

(c) Coastal ecosystem services sustain livelihoods, support food security, and aid recovery after climate disasters.
(d) Ecosystem-based Adaptation delivers key co-benefits such as biodiversity conservation and carbon sequestration.
(e) Nature-based solutions are often more flexible, adaptive, and cost-effective than hard engineering approaches.
(f) Successful Ecosystem-based Adaptation depends on policy integration, community participation, and long-term governance.

References

Abdalqadir, M., Hughes, D., Gomari, S. R., et al. (2024). A state of the art of review on factors affecting the enhanced weathering in agricultural soil: Strategies for carbon sequestration and climate mitigation. *Environmental Science and Pollution Research, 31*, 19047–19070. https://doi.org/10.1007/s11356-024-32498-5

Almutairi, A., Mourshed, M., & Ameen, R. F. M. (2020). Coastal community resilience frameworks for disaster risk management. *Natural Hazards, 101*, 595–630. https://doi.org/10.1007/s11069-020-03875-3

Atmaja, T., Yulianti, W., Arini, E. Y., et al. (2025). Scaling Nature-based solutions (NbS): Lessons from global progress and Indonesia's path to sustainability. *Earth Systems and Environment*. https://doi.org/10.1007/s41748-025-00842-5

Claudet, J., Blythe, J., Gill, D. A., Bennett, N. J., Naggea, J., et al. (2024). Advancing ocean equity at the nexus of development, climate and conservation policy. Nature Ecology & Evolution. https://doi.org/10.1038/s41559-024-02417-5

Donatti, C. I., Harvey, C. A., Hole, D., et al. (2020). Indicators to measure the climate change adaptation outcomes of ecosystem-based adaptation. *Climatic Change, 158*, 413–433. https://doi.org/10.1007/s10584-019-02565-9

Ferreira, Ó. (2022). Modelling risk reduction measures to minimise future impacts of storms at coastal areas. In C. Kondrup et al. (Eds.), *Climate adaptation modelling*. Springer Climate. https://doi.org/10.1007/978-3-030-86211-4_8

Funk, A., O'Higgins, T. G., Borgwardt, F., Trauner, D., & Hein, T. (2020). Ecosystem-based management to support conservation and restoration efforts in the Danube Basin. In T. O'Higgins, M. Lago, & T. DeWitt (Eds.), *Ecosystem-based management, ecosystem services and aquatic biodiversity*. Springer, Cham. https://doi.org/10.1007/978-3-030-45843-0_22

Gauthier, S., May, B., & Vasseur, L. (2021). Ecosystem-based adaptation to protect avian species in coastal communities in the greater Niagara region, Canada. *Climate, 9*(6), 91. https://doi.org/10.3390/cli9060091

Giardino, A., Nederhoff, K., & Vousdoukas, M. (2018). Coastal hazard risk assessment for small islands: Assessing the impact of climate change and disaster reduction measures on Ebeye (Marshall Islands). *Regional Environmental Change, 18*, 2237–2248. https://doi.org/10.1007/s10113-018-1353-3

Gonçalves, C., & Pinho, P. (2025). The governance of the coastal region: Evolutionary changes in the conceptualisation and integration of landscape in Portuguese coastal planning institutions. *Landscape Ecology, 40*, 41. https://doi.org/10.1007/s10980-025-02053-9

Hill, B., Jiang, J., Bosher, L., & Liang, Q. (2025). Ecosystems-based approaches for disaster risk reduction. In R. Shaw (Eds.), *Encyclopedia of disaster risk reduction*. Springer. https://doi.org/10.1007/978-981-97-4547-0_216-1

Hoffmann, S. (2022). Challenges and opportunities of area-based conservation in reaching biodiversity and sustainability goals. *Biodiversity and Conservation, 31*, 325–352. https://doi.org/10.1007/s10531-021-02340-2

Hoque, S. F., Quinn, C., & Sallu, S. (2018). Differential livelihood adaptation to social-ecological change in coastal Bangladesh. *Regional Environmental Change, 18*, 451–463. https://doi.org/10.1007/s10113-017-1213-6

Huynh, L. T. M., Su, J., & Gasparatos, A. (2025). Differentiated trajectories of ecosystem-based adaptation for urban coastal defence in the Asian-Pacific region: A biodiversity–climate–society nexus perspective. *Ocean & Coastal Management, 270*, Article 107799. https://doi.org/10.1016/j.ocecoaman.2025.107799

Jiménez-Navarro, I. C., López-Ballesteros, A., Mesman, J. P., et al. (2025). Modeling climate impacts on ecosystem services in an anthropized coastal lagoon for effective planning and adaptation. *Climatic Change, 178*, 92. https://doi.org/10.1007/s10584-025-03926-3

Kamal, M. Z. U., Akter, M., Binte, B. I., & Mina, K. K. (2024). Carbon sequestration and climate change mitigation. In M. M. Rahman, J. C.

Biswas, & R. S. Meena (Eds.), *Climate change and soil-water-plant Nexus*. Springer. https://doi.org/10.1007/978-981-97-6635-2_15

Khan, Y. D. I., Goswami, M., & Nautiyal, S. (Eds.) (2025). Ecosystem-based approaches for resilience building in Himalayan landscapes. In *Insights for climate change and disaster risk reduction*. Springer. ISBN: 978-981-95-2006-0. https://doi.org/10.1007/978-981-95-2007-7

Kriegl, M., Kluger, L. C., Gorris, P., et al. (2022). Coastal livelihood resilience to abrupt environmental change: The role of social capital in a Peruvian bay. *Regional Environmental Change, 22*, 103. https://doi.org/10.1007/s10113-022-01959-3

Lawlor, P., & Jackson, D. W. T. (2022). A nature-based solution for coastal foredune restoration: The case study of Maghery, County Donegal, Ireland. In I. Misiune, D. Depellegrin, L. E. Vigl (Eds.), *Human-nature interactions*. Springer. https://doi.org/10.1007/978-3-031-01980-7_32

Luo, C., Ma, X., Wang, Y., Yan, W., Han, Y., Yu, W., & Fan, B. (2025). Ecosystem services thresholds and interconnected feedback loops in the vulnerable Tarim River Basin: Confronting climate and vegetation transformations. *Global Ecology and Conservation, 59*, Article e03529. https://doi.org/10.1016/j.gecco.2025.e03529

Malakar, K. D. (2026). Geographies of Disaster Risk Reduction Practices: New Approaches, Progress, and Contemporary Challenges. Emerald Publishing Limited. ISBN: 9781836620112. https://doi.org/10.1108/978-1-83662-010-5

Mukherjee, M., & Shaw, R. (Eds.). (2021). *Ecosystem-based disaster and climate resilience: Integration of blue-green infrastructure in sustainable development*. Springer. ISBN: 978-981-16-4814-4. https://doi.org/10.1007/978-981-16-4815-1

Munang, R., Thiaw, I., Alverson, K., Liu, J., & Han, Z. (2013). The role of ecosystem services in climate change adaptation and disaster risk reduction. *Current Opinion in Environmental Sustainability, 5*(1), 47–52. https://doi.org/10.1016/j.cosust.2013.02.002

Nagaraju, T. V., & Bala, G. S. (2025). Climate change and coastal hazards in delta ecosystems. In T. V. Nagaraju, & G. S. Bala (Eds.), *Delta ecosystems. Environmental contamination remediation and management*. Springer. https://doi.org/10.1007/978-3-031-97920-0_2

Nair, P. K. R., Kumar, B. M., & Nair, V. D. (2021). Carbon sequestration and climate change mitigation. In *An Introduction to agroforestry*. Springer. https://doi.org/10.1007/978-3-030-75358-0_20

O'Higgins, T., O'Higgins, L., O'Hagan, A. M., & Ansong, J. O. (2019). Challenges and opportunities for ecosystem-based management and marine spatial planning in the Irish Sea. In J. Zaucha, & K. Gee (Eds.), *Maritime spatial planning*. Palgrave Macmillan. https://doi.org/10.1007/978-3-319-98696-8_3

Parvin, G. A., Ara, S., & Shumi, K. F. (2025). Importance, vulnerabilities, initiatives, and finance for coastal ecosystem management. In R. Shaw (Eds.), *Encyclopedia of disaster risk reduction.* Springer. https://doi.org/10.1007/978-981-97-4547-0_33-1

Peng, Y., Welden, N. A., & Renaud, F. G. (2025). Ecosystem-based adaptation strategies to multi-hazard risk reduction and policy implications in the Pearl River and Yangtze River deltas, China. *International Journal of Disaster Risk Reduction, 116*, Article 105053. https://doi.org/10.1016/j.ijdrr.2024.105053

Prato, J., et al. (2025). Marine ecosystem services for climate change adaptation and mitigation strategies in the seaflower biosphere reserve: Coastal protection and fish biodiversity refuge at Caribbean insular territories. In J. E. M. Pineda, A. F. Osorio, C. Toro, & C. S. Velásquez-Calderón (Eds.), *Climate change adaptation and mitigation in the seaflower biosphere reserve. Disaster risk reduction.* Springer. https://doi.org/10.1007/978-981-97-6663-5_8

Riera-Spiegelhalder, M., Campos-Rodrigues, L., Ensenado, E. M., Dekker-Arlain, J. D., Papadopoulou, O., Arampatzis, S., & Vervoort, K. (2023). Socio-economic assessment of ecosystem-based and other adaptation strategies in coastal areas: A systematic review. *Journal of Marine Science and Engineering, 11*(2), 319. https://doi.org/10.3390/jmse11020319

Sales, V. (2026). Sea level rise: Scientific evidence, socio-economic realities, and adaptation challenges for coastal communities. In E. Fornalé (Eds.), *Sea level rise*. Palgrave Macmillan. https://doi.org/10.1007/978-3-031-89171-7_2

Saroar, M. M., Rahman, M. M., Bahauddin, K. M., & Rahaman, M. A. (2019). Ecosystem-based adaptation: Opportunities and challenges in coastal Bangladesh. In S. Huq, J. Chow, A. Fenton, C. Stott, J. Taub, & H. Wright (Eds.), *Confronting climate change in Bangladesh.* The Anthropocene: Politik—Economics—Society—Science (Vol. 28). Springer. https://doi.org/10.1007/978-3-030-05237-9_5

Scarano, F. R. (2017). Ecosystem-based adaptation to climate change: Concept, scalability and a role for conservation science. *Perspectives in Ecology and Conservation, 15*(2), 65–73. https://doi.org/10.1016/j.pecon.2017.05.003

Setiawan, T., Myrna, R., Isnawati, N. W., et al. (2025). The environmental dynamics in shaping ecological governance for sustainable development in coastal areas. *Discover Sustainability, 6*, 925. https://doi.org/10.1007/s43621-025-01849-8

Setiawati, M. D., Nandika, M. R., Hernawan, U. E., et al. (2024). Application of coastal hazard index to advance nature-based protection for coastal communities in the small islands. *Discover Applied Science, 6*, 462. https://doi.org/10.1007/s42452-024-06164-x

Siddiqui, M. R., & Ishtiaque, A. (2025). Roles and nature of public participation in climate change adaptation in coastal South Asia. *Discover Environment, 3*, 166. https://doi.org/10.1007/s44274-025-00360-1

Sierra-Correa, P. C., & Kintz, J. R. C. (2015). Ecosystem-based adaptation for improving coastal planning for sea-level rise: A systematic review for mangrove coasts. *Marine Policy, 51*, 385–393. https://doi.org/10.1016/j.marpol.2014.09.013

Singh, A. et al. (Eds.). (2020). *Development in coastal zones and disaster management*. Palgrave Macmillan. ISBN: 978-981-15-4293-0. https://doi.org/10.1007/978-981-15-4294-7

Spencer, T., Munera, C., Pizarro, V., Lozano-Rivera, P., Esquivel, C., & Cardona, O. D. (2024). Coastal ecosystems contribution to climate adaptation and disasters risk management in the tropical Americas. *Nature-Based Solutions, 5*, Article 100112. https://doi.org/10.1016/j.nbsj.2024.100112

Thupalli, R., & Deen, T. A. (2018). An investment strategy for reducing disaster risks and coastal pollution using nature based solutions. In N. Nagabhatla & C. Metcalfe (Eds.), *Multifunctional wetlands. Environmental contamination remediation and management*. Springer. https://doi.org/10.1007/978-3-319-67416-2_5

Turner-Walker, S. (2023). Local resource governance: Strategies for adapting to change. In A. Triyanti, M. Indrawan, L. Nurhidayah, & M. A. Marfai (Eds.), *Environmental governance in Indonesia. Environment & policy* (Vol. 61). Springer. https://doi.org/10.1007/978-3-031-15904-6_22

Ussiri, D. A. N., & Lal, R. (2017). *Carbon sequestration for climate change mitigation and adaptation*. Springer. ISBN: 978-3-319-53843-3. https://doi.org/10.1007/978-3-319-53845-7

Vasseur, L. (2021). How ecosystem-based adaptation to climate change can help coastal communities through a participatory approach. *Sustainability, 13*(4), 2344. https://doi.org/10.3390/su13042344

Yáñez-Arancibia, A., Day, J. W., & Reyes, E. (2013). Understanding the coastal ecosystem-based management approach in the Gulf of Mexico. *Journal of Coastal Research, 63*, 244–262. https://doi.org/10.2307/23486516

6

Climate Disasters and Enhancing Coastal Resilience: Applications of Community-Based Adaptation and Participatory Approaches

This chapter explores climate disasters through the lens of coastal resilience, with a strong focus on the role of communities in shaping effective responses to climate change. Coastal areas are increasingly exposed to hazards such as cyclones, storm surges, coastal erosion, and salinity intrusion, all of which place serious pressure on livelihoods, ecosystems, and everyday life. In such settings, resilience cannot depend only on engineering solutions or physical infrastructure. It also depends on people, their knowledge, social relationships, and ability to act together in times of risk. The chapter adopts a community-based and participatory perspective to understand how coastal populations respond to climate threats and adapt over time. It discusses key themes such as community engagement, local and indigenous knowledge, social networks, collaborative governance, capacity building, communication, and participatory monitoring. These elements show how locally driven processes contribute to stronger and more sustainable resilience outcomes. The chapter also reflects on practical challenges that affect community participation, including social inequalities, power imbalances, and institutional limitations. Through this balanced approach, the

K. D. Malakar et al., *Coastal Ecologies at Risk*, Palgrave Studies in Environmental Sustainability, https://doi.org/10.1007/978-3-032-26081-9_6

chapter offers insights into how coastal communities can reduce vulnerability and work toward more adaptive and resilient futures under growing climate uncertainty.

Key Points of the Chapter

- Describes how coastal resilience is stronger when communities actively shape decisions rather than remain passive recipients.
- Explains how combining indigenous, local, and scientific knowledge leads to more relevant and effective climate responses.
- Highlights the role of collective action and social networks in enabling quicker response, recovery, and long-term adaptation.
- Examines how participatory learning, communication, and capacity building help resilience efforts evolve over time.
- Highlights the importance of inclusive and equitable governance in reducing vulnerability and supporting sustainable coastal futures.

6.1 Importance of Community Engagement in Climate Resilience

Community engagement plays a central role in building climate resilience, especially in coastal regions where climate hazards intersect with high population density, sensitive ecosystems, and livelihoods that depend on natural resources. Within community-based adaptation and participatory approaches, the involvement of local people is not only a matter of principle but also a practical necessity. Coastal communities possess long-term knowledge of local weather patterns, shoreline changes, and past disaster experiences that are often overlooked in externally driven planning processes. When this knowledge is incorporated into resilience planning, adaptation measures become more relevant to local conditions, culturally appropriate, and better aligned with the actual risks faced by different households and social groups. Community engagement thus shifts climate resilience away from purely technical solutions toward processes rooted in social realities. Participation also strengthens the acceptance and durability of adaptation strategies. Interventions such as coastal protection measures, early warning systems, planned relocation, and livelihood diversification depend on continued cooperation from local residents. Involvement in identifying risks and shaping solutions encourages a sense of ownership among community members. This sense of responsibility supports long-term maintenance, cooperation, and commitment, particularly in settings where institutional capacity is limited and adaptation is an ongoing effort rather than a one-time activity. Participatory processes further help reduce conflict by ensuring that interventions are sensitive to local priorities and social arrangements. Community engagement is equally important for addressing the social dimensions of vulnerability that influence how climate disasters are experienced. Exposure to coastal hazards is shaped not only by physical location but also by income, gender, caste, age, and access to information. Participatory approaches help make these differences visible by creating spaces where diverse voices can be expressed. Inclusive engagement allows adaptation strategies to respond more effectively to the needs of groups facing higher risks, including small-scale fishers, informal settlers, elderly residents, and women whose livelihoods and care responsibilities are often most affected during climate shocks. In this way, community-based adaptation contributes to resilience while also promoting social equity (Monckeberg & Gómez, 2025; Shammin et al., 2022).

Engaged communities are better positioned to act as active contributors to resilience rather than passive recipients of external assistance. Local institutions, community groups, and informal networks often mobilize resources, organize collective action, and share information more quickly than outside agencies. During climate-related disasters, these networks frequently provide early response through evacuation support, shelter management, and initial recovery efforts. Strengthening community capacity through participatory planning enhances adaptive capacity and enables coastal societies to anticipate, absorb, and recover from climate stresses more effectively. Community engagement also supports the long-term sustainability of climate resilience initiatives. Projects developed through participatory processes are more likely to be maintained, adjusted, and continued because they reflect local priorities and institutional conditions. As climate risks intensify and uncertainty increases, the ability of communities to learn, adapt, and refine their strategies becomes increasingly important. Within the focus of this chapter on climate disasters and coastal resilience, community engagement remains a foundational element that connects policy and technical interventions with the everyday experiences of those most affected by climate change (Islam & Islam, 2025; Almutairi et al., 2020).

6.2 Participatory Approaches in Climate Disaster Preparedness

Participatory approaches to climate disaster preparedness place coastal communities at the center of planning, decision-making, and action. In regions frequently exposed to cyclones, storm surges, sea-level rise, and flooding, preparedness efforts remain limited when they are shaped only by external technical agencies. Coastal residents live with risk on a daily basis and possess practical knowledge formed through repeated experiences of disasters. This includes familiarity with safe locations, evacuation routes, seasonal weather behavior, and informal coping practices that support survival during crises. When such experience is incorporated into formal preparedness systems, disaster planning reflects local realities rather than distant assessments of risk. A key contribution of participatory preparedness lies in identifying and mapping local risks and resources. Methods such as participatory hazard mapping, seasonal calendars, and vulnerability assessments enable community members to document high-risk zones, shelters, infrastructure, water sources, and livelihood assets. These exercises often reveal small-scale vulnerabilities, including informal housing, isolated households, or poorly drained areas, that are frequently overlooked in centralized planning. Preparedness plans developed through community-led mapping tend to be more accurate, socially informed, and easier to put into practice during emergencies.

Participatory approaches also strengthen early warning and communication systems. Warnings are only useful when they are trusted, clearly understood, and acted upon. In many coastal settings, official alerts may not reach all households or may be ignored due to unclear messages or earlier experiences with false alarms. Community-based communication networks involving local leaders, volunteers, women's groups, and youth organizations help translate technical information into familiar language and ensure that warnings reach marginalized groups. These locally rooted channels improve response time and encourage timely action when hazards are approaching. Capacity building at the community level is another important element of participatory disaster preparedness. Training local volunteers in first aid, search and rescue, shelter management, and damage assessment creates a group of responders who can act immediately after disasters occur. This is especially important in coastal areas where damaged infrastructure or severe weather can delay outside assistance. Such training builds practical skills alongside confidence, leadership, and shared responsibility, enabling communities to take an active role in disaster response (Ravazzoli et al., 2025; Malakar & Roy, 2024).

Participatory preparedness also supports social cohesion and collective action. Inclusive planning processes strengthen trust and cooperation among residents, which become vital during emergencies. Strong social ties help communities coordinate evacuations, share resources, and support vulnerable members. In coastal areas where livelihoods are closely connected, collective action plays an important role in recovery and continuity. Equity remains a central concern within participatory approaches. Climate hazards affect people differently, and preparedness strategies that overlook social differences can deepen existing inequalities. Participatory forums create opportunities for women, small-scale fishers, informal workers, and other vulnerable groups to express their specific needs related to shelter access, livelihood protection, and care responsibilities. This leads to preparedness plans that are more inclusive and socially responsive. Key participatory strategies for climate disaster preparedness include community-based risk assessment, participatory early warning systems, local disaster management committees, regular training and drills, and inclusive planning processes that involve marginalized groups. Within the wider context of coastal resilience, participatory preparedness links policy objectives with everyday experience. It allows scientific knowledge, institutional support, and community insight to work together in developing preparedness systems that can adapt over time

(Fig. 6.1). As climate-related disasters become more frequent and intense, preparedness will increasingly depend on how well communities are supported to anticipate risks, organize collectively, and respond with confidence (Rajeswari et al., 2025; Ross et al., 2015).

Fig. 6.1 Participatory strategies in climate disaster preparedness (Prepared by the authors)

6.3 Community-Based Adaptation and Climate Disaster Response

Community-based adaptation (CBA) provides a people-centered way of responding to climate risks, especially in coastal regions where environmental exposure is closely connected to livelihoods, settlement patterns, and social vulnerability (Fig. 6.2). In the context of climate disasters, CBA treats affected communities as active participants rather than passive recipients of external assistance. Coastal populations possess practical knowledge shaped by daily interaction with their environment, social relationships, and cultural practices. When this knowledge is used in planning and response, disaster management becomes more grounded in local realities and better suited to the needs of everyday life in coastal areas. A major contribution of community-based adaptation lies in linking immediate disaster response with longer-term resilience building. Coastal communities often face repeated hazards such as cyclones, storm surges, flooding, and erosion. These conditions require quick emergency action alongside gradual adjustments in housing, livelihoods, and resource use. CBA allows communities to reflect on disaster experiences and use them as opportunities for learning and change. Rebuilding houses with locally appropriate designs, restoring mangroves after storm damage, or diversifying income sources following disruptions to fishing are examples of how recovery activities can support future adaptation. This continuity between response and adaptation helps reduce recurring losses and strengthens resilience over time (Baten et al., 2025; Niu et al., 2023).

Community-based adaptation also improves the speed and effectiveness of disaster response. In many coastal areas, formal emergency services may be limited or delayed, particularly during extreme weather events. Local institutions such as village committees, self-help groups, fisher associations, and youth groups often provide the first response. These groups are able to organize evacuations, manage shelters, distribute relief, and identify households in urgent need more quickly than external agencies. Their close ties within the community allow them to draw on detailed knowledge of household locations, vulnerable individuals, and safe routes, which is especially important during fast-moving disaster

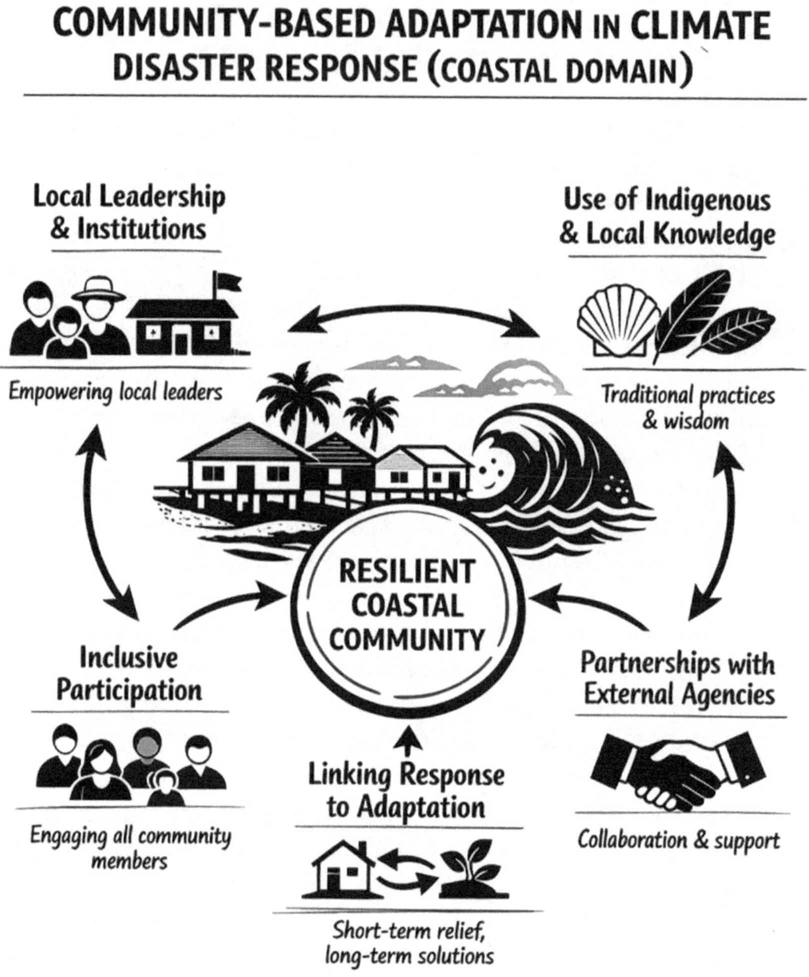

Fig. 6.2 Community-based adaptation in climate disaster response (Prepared by the authors)

situations. Equity and inclusion are strengthened through community-based adaptation during disaster response. Climate disasters tend to deepen existing social inequalities and often have greater impacts on women, older people, persons with disabilities, and low-income groups. Participatory CBA processes help make these differences visible and avoid uniform responses that overlook specific needs. When marginalized groups are involved in planning and decision-making, relief and recovery measures become more responsive. Attention to safety and privacy in shelters, accessibility for persons with disabilities, and livelihood recovery for small-scale fishers and informal workers contribute to fairer outcomes. Community-based adaptation also supports social and psychological recovery after disasters. Climate-related shocks often bring loss, stress, and uncertainty, affecting both individuals and community relationships. Community-led response and recovery activities encourage mutual support and shared problem-solving. Rebuilding together, sharing resources, and taking part in local recovery initiatives help restore confidence and strengthen social bonds. These social connections play an important role in helping communities cope with future climate stresses. Key features of community-based adaptation in climate disaster response include strong local leadership and institutions, the use of indigenous and local knowledge, links between recovery efforts and future risk reduction, inclusive participation of women and marginalized groups, and partnerships with external agencies for technical and financial support. Within broader efforts to enhance coastal resilience, community-based adaptation and disaster response remain closely connected. Disaster response shapes future vulnerability as much as it protects lives and assets in the short term. Placing communities at the center of both response and adaptation supports learning, innovation, and empowerment that extend beyond individual disaster events. As climate change intensifies coastal hazards, community-based adaptation continues to hold a central place in building resilient and sustainable coastal development (Babapoorkamani & Ricci, 2025; Busayo & Kalumba, 2020).

6.4 Indigenous and Traditional Knowledge Systems

Indigenous and traditional knowledge systems form a strong foundation for building coastal resilience in the face of climate disasters. Over long periods, coastal and island communities have developed detailed understandings of ocean currents, wind behavior, tidal movements, soil conditions, and ecosystem responses through close observation and everyday experience. This knowledge is closely linked to cultural practices, livelihood activities, oral traditions, and local institutions (Fig. 6.3). Long before the use of formal scientific tools, such knowledge helped communities anticipate environmental change and cope with uncertainty. Within community-based adaptation and participatory approaches, indigenous and traditional knowledge provides locally grounded insights that are essential for shaping climate resilience strategies suited to specific coastal settings. A key strength of indigenous and traditional knowledge lies in its strong connection to place and its capacity to evolve over time. Rather than remaining fixed, this knowledge develops through repeated interaction with the local environment. Coastal fishers often observe changes in waves, winds, cloud formations, and marine life to anticipate storms and seasonal shifts. Farming and mangrove-dependent communities pay close attention to soil salinity, plant health, and tidal rhythms to guide cropping, harvesting, and resource use. This detailed environmental awareness allows communities to notice early signs of climate stress and respond in ways that are timely and sensitive to local ecological conditions (Hanazaki, 2024; Alexander et al., 2019a).

Traditional knowledge systems also promote ways of living that maintain balance with coastal and marine ecosystems, which is essential for long-term resilience. Practices such as conserving mangroves, managing fishing seasons, protecting sand dunes, and maintaining freshwater ponds are not only cultural traditions but also practical measures that reduce disaster risk. These practices help limit erosion, reduce the impact of storm surges, and support livelihoods during climate shocks. When incorporated into contemporary adaptation planning, they offer nature-based and low-cost solutions that complement engineering and technological measures. Giving space to indigenous and traditional knowledge within participatory processes strengthens climate governance. In many contexts, such knowledge has been ignored or undervalued, leading to adaptation strategies that fail to connect with local realities. Valuing traditional knowledge through participatory engagement helps build trust between communities, planners, and institutions.

Fig. 6.3 Indigenous and traditional knowledge systems in coastal climate resilience (Prepared by the authors)

This trust supports stronger involvement in disaster preparedness, early warning, and recovery activities, which improves the adaptive capacity of coastal regions. Indigenous and traditional knowledge also contributes to social resilience. These knowledge systems are passed on through shared practices, rituals, and learning across generations, reinforcing social bonds and collective responsibility. During climate disasters, strong social networks support cooperation, mutual assistance, and coordinated action. Elders often play an important role by drawing on past experiences of extreme events to guide community responses and ensure that valuable lessons are retained. Key contributions of indigenous and traditional knowledge to coastal climate resilience include early warning based on natural indicators, ecosystem-based practices that protect coastal environments, livelihood strategies that reduce climate risk, cultural continuity that strengthens social cohesion, and locally appropriate solutions that are affordable and environmentally sustainable. Within efforts to address climate disasters and strengthen coastal resilience, indigenous and traditional knowledge should be treated as complementary to scientific approaches. When combined with climate data, modeling, and policy frameworks, local knowledge improves the relevance and acceptance of adaptation strategies. Participatory spaces that bring together scientists, policymakers, and indigenous communities support shared learning and the development of solutions that are technically sound and socially meaningful. As climate change continues to reshape coastal environments, the well-being of vulnerable communities will increasingly depend on how effectively these diverse knowledge systems are respected, integrated, and applied in building resilient and sustainable coastal futures (Dorji et al., 2024; Malakar et al., 2023; Alexander et al., 2019b).

6.5 Community Resilience and Social Capital

Community resilience in coastal regions exposed to climate disasters is closely linked to the strength of social capital, understood as the relationships of trust, cooperation, shared norms, and mutual support that connect people within a community. Physical infrastructure and technical systems are important for reducing disaster risk, but social relationships often shape how well communities prepare for, respond to, and recover from climate shocks. In coastal settings where livelihoods are closely connected and exposure to hazards such as cyclones, storm surges, and flooding is frequent, social capital functions as a vital resource that supports collective action and adaptive capacity. Strong social networks support disaster preparedness by enabling the exchange of information and resources before hazards occur. Communities with high levels of trust and regular interaction are more likely to share warnings, coordinate evacuation plans, and assist vulnerable households, including older people, persons with disabilities, and residents of informal settlements.

Informal connections among neighbors, kin groups, and livelihood associations often operate more quickly and reliably than formal systems during emergencies. Through these networks, people know who needs help, where shelters are located, and how labor and supplies can be mobilized in time. During climate disasters, social capital becomes even more important. When infrastructure is damaged and outside support is delayed, communities rely on internal cooperation to manage immediate risks. Collective efforts help organize rescue activities, manage shelters, and distribute food and water. Communities with strong social ties are also better able to manage tension and uncertainty during crises, reducing the risk of conflict and social breakdown. In this way, social capital helps convert exposure to risk into shared capacity for action and response (Hall et al., 2023; Partelow, 2021).

Social capital also plays an important role in recovery and longer-term adaptation after disasters. Rebuilding homes, restoring livelihoods, and repairing damaged ecosystems require sustained cooperation. Communities with strong social bonds are better able to share labor, pool resources, and exchange knowledge, which can speed up recovery and reduce dependence on external assistance. Social networks also provide emotional and psychological support, helping individuals and families cope with loss, stress, and uncertainty after climate-related events. Within community-based adaptation and participatory approaches, social capital functions both as a foundation and as a result of resilience-building efforts. Participation in planning and decision-making strengthens trust among community members and between communities and institutions. This supports shared responsibility and continued engagement in adaptation activities such as ecosystem restoration, livelihood diversification, and disaster preparedness. Over time, these interactions deepen social ties, creating a cycle in which cooperation strengthens resilience and resilience further reinforces cooperation. In efforts to strengthen coastal resilience, attention to social capital is as important as investment in physical infrastructure. Policies and initiatives that support local organizations, encourage inclusive participation, and promote community leadership contribute to stronger disaster outcomes and more resilient coastal societies. As climate risks continue to intensify, the ability of communities to trust one another, act collectively, and mobilize shared resources will remain central to adapting to an uncertain environmental future (Abunyewah et al., 2023; Carmen et al., 2022).

6.6 Participatory Monitoring and Evaluation

Participatory monitoring and evaluation (PM&E) plays an important role in climate resilience initiatives in coastal regions by helping adaptation and disaster risk reduction efforts remain accountable, responsive, and closely connected to local realities. Within community-based adaptation and participatory approaches, PM&E moves evaluation away from externally driven assessments toward a shared process of learning that involves community members, local institutions, and implementing agencies. People who live with climate risks on a daily basis are well placed to judge whether interventions respond to real needs, reflect local priorities, and treat different groups fairly. In coastal settings where environmental conditions and climate hazards change quickly, PM&E supports continuous feedback and adjustment. Communities can observe and track changes such as shoreline erosion, mangrove condition, flooding patterns, livelihood security, and access to safe shelters. This ongoing observation helps identify early signs of progress as well as emerging problems. When local residents take part in data collection and discussion, monitoring becomes an active and meaningful process rather than a one-time exercise. This flexibility is especially important in disaster-prone areas, where plans may need frequent revision as risks evolve. Participatory monitoring also strengthens transparency and local ownership of climate resilience initiatives. Involvement in setting goals, choosing indicators, and reviewing progress allows communities to better understand how decisions are made and how resources are used. This reduces the risk of exclusion, misuse of funds, or interventions that fail to address the needs of vulnerable groups. PM&E also provides communities with a space to engage with external agencies, government bodies, and non-governmental organizations, supporting relationships built on trust and shared responsibility (Malakar et al., 2025; Nurzaman et al., 2020).

PM&E contributes to learning and capacity building at the community level. Regular reflection meetings, participatory reviews, and simple evaluation tools help residents develop skills in observation, analysis, and collective decision-making. These skills remain valuable beyond individual projects and support longer-term adaptive capacity. Communities learn not only which actions reduce disaster risk but also why certain approaches perform better than others, allowing them to improve future responses to climate threats. PM&E also supports inclusion and social resilience by making diverse experiences visible. Climate disasters affect people differently depending on gender, age, livelihood, and social position. Participatory evaluation methods create space for these differences to be expressed and considered. Including local perspectives alongside numerical information provides a fuller understanding of resilience that reflects wellbeing, safety, and dignity in addition to physical protection (House et al., 2023; Martin et al., 2022).

In efforts to strengthen coastal resilience, participatory monitoring and evaluation connects action with learning. It encourages movement away from fixed project cycles toward approaches that value reflection, adaptation, and accountability. As climate change brings greater uncertainty to coastal systems, the ability to learn collectively and adjust strategies becomes increasingly important. PM&E therefore serves not only as a practical method but also as a participatory practice that strengthens the effectiveness and credibility of climate resilience efforts.

6.7 Collaborative Decision-Making and Governance

Collaborative decision-making and governance play a central role in building climate resilience in disaster-prone coastal regions, where environmental risks, livelihood systems, and institutional responsibilities are closely connected. Climate hazards such as cyclones, coastal flooding, and erosion extend beyond administrative boundaries and sectoral divisions, making isolated and top-down responses inadequate. Within community-based adaptation and participatory approaches, collaborative governance brings together communities, local governments, civil society organizations, and scientific institutions to jointly identify problems, set priorities, and design responses that reflect both technical understanding and local experience. An important feature of collaborative decision-making is the recognition of coastal communities as legitimate participants in climate governance. People living along the coast have direct experience of shoreline change, shifts in fish stocks, salinity intrusion, and storm patterns. This experiential knowledge adds depth to scientific data and strengthens planning processes. When local perspectives are included in decision-making, policies and projects are more closely aligned with local conditions and more likely to gain community acceptance. Participatory platforms such as village councils, local committees, and multi-stakeholder forums allow community members to influence decisions related to resource allocation, infrastructure development, and adaptation priorities. Shared decision-making reduces the risk of poorly suited interventions and supports the long-term sustainability of resilience efforts. Collaborative governance also improves accountability and transparency in climate-related actions. Inclusive decision-making processes make it easier for stakeholders to understand how priorities are determined and how resources are distributed. This openness helps build trust between communities and institutions, which is especially important in coastal areas where past development or disaster responses may have created tension. Trust supports cooperation during emergencies and encourages community involvement in preparedness, evacuation, and recovery activities (Rölfer et al., 2024; Woldesenbet, 2018).

Managing uncertainty is another important role of collaborative governance. Climate change introduces risks that are dynamic and unpredictable, requiring governance systems that can adjust over time. Bringing together actors with different types of knowledge and responsibility allows decision-making to remain flexible. Communities can share timely observations of environmental change and the effects of interventions, while government agencies and technical experts can provide policy direction, funding, and regulatory support. Ongoing interaction among these actors allows adaptation strategies to be revised as conditions evolve. Collaborative governance also contributes to equity in climate adaptation. Climate disasters often deepen existing social inequalities, leaving marginalized groups more exposed and less able to recover. Participatory decision-making creates opportunities for these groups to express their needs and influence outcomes. This may include decisions about the location of shelters, the design of livelihood support programs, or the restoration of coastal ecosystems. Inclusive governance helps ensure that resilience-building efforts respond to those facing the greatest risks. Within the wider discussion of climate disasters and coastal resilience, collaborative decision-making and governance provide an institutional foundation for community-based adaptation. They support adaptation processes shaped through dialogue, negotiation, and shared responsibility rather than external direction alone. As climate risks continue to increase in scale and complexity, the capacity of coastal societies to govern collectively, listen to diverse perspectives, and act together will play a key role in shaping resilient and sustainable futures (Asio et al., 2026; Basco-Carrera et al., 2017).

6.8 Communication and Awareness

Communication and awareness-raising play an important role in strengthening coastal resilience to climate disasters because they shape how risks are understood and how people respond to them. In coastal regions exposed to cyclones, storm surges, flooding, and sea-level rise, timely and reliable information can directly affect safety and survival. Within community-based adaptation and participatory approaches, communication extends beyond one-way delivery of technical messages. It operates as a shared process through which communities, institutions, and experts exchange knowledge, experiences, and concerns related to climate risk. Communication supports disaster preparedness by helping people understand the nature of climate threats and the actions needed to reduce harm. Scientific information related to weather forecasts, early warnings, and evacuation procedures becomes more useful when it is explained in familiar language and shared through trusted local channels. Local leaders, teachers, women's groups, fishers' associations, and youth networks play an important role in spreading information across communities. These actors help reach households that may not depend on formal media or digital platforms. Communication grounded in local relationships increases trust and improves the likelihood that warnings are taken seriously and acted upon in time (Höllermann & Heidenreich, 2025; Bracken et al., 2016).

Awareness-raising also contributes to longer-term climate adaptation by shaping everyday attitudes and behavior. Many climate risks develop gradually, such as coastal erosion, salinity intrusion, and ecosystem decline, and may not always be seen as immediate dangers. Participatory awareness activities, including community discussions, school programs, storytelling, and visual tools, help connect everyday environmental changes with broader climate processes. This shared understanding encourages preventive actions such as mangrove protection, changes in housing design, and livelihood diversification, rather than reliance on emergency response alone. Communication processes also strengthen social cohesion and shared responsibility. Open discussion of climate risks and adaptation options helps communities develop a common understanding of vulnerability and resilience. Shared dialogue reduces misinformation, fear, and feelings of helplessness, while supporting a sense of collective agency. In coastal areas where safety and livelihoods are closely connected, shared awareness promotes cooperation in maintaining evacuation routes, managing shelters, and restoring ecosystems that offer natural protection. Inclusive communication is essential for equitable resilience building. Climate information is often unevenly distributed, with women, older people, and marginalized groups having limited access to formal communication systems. Participatory approaches intentionally engage these groups using culturally appropriate methods and local languages. Inclusive communication reduces the risk of exclusion during emergencies and ensures that adaptation strategies reflect diverse needs and perspectives. Within efforts to strengthen coastal resilience, communication serves both practical and social purposes. It links scientific knowledge with lived experience, connects institutions with communities, and turns climate risks into shared concerns that guide collective action. As climate change continues to intensify coastal hazards, the ability to communicate clearly, listen carefully, and build awareness together remains central to people-centered and sustainable climate resilience (Areia et al., 2023; Bruno et al., 2021).

6.9 Capacity Building and Skills Development

Capacity building and skills development form an important part of climate resilience in disaster-prone coastal regions, where communities face growing uncertainty and repeated exposure to climate risks. Within community-based adaptation and participatory approaches, capacity building involves more than sharing technical information. It includes strengthening local institutions, leadership, and the ability of communities to identify problems and act collectively. Coastal resilience depends not only on physical infrastructure or external assistance but also on the skills and confidence of local people to anticipate hazards, respond effectively, and recover in ways that reduce future risk. At the community level, skills development improves preparedness and emergency response. Training related to early warning interpretation, evacuation procedures, first aid, search and rescue, and shelter management allows local volunteers and households to respond quickly during cyclones, floods, and storm surges. These skills are particularly important in remote or densely populated coastal areas where formal emergency services may be delayed or stretched. When communities develop practical response abilities, they are better able to protect lives and property and become less dependent on outside support during emergencies. Capacity building also supports longer-term adaptation by strengthening livelihoods and environmental management. Many coastal communities depend on climate-sensitive activities such as fishing, aquaculture, agriculture, and tourism. Skills related to sustainable resource use, climate-resilient farming practices, alternative livelihood options, and small-scale enterprise development help households reduce exposure to climate shocks. Training in ecosystem-based practices, including mangrove restoration and dune stabilization, enables communities to protect shorelines while sustaining local livelihoods (Temali, 2026; Khan et al., 2022).

Participatory approaches to capacity building encourage shared learning and collective action. When training programs are planned and implemented with community involvement, they reflect local priorities and build on existing knowledge. Peer learning, community workshops, and practical demonstrations create opportunities for people to exchange experiences, test new practices, and adapt ideas to local conditions. This shared learning process strengthens trust and cooperation, which are essential during disaster response and recovery. Capacity building also contributes to more inclusive and equitable resilience. Women, youth, and marginalized groups often experience greater impacts from climate hazards but have fewer opportunities to gain skills or participate in decision-making. Targeted training that supports leadership, technical skills, and organizational capacity among these groups helps reduce inequality and broaden participation in resilience efforts. When a wider range of community members gain skills and confidence, the adaptive capacity of coastal communities increases. Within the wider discussion of climate disasters and coastal resilience, capacity building and skills development provide a lasting human foundation for community-based adaptation. Infrastructure may be damaged and external support may vary, but skills, knowledge, and strong local institutions remain within communities. Investing in people's ability to understand risks, innovate, and work together supports coastal societies in facing climate challenges with greater confidence and collective strength (McEwen et al., 2022; Brown et al., 2022).

6.10 Challenges and Opportunities in Community Engagement

Community engagement is widely viewed as a central element of climate disaster management and coastal resilience, but its practice is shaped by a range of challenges and opportunities. Coastal communities face high exposure to hazards such as cyclones, flooding, erosion, and salinity intrusion, which makes their involvement in adaptation and disaster response both necessary and urgent. Turning the idea of participation into meaningful and sustained engagement requires careful attention to social conditions, institutional structures, and local contexts. A major challenge in community engagement lies in unequal access to power and resources within and between communities. Social differences related to income, gender, caste, age, or occupation often shape who is able to participate and whose views are considered. Local elites may dominate decision-making, while poorer households, women, informal workers, and marginalized groups remain underrepresented. Limited literacy, restricted access to information, and time constraints further reduce participation, particularly for those balancing livelihood activities and household responsibilities. Under such conditions, participation risks becoming symbolic rather than genuinely inclusive (Parvin et al., 2025; Park et al., 2024).

Institutional and governance-related constraints also affect the quality of community engagement. In many coastal regions, climate adaptation and disaster management are guided by centralized agencies with limited mechanisms for local input. Short project timelines, rigid procedures, and fragmented responsibilities across departments can weaken long-term collaboration with communities. Past experiences of poorly designed or top-down interventions may also create mistrust, making communities hesitant to engage fully or invest effort in participatory processes. Despite these challenges, community engagement offers important opportunities to strengthen coastal resilience. Local knowledge and lived experience provide insights into environmental change, hazard patterns, and coping practices that are difficult to capture through technical assessments alone. Involving communities in risk mapping, preparedness planning, and recovery activities helps ensure that adaptation strategies reflect local conditions and social needs. This strengthens the relevance, acceptance, and durability of resilience efforts. Community engagement also supports the growth of social capital and local leadership. Participatory processes encourage dialogue, cooperation, and shared problem-solving, which strengthen relationships within communities and between communities and institutions. These connections are especially valuable during climate disasters, when coordinated action and mutual support can reduce losses and protect lives. Over time, participation helps build organizational capacity, allowing communities to express their needs more clearly and engage more effectively with external agencies (Botero et al., 2025; O'Riordan, 2005).

Therefore, the context of climate disasters and coastal resilience, community engagement creates the possibility of moving communities from passive recipients of assistance to active partners in governance. Investment in inclusive participation, transparent decision-making, and long-term capacity building can gradually address existing barriers. Although challenges related to inequality, institutional rigidity, and limited resources remain, the potential outcomes include more responsive policies, stronger social cohesion, and coastal societies better equipped to adapt to increasing climate risks.

6.11 Conclusion and Arguments

The discussion on community engagement and the challenges and opportunities of participation shows that coastal climate resilience is shaped by social processes as much as environmental conditions. Climate disasters do not only damage physical infrastructure; they disrupt livelihoods, social relationships, and local governance systems that support everyday life in coastal regions. Community engagement helps ensure that adaptation and disaster risk reduction grow from within communities rather than being imposed from outside. When local people take part in planning and action, responses reflect lived experience, local priorities, and existing capacities. Through participatory planning, the use of indigenous and traditional knowledge, collaborative governance, and community-based adaptation, resilience becomes closely tied to local realities rather than remaining limited to formal policy frameworks.

These approaches strengthen climate action by recognizing communities as active contributors to knowledge and solutions. Participatory preparedness, monitoring, and decision-making allow coastal societies to learn from climate events and adjust strategies as risks change over time. Such processes support ongoing learning instead of fixed responses. Social capital, awareness, and capacity building further strengthen the human and institutional foundations that enable communities to act together during emergencies and recovery. These elements support cooperation, trust, and shared responsibility, which are essential during periods of crisis. Challenges linked to inequality, weak institutions, and limited resources continue to shape participation. Social divisions can restrict whose voices are heard, and institutional limitations may reduce opportunities for long-term engagement. Even so, meaningful community involvement creates strong possibilities for positive change.

Greater local ownership, attention to equity, space for innovation, and improved trust between communities and institutions contribute to more durable outcomes. These gains reduce reliance on external interventions and support locally driven adaptation efforts. As climate uncertainty increases, coastal resilience will depend less on the amount of infrastructure built and more on the ability of people to work together, draw on their knowledge, and guide their own responses to risk. Communities that are empowered to act collectively are better positioned to anticipate change, respond to challenges, and shape adaptive pathways that address both present needs and future pressures. The chapter discussions lead to the following arguments:

(a) Community engagement grounds climate resilience in local knowledge, experience, and cultural practices.
(b) Participatory approaches build trust and encourage long-term acceptance of resilience measures.
(c) Indigenous and traditional knowledge offers low-cost and locally suitable adaptation solutions.
(d) Social capital supports rapid response, mutual support, and sustained recovery after disasters.
(e) Participatory monitoring improves accountability, learning, and adaptive decision-making.
(f) Collaborative governance enhances coordination, transparency, and equity across institutions.
(g) Communication and awareness reduce misinformation and strengthen preparedness.
(h) Capacity building enables communities to manage climate risks with greater independence.
(i) Community engagement, despite challenges, supports inclusive and sustainable coastal resilience.

References

Abunyewah, M., Erdiaw-Kwasie, M. O., Okyere, S. A., Thayaparan, G., Byrne, M., Lassa, J., Zander, K. K., Fatemi, M. N., & Maund, K. (2023). Influence of personal and collective social capital on flood preparedness and community resilience: Evidence from Old Fadama, Ghana. *International Journal of Disaster Risk Reduction, 94*, 103790. https://doi.org/10.1016/j.ijdrr.2023.103790

Alexander, S. M., Provencher, J. F., Henri, D. A., et al. (2019a). Bridging Indigenous and science-based knowledge in coastal-marine research, monitoring, and management in Canada: A systematic map protocol. *Environmental Evidence, 8*, 15. https://doi.org/10.1186/s13750-019-0159-1

Alexander, S. M., Provencher, J. F., Henri, D. A., et al. (2019b). Bridging Indigenous and science-based knowledge in coastal and marine research, monitoring, and management in Canada. *Environmental Evidence, 8*, 36. https://doi.org/10.1186/s13750-019-0181-3

Almutairi, A., Mourshed, M., & Ameen, R. F. M. (2020). Coastal community resilience frameworks for disaster risk management. *Natural Hazards, 101*, 595–630. https://doi.org/10.1007/s11069-020-03875-3

Areia, N. P., Tavares, A. O., & Costa, P. J. (2023). Public perception and preferences for coastal risk management: Evidence from a convergent parallel mixed-methods study. *Science of the Total Environment, 882*, Article 163440. https://doi.org/10.1016/j.scitotenv.2023.163440

Asio, M. V. S., Cvitanovic, C., Lau, J., & Diedrich, A. (2026). Collaborative governance in marine and coastal management: A systematic review. *Ocean & Coastal Management, 272*, Article 108020. https://doi.org/10.1016/j.ocecoaman.2025.108020

Babapoorkamani, A., & Ricci, L. (2025). Decision-making strategies for climate change adaptation in coastal regions of Africa. *Environmental Development, 101196*. https://doi.org/10.1016/j.envdev.2025.101196

Basco-Carrera, L., van Beek, E., Jonoski, A., et al. (2017). Collaborative modelling for informed decision making and inclusive water development. *Water Resources Management, 31*, 2611–2625. https://doi.org/10.1007/s11269-017-1647-0

Baten, M. A., Al Zabir, A., Hossain, M. M., et al. (2025). Community driven climate change adaptation strategies in Ephemeral wetlands of northeastern Bangladesh. *Discover Environment, 3*, 176. https://doi.org/10.1007/s44274-025-00357-w

Botero, C. M., Suman, D. O., & Milanes, C. B. (2025). The multiple challenges faced by coastal and marine governance. *Water, 17*(15), 2322. https://doi.org/10.3390/w17152322

Bracken, L. J., Oughton, E. A., Donaldson, A., et al. (2016). Flood risk management, an approach to managing cross-border hazards. *Natural Hazards, 82*(Suppl 2), 217–240. https://doi.org/10.1007/s11069-016-2284-2

Brown, S., Hanson, S. E., Sear, D., et al. (2022). Assessing hazards and disaster risk on the coast for Pacific small island developing States: The need for a data-driven approach. *Anthropocene Coasts, 5*, 5. https://doi.org/10.1007/s44218-022-00005-3

Bruno, M. F., Motta Zanin, G., Barbanente, A., & Damiani, L. (2021). Understanding the cognitive components of coastal risk assessment. *Journal of Marine Science and Engineering, 9*(7), 780. https://doi.org/10.3390/jmse9070780

Busayo, E. T., & Kalumba, A. M. (2020). Coastal climate change adaptation and disaster risk reduction: A review of policy, programme and practice for sustainable planning outcomes. *Sustainability, 12*(16), 6450. https://doi.org/10.3390/su12166450

Carmen, E., Fazey, I., Ross, H., et al. (2022). Building community resilience in a context of climate change: The role of social capital. *Ambio, 51*, 1371–1387. https://doi.org/10.1007/s13280-021-01678-9

Dorji, T., Rinchen, K., Morrison-Saunders, A., et al. (2024). Understanding how indigenous knowledge contributes to climate change adaptation and resilience: A systematic literature review. *Environmental Management, 74*, 1101–1123. https://doi.org/10.1007/s00267-024-02032-x

Hall, C. E., Wehling, H., Stansfield, J., et al. (2023). Examining the role of community resilience and social capital on mental health in public health emergency and disaster response: A scoping review. *BMC Public Health, 23*, 2482. https://doi.org/10.1186/s12889-023-17242-x

Hanazaki, N. (2024). Local and traditional knowledge systems, resistance, and socioenvironmental justice. *Journal of Ethnobiology and Ethnomedicine, 20*, 5. https://doi.org/10.1186/s13002-023-00641-0

Höllermann, B., & Heidenreich, A. (2025). Certainly uncertain: The role of uncertainty perception for flood risk preparedness and response. *Natural Hazards, 121*, 22027–22048. https://doi.org/10.1007/s11069-025-07675-5

House, J., Kleiber, D., Steenbergen, D. J., et al. (2023). Participatory monitoring in community-based fisheries management through a gender lens. *Ambio, 52*, 300–318. https://doi.org/10.1007/s13280-022-01783-3

Islam, M. M., & Islam, A. R. M. T. (2025). Strengthening climate resilience in coastal Bangladesh: Analyzing the role of NGOs in adaptation governance from an intersectionality lens. *Progress in Disaster Science, 100498*. https://doi.org/10.1016/j.pdisas.2025.100498

Khan, A. S., Kumar, M. S., & Chella, R. S. (2022). Risk communication and capacity-building: A case study on framing CBA strategies of artisanal fishing communities to sea-level rise using BASIEC. *Climate Services, 26*, Article 100299. https://doi.org/10.1016/j.cliser.2022.100299

Malakar, K. D., Kumar, M., Anand, S., & Kuzur, G. (2023). Climate Change and Socio-Ecological Transformation: Applications and Field Narratives. In: *Climate Change and Socio-Ecological Transformation. Advances in Geographical and Environmental Sciences*. Springer, Singapore. https://doi.org/10.1007/978-981-99-4390-6_7

Malakar, K. D., & Roy, S. (2024). Applications of participatory GIS: A socio-ecological approaches and mapping. In *Mapping geospatial citizenship*. SpringerBriefs in GIS. Springer. https://doi.org/10.1007/978-3-031-63107-8_3

Malakar, K. D., Roy, S., & Kumar, M. (2025). Participatory GIS: A solution to traditional GIS challenges. In *Geospatial technologies in coastal ecologies monitoring and management*. Advances in Geographic Information Science. Springer. https://doi.org/10.1007/978-3-031-92017-2_9

Martin, D. M., Specht, J. A., Canick, M. R., et al. (2022). Using decision analysis to integrate habitat and community values for coastal resilience planning. *Estuaries and Coasts, 45*, 331–344. https://doi.org/10.1007/s12237-021-00970-x

McEwen, L., Roberts, L., Holmes, A., et al. (2022). Building local capacity for managing environmental risk: A transferable framework for participatory, place-based, narrative-science knowledge exchange. *Sustainability Science, 17*, 2489–2511. https://doi.org/10.1007/s11625-022-01169-0

Monckeberg, E., & Gómez, S. (2025). Exploring the potential of coastal cities to address climate change towards an inclusive, equitable and politically engaged orientation. *Anthropocene Coasts, 8*, 33. https://doi.org/10.1007/s44218-025-00099-5

Niu, Y., Abdullayev, V., Alyar, A. V., & Kamran, A. T. (2023). Resilience and adaptation to climate change: Community-based strategies in coastal regions. *Estidamaa, 2023*, 36–43. https://doi.org/10.70470/ESTIDAMAA/2023/005

Nurzaman, A., Shaw, R., & Roychansyah, M. S. (2020). Measuring community resilience against coastal hazards: Case study in Baron Beach,

Gunungkidul Regency. *Progress in Disaster Science, 5*, Article 100067. https://doi.org/10.1016/j.pdisas.2020.100067

O'Riordan, T. (2005). Inclusive and community participation in the coastal zone: Opportunities and dangers. In J. Vermaat, W. Salomons, L. Bouwer, & K. Turner (Eds.), *Managing European Coasts*. Environmental Science. Springer. https://doi.org/10.1007/3-540-27150-3_9

Park, H., Nam, K., & Egawa, S. (2024). The gaps between institutional and practical disaster risk management measures on coastal flood risks in South Korea's coastal communities. *International Journal of Disaster Risk Science, 15*, 594–607. https://doi.org/10.1007/s13753-024-00579-1

Partelow, S. (2021). Social capital and community disaster resilience: Post-earthquake tourism recovery on Gili Trawangan, Indonesia. *Sustainability Science, 16*, 203–220. https://doi.org/10.1007/s11625-020-00854-2

Parvin, G. A., Ara, S., & Shumi, K. F. (2025). Importance, vulnerabilities, initiatives, and finance for coastal ecosystem management. In R. Shaw (Eds.), *Encyclopedia of disaster risk reduction*. Springer. https://doi.org/10.1007/978-981-97-4547-0_33-1

Rajeswari, D., Jayaraman, R., Celina, D., et al. (2025). Community change agents and disaster preparedness among women in coastal areas. *BMC Emergency Medicine, 25*, 234. https://doi.org/10.1186/s12873-025-01393-8

Ravazzoli, E., Lavarello-Schettini, R., Oberti, B., et al. (2025). Community-based approaches in disaster risk reduction and climate change adaptation: An analysis of applied participatory processes. *Climatic Change, 178*, 225. https://doi.org/10.1007/s10584-025-04078-0

Rölfer, L., Celliers, L., Fernandes, M., et al. (2024). Assessing collaboration, knowledge exchange, and stakeholder agency in coastal governance to enhance climate resilience. *Regional Environmental Change, 24*, 6. https://doi.org/10.1007/s10113-023-02163-7

Ross, H., Shaw, S., Rissik, D., et al. (2015). A participatory systems approach to understanding climate adaptation needs. *Climatic Change, 129*, 27–42. https://doi.org/10.1007/s10584-014-1318-6

Shammin, M. R., Haque, A. K. E., & Faisal, I. M. (2022). A framework for climate resilient community-based adaptation. In A. K. E. Haque, P. Mukhopadhyay, M. Nepal, & M. R. Shammin (Eds.), *Climate change and community resilience*. Springer. https://doi.org/10.1007/978-981-16-0680-9_2

Temali, I. K. (2026). Climate change adaptation green technologies for coastal risk prevention: Case studies summary from coastal countries. In E. Berisha, J. Moodie, L. Allkja, & M. Jeftić (Eds.), *The Palgrave handbook of just green transitions in the Western Balkans and beyond*. Palgrave Macmillan. https://doi.org/10.1007/978-3-031-95075-9_19

Woldesenbet, W. G. (2018). Collaborative governance: Assessing the problem of weak cross-sectoral collaborations for the governance of Addis Ababa Rivers. *Applied Water Science, 8*, 116. https://doi.org/10.1007/s13201-018-0763-1

7

Climate-Induced Disaster Risk Reduction and Integrated Coastal Zone Management

Coastal areas play a vital role in supporting human life, natural ecosystems, and economic activities. At the same time, they are among the most vulnerable regions to climate-related disasters. Rising sea levels, stronger cyclones, coastal erosion, flooding, and the loss of coastal ecosystems are no longer occasional events. They are becoming frequent and severe, placing growing pressure on communities and environments along the coast. These challenges cannot be managed effectively through traditional disaster response methods alone. This chapter introduces climate disasters and coastal ecologies by placing Disaster Risk Reduction within the wider context of Integrated Coastal Zone Management. It explains why preventive, adaptive, and coordinated approaches are essential for reducing climate-related risks in coastal regions. The chapter explores how climate change alters coastal hazards and increases vulnerability across both social and ecological systems. It highlights the value of understanding local conditions, using scientific tools, and strengthening community participation in risk reduction efforts. The chapter also emphasizes the role of governance, planning, and cooperation at local, national, and international levels. By connecting policy, science, and

K. D. Malakar et al., *Coastal Ecologies at Risk*, Palgrave Studies in Environmental Sustainability, https://doi.org/10.1007/978-3-032-26081-9_7

practical action, it sets the foundation for building safer, more resilient coastal environments in a changing climate.

Key Points of the Chapter

- Describes climate-related Disaster Risk Reduction as an integral part of Integrated Coastal Zone Management for addressing coastal risks.
- Explains the role of hazard, vulnerability, and geospatial assessments in supporting informed coastal planning.
- Highlights the combined role of ecosystems, communities, and incentives in strengthening disaster risk reduction efforts.
- Examines the importance of early warning, preparedness, and resilient recovery in coastal disaster management.

7.1 Climate Disaster Risk Reduction and Coastal Communities

Climate-related Disaster Risk Reduction (DRR) refers to a systematic approach to identifying, assessing, and reducing risks associated with climate-induced hazards such as cyclones, floods, storm surges, sea-level rise, heatwaves, and coastal erosion. Unlike conventional disaster management practices that mainly focus on post-disaster response and relief, climate-related DRR emphasizes anticipatory, preventive, and resilience-building measures. These measures address both present and future risks under changing climatic conditions. Climate-related DRR combines scientific knowledge of climate variability and climate change with social, economic, institutional, and environmental considerations. The aim is to reduce the vulnerability and exposure of human communities, ecosystems, and infrastructure. Within this broader framework, attention extends beyond individual hazards to include multiple and interconnected risks. Climate change can increase the frequency, intensity, and spatial extent of disasters, particularly in fragile coastal areas where human and natural systems are closely linked. As a result, climate-related DRR is closely connected with climate adaptation strategies, sustainable development principles, and ecosystem-based approaches.

Risk reduction is therefore treated as part of long-term planning and governance rather than as isolated or short-term actions (Tasnuva et al., 2024; Busayo & Kalumba, 2020).

The importance of climate-related Disaster Risk Reduction has increased considerably in the context of accelerating climate change and rapid socio-economic change. Climate-induced disasters pose serious threats to lives, livelihoods, infrastructure, and ecosystems, often reversing development gains and increasing poverty and inequality. Coastal regions face particular risks due to high population density, strong dependence on climate-sensitive economic activities, and growing exposure to sea-level rise, extreme weather events, and saline intrusion. In this setting, climate-related DRR plays a vital role in reducing disaster losses by strengthening early warning systems, promoting climate-resilient infrastructure, improving the adaptive capacity of communities, and protecting natural buffers such as mangroves, wetlands, and coastal dunes. A shift from reactive disaster response to proactive risk governance helps reduce long-term social and economic losses linked to disasters. From a policy and planning perspective, climate-related DRR provides a strong basis for integrated and coordinated coastal management. It supports the connection between disaster risk reduction, Integrated Coastal Zone Management (ICZM), land-use planning, climate adaptation, and environmental conservation. The inclusion of risk assessments in development planning helps prevent maladaptation and avoids the creation of new risks. Climate-related DRR also supports inclusive and participatory approaches that value local knowledge, address uneven vulnerabilities, and strengthen social resilience. At a broader level, it contributes to sustainable development by enhancing the ability of societies to manage uncertainty, adapt to climate change, and follow development pathways that are resilient, equitable, and environmentally sound (Kautsary et al., 2025; Mal et al., 2018).

Climate change has altered the nature, scale, and intensity of disasters across the world, with coastal regions becoming some of the most vulnerable areas. Rising global temperatures influence ocean and atmospheric processes, leading to sea-level rise, higher sea surface temperatures, changing rainfall patterns, and more frequent and intense extreme weather events such as cyclones, storm surges, floods, and heatwaves.

These climate-induced hazards have greater impacts on coastal zones because of their physical exposure, ecological sensitivity, and concentration of population and economic activities. Climate change therefore acts as a factor that increases existing risks and creates new forms of disaster risk for coastal communities. Coastal communities are located at the boundary between land and sea and depend heavily on climate-sensitive natural resources and ecosystems. Livelihoods such as fisheries, aquaculture, agriculture, salt production, tourism, and port-related activities are directly influenced by climatic conditions. Sea-level rise and coastal erosion threaten settlements, infrastructure, and freshwater sources, while saline intrusion reduces agricultural productivity and affects drinking water quality. Extreme weather events disrupt local economies, damage housing and public infrastructure, and increase health risks and displacement. In many developing regions, limited access to resources, weak institutional support, and social marginalization further increase the impacts of climate-induced disasters.

The relationship between climate change and disasters in coastal areas is complex and cumulative. Slow processes such as sea-level rise and ocean warming interact with sudden events like cyclones and storm surges, leading to combined and cascading impacts. Higher sea levels increase the destructive potential of storm surges, while the degradation of coastal ecosystems reduces natural protection against flooding and wave action. Changes in hazard patterns over time and space expose new areas to risk and challenge traditional coping practices. These conditions highlight the need to move beyond static hazard management toward adaptive and forward-looking disaster risk reduction approaches. Coastal communities are not passive recipients of disaster impacts. Their vulnerability and resilience are shaped by social, economic, cultural, and institutional conditions. Poverty, insecure land tenure, inadequate housing, limited access to early warning systems, and exclusion from decision-making processes increase disaster risk. At the same time, local knowledge, social networks, and customary practices often play an important role in coping with and recovering from disasters. Climate change can exceed these traditional capacities by increasing the frequency and severity of hazards beyond past experience.

Strengthening community-level resilience therefore requires approaches that combine scientific knowledge with local practices.

The connection between climate change, disasters, and coastal communities forms a core element of climate-induced Disaster Risk Reduction and Integrated Coastal Zone Management (ICZM). Disaster Risk Reduction provides tools to assess and reduce risk, while ICZM offers a coordinated framework for managing coastal resources, land use, and development. Integrating climate change considerations into DRR and ICZM allows planners and policymakers to address both sudden disasters and slow environmental changes, reduce vulnerability and exposure, and improve the adaptive capacity of coastal systems. Such integration helps avoid fragmented interventions that may reduce risk in one sector while increasing it in another (Jiang et al., 2023; de Lima et al., 2022). Following are the key dimensions of this linkage include the following:

(a) Climate change intensifies existing coastal hazards and generates new risks through sea-level rise, extreme events, and ecosystem degradation.
(b) Disasters affect coastal communities through loss of lives, livelihoods, infrastructure, and ecosystems, along with long-term social and economic disruption.
(c) Vulnerability differs across communities depending on socio-economic conditions, gender, occupation, and access to resources and institutions.
(d) Healthy coastal ecosystems such as mangroves, coral reefs, and wetlands reduce disaster risk, while ecosystem degradation increases exposure.
(e) Integrated approaches linking climate adaptation, Disaster Risk Reduction, and ICZM are essential for sustainable and resilient coastal development.

In conclusion, the relationship between climate change, disasters, and coastal communities' points to the need for integrated, multi-level, and participatory approaches to risk reduction. Climate-induced disasters threaten not only physical assets but also the social and economic foundations of coastal societies. Addressing these challenges requires a move away from reactive responses toward proactive planning that incorporates climate science, risk assessment, ecosystem-based solutions, and community participation. Within the broader scope of this chapter, this linkage provides a clear foundation for advancing climate-induced Disaster Risk Reduction and Integrated Coastal Zone Management as complementary approaches for building resilient coastal futures.

7.2 Coastal Hazards and Vulnerability Assessment

Coastal zones are dynamic and complex socio-ecological systems where natural processes, human activities, and climatic forces interact continuously. In the context of climate-induced Disaster Risk Reduction and Integrated Coastal Zone Management, the identification of coastal hazards and the assessment of vulnerability provide a basic foundation for effective planning, policy formulation, and risk reduction. Climate change has intensified both sudden and gradual coastal hazards, increasing the exposure of populations, infrastructure, and ecosystems. For this reason, systematic hazard identification, careful assessment of hazard characteristics, and detailed vulnerability analysis are necessary to reduce disaster risk and strengthen coastal resilience. Coastal hazards are physical events or processes that can cause damage in coastal areas. These hazards may originate in coastal and marine environments or become more severe at the interface between land and sea. They may be climatic, oceanographic, geological, or human-induced in nature, and their impacts are often increased by climate change. The identification of coastal hazards involves recognizing the full range of threats affecting a particular coastal region, considering both past events and emerging risks under future climate conditions. Major climate-induced coastal hazards include tropical cyclones, storm surges, coastal flooding, sea-level rise, coastal erosion, saline intrusion, heat stress, and changes in wave and current patterns. Secondary and cascading impacts such as infrastructure damage, ecosystem degradation, and public health problems may also follow major hazard events (Lopes et al., 2022; Tanim et al., 2022).

Hazard identification goes beyond listing events and requires an understanding of when and where hazards occur. This includes examining their frequency, magnitude, duration, seasonal variation, and spatial extent. Climate change complicates this process by altering baseline conditions and increasing uncertainty. Gradual sea-level rise can turn occasional flooding into repeated or permanent inundation, while increasing sea surface temperatures may intensify cyclonic activity. Hazard identification therefore needs to be continuous and adaptive, combining historical records, climate projections, and local knowledge to reflect changing risk patterns in coastal areas. Once hazards are identified, hazard assessment is required to evaluate their potential impacts and associated risk levels. Coastal hazard assessment focuses on the likelihood of hazard occurrence and the severity of expected impacts on physical structures, ecosystems, and human populations. This process uses both quantitative and qualitative methods, including hazard mapping, modeling, remote sensing, geographic information systems, and scenario-based analysis. Maps showing flood-prone zones, erosion areas, storm surge extents, and locations vulnerable to sea-level rise are important tools for visualizing risk and supporting land-use planning. Climate-induced hazard assessment must account for long-term change and uncertainty. Assessments based only on past conditions are no longer sufficient, as future hazards may exceed historical experience. The use of climate change scenarios allows planners to examine possible future changes in hazard intensity and frequency and supports early risk reduction efforts. In coastal regions, interactions between multiple hazards are particularly important. Sea-level rise can increase storm surge impacts, while coastal erosion can weaken natural and built defenses and raise flood risk. Comprehensive hazard assessment therefore requires an integrated approach that considers multiple hazards together and aligns with the principles of disaster risk reduction and integrated coastal zone management. Vulnerability refers to the likelihood that exposed people, assets, and systems will suffer harm when affected by hazards. In coastal settings, vulnerability is shaped by physical, social, economic, ecological, and institutional factors. Climate change increases vulnerability by expanding exposure to hazards and placing additional pressure on the ability of coastal systems to cope and adapt. Vulnerability changes over time in response to development patterns, governance arrangements, environmental change, and social conditions. A clear understanding of vulnerability therefore requires attention to people, livelihoods, and systems rather than a focus on hazards alone (Panda et al., 2022; Haq & Milliman, 1996).

Coastal vulnerability has several dimensions. Physical vulnerability relates to the sensitivity of housing, infrastructure, and essential facilities to hazard impacts and is influenced by location, construction quality, and maintenance. Social vulnerability reflects characteristics such as income level, age, gender, health, education, and social inclusion, which affect the capacity of individuals and communities to prepare for, respond to, and recover from disasters. Economic vulnerability arises from dependence on climate-sensitive livelihoods, limited income options, and lack of financial resources or insurance. Ecological vulnerability concerns the condition of coastal ecosystems such as mangroves, coral reefs, wetlands, and dunes, which provide natural protection from hazards but are increasingly stressed by climate change and human activity. Vulnerability analysis involves the systematic examination of factors that increase susceptibility and limit adaptive capacity in coastal regions. It seeks to identify who and what is vulnerable, the reasons for this vulnerability, and how vulnerability may change under different climate and development conditions. Exposure, sensitivity, and adaptive capacity are commonly used as key components of vulnerability analysis. Exposure refers to the presence of people, infrastructure, and ecosystems in areas prone to hazards. Sensitivity describes the extent to which these elements are affected by hazards. Adaptive capacity refers to the ability to adjust, cope, and recover from impacts. Coastal vulnerability assessment uses a range of methods, including indicator-based indices, participatory approaches, household surveys, and spatial analysis. Composite indices combine multiple indicators to compare vulnerability across locations or social groups. Participatory methods help document local knowledge, lived experience, and community priorities that may not be captured through technical analysis alone. The combination of qualitative and quantitative approaches improves the usefulness and credibility of vulnerability assessments, particularly in diverse coastal environments (Malakar et al., 2023a; Tanim et al., 2022).

Climate change adds a time dimension to vulnerability analysis by introducing long-term trends and uncertainty. Vulnerability may increase in the future due to sea-level rise, population growth, unplanned urban expansion, and ecosystem degradation, even where present risk levels appear manageable. Vulnerability assessment therefore needs to be forward-looking and scenario-based. This supports the identification of priority areas and population groups and helps design risk reduction strategies that address underlying causes of vulnerability rather than only the immediate impacts of hazards. Coastal hazard and vulnerability assessments are central to climate-induced Disaster Risk Reduction and Integrated Coastal Zone Management. Disaster Risk Reduction frameworks view risk as a combination of hazard, exposure, and vulnerability, while Integrated Coastal Zone Management focuses on balancing development, environmental protection, and risk management. Integrating hazard identification, hazard assessment, and vulnerability analysis within coastal management processes improves decision-making and reduces conflict between development goals and safety concerns. This approach also helps avoid practices such as infrastructure development in high-risk areas that can increase long-term vulnerability. Vulnerability assessment also contributes to equity by highlighting differences in risk across social groups and geographic areas. Recognizing that disaster risk is shaped by social and economic conditions allows policymakers to design interventions that strengthen livelihoods, improve access to services, enhance governance, and restore degraded ecosystems. Ecosystem-based approaches, climate-resilient infrastructure, land-use planning, and community-based adaptation measures all rely on sound hazard and vulnerability assessments. In the context of climate change, these assessments support a shift from reactive disaster response to preventive and planned risk reduction.

Coastal hazard and vulnerability assessment forms a key component of climate-induced Disaster Risk Reduction and Integrated Coastal Zone Management. Systematic identification and assessment of hazards, combined with detailed vulnerability analysis, make it possible to understand the changing risk conditions of coastal areas. Climate change has increased the urgency of this work by intensifying hazards, expanding exposure, and challenging existing coping practices. Reducing disaster risk in coastal zones therefore requires integrated, multi-hazard, and forward-looking assessments that consider physical processes, social and economic conditions, ecosystem health, and institutional capacity. Incorporating hazard and vulnerability assessments into coastal management frameworks can strengthen resilience, reduce disaster losses, and support development paths that are sustainable, inclusive, and responsive to climate change (Anjita et al., 2025; Basheer Ahammed et al., 2019).

7.2.1 Application of Geospatial Technologies for Hazard/Disasters Mapping and Vulnerability Risk Reduction

The increasing frequency and intensity of climate-induced hazards have emphasized the need for accurate, timely, and spatially detailed information to support Disaster Risk Reduction and Integrated Coastal Zone Management. Geospatial technologies, including Remote Sensing, Geographic Information Systems, Global Navigation Satellite Systems, and spatial modeling tools, now play an important role in understanding hazards, assessing vulnerability, and guiding risk reduction efforts. These technologies support the integration of information from multiple sources and help examine the relationships between climate change, coastal processes, human activities, and disaster risk (Fig. 7.1). Their use has contributed to a shift in disaster management practices from emergency response toward planning, prevention, and preparedness. Geospatial technologies support hazard and disaster mapping by providing spatially continuous and regularly updated information across large and often difficult-to-access coastal areas. Satellite-based remote sensing provides data on atmospheric, oceanic, and land conditions related to climate-induced hazards. These include sea surface temperature, changes in sea level, rainfall intensity, shoreline movement, land cover change, and ecosystem condition. Such information helps identify areas exposed to hazards, monitor changes over time, and detect early signs of extreme events. Geographic Information Systems function as a central platform for storing, analyzing, visualizing, and modeling spatial data, supporting multi-hazard assessments that are essential for integrated coastal zone management (Ineza et al., 2025; Kaur et al., 2018).

Geospatial technologies are widely used to identify and map climate-induced hazards. In coastal areas, hazards such as cyclones, storm surges,

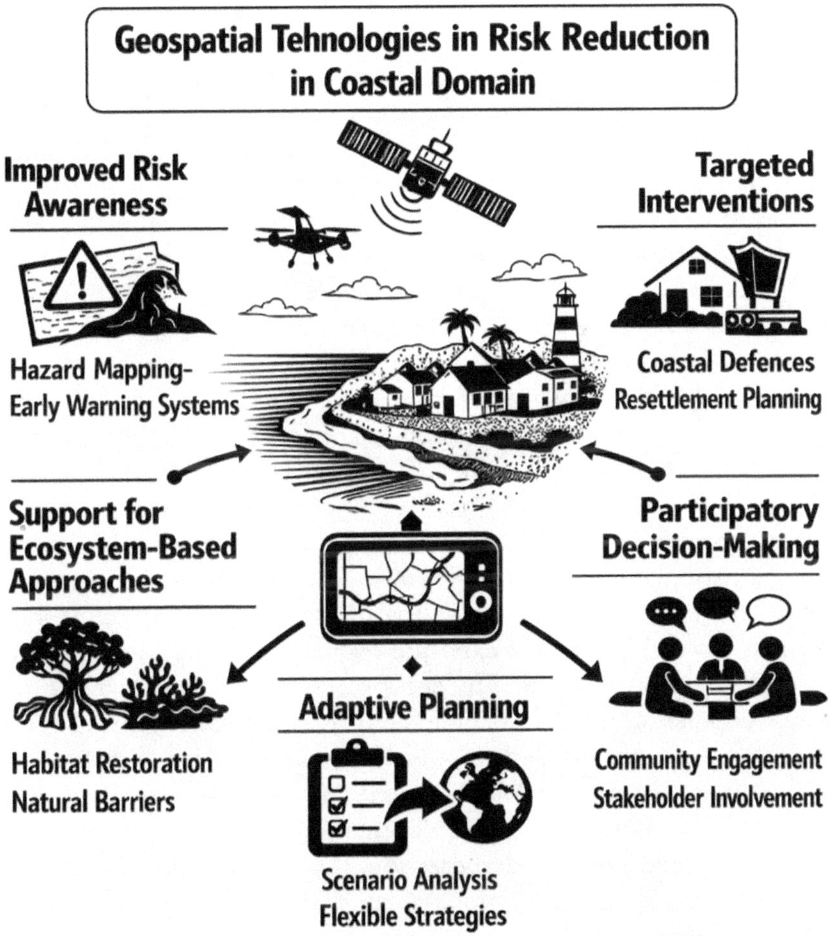

Fig. 7.1 Geospatial technologies in risk reduction (Prepared by the authors)

coastal flooding, sea-level rise, erosion, and saline intrusion vary significantly across space. Remote sensing data from optical, radar, and thermal sensors help detect and map areas affected by hazards, even during adverse weather conditions. Synthetic Aperture Radar imagery is particularly useful for mapping floods during cyclonic events because it can collect data through cloud cover and at any time of day. Long-term satellite observations also support the analysis of gradual changes such as shoreline retreat, wetland loss, and land degradation, which are important for understanding slow-onset climate hazards. GIS-based hazard mapping combines information on hazard intensity, frequency, and spatial extent to produce maps that show different levels of risk. These maps are important tools for land-use planning, zoning decisions, and infrastructure development in coastal regions. Visual representation of hazard-prone areas helps identify locations where development should be limited, modified, or supported by protective measures. Hazard maps can be revised as new data become available, allowing them to remain useful under changing environmental conditions (Weday et al., 2023; Shafapourtehrany et al., 2023).

Geospatial technologies also play a key role in vulnerability and exposure assessment, which together shape disaster risk. Vulnerability varies across space due to differences in social conditions, economic activities, infrastructure quality, and ecosystem health. Geographic Information Systems allow demographic data, socio-economic indicators, land-use patterns, infrastructure locations, and ecological features to be combined with hazard maps. This spatial overlay approach helps identify areas where high hazard exposure overlaps with high vulnerability. In coastal settings, vulnerability assessments often include indicators such as population density, income levels, housing conditions, access to services, dependence on climate-sensitive livelihoods, and distance from the shoreline. Ecosystem-related vulnerability can also be assessed using remote sensing, particularly where degradation of mangroves, coral reefs, or wetlands reduces natural protection. These integrated assessments are valuable for Integrated Coastal Zone Management, as they support decisions that consider human safety, economic needs, and ecosystem protection together (Sajjad et al., 2024; Hoque et al., 2021).

Geospatial technologies support risk modeling by combining information on hazards, exposure, and vulnerability within spatial analysis frameworks. Spatial models can be used to examine potential impacts under different hazard conditions, including future changes related to sea-level rise or increased storm intensity. Scenario analysis allows planners and decision-makers to compare the outcomes of different development and risk reduction options. For example, GIS-based models can assess how disaster risk may change under unplanned coastal expansion compared to regulated development combined with ecosystem protection. In climate-induced disaster risk reduction, this type of modeling supports early planning by identifying areas that may face higher risk in the future, even if present conditions appear manageable. This approach helps avoid inappropriate development and supports long-term investment decisions that are better suited to changing climate conditions. Scenario analysis strengthens the connection between disaster risk reduction and integrated coastal zone management by incorporating risk considerations into long-term planning (Hassani, 2016; Ergashev, 2012).

Geospatial technologies are also essential for early warning systems and disaster response. Satellite observations of weather patterns, ocean conditions, and land surface changes support forecasting of extreme events such as cyclones, storm surges, and floods. GIS-based decision-support systems combine real-time hazard information with data on population distribution, evacuation routes, shelters, and critical infrastructure. This integration allows warnings to be issued more quickly and evacuation planning to be more effective. Remote sensing also supports rapid assessment after disasters by mapping flooded areas, damaged infrastructure, and environmental impacts. This information helps guide relief, recovery, and reconstruction activities. In coastal regions, where hazards can develop rapidly and cause severe damage, the speed and spatial accuracy of geospatial technologies are especially important. These tools improve coordination among agencies and support decisions that reduce loss of life and property (Yadav et al., 2026; Singh & Zommers, 2014).

Geospatial technologies contribute directly to vulnerability risk reduction by supporting focused and well-informed interventions. Identification of high-risk locations and vulnerable populations helps guide actions such as climate-resilient infrastructure development, ecosystem restoration, livelihood diversification, and community-based adaptation. Within Integrated Coastal Zone Management processes, geospatial data support coordination among Disaster Risk Reduction, coastal resource management, urban planning, and environmental conservation. Key contributions of geospatial technologies include improved awareness of risk through maps and visual analysis, identification of priority areas for intervention, support for ecosystem-based approaches, and assistance in developing adaptive planning strategies. Participatory mapping and community-level GIS approaches also allow local knowledge to be included in planning processes, improving the relevance and acceptance of risk reduction measures. Several challenges affect the use of geospatial technologies, particularly in developing coastal regions. Common issues include limited access to high-resolution or updated data, shortages of trained personnel, and weak coordination among institutions. Uncertainty related to climate projections and modeling assumptions can also influence the interpretation of results. Addressing these challenges requires investment in technical capacity, improved data sharing, stronger institutional coordination, and the combination of geospatial analysis with field observations and social knowledge (Lanlan et at., 2024; Ballesteros & Esteves, 2021; Ramieri et al., 2019).

Therefore, the geospatial technologies have become important tools for hazard mapping and vulnerability risk reduction in the context of climate change. Their ability to capture spatial and temporal change, integrate different types of data, and support scenario analysis makes them suitable for addressing the complex risks faced by coastal areas. Within climate-induced Disaster Risk Reduction and Integrated Coastal Zone Management, these technologies provide a sound basis for informed decision-making, preventive planning, and sustainable coastal development. By linking climate information, spatial analysis, and policy action, geospatial technologies support efforts to reduce vulnerability, strengthen resilience, and guide coastal communities toward safer and more adaptive futures.

7.3 Integrated Coastal Zone Management (ICZM)

Coastal zones are among the most dynamic and productive regions of the world. They function as areas where land and sea systems interact while supporting large populations, critical infrastructure, and diverse ecosystems. At the same time, coastal areas are increasingly exposed to climate-induced hazards such as sea-level rise, coastal erosion, storm surges, cyclones, saline intrusion, and flooding. These growing risks have created serious challenges for coastal development and environmental sustainability. In response, Integrated Coastal Zone Management has emerged as a comprehensive approach to address development pressures, environmental degradation, and disaster risk. Within the broader framework of climate-induced Disaster Risk Reduction and Integrated Coastal Zone Management, ICZM provides an important governance and planning approach to balance development needs with risk reduction and environmental protection.

Integrated Coastal Zone Management is a holistic and participatory process that aims to support sustainable management of coastal areas by balancing environmental, economic, social, and institutional goals. The central idea of ICZM is integration across sectors, disciplines, spatial scales, and levels of governance. This approach helps overcome fragmented and sector-based planning practices that often limit effective coastal management. Coastal issues are addressed in an interconnected manner, linking land-based and marine activities, natural processes and human actions, and short-term development needs with long-term sustainability goals. ICZM functions both as a process and as a framework. As a process, it involves continuous planning, implementation, monitoring, and adjustment based on changing scientific knowledge and socio-economic conditions. As a framework, it provides guiding principles, institutional arrangements, and policy tools that support coordination among government agencies, local communities, private sector actors, and civil society organizations. Adaptability is a key feature of ICZM, as it recognizes uncertainty and change, particularly in the context of climate variability and long-term climate change.

Climate change has intensified coastal hazards and expanded the scale of risk, making conventional coastal management approaches less effective. ICZM addresses this challenge by integrating disaster risk reduction and climate adaptation into coastal planning and management. The focus moves from responding to disasters after they occur toward reducing exposure and vulnerability through planned and preventive action. The integration of hazard assessment, vulnerability analysis, land-use planning, ecosystem management, and community participation allows ICZM to address both sudden disasters and gradual climatic changes. In climate-induced disaster risk reduction, ICZM plays an important role in aligning development decisions with risk considerations. Measures such as regulating coastal land use based on hazard zones, promoting climate-resilient infrastructure, and conserving natural buffers such as mangroves and wetlands form key elements of ICZM. This approach helps prevent development practices that may increase future disaster risk and instead supports coastal development pathways that are resilient and sustainable over time (Farinós-Dasí et al., 2025; Abdou et al., 2025; Mall et al., 2019).

The effectiveness of ICZM depends on a set of guiding principles that shape planning, implementation, and evaluation. These principles provide a foundation for managing coastal areas in a balanced and sustainable way. Integration is a fundamental principle of ICZM. Coordination across sectors such as fisheries, tourism, urban development, agriculture, disaster management, and environmental protection is essential. Integration also applies across governance levels, connecting local, regional, and national institutions. Scientific knowledge is linked with policy and practice, while traditional knowledge is combined with formal planning tools. Without such integration, coastal management efforts remain fragmented and less effective in addressing climate-related risks. Ecosystem-based management is another key principle. Coastal ecosystems such as mangroves, coral reefs, salt marshes, dunes, and seagrass beds provide important services including shoreline protection, reduction of wave energy, flood regulation, and livelihood support. ICZM emphasizes the conservation, restoration, and sustainable use of these ecosystems as effective responses to climate-induced hazards. This principle supports nature-based approaches and strengthens resilience in both human and natural systems. Participation and stakeholder engagement are central to ICZM. Coastal areas are used and inhabited by diverse groups with different needs and levels of vulnerability. ICZM promotes inclusive decision-making processes that involve local communities, indigenous groups, women, and marginalized populations alongside technical experts and policymakers. Participation improves fairness, strengthens local ownership, and increases the effectiveness of management actions. In Disaster Risk Reduction, community involvement supports the use of local knowledge, improves risk awareness, and enhances preparedness.

Precaution and prevention are particularly important under conditions of climate uncertainty. ICZM supports early action to reduce risk, even when scientific information is incomplete. Measures such as limiting development in high-risk areas, enforcing building standards, and protecting coastal ecosystems help prevent future losses. Prevention is emphasized as a more effective and sustainable strategy than reliance on post-disaster response. Adaptability and learning are essential components of ICZM because coastal systems and climate impacts continue to change. ICZM is designed as a flexible and ongoing process that includes monitoring, evaluation, and feedback. This allows management strategies to be adjusted as conditions evolve or new information becomes available. Adaptability is especially important for addressing long-term risks such as sea-level rise and changing hazard patterns. Equity and social justice are also important principles of ICZM. Coastal risks and benefits are not evenly distributed, and climate-induced disasters often affect poorer and more vulnerable groups more severely. ICZM seeks to reduce these inequalities by improving access to resources, reducing vulnerability, and protecting livelihoods. Attention to social dimensions of risk supports inclusive and sustainable coastal development (Li et al., 2026; Ibrahim & Shaw, 2012).

The principles of ICZM are applied through a range of policy instruments, planning tools, and institutional arrangements. These include coastal zoning regulations, setback policies, integrated land and sea planning, environmental impact assessment, disaster risk assessment, and coordination mechanisms among agencies. Scientific inputs such as hazard mapping, vulnerability assessment, and climate projections support informed decision-making. Capacity building and public awareness are also important, as they enable stakeholders to participate meaningfully in ICZM processes. ICZM provides a structured framework for managing development pressures while respecting environmental limits and disaster risk considerations. Integrated planning, ecosystem conservation, participatory governance, and adaptive management support sustainable livelihoods, protect ecosystems, and reduce vulnerability to climate-induced hazards. The relevance of ICZM extends beyond risk reduction to broader development goals such as economic growth, food security, and environmental sustainability. Integrated Coastal Zone Management offers a comprehensive approach to managing the challenges faced by coastal areas under changing climatic conditions. As a concept and a guiding framework, ICZM emphasizes integration, ecosystem-based management, participation, precaution, adaptability, and equity. Within the framework of climate-induced Disaster Risk Reduction and Integrated Coastal Zone Management, ICZM connects climate knowledge, risk reduction, and sustainable development. By incorporating risk considerations into coastal planning and governance, ICZM strengthens resilience, reduces disaster losses, and supports development pathways that are environmentally sound, socially inclusive, and responsive to climate change.

7.3.1 Participatory Approaches in ICZM

Integrated Coastal Zone Management is widely recognized as a holistic and adaptive framework for managing coastal areas that face increasing pressure from development, environmental degradation, and climate-induced disasters. Coastal zones are complex and contested spaces where different interests and uses often overlap. For this reason, effective ICZM cannot rely only on top-down planning approaches. Participatory methods that involve local communities and diverse stakeholders in decision-making have become a central feature of contemporary ICZM. Within the broader framework of climate-induced Disaster Risk Reduction and Integrated Coastal Zone Management, participation is particularly important because it strengthens local resilience, improves awareness of risk, and supports disaster risk reduction strategies that are socially acceptable, locally relevant, and sustainable.

Participatory approaches in ICZM refer to structured processes through which stakeholders such as coastal communities, local governments, resource users, civil society organizations, scientists, and private sector actors are involved in planning, implementation, and monitoring of coastal management activities. Participation is based on the recognition that coastal systems are socio-ecological in nature and that local populations hold valuable knowledge developed through long-term interaction with coastal environments. In the context of climate change, where uncertainty and variability are high, including diverse perspectives helps improve the flexibility and effectiveness of management strategies. Participation also helps address governance challenges commonly found in coastal zones, including overlapping responsibilities, sector-based planning, and conflicts over resource use. Dialogue and cooperation among stakeholders improve coordination across sectors and levels of governance and help build trust. From a disaster risk reduction perspective, participation shifts attention from externally driven actions toward locally grounded risk governance. This approach allows communities to take an active role in reducing their own vulnerability (Hietala et al., 2021; Rockloff et al., 2004; Kannen, 2002).

Climate-induced hazards such as cyclones, storm surges, coastal flooding, erosion, and sea-level rise often affect coastal communities with limited resources and adaptive capacity. Participatory approaches in ICZM support disaster risk reduction by bringing local knowledge of hazards, past events, and coping practices into formal planning processes. Communities often have detailed understanding of flood-prone areas, shoreline changes, evacuation routes, and ecosystem conditions that may not be fully captured through technical assessments. Combining this knowledge with scientific information and spatial analysis improves the relevance and accuracy of risk assessments. Participation also strengthens preparedness and response by increasing awareness of risk and encouraging shared responsibility for disaster risk reduction actions. Community involvement in hazard mapping, early warning communication, and contingency planning improves the effectiveness of these systems and ensures they are accessible and culturally appropriate. Through these processes, participatory ICZM reduces exposure and vulnerability while strengthening social networks that support recovery and resilience (Wen et al., 2023).

Participatory approaches in ICZM take different forms and operate at various levels, from information sharing to shared decision-making and co-management (Fig. 7.2). The quality and depth of participation strongly influence outcomes. Common forms of participation include consultative participation, where stakeholders provide input through meetings, workshops, and public consultations on coastal issues and risk reduction measures. Collaborative participation involves joint analysis, planning, and problem-solving through multi-stakeholder platforms or committees. Community-based management places local communities at the center of managing coastal resources and implementing risk reduction actions with institutional support. Co-management and shared governance involve shared authority and responsibility between government agencies and local stakeholders, supporting accountability and long-term commitment. These forms often exist together within ICZM processes. Effective participation requires meaningful involvement that influences decisions and outcomes rather than symbolic engagement (Veloso-Gomes et al., 2008; Stead, 2005).

A range of tools and methods support participation in ICZM. Participatory rural appraisal, focus group discussions, stakeholder mapping, and community consultations are commonly used to understand local perceptions, priorities, and vulnerabilities. Participatory mapping and community-based geographic information systems allow stakeholders to identify hazard-prone areas, important resources, and safe locations, creating a shared understanding of risk. Scenario-building and visioning exercises help communities explore future climate and development pathways and support adaptive planning. Capacity building is an essential part of participatory ICZM. Meaningful engagement depends on access to information, skills, and institutional support. Training on climate change, disaster preparedness, and coastal governance improves the ability of communities and local institutions to participate effectively. These efforts are especially important for marginalized groups

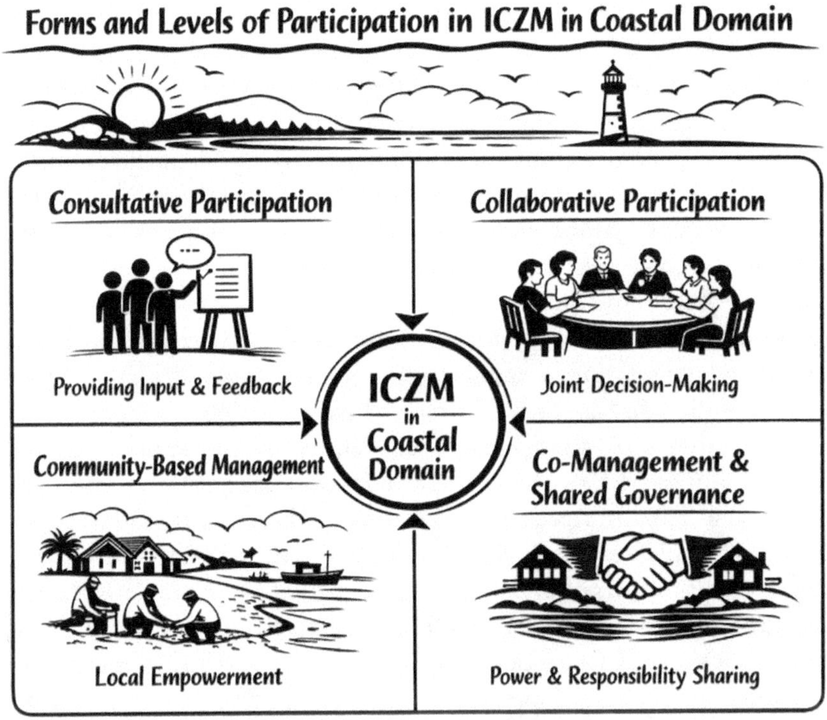

Fig. 7.2 Forms and levels of participation in ICZM (Prepared by the authors)

whose voices are often excluded from formal decision-making. Participation in ICZM must address social differences within coastal communities. Vulnerability to climate-induced disasters varies according to gender, age, occupation, income, and social position, shaping access to resources, information, and decision-making power. Participatory processes that ignore these differences risk reinforcing existing inequalities and excluding those most at risk. Inclusive participation emphasizes representation of women, indigenous peoples, small-scale fishers, informal settlers, and other marginalized groups. Safe spaces for dialogue, recognition of customary rights, and transparency in decision-making strengthen trust and accountability. Attention to social dimensions of risk improves fairness and effectiveness in disaster risk reduction (Miao et al., 2025; Malakar & Roy, 2024; Aggrey et al., 2021).

Integrating participation into ICZM has important implications for coastal governance. Institutional arrangements must be flexible, transparent, and responsive to stakeholder input. Decentralized decision-making, coordination among agencies, and legal recognition of community roles are often required to support meaningful participation. In climate-induced disaster risk reduction, institutionalized participation helps align local actions with national and regional strategies. Participatory governance also improves policy acceptance and compliance, as stakeholders are more likely to support decisions they have helped shape. This is particularly relevant for land-use regulation, ecosystem conservation, and adaptation strategies that involve trade-offs and difficult choices (Frølich & Caspersen, 2015).

Participatory ICZM faces several challenges, including unequal power relations, limited local capacity, participation fatigue, and conflicting interests. Participatory processes may require time and resources that are not always available, especially in disaster situations. Translating participatory inputs into policy action also depends on strong institutional commitment and coordination. Clear objectives, well-designed participatory processes, capacity building, and conflict resolution mechanisms are needed to address these challenges. Combining participation with scientific analysis helps ensure that decisions are both inclusive and technically sound. Participatory approaches strengthen ICZM by building adaptive capacity at community and institutional levels. Integration of local knowledge, shared responsibility, and strong social networks supports community-based and ecosystem-based adaptation strategies. Participation also encourages learning and adjustment of management actions as risks change over time. Key contributions of participatory ICZM include improved understanding of local hazards and vulnerabilities, better design and acceptance of risk reduction measures, stronger preparedness and response capacity, greater equity in coastal governance, and increased resilience of coastal socio-ecological systems (Malakar & Roy, 2024).

Therefore, the participatory approaches are essential for effective Integrated Coastal Zone Management under conditions of climate change and increasing disaster risk. Within the framework of climate-induced Disaster Risk Reduction and Integrated Coastal Zone Management, participation makes coastal management more socially grounded and adaptive. Active involvement of communities and stakeholders improves risk awareness, reduces vulnerability, and strengthens resilience to climate-induced hazards. Despite ongoing challenges, inclusive and well-designed participatory approaches provide a practical pathway toward sustainable, equitable, and resilient coastal development.

7.3.2 Planning and Governance Frameworks

Integrated Coastal Zone Management has developed as a comprehensive planning and governance framework to respond to the increasing complexity of coastal regions affected by climate change, development pressures, and disaster risk. Coastal zones experience strong interactions between natural processes and human activities and are often governed by multiple institutions with overlapping responsibilities. Within the framework of climate-induced Disaster Risk Reduction and Integrated Coastal Zone Management, effective planning and governance systems are necessary to translate ICZM principles into practical actions that reduce vulnerability, manage risk, and support sustainable coastal development. ICZM planning frameworks aim to bring together environmental protection, economic development, social equity, and disaster risk reduction within a coordinated process. Unlike sector-based or project-focused approaches, ICZM follows a long-term and strategic perspective that considers cumulative impacts, interactions between land and sea, and future climate conditions. Planning under ICZM is designed as an adaptive process, allowing policies and actions to change in response to new scientific knowledge, socio-economic developments, and emerging risks. This adaptive approach is particularly important in coastal areas where climate-induced hazards such as sea-level rise, storm surges, coastal erosion, and flooding are becoming more frequent and severe. Integration of disaster risk reduction and climate change adaptation forms a central part of ICZM planning. Coastal plans and policies incorporate hazard assessment, vulnerability analysis, and risk-sensitive land-use planning. Planning tools such as coastal zoning, setback regulations, and development controls are widely used to reduce exposure in high-risk areas. Aligning development decisions with risk information helps avoid the creation of new vulnerabilities and reduces long-term disaster losses. Prevention and preparedness are emphasized rather than reliance on post-disaster response, making coastal planning more effective and sustainable over time (Ahlhorn, 2018).

Governance frameworks provide the institutional structure that supports ICZM planning. Coastal governance involves multiple levels of authority, including local governments, national agencies, and regional bodies, along with stakeholders such as coastal communities, private sector actors, and civil society organizations. ICZM governance frameworks address this complexity through coordination, integration, and collaboration. Coordination across sectors such as environment, fisheries, tourism, urban development, disaster management, and infrastructure helps prevent conflicting policies. Coordination across governance levels ensures that national strategies reflect local needs and that local actions support broader policy goals. Institutional arrangements for ICZM governance differ across regions but often include inter-agency committees, coastal authorities, or dedicated ICZM units that support coordination and decision-making. These institutions play an important role in aligning policies, sharing information, resolving conflicts, and overseeing implementation. In climate-induced disaster risk reduction, strong connections between ICZM institutions and disaster management agencies, climate adaptation bodies, and planning authorities are essential. Such coordination supports effective responses to both sudden disasters and gradual climate impacts. Legal and policy frameworks are another key component of ICZM governance. Effective ICZM requires laws and policies that clearly define roles and responsibilities, support integrated planning, and allow enforcement of coastal regulations. Coastal legislation often reflects principles such as sustainability, precaution, ecosystem protection, and public participation. Alignment of legal frameworks with disaster risk reduction and climate adaptation policies strengthens their role in managing coastal risk and guiding development in hazard-prone areas. Gaps or inconsistencies in laws and policies can weaken ICZM implementation, highlighting the need for legal clarity and consistency (Malakar et al., 2023b).

Scientific knowledge and information systems support decision-making in ICZM planning and governance. Coastal planning relies on data related to hazards, ecosystems, land use, socio-economic conditions, and climate projections. Use of this information improves transparency and supports evidence-based decisions. Scenario-based planning is increasingly applied to explore future risk conditions under different climate and development pathways. This approach supports forward-looking governance by helping policymakers anticipate change and design strategies that remain effective under uncertain conditions. Stakeholder participation is an important governance principle within ICZM planning frameworks. Inclusive governance improves the legitimacy, accountability, and effectiveness of coastal management decisions. Involvement of local communities, resource users, and other stakeholders ensures that local knowledge and social priorities are reflected in policies and actions. In Disaster Risk Reduction, participatory governance increases risk awareness, strengthens preparedness, and supports acceptance of risk-sensitive measures such as land-use controls or relocation when required. Decentralization and multi-level governance are also important features of ICZM frameworks. National policies provide overall direction and regulatory support, while local governments and community institutions are often best placed to implement and enforce coastal management measures. Decentralized governance allows solutions that reflect local environmental and socio-economic conditions. Coordination mechanisms remain necessary to ensure consistency, equity, and effective management of shared or transboundary coastal issues. ICZM planning and governance frameworks face several challenges. Institutional fragmentation, limited technical and financial capacity, stakeholder conflicts, and pressure for short-term development can hinder effective implementation. Climate change adds further complexity by increasing uncertainty and requiring planning horizons that extend beyond conventional policy cycles. Addressing these challenges requires long-term political commitment, capacity building, and continuous learning within institutions. Monitoring, evaluation, and accountability strengthen ICZM governance. Regular review of planning outcomes and governance performance helps identify gaps and improve future actions. Adaptive governance systems that include feedback mechanisms allow ICZM frameworks to respond to changing risk conditions and evolving social priorities. In climate-induced disaster risk reduction, such adaptive governance is essential for managing dynamic and uncertain coastal environments (Malakar et al., 2023c).

Therefore, the planning and governance frameworks form the foundation of Integrated Coastal Zone Management and shape its effectiveness in addressing climate-induced disaster risk. Integration across sectors and governance levels, incorporation of risk reduction into development planning, and support for participatory and adaptive governance make ICZM a practical approach for managing coastal zones under climate change. Within the broader context of climate-induced Disaster Risk Reduction and Integrated Coastal Zone Management, these frameworks provide the institutional basis for reducing vulnerability, strengthening resilience, and guiding sustainable and equitable coastal development under uncertain climatic conditions.

7.4 Climate Change Adaptation Strategies for Coastal Areas

7.4.1 Ecosystem-Based Adaptation (EbA) and Disaster Risk Reduction (DRR)

Coastal areas are among the most vulnerable regions to climate change because of their exposure to sea-level rise, coastal erosion, saline intrusion, cyclones, storm surges, flooding, and changing ocean conditions. These climate-related pressures interact with rapid urban growth, strong dependence on coastal resources for livelihoods, and ongoing environmental degradation, leading to higher disaster risk. Within the framework of climate-induced Disaster Risk Reduction and Integrated Coastal Zone Management, climate change adaptation strategies are essential for reducing vulnerability, strengthening resilience, and supporting sustainable coastal development. Ecosystem-based Adaptation and Disaster Risk Reduction are widely recognized as complementary approaches for addressing present and future climate risks in coastal zones. Climate change adaptation in coastal areas refers to adjustments in natural, social, and economic systems in response to actual or expected climate impacts. Adaptation differs from short-term coping actions by focusing on long-term resilience, flexibility, and learning. Coastal adaptation requires integrated approaches that address environmental processes, social and economic conditions, and governance systems together. Ecosystem-based Adaptation and Disaster Risk Reduction provide two closely linked pathways for achieving this integration. Ecosystem-based Adaptation focuses on the role of ecosystems in supporting adaptation, while Disaster Risk Reduction emphasizes reducing exposure, vulnerability, and hazard impacts through prevention and preparedness (Peng et al., 2025; Hill et al., 2025).

Ecosystem-based Adaptation uses biodiversity and ecosystem services to help people adjust to the impacts of climate change. In coastal settings, this approach recognizes that healthy ecosystems such as mangroves, coral reefs, seagrass meadows, salt marshes, dunes, and coastal wetlands provide important regulating services that reduce climate-related risks. These ecosystems act as natural buffers by reducing wave energy, stabilizing shorelines, limiting erosion, regulating floods, and supporting livelihoods that strengthen social and economic resilience. Ecosystem-based Adaptation emphasizes conservation, restoration, and sustainable management of coastal ecosystems as practical and flexible adaptation options. Mangrove restoration, for example, can reduce storm surge impacts and coastal flooding while also supporting fisheries, storing carbon, and providing income opportunities. Coral reefs and seagrass beds reduce wave action and protect shorelines, while coastal wetlands absorb floodwaters and improve water quality. Maintaining healthy ecosystems strengthens the ability of coastal systems to adjust to climate variability and long-term change. A key strength of Ecosystem-based Adaptation is its ability to deliver multiple benefits at the same time. Ecosystem-based measures support biodiversity conservation, livelihoods, climate mitigation, and cultural values in addition to reducing disaster risk. These measures are adaptive in nature, as ecosystems can respond to gradual changes such as sea-level rise when sufficient space and protection are provided. Effective implementation depends on long-term planning, community involvement, and governance systems that prevent ecosystem degradation and inappropriate development (Nandy et al., 2013; Ahammad et al., 2013).

Disaster Risk Reduction aims to reduce disaster losses by lowering exposure, vulnerability, and risk. In coastal areas, this approach addresses sudden hazards such as cyclones and storm surges, as well as slow processes such as erosion and sea-level rise. Climate change adaptation and Disaster Risk Reduction are closely linked because both focus on anticipating future risks and reducing impacts before disasters occur. Common Disaster Risk Reduction strategies in coastal areas include hazard and vulnerability assessment, risk-sensitive land-use planning, climate-resilient infrastructure, early warning systems, preparedness planning, and community-based disaster management. Coastal zoning and setback rules help restrict development in high-risk locations, while resilient buildings and infrastructure reduce damage during extreme events. Early warning systems combined with evacuation planning and awareness programs reduce loss of life and property. Disaster Risk Reduction also addresses underlying drivers of vulnerability, including poverty, weak governance, and limited access to resources. From an adaptation perspective, Disaster Risk Reduction strengthens the capacity of coastal communities and institutions to cope with climate extremes that are expected to increase in frequency and intensity. Incorporating climate projections into risk reduction planning helps ensure that measures remain effective under future conditions rather than relying only on past hazard patterns (Wen et al., 2023; Ahammad et al., 2013).

Ecosystem-based Adaptation and Disaster Risk Reduction work best when applied together. Ecosystem-based measures reduce hazard intensity and exposure through natural protection, while Disaster Risk Reduction provides planning, institutional support, and preparedness mechanisms to manage remaining risk. Combined application supports a balanced strategy that links nature-based solutions with structural, social, and institutional measures. Integration of these approaches is especially important in coastal areas, where ecosystem loss often increases disaster risk. Degradation of mangroves and dunes can intensify storm surge impacts and increase dependence on costly engineered defenses. Restoration of ecosystems alongside Disaster Risk Reduction measures such as land-use controls, building standards, and early warning systems improves resilience and long-term sustainability. In this way, the adaptation strategies combining ecosystem-based adaptation and disaster risk reduction;

(a) Conservation and restoration of coastal ecosystems such as mangroves, wetlands, coral reefs, and dunes to reduce erosion, flooding, and wave impacts while supporting livelihoods.
(b) Risk-sensitive coastal planning that integrates ecosystem protection with hazard and vulnerability assessment.
(c) Hybrid approaches that combine ecosystem-based measures with engineered structures to strengthen protection.
(d) Community-based adaptation and Disaster Risk Reduction through local involvement in ecosystem management, hazard mapping, preparedness, and monitoring.
(e) Early warning and preparedness systems that link ecosystem monitoring with hazard forecasting and community response.
(f) Livelihood diversification that supports climate-resilient and ecosystem-based economic activities.
(g) Role of Integrated Coastal Zone Management in Integrating Ecosystem-based Adaptation and Disaster Risk Reduction.

So, the Integrated Coastal Zone Management provides the planning and governance framework needed to implement Ecosystem-based Adaptation and Disaster Risk Reduction in a coordinated manner. This approach supports cross-sector coordination, stakeholder participation, and adaptive management, helping align ecosystem conservation, disaster risk reduction, and development objectives. Through Integrated Coastal Zone Management, both approaches can be incorporated into coastal policies, development plans, and governance systems, reducing fragmentation and avoiding maladaptive outcomes. Integrated Coastal Zone Management also supports coordination across scales by linking local ecosystem initiatives with national disaster management and climate adaptation strategies. Embedding Ecosystem-based Adaptation and Disaster Risk Reduction within this framework improves coherence, equity, and effectiveness in addressing long-term climate risks.

And the implementation of Ecosystem-based Adaptation and Disaster Risk Reduction in coastal areas faces several challenges. Ecosystem-based measures often require time to deliver protective benefits, which may not align with short planning cycles. Land-use pressure, competing development interests, and weak enforcement can reduce effectiveness. Disaster Risk Reduction efforts may be limited by financial constraints, technical capacity, and institutional fragmentation. Uncertainty in climate projections also complicates long-term planning. Addressing these challenges depends on strong governance, inclusive participation, capacity building, and sustained political support. Linking scientific knowledge with local experience and maintaining flexible planning approaches are essential for managing uncertainty and change. Climate change adaptation in coastal areas requires integrated strategies that address both environmental processes and human vulnerability. Ecosystem-based Adaptation and Disaster Risk Reduction provide complementary pathways for strengthening coastal resilience. Ecosystem-based Adaptation enhances the protective role of ecosystems, while Disaster Risk Reduction reduces exposure and vulnerability through planning, preparedness, and governance. Within the framework of climate-induced Disaster Risk Reduction and Integrated Coastal Zone Management, combining these approaches offers a practical and sustainable way to manage climate risks. Nature-based solutions combined with proactive risk reduction measures can reduce disaster losses, protect ecosystems, and support development pathways that are resilient, inclusive, and adaptable to a changing climate (Fig. 7.3).

7.4.2 Incentive-Based Adaptation (IbA) and Disaster Risk Reduction (DRR)

Coastal areas are increasingly exposed to climate change impacts such as sea-level rise, stronger cyclones, storm surges, coastal erosion, saline intrusion, and flooding. These hazards pose serious threats to lives, livelihoods, infrastructure, and fragile coastal ecosystems. The risks are especially high in regions where poverty, dense populations, and dependence on climate-sensitive resources are common. Within the framework

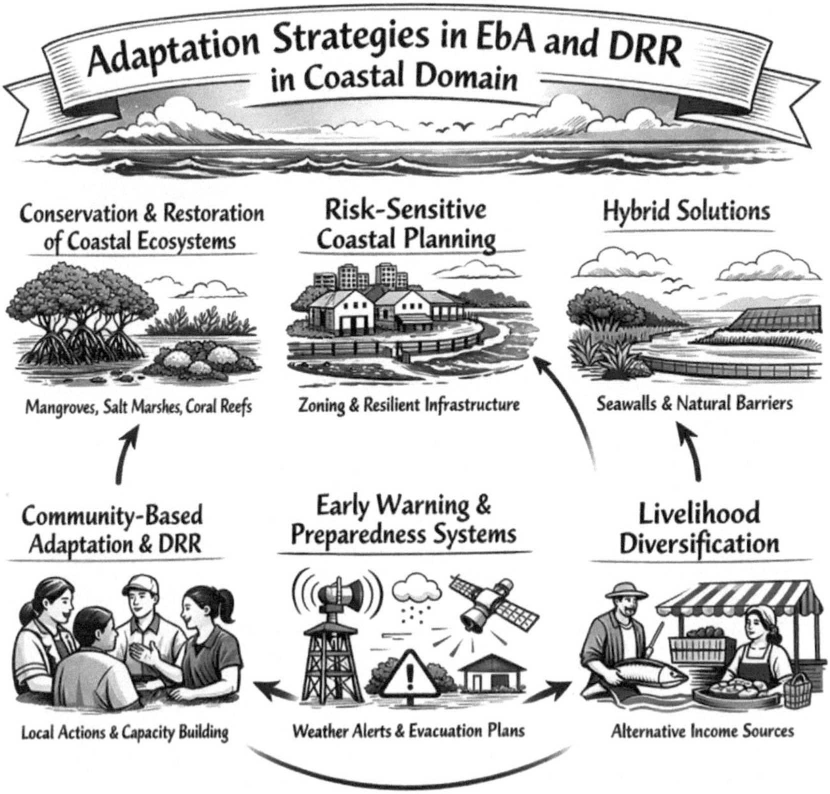

Fig. 7.3 Adaptation strategies in EbA and DRR (Prepared by the authors)

of climate-induced Disaster Risk Reduction and Integrated Coastal Zone Management, climate change adaptation strategies aim to reduce vulnerability, strengthen resilience, and support the ability of coastal communities and institutions to cope with both present and future climate risks. Along with ecosystem-based and structural measures, Incentive-based Adaptation combined with Disaster Risk Reduction has gained attention as a practical approach to influence behavior, guide development choices, and integrate risk reduction into coastal development processes. Climate change adaptation in coastal areas involves more than technical solutions. Economic, institutional, and behavioral changes are also required. Incentive-based Adaptation focuses on providing positive incentives such as financial support, policy benefits, or institutional rewards that encourage individuals, communities, and private actors to adopt climate-resilient and risk-reducing practices. Disaster Risk Reduction offers a structured approach for identifying hazards, assessing vulnerability, and applying measures to reduce disaster risk. When combined, Incentive-based Adaptation and Disaster Risk Reduction connect economic decision-making with long-term resilience and sustainable coastal management.

Incentive-based Adaptation refers to the use of economic and policy incentives to motivate stakeholders to adopt practices that improve climate resilience and reduce disaster risk. In coastal regions, decisions are often shaped by short-term economic needs, even when these choices increase long-term vulnerability. Incentives can help change these patterns by reshaping costs and benefits. This encourages risk-sensitive land use, safer construction practices, livelihood diversification, and early adaptation action. Incentive-based Adaptation tools include subsidies for climate-resilient housing, tax benefits for moving away from high-risk coastal areas, insurance schemes linked to risk reduction actions, conditional grants for local governments, and livelihood incentives that reduce dependence on climate-sensitive activities. These tools are particularly useful in coastal areas where development pressure is strong and regulation alone may not be effective. Aligning economic incentives with adaptation and Disaster Risk Reduction goals helps translate policy objectives into practical action.

In coastal settings, Incentive-based Adaptation supports action at different levels. At the household level, incentives can promote safer housing designs, raised structures, and improved sanitation in flood-prone areas. At the community level, incentives can encourage shared efforts such as maintaining drainage systems, participating in preparedness programs, and adopting safer livelihood practices. At the level of local government, performance-based funding can support the inclusion of climate risk considerations in coastal planning and infrastructure development. The private sector also plays an important role in coastal areas through tourism, ports, fisheries, and urban development. Insurance discounts for risk-reducing actions, affordable finance for resilient infrastructure, and regulatory benefits for sustainable investment can encourage private sector involvement in adaptation. When well designed, Incentive-based Adaptation helps reduce disaster risk while also supporting innovation and long-term economic stability in coastal regions (Kuhl et al., 2020).

Disaster Risk Reduction provides the analytical and institutional base for effective Incentive-based Adaptation. Disaster Risk Reduction defines risk in terms of hazard, exposure, and vulnerability and places emphasis on prevention, preparedness, and mitigation rather than post-disaster response. In coastal areas, Disaster Risk Reduction strategies include hazard and vulnerability assessment, early warning systems, risk-sensitive land-use planning, resilient infrastructure, and community preparedness. Incentive-based Adaptation strengthens Disaster Risk Reduction by encouraging voluntary action and early investment in safety. Risk-based insurance schemes can motivate households to relocate from highly exposed areas or invest in safer construction. Subsidies and grants can support adoption of risk reduction measures that might otherwise be considered too costly. Integrating incentives into Disaster Risk Reduction frameworks improves acceptance and economic feasibility of adaptation actions. Combining Incentive-based Adaptation and Disaster Risk Reduction improves climate adaptation in coastal areas. Disaster Risk Reduction identifies where risks exist and why they matter, while Incentive-based Adaptation shapes how people and institutions respond to those risks. Together, these approaches address both structural drivers of vulnerability and behavioral aspects of adaptation. Key areas of synergy include risk-sensitive development, preparedness, and long-term resilience. Land-use regulations supported by relocation incentives reduce exposure in high-risk coastal zones. Insurance and social protection linked to preparedness measures strengthen household resilience. Incentives for local governments to invest in risk reduction improve institutional capacity and governance.

(a) Risk-based insurance and social protection schemes that reward risk-reducing behavior and support vulnerable groups.
(b) Subsidies and financial incentives for climate-resilient housing, infrastructure improvement, and livelihood diversification.

(c) Performance-based grants that encourage local governments to integrate disaster risk reduction and climate adaptation into coastal planning.
(d) Relocation and land-use incentives that support voluntary movement away from highly exposed coastal zones.
(e) Private sector incentives such as tax benefits, affordable finance, and regulatory support for resilient coastal investment.
(f) Role of Integrated Coastal Zone Management in Integrating Incentive-based Adaptation and Disaster Risk Reduction.

Therefore, the Integrated Coastal Zone Management provides the planning and governance framework needed to combine Incentive-based Adaptation and Disaster Risk Reduction in a coordinated way. This approach supports coordination across sectors and governance levels and ensures that incentives align with long-term sustainability and risk reduction goals. Hazard and vulnerability assessments within Integrated Coastal Zone Management help target incentives to appropriate locations and reduce the risk of maladaptation. Integrated Coastal Zone Management also supports stakeholder participation and transparency in the design of incentive mechanisms, which strengthens trust and effectiveness. Embedding Incentive-based Adaptation and Disaster Risk Reduction within coastal planning and governance structures makes adaptation strategies more coherent, equitable, and responsive to changing climate conditions. Several challenges affect the application of Incentive-based Adaptation and Disaster Risk Reduction in coastal areas. Poorly designed incentives may encourage continued settlement in high-risk locations. Unequal access to incentives can exclude vulnerable groups and increase inequality. Disaster Risk Reduction efforts may face limits related to funding, institutional coordination, and enforcement. Uncertainty in climate projections also makes long-term planning more difficult. Addressing these challenges requires careful design of incentives, strong risk assessment, effective governance, and regular monitoring and evaluation. Linking scientific knowledge with local experience and maintaining flexible planning approaches are important for managing uncertainty and change. Incentive-based Adaptation and Disaster Risk Reduction provide complementary approaches for climate change adaptation

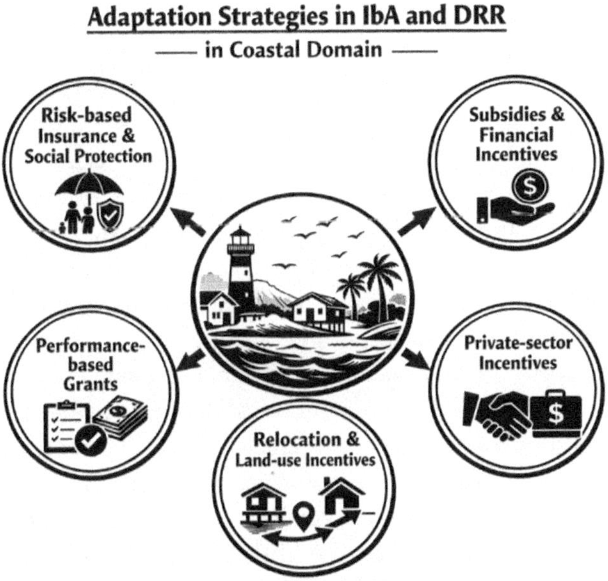

Fig. 7.4 Adaptation strategies in IbA and DRR (Prepared by the authors)

in coastal areas. Disaster Risk Reduction offers the framework for understanding and reducing risk, while Incentive-based Adaptation influences behavior and development choices by aligning economic incentives with resilience goals. Within the framework of climate-induced Disaster Risk Reduction and Integrated Coastal Zone Management, combining these approaches supports proactive, inclusive, and sustainable coastal adaptation. By guiding decisions, strengthening preparedness, and reducing long-term vulnerability, these strategies contribute to coastal systems that are better able to face the growing challenges of climate change (Fig. 7.4).

7.4.3 Community-Based Adaptation (CbA) and Disaster Risk Reduction (DRR)

Coastal areas face some of the most serious impacts of climate change. Rising sea levels, coastal erosion, cyclones, storm surges, flooding, and saline intrusion are increasing risks to lives, livelihoods, infrastructure, and coastal ecosystems. These risks are especially severe in communities whose economies depend on climate-sensitive resources. Within the framework of climate-induced Disaster Risk Reduction and Integrated Coastal Zone Management, adaptation strategies that place people at the center of planning and action are particularly important. Community-based Adaptation, when linked with Disaster Risk Reduction, provides a locally grounded and participatory approach to reducing vulnerability and strengthening resilience in coastal regions. Climate change adaptation in coastal areas needs to reflect local environmental conditions, knowledge systems, and social and economic realities. Large infrastructure projects and policy measures are important, but they often do not address the everyday risks faced by coastal populations. Community-based Adaptation focuses on the direct involvement of communities in identifying risks, designing solutions, and carrying out adaptation and risk reduction actions. Disaster Risk Reduction complements this approach by offering structured methods to understand hazards, reduce exposure, and improve preparedness. Together, these approaches connect local action with wider coastal planning and governance systems such as Integrated Coastal Zone Management (Bera, 2025; Cruickshank et al., 2024).

Community-based Adaptation refers to adaptation initiatives that are designed and implemented with active participation of local communities, especially those most exposed to climate change impacts. In coastal areas, this approach recognizes that communities hold valuable knowledge of coastal processes, hazard patterns, and coping practices developed through long-term interaction with their environment. This local knowledge is essential for identifying risks, setting priorities, and ensuring that adaptation measures are socially acceptable and culturally appropriate. Community-based Adaptation aims to strengthen local adaptive capacity by addressing both immediate climate risks and underlying causes of vulnerability such as poverty, social exclusion, and limited access to resources. In coastal settings, initiatives may include strengthening housing against floods and cyclones, diversifying livelihoods, improving access to early warning information, restoring local ecosystems, and strengthening community organizations. By supporting community ownership of adaptation processes,

Community-based Adaptation contributes to long-term resilience and sustainability.

Disaster Risk Reduction offers a systematic and preventive approach to reducing disaster losses by lowering exposure, vulnerability, and risk. In coastal areas, Disaster Risk Reduction addresses sudden events such as cyclones and storm surges as well as gradual processes such as erosion and sea-level rise. Common measures include hazard and vulnerability assessments, early warning systems, evacuation planning, resilient infrastructure, and preparedness planning. From a climate adaptation perspective, Disaster Risk Reduction helps anticipate future risks and prevent disasters rather than responding after damage has occurred. Integrating climate change considerations into Disaster Risk Reduction ensures that risk reduction measures remain effective as hazard patterns change. In coastal regions, this approach also focuses on protecting critical infrastructure and essential services that support community safety and recovery. Community-based Adaptation and Disaster Risk Reduction work best when applied together. Disaster Risk Reduction provides technical and institutional tools for assessing risk and improving preparedness, while Community-based Adaptation ensures that these tools are shaped by local experience and community participation. Integration of these approaches shifts disaster management away from a response-focused model toward prevention and resilience.

In coastal areas, integration allows communities to take part in hazard mapping, vulnerability assessment, and preparedness planning. Community participation improves the quality of risk assessments by adding local knowledge and historical experience. It also builds trust and shared responsibility, which are essential for effective early warning systems, evacuation planning, and risk-sensitive land-use practices. Through Community-based Adaptation, Disaster Risk Reduction measures become more inclusive and sustainable.

(a) Participatory risk assessment through local consultations and mapping to identify hazards, vulnerable groups, and critical assets.
(b) Community-based early warning systems that use locally appropriate communication methods and trained volunteers.

(c) Strengthening climate-resilient livelihoods through income diversification and safer livelihood practices.
(d) Resilient housing and infrastructure through community-led improvements to homes, shelters, and local facilities.
(e) Local ecosystem stewardship through protection and restoration of mangroves, dunes, and wetlands that reduce hazard impacts.
(f) Preparedness and response planning through local evacuation plans, drills, and community disaster committees.
(g) Role of Integrated Coastal Zone Management in Supporting Community-based Adaptation and Disaster Risk Reduction

Integrated Coastal Zone Management provides the planning and institutional framework needed to support and expand Community-based Adaptation and Disaster Risk Reduction initiatives. This framework promotes coordination across sectors and governance levels and helps align community actions with regional and national coastal policies. Through Integrated Coastal Zone Management, community-based efforts can be incorporated into coastal development planning, land-use regulation, and ecosystem management, reducing fragmentation. Integrated Coastal Zone Management also supports capacity building and knowledge exchange among communities, local authorities, and technical institutions. Recognizing communities as key actors in coastal governance strengthens participation and improves the legitimacy of adaptation strategies. This alignment is important for managing long-term climate risks that require sustained collective action. Community-based Adaptation places strong emphasis on equity and inclusion. Climate-related disasters often have greater impacts on marginalized groups such as women, older people, small-scale fishers, and informal settlers in coastal areas. Community-based approaches aim to involve these groups directly in adaptation and Disaster Risk Reduction processes so that their needs and priorities are reflected. Reducing social vulnerability and strengthening social networks contribute to fairer and more effective risk reduction outcomes. Community-based Adaptation and Disaster Risk Reduction face several challenges in coastal contexts. Limited financial and technical resources can restrict local initiatives, and power differences within communities may affect participation.

Community-led actions alone may not be sufficient to manage large-scale or shared coastal risks, which require support from higher levels of governance. Climate uncertainty also makes long-term planning difficult and requires continuous learning. Addressing these challenges depends on strong institutional support, capacity building, and close integration of community efforts with wider Integrated Coastal Zone Management and Disaster Risk Reduction frameworks. Combining local knowledge with scientific information improves the effectiveness and reach of adaptation strategies. Community-based Adaptation and Disaster Risk Reduction together offer a people-centered approach to climate change adaptation in coastal areas. Disaster Risk Reduction provides structured methods for understanding and reducing risk, while Community-based Adaptation ensures that these methods are locally relevant and inclusive. Within the framework of climate-induced Disaster Risk Reduction and Integrated Coastal Zone Management, integration of these approaches strengthens resilience, reduces vulnerability, and improves the ability of coastal communities to cope with climate-related hazards. Placing communities at the core of adaptation and risk reduction supports sustainable and resilient coastal development under changing climate conditions (Fig. 7.5).

7.5 Disaster Risk Reduction Measures in Coastal Zones

7.5.1 Early Warning Systems and Preparedness Planning

Coastal zones are increasingly exposed to climate-induced hazards such as cyclones, storm surges, coastal flooding, tsunamis, and extreme rainfall events. Climate change has increased the frequency and severity of these hazards, while sea-level rise and rapid coastal development have expanded exposure and vulnerability. Within the framework of climate-induced Disaster Risk Reduction and Integrated Coastal Zone Management, disaster risk reduction measures that focus on early warning systems and preparedness planning are essential for reducing loss of life, limiting

Fig. 7.5 Adaptation strategies in CbA and DRR (Prepared by the authors)

damage to property and livelihoods, and strengthening the resilience of coastal communities. These measures represent preventive and forward-looking approaches that support long-term adaptation and sustainable coastal management. Early warning systems are a core element of disaster risk reduction in coastal areas. An effective early warning system goes beyond hazard forecasting and functions as an integrated process that combines risk knowledge, monitoring and forecasting, communication, and response capacity. In coastal settings, early warning systems are designed to detect and predict hazards such as cyclones, storm surges, high waves, and floods and to provide timely and useful information to people at risk. Improvements in weather observation, ocean monitoring, remote sensing, and forecasting techniques have increased the accuracy and lead time of warnings, allowing coastal communities to prepare more effectively for extreme events.

Risk knowledge forms the foundation of early warning systems in coastal zones. This involves identifying hazards, exposure, and vulnerability through hazard mapping, historical analysis, and community-level assessment. Clear knowledge of who is at risk, where they are located, and why they are vulnerable helps ensure that warnings are relevant and targeted. In coastal areas, risk knowledge must consider both sudden hazards such as cyclones and slow processes such as erosion and sea-level rise that can worsen disaster impacts. Combining scientific information with local experience improves the usefulness and acceptance of warning messages. Monitoring and forecasting are the technical components of early warning systems. Coastal early warning relies on meteorological stations, satellites, ocean buoys, tide gauges, and radar systems to track weather patterns and ocean conditions in real time. These observations feed into forecasting models that estimate hazard movement, intensity, and possible impacts. Although technical capacity has improved, effective monitoring and forecasting depend on continued investment, data sharing, and coordination among institutions, especially in coastal regions with limited resources.

Communication and dissemination link forecasting with community action. In coastal zones, warnings must reach diverse groups, including fishing communities at sea, informal settlements, and remote coastal villages. Use of multiple communication channels such as mobile alerts, radio broadcasts, sirens, community volunteers, and local institutions helps ensure that warnings reach all groups. Clarity of messages is equally important, as warnings should explain the nature of the hazard, expected impacts, and recommended actions. Trust in warning sources and familiarity with warning procedures strongly influence how communities respond. Preparedness planning supports early warning systems by ensuring that individuals, communities, and institutions are ready to act when warnings are issued. Preparedness planning includes the development and regular updating of contingency plans, evacuation routes, shelters, and response arrangements suited to coastal risk conditions. In coastal areas, preparedness must address challenges such as rapid flooding, limited evacuation time, and access constraints. Well-planned preparedness measures help convert early warning information into timely and effective action.

At the community level, preparedness planning focuses on awareness, training, and organization. Community disaster management committees, evacuation drills, and education programs help translate risk information into practical knowledge and skills. Measures such as household emergency plans, safe shelters, and protection of assets and livelihoods improve coping capacity and reduce confusion during disasters. Preparedness planning should be inclusive and responsive to the needs of women, children, older people, persons with disabilities, and livelihood groups such as fishers. At the institutional level, preparedness planning requires coordination among coastal management authorities, disaster management agencies, local governments, and service providers. Clear assignment of roles, standard procedures, and resource arrangements support effective response. Within Integrated Coastal Zone Management, preparedness planning should align with land-use planning, infrastructure development, and ecosystem management so that emergency actions and long-term risk reduction goals remain consistent. Key disaster risk reduction measures related to early warning and preparedness in coastal zones include hazard-specific early warning systems for cyclones, storm surges, floods, and tsunamis; risk-informed communication that delivers clear and timely messages through multiple channels; community-based preparedness through local committees, drills, and awareness programs; evacuation planning supported by safe routes and shelters accessible to vulnerable groups; and strong institutional coordination that links early warning and preparedness with coastal governance systems (Rajeswari et al., 2025; Šakić Trogrlić et al., 2022).

Early warning systems and preparedness planning face several challenges in coastal areas. These include technology gaps, weak last-mile communication, limited awareness of risk, and social and economic barriers to preparedness and evacuation. Climate change adds further complexity by increasing uncertainty and reducing available response time. Addressing these challenges requires continuous capacity building, strong community engagement, and better integration of early warning and preparedness within broader Disaster Risk Reduction and Integrated Coastal Zone Management frameworks. Therefore, the early warning systems and preparedness planning are essential disaster risk reduction measures for coastal zones affected by climate change. By supporting timely action and informed decision-making, they significantly reduce loss of life and damage from climate-induced hazards (Fig. 7.6). Within the framework of climate-induced Disaster Risk Reduction and Integrated Coastal Zone Management, strengthening early warning and preparedness contributes to proactive risk governance, improved resilience, and sustainable coastal development under changing climatic conditions.

7.5.2 Resilient Infrastructure and Coastal Defences

Coastal zones are highly exposed to climate-induced hazards such as cyclones, storm surges, coastal flooding, erosion, and sea-level rise. Rapid urbanization, concentration of critical infrastructure, and strong dependence on climate-sensitive livelihoods further increase disaster risk in these areas. Within the framework of climate-induced Disaster Risk Reduction and Integrated Coastal Zone Management, building resilient infrastructure and effective coastal defenses is a key element of disaster risk reduction. These measures aim to reduce exposure, limit damage during extreme events, and support faster recovery, while also contributing to long-term sustainability and adaptation to climate change. Resilient coastal infrastructure refers to physical structures that are planned, designed, built, and maintained to withstand climate-related hazards and changing environmental conditions. In coastal areas, this includes housing, roads, ports, embankments, drainage systems, water supply networks, energy systems, and critical facilities such as hospitals, shelters, and communication services. Climate change requires a move away from design practices based only on past conditions toward risk-informed approaches that consider stronger hazards, uncertainty, and long-term trends such as sea-level rise. Elevation of structures, use of flood- and corrosion-resistant materials, flexible design standards, and

Fig. 7.6 Early warning and preparedness in coastal zones disaster risk reduction (Prepared by the authors)

backup systems for essential services are important features of resilient coastal infrastructure.

Coastal defenses are built to protect shorelines, settlements, and assets from marine and coastal hazards. Common hard engineering measures include seawalls, revetments, groynes, breakwaters, and embankments, which are widely used to reduce wave energy, control erosion, and limit flooding. These structures can provide immediate protection, but their performance depends on careful design, regular maintenance, and consistency with natural coastal processes. Poorly planned defenses can shift risk to nearby areas, disturb sediment movement, or fail during extreme events that are becoming more intense under climate change. Experience from many coastal regions shows the value of combining hard structures with soft and hybrid approaches. Nature-based and hybrid coastal defenses link engineered structures with natural features such as dunes, mangroves, wetlands, and beaches. These approaches improve protection while supporting ecological functions and natural adaptation processes. From a Disaster Risk Reduction perspective, combining engineered measures with ecosystem-based solutions reduces reliance on a single line of defense and improves overall system performance. Building resilient infrastructure and coastal defenses involves more than technical design. Planning and governance play a major role in determining where infrastructure should be located and how much protection is needed. Risk-sensitive land-use planning helps guide development away from high-risk coastal areas. Limiting development in such zones is often more effective and less costly than constructing large protective structures. Where development cannot be avoided, design standards should be based on hazard assessments and climate projections. Integrated Coastal Zone Management provides a framework to coordinate infrastructure development, coastal defense planning, environmental protection, and disaster risk reduction.

Social and equity concerns are central to resilient infrastructure development in coastal zones. Poor and marginalized communities often live in the most hazard-prone locations and have limited access to safe housing and protective infrastructure. Disaster Risk Reduction strategies should prioritize inclusive investments that protect vulnerable populations and ensure access to safe shelters, evacuation routes, and basic services. Involving communities in planning, construction, and maintenance strengthens local ownership and improves the long-term effectiveness of coastal defenses. Therefore, the resilient infrastructure and coastal defenses are essential disaster risk reduction measures for coastal zones facing climate change. By reducing exposure, limiting damage, and supporting quicker recovery, these measures help safeguard lives, livelihoods, and development gains. Within the framework of climate-induced Disaster Risk Reduction and Integrated Coastal Zone Management, resilient infrastructure and coastal defenses should be guided by risk information, combined with ecosystem-based approaches, and supported by inclusive and adaptive planning to achieve sustainable and resilient coastal development.

7.5.3 Disaster Response and Recovery Strategies Specific to Coastal Disasters

Coastal disasters such as cyclones, storm surges, coastal flooding, tsunamis, and erosion-related failures present serious challenges because of their rapid onset, wide spatial spread, and combined land and sea impacts. Climate change has increased the intensity of these hazards and added uncertainty, leading to greater damage to infrastructure, livelihoods, and ecosystems. Within the framework of climate-induced Disaster Risk Reduction and Integrated Coastal Zone Management, disaster response and recovery strategies need to be designed for coastal conditions and linked with long-term resilience building and sustainable coastal governance. Response to coastal disasters focuses first on saving lives in situations where access is limited, communication systems are damaged, and environments are unsafe. Rapid evacuation, search and rescue, emergency medical support, and provision of safe shelters are essential, particularly where flooding and storm surges isolate settlements. Response operations must consider sea conditions, tidal patterns, and debris-filled waters that limit movement and complicate rescue efforts. Close coordination among coastal authorities, disaster management agencies, maritime services, and local governments supports timely action and effective use of resources.

Relief delivery and logistics require special planning during coastal disasters. Damage to ports, jetties, roads, and bridges can interrupt supply chains and delay assistance. Alternative transport options such as boats, amphibious vehicles, and air support are often needed. Early restoration of basic services including electricity, water supply, sanitation, and communication reduces health risks and supports emergency response. Clear command structures and reliable communication among agencies improve awareness and decision-making during response operations.

Coastal livelihoods such as fishing, aquaculture, tourism, and salt production are often severely disrupted during disasters. Response efforts should include immediate support for affected households through emergency cash assistance, food distribution, and protection of productive assets such as boats, nets, and tools. Community-based response systems, including trained volunteers and local committees, improve outreach, trust, and cultural relevance, especially in information sharing and shelter management. Special arrangements are required for vulnerable groups including women, children, older people, persons with disabilities, and migrant or informal coastal populations. Inclusive shelter design, targeted health services, and protection measures help reduce inequality and prevent additional social risks during emergencies. Recovery after coastal disasters involves more than physical rebuilding. It also includes restoring livelihoods, ecosystems, and governance systems. Early recovery activities focus on clearing debris, providing temporary housing, and restoring essential services. Medium- and long-term recovery emphasizes resilient reconstruction and economic recovery. Recovery planning in coastal areas should be guided by hazard assessments and climate projections to avoid rebuilding in unsafe locations. Building back better is especially important in coastal regions. Recovery measures include relocating or redesigning infrastructure in high-risk areas, raising structures, improving drainage, and using stronger construction materials and standards. Restoration of coastal ecosystems such as mangroves, dunes, and wetlands is equally important because these ecosystems provide natural protection and support local livelihoods. When ecosystem restoration reflects community needs, it speeds recovery and strengthens future risk reduction.

So, the recovery in coastal areas depends on alignment with Integrated Coastal Zone Management to ensure coordination across sectors and governance levels. Recovery investments should support risk-sensitive land-use planning, ecosystem protection, and participatory decision-making. Transparent processes, community engagement, and monitoring improve accountability and long-term outcomes. Coordination with climate adaptation strategies further supports resilience and sustainable coastal development. Disaster response and recovery strategies in coastal areas must be designed for local conditions, coordinated across institutions, and focused on future risk. Combining timely emergency response with resilient recovery guided by Integrated Coastal Zone Management principles reduces losses, restores livelihoods and ecosystems, and strengthens preparedness for climate-induced hazards. This approach supports safer, more resilient coastal communities under changing climatic conditions.

7.6 Coastal Resilience and Climate Change Mitigation

7.6.1 Coastal Ecosystems in Climate Change Mitigation

Coastal resilience refers to the ability of coastal systems, including communities, infrastructure, and ecosystems, to anticipate climate-related stress, absorb impacts, recover after disturbance, and adapt to changing conditions. In a changing climate, resilience in coastal zones is closely linked to disaster risk reduction, climate adaptation, and climate change mitigation. Coastal ecosystems play an important dual role by reducing exposure to climate-related hazards while also supporting mitigation through carbon storage. Within the framework of climate-induced Disaster Risk Reduction and Integrated Coastal Zone Management, strengthening the mitigation role of coastal ecosystems is essential for resilient and sustainable coastal development. Coastal ecosystems such as mangroves, salt marshes, seagrass meadows, tidal flats, and coastal wetlands function as highly effective natural carbon sinks. These ecosystems capture atmospheric carbon dioxide and store it in vegetation and sediments over long periods. Coastal wetlands can continue storing carbon for centuries if they remain intact and are able to respond naturally to sea-level rise. This capacity makes coastal ecosystems an important nature-based option for climate change mitigation. Mangrove forests are especially significant because of their high productivity and large carbon storage both above and below the ground. Carbon stored in mangrove sediments often exceeds that found in many terrestrial forests. Seagrass meadows also store substantial amounts of carbon through dense root systems that trap organic matter and stabilize sediments. Salt marshes contribute through continuous sediment accumulation. Protection and restoration of these ecosystems therefore support climate mitigation while also contributing to adaptation and disaster risk reduction.

The mitigation role of coastal ecosystems is closely linked with coastal resilience. Healthy ecosystems reduce the intensity of climate-related hazards while providing carbon storage benefits. Mangroves and salt marshes reduce wave energy, lower storm surge impacts, and stabilize shorelines. Seagrasses and coral reefs help reduce erosion and improve water quality. By lowering hazard impacts, these ecosystems reduce disaster losses and recovery costs, strengthening overall coastal resilience. Human pressure and climate change increasingly threaten the mitigation potential of coastal ecosystems. Coastal development, land reclamation, expansion of aquaculture, pollution, and changes in sediment flow have led to widespread degradation of blue carbon ecosystems. When these systems are damaged or destroyed, stored carbon is released into the atmosphere, turning natural sinks into sources of emissions. This process weakens mitigation efforts and increases vulnerability to coastal hazards, creating a cycle of rising risk and declining resilience. Integrated Coastal Zone Management offers an effective platform to protect and enhance the mitigation role of coastal ecosystems. Coordinated planning across sectors and governance levels helps align ecosystem conservation with development, disaster risk reduction, and climate policies. Including carbon storage considerations in coastal planning, land-use regulation, and environmental assessments supports ecosystem protection and guides restoration efforts. Through Integrated Coastal

Zone Management, ecosystem-based approaches become part of routine coastal governance rather than isolated actions.

Coastal ecosystems support mitigation and resilience in several ways. They store carbon in vegetation and sediments over long periods, prevent emissions by avoiding ecosystem loss, reduce disaster risk by weakening storm surges and erosion, support livelihoods through fisheries and tourism, and adapt to sea-level rise when space and sediment supply are available. Restoring degraded coastal ecosystems provides a strong link between mitigation, adaptation, and disaster risk reduction. Mangrove planting, seagrass restoration, and wetland rehabilitation improve carbon storage while restoring protective ecosystem services. Community involvement and supportive governance strengthen local stewardship and social resilience, improving the long-term success of restoration efforts. Several challenges limit the integration of coastal ecosystems into climate mitigation strategies. Awareness of carbon storage benefits remains limited, ecosystem services are often undervalued, land ownership conflicts persist, and institutional coordination is weak. Climate impacts such as sea-level rise and temperature stress may also affect ecosystem performance when adaptive space is restricted. Addressing these challenges requires science-based planning, supportive policies, and long-term investment through Integrated Coastal Zone Management and climate governance systems. Therefore, the coastal ecosystems are central to both climate mitigation and coastal resilience. Their capacity to store carbon, reduce disaster risk, and support livelihoods makes them essential components of integrated coastal strategies. Within the framework of climate-induced Disaster Risk Reduction and Integrated Coastal Zone Management, protecting and restoring coastal ecosystems offers a practical and adaptive pathway to address climate change, disaster risk, and sustainable development together (Fig. 7.7).

7.6.2 Sustainable Land Use and Coastal Development Practices

Coastal zones are among the most heavily used and rapidly changing regions, supporting settlements, ports, tourism, fisheries, industry, and

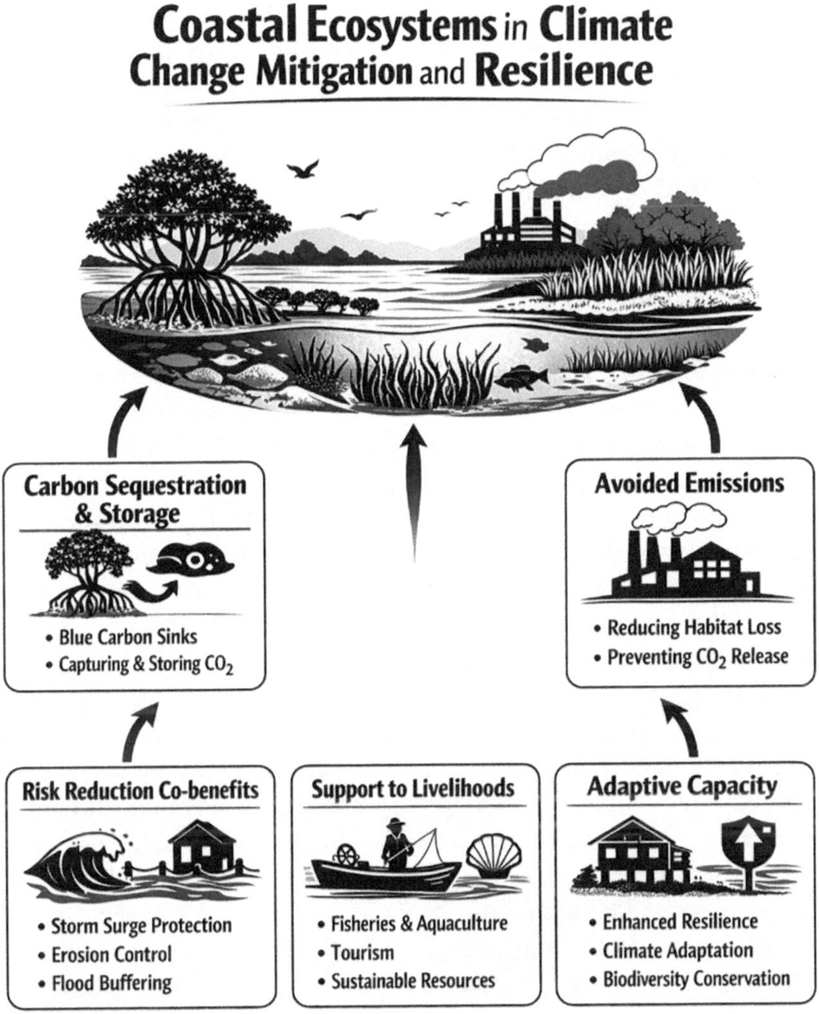

Fig. 7.7 Coastal ecosystems in climate change mitigation and resilience (Prepared by the authors)

critical infrastructure. These areas also face high exposure to climate-related hazards such as sea-level rise, storm surges, coastal flooding, erosion, and saline intrusion. Unsustainable land use and poorly planned development have increased disaster risk by expanding exposure and weakening natural protection systems. Within the framework of climate-induced Disaster Risk Reduction and Integrated Coastal Zone Management, sustainable land use and coastal development practices are essential for reducing vulnerability and supporting long-term environmental and economic stability. Sustainable land use in coastal areas involves managing land in ways that balance development needs with environmental limits and disaster risk considerations. This approach emphasizes alignment between land-use decisions, coastal processes, ecosystem functions, and climate projections. Development that focuses only on short-term economic gains often leads to repeated disaster losses and long-term environmental damage. Climate change has increased uncertainty and hazard intensity, making precautionary and long-term planning approaches increasingly important.

Risk-sensitive planning is a core element of sustainable coastal land use. Hazard and vulnerability assessments guide decisions on where development should occur and where it should be limited. Coastal zoning and setback regulations help restrict development in flood-prone areas, erosion zones, and low-lying land exposed to sea-level rise. By reducing exposure, these measures lower the need for costly protection structures and repeated emergency response. Avoiding risk through land-use planning is often more effective than managing hazards after development has occurred. Sustainable coastal development also depends on conserving ecosystems that provide natural protection and regulatory services. Mangroves, dunes, wetlands, beaches, and coastal forests act as buffers against waves and flooding while supporting biodiversity and livelihoods. Development that damages these ecosystems reduces both environmental sustainability and disaster resilience. Integrating ecosystem conservation into land-use planning through protected areas, buffer zones, and nature-based approaches strengthens adaptive capacity and supports disaster risk reduction. Urban growth and infrastructure development pose major challenges in coastal areas. Rapid and unplanned expansion has increased population density in high-risk locations and placed pressure on land and services. Sustainable coastal urban development promotes planned growth, resilient infrastructure, and adequate drainage, sanitation, and transport systems. Climate-resilient building codes, elevation standards, and infrastructure design help ensure safety under future hazard conditions. Consideration of cumulative impacts is essential, as multiple projects can collectively increase risk.

On the other hand, governance is central to sustainable land use and coastal development. Integrated planning across land use, disaster management, environmental protection, and economic development reduces conflict and risk. Integrated Coastal Zone Management provides a framework to coordinate these sectors and ensure consistent policies. Public participation strengthens fairness and acceptance of land-use decisions, especially when restrictions, relocation, or trade-offs are required. So, the sustainable land use and coastal development practices include risk-sensitive planning that incorporates climate projections, zoning and setback regulations in high-risk and sensitive areas, ecosystem-based planning that protects natural buffers, climate-resilient urban design, managed retreat in areas of unavoidable risk, and participatory planning that reflects local needs and priorities. Several challenges affect the implementation of sustainable land use in coastal regions. Economic pressure, land scarcity, weak enforcement, and rapid informal development often undermine planning efforts. Climate change adds long-term uncertainty that extends beyond standard planning timeframes. Addressing these issues requires strong institutions, political commitment, and continued use of scientific information in decision-making. Therefore, the sustainable land use and coastal development practices are fundamental to reducing disaster risk and building resilience in coastal areas affected by climate change. Aligning development with coastal processes, ecosystem functions, and hazard conditions helps prevent new risks while supporting livelihoods and economic growth. Within the framework of climate-induced Disaster Risk Reduction and Integrated Coastal Zone Management, sustainable land use functions as a preventive and transformative approach that protects both coastal communities and the environments on which they depend.

7.7 Policy and International Cooperation

7.7.1 International Agreements and Frameworks

Climate-induced risks in coastal zones extend beyond administrative and national boundaries because oceans, coastal ecosystems, and climate processes are shared across regions. Disaster Risk Reduction and Integrated Coastal Zone Management therefore depend on coordinated policy frameworks and strong international cooperation. International agreements provide shared principles, goals, and mechanisms that guide national and local action. These frameworks support knowledge exchange, mobilize financial resources, and help align coastal risk reduction with climate adaptation and sustainable development goals.

Global climate policy frameworks shape responses to the causes of climate-related coastal risks. The Paris Agreement sets a common commitment to limit global temperature rise and strengthen adaptation and resilience. Its adaptation focus is especially important for coastal areas, as it emphasizes adaptive capacity, resilience building, and reduced vulnerability. Nationally Determined Contributions and National Adaptation Plans increasingly address coastal priorities such as sea-level rise, ecosystem protection, resilient infrastructure, and community-based adaptation. These policy instruments create opportunities to link Disaster Risk Reduction with Integrated Coastal Zone Management at the national level. International disaster policy is guided by the Sendai Framework for Disaster Risk Reduction, which promotes a shift from disaster response toward prevention and risk governance. Climate change is recognized as a major driver of disaster risk, with emphasis placed on risk assessment, governance strengthening, investment in resilience, and preparedness for response and recovery. In coastal areas, these priorities support actions such as multi-hazard risk assessment, early warning systems, risk-sensitive land-use planning, and resilient infrastructure, all of which are core components of Integrated Coastal Zone Management. Global development agendas further support the integration of climate action, disaster risk reduction, and coastal management. The Sustainable Development Goals provide a broad framework linking climate resilience, ocean protection, and inclusive development. Goals related to climate action, sustainable cities, marine ecosystems, and resilient infrastructure highlight the need for integrated coastal policies that reduce disaster risk while supporting livelihoods and ecosystem health. These linkages encourage coordination across sectors and governance levels and help align national coastal strategies with global development commitments.

International cooperation also plays an important role in addressing shared coastal challenges through regional and multilateral arrangements. Regional seas agreements, coastal resilience initiatives, and transboundary risk management platforms enable countries to coordinate responses to sea-level rise, cyclones, erosion, and ecosystem degradation. Cooperation at this scale supports data sharing, common standards, joint monitoring, and coordinated early warning systems, which are especially important for shared coastlines and river deltas. International agreements and cooperative mechanisms support coastal Disaster Risk Reduction and Integrated Coastal Zone Management in several ways. They promote policy alignment across climate adaptation, disaster risk reduction, and coastal planning. They provide guidance through shared principles, targets, and best practices. They help mobilize finance and support access to resilient technologies in vulnerable coastal regions. They encourage knowledge exchange through research collaboration and capacity building. They also emphasize equity by focusing support on countries and communities facing the highest coastal risks. Challenges remain in translating international commitments into national and local action. Limited technical capacity, fragmented institutions, and competing development priorities often slow implementation. Strong governance systems, inclusive participation, and long-term political commitment are needed to address these gaps. Integrated Coastal Zone Management offers a practical pathway by applying international principles through spatial planning, ecosystem-based approaches, and participatory decision-making.

Therefore, the international agreements and cooperative frameworks play an important role in shaping climate-induced disaster risk reduction and integrated coastal management. By offering shared direction, resources, and accountability, they allow countries to address coastal risks collectively while adapting actions to local contexts. Stronger alignment between global frameworks and Integrated Coastal Zone Management is essential for building resilient, sustainable, and equitable coastal futures under changing climate conditions.

7.7.2 Role of National and Local Policies in Resilience

Resilience to climate-induced hazards in coastal zones depends strongly on the quality and coordination of national and local policies. Climate change has increased risks such as sea-level rise, storm surges, flooding, erosion, and saline intrusion, making isolated or sector-based responses insufficient. Within the framework of climate-induced Disaster Risk Reduction and Integrated Coastal Zone Management, policies at national and local levels create the conditions through which resilience efforts are planned, financed, coordinated, and sustained. Clear policy direction helps translate scientific knowledge and global commitments into practical actions that reduce vulnerability and strengthen adaptive capacity across scales.

National policies provide strategic guidance for resilience efforts. They define long-term goals, legal mandates, and institutional arrangements related to disaster risk reduction, climate adaptation, and coastal management. Through legislation, strategies, and action plans, governments can integrate resilience into development planning, infrastructure investment, land-use regulation, and environmental management. In coastal areas, national policies are particularly important for addressing large-scale and cross-sector concerns such as coastal zoning, protection of key ecosystems, climate-resilient infrastructure standards, and inclusion of disaster risk considerations in economic planning. Shared standards and priorities at the national level help limit fragmentation and promote consistency across regions. Local policies and governance systems play an equally important role because resilience is shaped at the community and landscape level. Local governments are often responsible for early response during climate-related disasters and manage land-use planning, building regulation, service delivery, and community engagement. Local policies allow national frameworks to be applied in ways that reflect local hazard conditions, social and economic realities, and cultural practices. In coastal zones, planning tools such as development plans, zoning rules, and building codes directly influence exposure and vulnerability by shaping settlement patterns and investment decisions. Coordination between national and local policies strongly influences the success of resilience efforts. Alignment across governance levels allows national strategies to be carried out through local action while local experience informs policy improvement at higher levels. Integrated Coastal Zone Management supports this coordination by linking national objectives with local implementation through participatory planning, spatial tools, and adaptive management. When policies are aligned across levels, resilience measures tend to be more coherent, efficient, and socially inclusive.

Policy frameworks also support the integration of disaster risk reduction and climate adaptation into everyday decision-making. Policies can require risk and vulnerability assessments, include climate considerations in development approvals, and promote ecosystem-based approaches. They also guide allocation of financial resources for resilience investments such as early warning systems, resilient infrastructure, ecosystem restoration, and capacity development. By promoting prevention and preparedness, policies shift attention away from short-term relief toward long-term risk reduction and adaptation. National and local policies contribute to resilience-building in several ways. They integrate disaster risk reduction and climate adaptation into development and sector planning. They guide risk-sensitive land use through zoning, setback rules, and building standards. They clarify institutional roles and responsibilities across agencies and governance levels. They support public investment and incentives for resilient infrastructure, livelihoods, and ecosystem protection. They also promote inclusive decision-making and address differences in vulnerability among social groups. Implementation challenges remain across many coastal regions. Limited technical and financial capacity, weak enforcement, political pressure, and competing development interests often reduce policy effectiveness. Rapid urban growth and informal development in coastal areas frequently outpace regulatory systems. Climate change adds further difficulty by introducing uncertainty and long-term risks that extend beyond standard planning timeframes. Strengthening institutions, improving capacity, and encouraging continuous learning are essential to address these issues.

Therefore, the national and local policies are central to resilience in coastal zones under climate change. Strategic direction, regulatory tools, and institutional coordination create the foundation for disaster risk reduction and adaptation. Within the framework of climate-induced Disaster Risk Reduction and Integrated Coastal Zone Management, well-aligned policies help turn resilience goals into practical action and support coastal development pathways that are safer, more inclusive, and responsive to a changing climate.

7.7.3 Importance of International Collaboration and Knowledge

Climate-induced risks in coastal zones extend across national boundaries because oceans, atmospheric processes, ecosystems, and socio-economic systems are shared. Sea-level rise, cyclones, storm surges, coastal erosion, and ecosystem degradation affect coastal regions regardless of political borders. National responses on their own are therefore limited in scope. Within the framework of Climate-induced Disaster Risk Reduction and Integrated Coastal Zone Management, international collaboration and shared knowledge play an important role in understanding, managing, and reducing coastal disaster risks under climate change.

International collaboration provides shared direction and guidance for responding to climate-related coastal risks. Global frameworks such as the Paris Agreement highlight the importance of strengthening adaptive capacity, resilience, and reduced vulnerability, with special attention to coastal and marine systems. The Sendai Framework for Disaster Risk Reduction also emphasizes prevention, preparedness, and resilience, while recognizing climate change as a major driver of disaster risk. These frameworks help align national coastal policies with common global priorities and support coordination between climate adaptation, disaster risk reduction, and integrated coastal management. Knowledge exchange forms a central part of international cooperation. Climate change introduces uncertainty, complex interactions, and changing risk patterns that require continuous learning. International scientific collaboration improves understanding of coastal processes, hazard behavior, ecosystem services, and social vulnerability through shared research, data, and methods. Progress in climate modelling, sea-level projections, early warning systems, and ecosystem-based approaches often results from joint international efforts and is later adapted to local coastal conditions. Knowledge sharing reduces duplication, improves risk assessment, and supports informed decision-making within Integrated Coastal Zone Management.

International cooperation also strengthens capacity and supports equity in resilience efforts. Many coastal regions facing high climate risk have limited technical skills, financial resources, and institutional capacity. Multilateral cooperation helps improve access to training, technology transfer, and climate finance, supporting implementation of Disaster Risk Reduction and climate adaptation measures. This approach reflects shared responsibility and differences in capacity, principles also reflected in global development agendas such as the Sustainable Development Goals, which link climate action, ocean protection, and inclusive development.

Regional and transboundary cooperation plays an important role in managing shared coastal ecosystems, river deltas, and early warning systems. Joint monitoring, coordinated planning, and common standards improve preparedness and help reduce risks that spread across borders. Cooperation of this kind supports trust and collective action, which are necessary for addressing long-term climate challenges in coastal areas. In summary, international collaboration and knowledge sharing are central to climate-induced disaster risk reduction and integrated coastal zone management. Shared frameworks, scientific cooperation, capacity support, and attention to equity help coastal societies respond more effectively to climate risks. In a climate future marked by growing interconnection and uncertainty, collective learning and coordinated action remain essential for achieving resilient and sustainable coastal systems.

7.8 Conclusion and Arguments

This chapter is the progression from the core concepts of climate-related Disaster Risk Reduction to policy frameworks and international cooperation, highlighting the interconnected nature of managing climate-induced risks in coastal zones. Climate change has reshaped coastal hazard conditions by increasing the intensity of sudden disasters and accelerating slow processes such as sea-level rise and coastal erosion. These changes make it clear that Disaster Risk Reduction cannot remain a stand-alone or reactive activity. It needs to be embedded within Integrated Coastal Zone Management frameworks that connect risk reduction with climate adaptation, development planning, and environmental governance. Consideration of hazard identification, vulnerability assessment, geospatial tools, participatory approaches, ecosystem-based and community-based adaptation, resilient infrastructure, preparedness, response, and recovery points to the need for integrated and forward-looking strategies. These elements reinforce one another when applied in a coordinated manner. Governance plays a central role, as national and local policies translate global commitments into actions suited to local coastal contexts. International cooperation, shared knowledge, and capacity development further support this process by strengthening coordination and learning across regions. Coastal resilience emerges not only from technical measures but also from social, ecological, and institutional processes. Coordination across sectors, governance levels, and stakeholder groups is essential for responding to the diverse and changing risks faced by coastal zones. Integration of Disaster Risk Reduction with Integrated Coastal Zone Management provides a clear pathway to reduce vulnerability, protect ecosystems and livelihoods, and support sustainable coastal development in a changing climate. The chapter discussions lead to the following arguments:

(a) Climate change increases coastal risks, making integrated and preventive disaster risk reduction essential.
(b) Integrated Coastal Zone Management provides a common framework linking disaster risk reduction, climate adaptation, land-use planning, and ecosystem management.

(c) Hazard and vulnerability assessments, supported by geospatial tools, guide risk-informed planning and decision-making.
(d) Participatory, community-based, and incentive-based approaches improve equity, local ownership, and long-term effectiveness of risk reduction measures.
(e) Ecosystem-based adaptation and resilient infrastructure together reduce hazard impacts while supporting sustainability.
(f) Early warning systems, preparedness planning, and context-specific response and recovery reduce disaster losses.
(g) National and local policies mainstream resilience, while international cooperation strengthens coordination, knowledge sharing, and capacity building.

References

Abdou, I., Yousry, A., & Osman, T. (2025). A hybrid methodological approach for assessing nature-based solutions within urban coastal resilience planning in the Nile Delta. *Journal of Engineering and Applied Science, 72*, 274. https://doi.org/10.1186/s44147-025-00798-3

Aggrey, J. J., Ros-Tonen, M. A. F., & Asubonteng, K. O. (2021). Using participatory spatial tools to unravel community perceptions of land-use dynamics in a mine-expanding landscape in Ghana. *Environmental Management, 68*, 720–737. https://doi.org/10.1007/s00267-021-01494-7

Ahammad, R., Nandy, P., & Husnain, P. (2013). Unlocking ecosystem based adaptation opportunities in coastal Bangladesh. *Journal of Coastal Conservation, 17*, 833–840. https://doi.org/10.1007/s11852-013-0284-x

Ahlhorn, F. (2018). *Integrated coastal zone management: Status, challenges and prospects*. Springer Vieweg Wiesbaden. ISBN: 978-3-658-17050-9. https://doi.org/10.1007/978-3-658-17052-3

Anjita N. A., Makhdumi, W., GS, D., et al. (2025). Coastal vulnerability assessment of the Kasaragod Coast in Kerala, West Coast of India. *Anthropocene Coasts, 8*, 38. https://doi.org/10.1007/s44218-025-00096-8

Ballesteros, C., & Esteves, L. S. (2021). Integrated assessment of coastal exposure and social vulnerability to coastal hazards in East Africa. *Estuaries and Coasts, 44*, 2056–2072. https://doi.org/10.1007/s12237-021-00930-5

Basheer Ahammed, K. K., & Pandey, A. C. (2019). Coastal social vulnerability and risk analysis for cyclone hazard along the Andhra Pradesh, East Coast of India. *KN-Journal of Cartography and Geographic Information, 69*, 285–303. https://doi.org/10.1007/s42489-019-00029-9

Bera, M. K. (2025). Community-based nature-based solution to disasters in Sundarban Islands in India. In W. Leal Filho, G. J. Nagy, & D. Y. Ayal (Eds.), *Handbook of nature-based solutions to mitigation and adaptation to climate change*. Springer. https://doi.org/10.1007/978-3-030-98067-2_99-1

Busayo, E. T., & Kalumba, A. M. (2020). Coastal climate change adaptation and disaster risk reduction: A review of policy, programme and practice for sustainable planning outcomes. *Sustainability, 12*(16), 6450. https://doi.org/10.3390/su12166450

Cruickshank, A., Moncada, S., & Thomas, A. (2024). A systematic review of community-based volcanic risk reduction and adaptation to the compound and cascading impacts of climate change in volcanic SIDS. *International Journal of Disaster Risk Reduction, 114*, Article 104995. https://doi.org/10.1016/j.ijdrr.2024.104995

de Lima, A. D. S., Scherer, M. E. G., Gandra, T. B. R., & Bonetti, J. (2022). Exploring the contribution of climate change policies to integrated coastal zone management in Brazil. *Marine Policy, 143*, Article 105180. https://doi.org/10.1016/j.marpol.2022.105180

Ergashev, B. A. (2012). A theoretical framework for incorporating scenarios into operational risk modeling. *Journal of Financial Services Research, 41*, 145–161. https://doi.org/10.1007/s10693-011-0105-z

Farinós-Dasí, J., Pinazo-Dallenbach, P., Peiró Sánchez-Manjavacas, E., et al. (2025). Disaster risk management, climate change adaptation and the role of spatial and urban planning: Evidence from European case studies. *Natural Hazards, 121*, 23479–23512. https://doi.org/10.1007/s11069-024-06448-w

Frølich, N., & Caspersen, J. (2015). Institutional governance structures. In J. Huisman, H. de Boer, D. D. Dill, & M. Souto-Otero (Eds.), *The Palgrave international handbook of higher education policy and governance*. Palgrave Macmillan. https://doi.org/10.1007/978-1-137-45617-5_21

Haq, B. U., & Milliman, J. D. (1996). Coastal vulnerability: Hazards and strategies. In J. D. Milliman & B. U. Haq (Eds.), *Sea-level rise and coastal subsidence: Coastal systems and continental margins* (Vol. 2). Springer. https://doi.org/10.1007/978-94-015-8719-8_20

Hassani, B. K. (2016). *Scenario analysis in risk management*. Springer Cham. ISBN: 978-3-319-25054-0. https://doi.org/10.1007/978-3-319-25056-4

Hietala, R., Ijäs, A., Pikner, T., et al. (2021). Data integration and participatory process in developing integrated coastal zone management (ICZM) in the northern Baltic Sea. *Journal of Coastal Conservation, 25*, 47. https://doi.org/10.1007/s11852-021-00833-4

Hill, B., Jiang, J., Bosher, L., & Liang, Q. (2025). Ecosystems-based approaches for disaster risk reduction. In R. Shaw (Eds.), *Encyclopedia of disaster risk reduction*. Springer. https://doi.org/10.1007/978-981-97-4547-0_216-1

Hoque, M., Pradhan, B., Ahmed, N., & Alamri, A. (2021). Drought vulnerability assessment using geospatial techniques in southern Queensland, Australia. *Sensors, 21*(20), 6896. https://doi.org/10.3390/s21206896

Ibrahim, H. S., & Shaw, D. (2012). Assessing progress toward integrated coastal zone management: Some lessons from Egypt. *Ocean & Coastal Management, 58*, 26–35. https://doi.org/10.1016/j.ocecoaman.2011.12.002

Ineza, W. B., Izere, E. C., Sonnenschein, R., Kukunda, C. B., Tumwebaze, F., Shumbusho, R., & Razak, K. A. (2025). Application of geospatial and ICT technologies for landslide disaster risk reduction in Rwanda. *Progress in Disaster Science*, 100466. https://doi.org/10.1016/j.pdisas.2025.100466

Jiang, L., Yang, T., Wang, X., Yu, J., Liu, J., & Zhang, K. (2023). Research on integrated coastal zone management from past to the future: A bibliometric analysis. *Frontiers in Marine Science, 10*, 1201811. https://doi.org/10.3389/fmars.2023.1201811

Kannen, A. (2002). Integrated Coastal Zone Management (ICZM) in the context of spatial planning. In G. Schernewski & U. Schiewer (Eds.), *Baltic coastal ecosystems*. Central and Eastern European Development Studies. Springer. https://doi.org/10.1007/978-3-662-04769-9_15

Kaur, H., Gupta, S., Parkash, S., & Thapa, R. (2018). Application of geospatial technologies for multi-hazard mapping and characterization of associated risk at local scale. *Annals of GIS, 24*(1), 33–46. https://doi.org/10.1080/19475683.2018.1424739

Kautsary, J., Andolia, T., & Ilma, D. (2025). Climate change disaster risk reduction in coastal settlements of Sayung Demak, with nature-based solutions. In *IOP Conference Series: Earth and Environmental Science* (Vol. 1543, No. 1, p. 012041). IOP Publishing. https://doi.org/10.1088/1755-1315/1543/1/012041

Kuhl, L., Van Maanen, K., & Scyphers, S. (2020). An analysis of UNFCCC-financed coastal adaptation projects: Assessing patterns of project design and contributions to adaptive capacity. *World Development, 127*, Article 104748. https://doi.org/10.1016/j.worlddev.2019.104748

Lanlan, J., Sarker, M. N. I., Ali, I., et al. (2024). Vulnerability and resilience in the context of natural hazards: A critical conceptual analysis. *Environment, Development and Sustainability, 26*, 19069–19092. https://doi.org/10.1007/s10668-023-03440-5

Li, Y., Zheng, W., Dong, T., Yang, Y., Wang, S., Kang, Y., Liu, Y., Chen, Y., & Liu, D. (2026). The evolution and development of integrated coastal zone management in China since the 1950s. *Coastal Management*, 1–24. https://doi.org/10.1080/08920753.2026.2623357

Lopes, N. D. R., Li, T., Matomela, N., et al. (2022). Coastal vulnerability assessment based on multi-hazards and bio-geophysical parameters. Case study - northwestern coastline of Guinea-Bissau. *Natural Hazards, 114*, 989–1013. https://doi.org/10.1007/s11069-022-05420-w

Mal, S., Singh, R. B., & Huggel, C. (Eds.). (2018). *Climate change, extreme events and disaster risk reduction: Towards sustainable development goals.* Springer Cham. ISBN: 978-3-319-56468-5. https://doi.org/10.1007/978-3-319-56469-2

Malakar, K. D., Kumar, M., Anand, S., & Kuzur, G. (2023a). Climate vulnerability and socio-ecological transformation. In *Climate change and socio-ecological transformation. Advances in geographical and environmental sciences.* Springer. https://doi.org/10.1007/978-981-99-4390-6_4

Malakar, K. D., Kumar, M., Anand, S., & Kuzur, G. (2023b). Disaster, policy instruments, and sustainability. In *Climate change and socio-ecological transformation. Advances in geographical and environmental sciences.* Springer. https://doi.org/10.1007/978-981-99-4390-6_6

Malakar, K. D., Kumar, M., Anand, S., & Kuzur, G. (2023c). Climate change and socio-ecological transformation: Applications and field narratives. In *Climate change and socio-ecological transformation. Advances in geographical and environmental sciences.* Springer. https://doi.org/10.1007/978-981-99-4390-6_7

Malakar, K. D., & Roy, S. (2024). Understanding of participatory GIS: Concepts and techniques. In *Mapping geospatial citizenship.* SpringerBriefs in GIS. Springer. https://doi.org/10.1007/978-3-031-63107-8_2

Mall, R. K., Srivastava, R. K., Banerjee, T., et al. (2019). Disaster risk reduction including climate change adaptation over South Asia: Challenges and ways forward. *International Journal of Disaster Risk Science, 10*, 14–27. https://doi.org/10.1007/s13753-018-0210-9

Miao, X., Yin, I. Z., Zhang, H., et al. (2025). Participatory research methods and tools for promoting children's psychological well-being: A systematic review. *Applied Research Quality Life, 20*, 1629–1664. https://doi.org/10.1007/s11482-025-10490-6

Nandy, P., Ahammad, R., Alam, M., & Islam, A. (2013). Coastal ecosystem based adaptation: Bangladesh experience. In R. Shaw, F. Mallick, & A. Islam (Eds.), *Climate change adaptation actions in Bangladesh*. Disaster Risk Reduction. Springer. https://doi.org/10.1007/978-4-431-54249-0_15

Panda, G. K., Chatterjee, U., Mishra, M., et al. (2022). Vulnerability assessment of environmental degradation due to coastal hazards using geospatial technology—a study of Puri coast along the Mahanadi Delta, India. *Arabian Journal of Geosciences, 15*, 1322. https://doi.org/10.1007/s12517-022-10578-5

Peng, Y., Welden, N. A., & Renaud, F. G. (2025). Ecosystem-based adaptation strategies to multi-hazard risk reduction and policy implications in the Pearl River and Yangtze River deltas, China. *International Journal of Disaster Risk Reduction, 116*, Article 105053. https://doi.org/10.1016/j.ijdrr.2024.105053

Rajeswari, D., Jayaraman, R., Celina, D., et al. (2025). Community change agents and disaster preparedness among women in coastal areas. *BMC Emergency Medicine, 25*, 234. https://doi.org/10.1186/s12873-025-01393-8

Ramieri, E., Bocci, M., & Markovic, M. (2019). Linking integrated coastal zone management to maritime spatial planning: The mediterranean experience. In J. Zaucha & K. Gee (Eds.), *Maritime spatial planning*. Palgrave Macmillan. https://doi.org/10.1007/978-3-319-98696-8_12

Rockloff, S. F., & Lockie, S. (2004). Participatory tools for coastal zone management: Use of stakeholder analysis and social mapping in Australia. *Journal of Coastal Conservation, 10*, 81–92. https://doi.org/10.1652/1400-0350(2004)010[0081:PTFCZM]2.0.CO;2

Sajjad, H., Rahaman, M. H., Masroor, M., Sharma, Y., Sharma, A., & Saha, T. K. (2024). Vulnerability assessment of forest ecosystem based on exposure, sensitivity and adaptive capacity in the Valmiki Tiger Reserve, India: A geospatial analysis. *Ecological Informatics, 80*, Article 102494. https://doi.org/10.1016/j.ecoinf.2024.102494

Šakić Trogrlić, R., van den Homberg, M., Budimir, M., McQuistan, C., Sneddon, A., & Golding, B. (2022). Early warning systems and their role in disaster risk reduction. In B. Golding (Eds.), *Towards the "perfect" weather warning*. Springer. https://doi.org/10.1007/978-3-030-98989-7_2

Shafapourtehrany, M., Batur, M., Shabani, F., Pradhan, B., Kalantar, B., & Özener, H. (2023). A comprehensive review of geospatial technology applications in earthquake preparedness, emergency management, and damage assessment. *Remote Sensing, 15*(7), 1939. https://doi.org/10.3390/rs1507 1939

Singh, A., & Zommers, Z. (Eds.). (2014). *Reducing disaster: Early warning systems for climate change*. Springer Dordrecht. ISBN: 978-94-017-8597-6. https://doi.org/10.1007/978-94-017-8598-3

Stead, S. M. (2005). A comparative analysis of two forms of stakeholder participation in European aquaculture governance: Self-regulation and integrated coastal zone management. In T. S. Gray (Eds.), *Participation in fisheries governance*. Reviews: Methods and Technologies in Fish Biology and Fisheries (Vol. 4). Springer. https://doi.org/10.1007/1-4020-3778-3_10

Tanim, A. H., Goharian, E., & Moradkhani, H. (2022). Integrated socio-environmental vulnerability assessment of coastal hazards using data-driven and multi-criteria analysis approaches. *Scientific Reports, 12*(1), 11625. https://doi.org/10.1038/s41598-022-15237-z

Tasnuva, A., Bari, Q. H., Islam, A. R. M. T., & Hassan, K. M. (2024). Developing a disaster risk index for coastal communities in southwest Bangladesh: Shifting from data-driven models to holistic approaches. *Ecological Indicators, 166*, Article 112381. https://doi.org/10.1016/j.ecolind.2024.112381

Veloso-Gomes, F., Barroco, A., Ramos Pereira, A., et al. (2008). Basis for a national strategy for integrated coastal zone management—in Portugal. *Journal of Coastal Conservation, 12*, 3–9. https://doi.org/10.1007/s11852-008-0017-8

Weday, M. A., Tabor, K. W., & Gemeda, D. O. (2023). Flood hazards and risk mapping using geospatial technologies in Jimma City, southwestern Ethiopia. *Heliyon, 9*(4). https://doi.org/10.1016/j.heliyon.2023.e14617

Wen, J., Wan, C., Ye, Q., et al. (2023). Disaster risk reduction, climate change adaptation and their linkages with sustainable development over the past 30 years: A review. *International Journal of Disaster Risk Science, 14*, 1–13. https://doi.org/10.1007/s13753-023-00472-3

Yadav, N., Thalakiya, R., Srivastava, S., et al. (2026). Scoping review of early warning systems in improving disaster risk outcomes: Perspective from evolution to usage in Southeast Asian countries. *BMC Public Health*. https://doi.org/10.1186/s12889-026-26292-w

Part III

Risk Reduction Practices and Socio-ecological Sustainability: Policy Frameworks and Future Challenges

8

Policy Frameworks and Governance: Enhancing Coastal Resilience

Coastal regions are some of the most active and valuable areas on Earth. They support diverse ecosystems large populations and many economic activities. At the same time these areas are facing growing pressure from climate change environmental degradation and rapid development. Rising sea levels stronger storms coastal erosion and unplanned growth are increasing the exposure of coastal regions to disasters and long-term risks. This chapter is written within this context of rising vulnerability and uncertainty. The chapter presents coastal resilience as more than a technical response to hazards. It explains resilience as a process shaped by policies institutions and governance systems that guide how coastal risks are managed. The discussion introduces key ideas related to coastal resilience and then explores policy tools, legal and regulatory frameworks, institutional arrangements, financing mechanisms, and participatory approaches. Attention is given to the need for coordination across local regional national and international levels and to the value of cooperation among different actors. The chapter also highlights the importance of monitoring, evaluation, and learning so that policies and governance systems can adjust to changing coastal conditions. It aims to support informed decision-making and provide practical insight

K. D. Malakar et al., *Coastal Ecologies at Risk*, Palgrave Studies in Environmental Sustainability, https://doi.org/10.1007/978-3-032-26081-9_8

for strengthening coastal governance in the face of climate change and development pressures.

Key Points of the Chapters

- Describes coastal resilience as an integrated governance process.
- Explains the need for coordinated coastal policy frameworks.
- Examines the role of community participation and multi-level governance.

Highlights the importance of international cooperation and capacity building, as well as adaptive governance through monitoring and learning.

8.1 Coastal Resilience and its Importance

Coastal resilience refers to the ability of coastal social and ecological systems to prepare for, respond to, and recover from disturbances while continuing to function and retain their core characteristics. In relation to policy frameworks and governance coastal resilience is not limited to emergency response or short-term recovery. It focuses on long-term planning coordination among institutions and the ability to adjust decisions as conditions change. This approach connects physical environmental social economic and governance aspects to address the complex risks faced by coastal areas particularly those linked to climate change sea level rise extreme weather events and increasing human activities. Coastal resilience is closely linked to systems thinking which views coastal zones as connected land and sea areas shaped by natural processes and human actions. The goal is not to fully control or eliminate hazards but to strengthen the capacity to adapt learn and adjust. Disturbances such as cyclones flooding erosion and salinity intrusion are common in coastal regions. The response of communities and institutions plays a key role in reducing long term impacts. This calls for a shift away from reactive and sector focused management toward integrated and forward-looking policy approaches.

The growing importance of coastal resilience is linked to the concentration of people infrastructure and economic activities along coastlines. Coastal areas support livelihoods such as fisheries tourism ports and agriculture and include sensitive ecosystems like mangroves wetlands and coral reefs. Building resilience helps protect these resources reduce disaster related losses and support sustainable development. From a governance perspective it also supports social equity by giving attention to vulnerable groups strengthening local capacities and promoting fair access to adaptation benefits. Several principles guide effective coastal resilience governance. Integration across sectors such as environment disaster management, climate adaptation, urban planning, and marine management is essential to avoid fragmented actions. Participation of local communities' indigenous groups civil society and the private sector improve relevance trust and shared responsibility. Flexibility and learning are equally important as policies must respond to uncertainty and change through monitoring, evaluation, and regular adjustment. Nature-based approaches support resilience by reducing risks while sustaining ecosystems and livelihoods. Coordination across local national and international levels supported by clear institutions adequate financing and capacity building is necessary to ensure long-term and inclusive coastal resilience (Almutairi et al., 2020; Nichols et al., 2025; Singh, 2020).

Coastal resilience has become an important focus in governance as coastal regions face increasing pressure from climate change and development. It refers to the ability of coastal systems, including ecosystems, communities, infrastructure, and institutions, to cope with climate-related stress and recover from shocks while continuing to support long-term development. Within policy and governance, coastal resilience provides a way to link climate adaptation, disaster risk reduction, and sustainable development so that actions taken to manage immediate risks do not weaken social, economic, or environmental conditions over time. Coastal zones are especially exposed to climate-related hazards such as sea level rise, warmer ocean temperatures, coastal erosion, saltwater intrusion, and stronger storms. These changes affect not only physical structures but also livelihoods, ecosystems, and social systems. Activities such as fisheries, tourism, and coastal agriculture are increasingly at risk as environmental conditions change. Coastal resilience responds to these challenges by encouraging adaptive actions that reduce vulnerability and strengthen the capacity of both communities and institutions to deal with uncertainty. This requires policies that are informed by climate science, risk assessment, and long-term planning rather than short-term responses.

The connection between coastal resilience and sustainable development is strong. Development in coastal areas depends on healthy ecosystems that provide protection, food, income, and other essential services. Climate change threatens these benefits and can undermine development progress if risks are not addressed. Governance approaches that focus on resilience help protect natural resources, support livelihoods, and ensure that development efforts remain effective under changing climate conditions. In this sense, resilience is a basic requirement for sustainable coastal development rather than an additional policy demand. Effective governance for coastal resilience depends on policies that connect climate adaptation with development planning. Fragmented and sector-focused approaches often fail to capture how risks build up over time and across different areas. More coordinated policy frameworks encourage long-term investment in adaptive infrastructure, ecosystem-based solutions, and social capacity building. Equity is also a key concern since climate impacts often affect vulnerable coastal groups more severely. Inclusive governance that supports participation and local knowledge strengthens resilience while also contributing to fair and sustainable development.

8.2 Policy and Governance Approaches for Coastal Resilience

8.2.1 National Policies and Strategies

National policies and strategies play a central role in strengthening coastal resilience by providing the legal institutional and financial basis for action across different sectors and levels. At the national scale these policies connect international climate commitments with the realities faced by coastal regions. They turn global goals on climate adaptation disaster risk reduction and sustainable development into national priorities rules and programs that reflect local coastal conditions. At the policy level coastal resilience is now commonly included in national climate policies disaster management frameworks and coastal or marine governance plans. This shows a clear move away from short-term and hazard-focused responses toward long-term planning that accounts for future risks. National adaptation plans climate strategies and development visions increasingly recognize coastal areas as both vulnerable and valuable spaces that require targeted investment. Integrating coastal resilience into national policies helps improve coordination, reduce overlaps among sectors, and align development planning with climate risks.

National strategies also help define institutional roles and coordination mechanisms. Coastal resilience involves many sectors including environment fisheries, urban development infrastructure, water resources, and disaster management. Clear national policies clarify responsibilities among agencies and encourage cooperation across ministries. This coordination is important for reducing fragmented decision making which often leads to inconsistent and ineffective actions along the coast. Regulatory and planning tools shaped by national strategies directly influence resilience outcomes. Land use zoning setback rules building standards and environmental assessments can limit exposure to hazards and guide safer development when climate risks are considered. National policies can also support ecosystem-based approaches such as mangrove restoration and wetland protection as practical and sustainable options alongside engineered measures. Financing and capacity building are key elements of national approaches. Building coastal resilience requires steady investment in data systems infrastructure upgrades and community level initiatives. National strategies help mobilize public resources attract international climate funding and encourage private sector participation. Strengthening technical skills institutions and links between science and policy further supports informed and effective decision making. Social inclusion and equity are also gaining attention in national coastal resilience policies. Coastal communities with fewer resources often face the greatest climate impacts. National strategies can support livelihoods, protect local rights, and promote participation in planning processes. Placing equity at the center of policy design helps ensure that resilience efforts contribute not only to risk reduction but also to long-term and inclusive development (Anderies & Barreteau, 2019; Chaudhuri, 2025).

8.2.2 ICZM Frameworks and Multi-Level Governance

Integrated Coastal Zone Management, commonly known as ICZM, is widely used as a policy and governance approach to strengthen coastal resilience in the face of climate change, environmental stress, and rising development pressure. Coastal zones function as closely linked social and ecological systems where land and sea processes affect one another over time. ICZM offers a structured way to manage these interactions and address fragmented decision making while supporting coordinated and sustainable coastal governance. ICZM brings together policies institutions and stakeholders from sectors such as environment fisheries urban development tourism, disaster management, and climate adaptation. This integrated approach is important because coastal hazards such as sea level rise storm surges erosion and saltwater intrusion do not follow administrative boundaries. Planning that considers the coast as a whole allows development goals to be balanced with ecosystem protection while reducing risk and strengthening the capacity of communities and infrastructure to adjust to change. ICZM encourages planning that looks beyond immediate hazards and focuses on long-term resilience (Krause & Stead, 2017; Ramieri et al., 2019).

Several principles within ICZM directly support coastal resilience. Ecosystem-based management emphasizes the protection and restoration of natural systems such as mangroves, wetlands dunes, and coral reefs which reduce hazard impacts and support livelihoods and biodiversity. Participation allows local communities and stakeholders to share knowledge and take part in decision making which improves relevance and acceptance of policies. Adaptive management supports learning over time through monitoring, evaluation, and the use of scientific and local knowledge to adjust policies as conditions change. Effective ICZM depends on coordination across different levels of governance.

Coastal risks often extend from local areas to national and regional scales and may cross political boundaries. Multi-level governance helps connect national policy direction with local action and ensures that local efforts receive the support they need. Clear roles shared responsibilities and regular coordination reduce gaps and overlaps in coastal management. Local institutions play an important role in putting ICZM into practice. Communities and local authorities are often the first to experience coastal hazards and have detailed knowledge of local conditions. Strengthening local governance through participation capacity building and shared responsibility improves the effectiveness of resilience measures. National governance provides legal frameworks coordination mechanisms and financial support while regional cooperation helps address shared ecosystems and cross boundary risks. ICZM supported by coordination across governance levels offers a practical path toward stronger coastal resilience. Integrated planning inclusive decision making and adaptive approaches help coastal regions manage climate risks reduce disaster impacts and pursue sustainable development over the long term.

8.3 Legal and Regulatory Frameworks for Coastal Resilience

8.3.1 Coastal Zone Regulations and Land-Use Planning

Legal and regulatory frameworks play a key role in supporting coastal resilience by translating policy goals into rules that guide development resource use and risk management in coastal areas. These frameworks influence where and how development takes place how ecosystems are protected and how climate related risks are addressed. Coastal zone regulations and land use planning are among the most important tools for reducing exposure to hazards managing competing uses and protecting sensitive coastal environments. Clear legal rules help ensure consistency accountability and transparency in decision making while allowing adjustments as coastal conditions change. Coastal zone regulations are designed to manage activities in fragile and hazard prone coastal areas.

They respond to risks such as sea level rise erosion storm surges and environmental degradation. Common measures include coastal setback limits development controls land use zoning and protection of ecosystems such as mangroves wetlands dunes and coral reefs. From a resilience perspective these rules aim to prevent unsafe development that increases long term vulnerability. Restricting construction in high-risk areas and encouraging appropriate land uses help reduce future disaster losses. Land use planning plays a central role in putting coastal resilience into practice. When climate risk information is included in planning decisions authorities can guide settlement growth infrastructure development and economic activities away from areas most exposed to hazards. Zoning based on risk levels elevation and coastal dynamics helps reduce exposure while supporting safer development. In this way land use planning acts as a long-term tool for resilience by aligning development patterns with changing coastal conditions (Sundar, 2022; Wu & Barrett, 2022).

The effectiveness of coastal regulations and land use planning depends on how well they fit within wider legal and institutional systems. Separate laws governing land environment fisheries and urban development can lead to overlapping responsibilities and gaps in regulation. Strong coastal governance seeks better alignment between coastal laws climate adaptation policies, disaster risk reduction measures, and environmental protection rules. Strategic and cumulative impact assessments further support informed decision making by considering long term and combined effects of development. Flexibility is increasingly important in legal frameworks for coastal resilience. Climate risks evolve and fixed regulations may become unsuitable over time. Updating setback lines revising zoning based on new hazard information and regularly reviewing policies help legal systems remain effective. These adaptive approaches allow regulations to respond to new scientific knowledge monitoring results and observed coastal change. Social equity is also an important concern in coastal regulation. Rules that shape land and resource use can strongly affect communities that depend on coastal areas for their livelihoods. Participatory planning transparent processes and protection of local and customary rights help ensure that resilience measures do not lead to displacement or inequality. Effective enforcement strong institutions and adequate monitoring are essential for turning legal provisions into real resilience outcomes.

8.3.2 EIA and Adaptive Management Frameworks for Dynamic Coastal Environments

Legal and regulatory frameworks play an important role in strengthening coastal resilience, particularly in coastal areas where natural processes climate variability and human activities interact in uncertain ways. Environmental impact assessment and adaptive management are two governance tools that work well together to support better planning informed decision making and learning over time. They help ensure that development in coastal zones is environmentally responsible socially acceptable and able to respond to present and future climate-related risks. Environmental impact assessment is widely used to examine the possible environmental and social effects of development projects. In coastal areas it is especially useful for identifying risks linked to infrastructure tourism ports aquaculture and urban growth in sensitive and hazard-prone locations. From a coastal resilience perspective EIA does more than predict impacts. It serves as a preventive planning tool by identifying exposure to sea level rise storm surges erosion and flooding before projects move forward. This helps avoid development choices that could increase long-term vulnerability.

The contribution of EIA to coastal resilience becomes stronger when climate change is included in assessment processes. Earlier assessments often depended on past environmental conditions which are no longer reliable as climate patterns change. More recent approaches encourage the use of climate projections long-term risk scenarios and analysis of combined impacts. This allows decision makers to consider how projects may perform under future conditions as well as their immediate effects. EIA therefore supports development that aligns with resilience goals and sustainable coastal management. Public participation adds further value to EIA in coastal settings. Local communities often have detailed knowledge of environmental change hazard patterns and ecosystem conditions.

When assessment processes include stakeholder engagement this knowledge can inform project design and mitigation planning. Inclusive assessments build trust reduce conflict and help ensure that resilience actions reflect local needs especially for vulnerable and resource dependent groups. Adaptive management addresses uncertainty at a broader governance level. Coastal environments change over time due to natural processes and climate impacts, and governance systems need the ability to adjust. Adaptive management allows policies standards and management practices to be updated based on monitoring results and new scientific evidence. This helps keep regulations effective as conditions evolve. Fixed rules can lose relevance in dynamic coastal environments. Adaptive management responds by supporting repeated cycles of action monitoring, evaluation, and adjustment. When supported by legal provisions this approach turns governance into a continuous learning process rather than a single decision. Linking environmental impact assessment with adaptive management strengthens coastal resilience by connecting early risk assessment with long-term monitoring and review. Strong institutions coordination and accountability are essential for these frameworks to work in practice. Clear responsibilities information sharing and cooperation among agencies improve implementation. Attention to fairness transparency and access to justice also helps protect livelihoods and rights. Together environmental impact assessment and adaptive management provide a practical and flexible foundation for strengthening coastal resilience under changing climate conditions (Roux et al., 2025; Valente et al., 2025).

8.4 Institutional Arrangements for Coastal Resilience

Institutional arrangements provide the practical foundation for coastal resilience governance by turning policies and laws into coordinated action on the ground. They shape how responsibilities are shared, how decisions are made across sectors and levels, and how resources are used to respond to changing coastal risks. Coastal resilience involves climate adaptation disaster risk reduction ecosystem management and sustainable development which makes well-organized institutions essential for coordination accountability and long-term effectiveness. Clear roles and responsibilities are a key part of effective institutional arrangements. Coastal resilience usually involves many public agencies responsible for environment climate change, disaster management, urban development, fisheries, water resources, and infrastructure. When responsibilities are unclear or overlap decision making becomes fragmented and resources are often misused. Strong institutional systems focus on clearly defined mandates supported by coordination mechanisms such as inter agency committees' coastal authorities or lead agencies for coastal management. These arrangements help align sector priorities coordinate policies and manage interactions between land and sea. National institutions play an important role by setting policy direction establishing standards and mobilizing financial and technical support for coastal resilience. They also connect international climate commitments with actions at regional and local levels. National policies funding programs and adaptation strategies create an enabling environment for resilience efforts. At the same time effective governance cannot rely only on central institutions. Regional and local bodies need authority resources and capacity to act.

Regional and local institutions are essential for applying resilience strategies in specific coastal settings. Local governments are often responsible for land use planning regulation enforcement infrastructure maintenance and emergency response. Their closeness to communities allows faster response to hazards and better use of local knowledge. Strengthening local institutions through training funding and technical support is therefore critical. Without this support national strategies often fail during implementation. Institutional arrangements increasingly support participation and stakeholder engagement. Coastal communities' indigenous groups civil society organizations and academic institutions contribute knowledge monitoring and public awareness. Inclusive governance builds trust improves decision making and helps ensure that resilience measures reflect social conditions and equity concerns. Participatory platforms and community-based management structures allow different actors to take part in planning and implementation. Cooperation between public and private actors has also become important for coastal resilience. Private actors are active in areas such as infrastructure tourism ports fisheries energy and insurance. When properly regulated partnerships can bring finance technology and management skills that support resilience goals. Clear rules transparency and accountability help ensure that private involvement supports long-term public interests rather than short-term gains. Coordination across sectors and governance levels remains essential. Coastal risks cross administrative boundaries and policy areas which makes cooperation necessary. Systems that support information sharing, joint planning, and conflict resolution reduce fragmentation and improve outcomes. Institutional learning and adaptability strengthen resilience over time. Coastal conditions change and institutions must be able to adjust policies and practices based on experience data and new knowledge. Review processes monitoring and feedback mechanisms help institutions improve performance and respond to emerging risks. Strong institutional arrangements help turn coastal resilience from policy intent into practical action. Clear roles coordination participation and adaptive capacity allow governance systems to manage climate risks and development pressures more effectively. When institutions are aligned, well-resourced, and flexible they support sustainable equitable and climate resilient coastal development (Owusu et al., 2024; Robinson et al., 2023).

8.5 Financing and Economic Instruments for Coastal Resilience

Financing and economic instruments play an important role in strengthening coastal resilience because they shape what actions can be planned funded and sustained over time. Coastal resilience often requires large and long-term investment in infrastructure ecosystem restoration data systems and institutional capacity. Without stable financing even well-planned policies and governance efforts are difficult to carry out. Financial tools therefore provide the support needed to turn resilience goals into practical action in coastal areas facing increasing climate and disaster risks. Funding for coastal resilience comes from many sources and operates at different levels. Public funding remains essential especially for coastal protection early warning systems risk reduction infrastructure and capacity building. Many governments now include resilience and climate adaptation within regular development budgets rather than treating them as separate spending areas. This helps ensure that public investment in coastal regions considers climate risks from the beginning. International climate finance development assistance and multilateral funds also play an important role particularly for countries with limited financial capacity and high exposure to coastal hazards.

New financing approaches are increasingly used to support public funding. Blended finance brings together public funds and private investment to support projects that improve resilience but may not attract private finance on their own. Insurance and risk transfer tools such as disaster insurance help spread financial risk and support recovery after extreme events. Dedicated resilience funds and climate adaptation programs within national financial systems also improve the continuity and reliability of funding for coastal resilience actions. Economic instruments influence how people and investors behave in coastal areas. Incentives such as subsidies tax relief and low interest loans can encourage climate resilient construction ecosystem restoration and sustainable livelihoods. Disincentives such as higher insurance costs or development charges in high-risk zones can discourage unsafe development. When these instruments are aligned with land use planning and regulations, they help guide economic activity toward safer and more resilient outcomes. Economic valuation supports better financing and policy decisions. Coastal ecosystems provide many benefits including protection from storms support for fisheries tourism income and carbon storage. These benefits are often overlooked in conventional economic assessments. Valuation methods help demonstrate the long-term value of ecosystems and show the cost effectiveness of nature-based solutions compared to hard infrastructure alone. Cost benefit analysis and avoided loss studies also help compare the cost of preventive action with the losses caused by repeated disasters. Including social impacts in these assessments helps address equity concerns. Financing and economic instruments are central to effective coastal resilience governance. Reliable funding supports long-term action while economic analysis helps prioritize investments and integrate resilience into development planning. When applied within strong policy frameworks these tools support fair efficient and lasting coastal resilience strategies and help coastal regions manage climate risks while pursuing sustainable development (Moaddine & Fosse, 2025; Rogge, 2025).

Insurance schemes and risk transfer mechanisms play an important role in coastal resilience governance as climate-related hazards become more frequent and economic exposure in coastal areas increases. These instruments support regulation planning and public investment by providing financial protection against losses and encouraging awareness of risk. Although they do not prevent hazards, they help manage remaining risks support recovery after disasters and reduce long-term economic disruption in vulnerable coastal regions. Coastal areas face growing impacts from cyclones storm surges flooding and erosion which place heavy financial pressure on households' businesses and public authorities. Insurance schemes help spread these risks across a wider group which reduces the burden on those directly affected. Property insurance, fisheries and crop insurance, and coverage for public infrastructure support recovery and help restore livelihoods after extreme events. In governance terms insurance also acts as a policy tool because pricing coverage conditions and exclusions reflect actual levels of risk. Risk transfer mechanisms extend beyond traditional insurance and include reinsurance catastrophe bonds and parametric insurance. These tools allow governments and insurers to shift part of the financial risk to wider markets which helps protect public budgets. Parametric insurance is especially useful in coastal areas because payments are triggered by measured events such as wind speed or water levels. This allows funds to be released quickly after disasters and supports timely response and recovery. Insurance schemes also shape long-term development choices in coastal zones. Premiums linked to risk levels and conditions attached to coverage can encourage safer construction compliance with coastal regulations and protection of natural buffers. When aligned with land use planning and building standards insurance tools support governance goals by discouraging development in high-risk areas. Poorly designed subsidies however can encourage unsafe development which shows the need for careful policy design. Public institutions have an important role in making insurance fair and accessible. Many coastal communities face barriers related to cost limited awareness and informal livelihoods. Public or mixed insurance schemes can address these challenges by subsidizing premiums, expanding coverage, and linking insurance with social protection programs. These approaches strengthen social resilience and support inclusive development. Insurance and risk transfer tools work best when combined with strong risk reduction and adaptation efforts. Without investment in prevention and ecosystem protection rising climate risks can make insurance unaffordable or unavailable. When integrated within broader policy frameworks these instruments support a balanced approach to managing coastal risks and building long-term resilience.

8.6 Climate Change Adaptation and Mitigation in Coastal Policies

Integrating climate change adaptation and mitigation into coastal policy frameworks is essential for strengthening coastal resilience under changing environmental conditions (Fig. 8.1). Coastal areas face direct climate risks such as sea level rise erosion, extreme weather events, and ecosystem loss. At the same time, they offer important opportunities to reduce emissions through healthy ecosystems and sustainable development practices. Bringing adaptation and mitigation together within coastal policies supports coordinated and long-term governance rather than isolated responses. Adaptation in coastal policy focuses on reducing vulnerability and improving the ability of coastal systems to cope with climate stress. Many governments now use climate risk assessments vulnerability mapping and future climate scenarios to guide planning and decisions. This allows policies to respond to expected changes rather than rely only on past climate patterns. Adaptation oriented policies promote safer land use planning resilient infrastructure ecosystem-based measures and livelihood diversification for vulnerable coastal communities. When these actions are included in formal policies adaptation becomes part of routine development planning. Mitigation is also important for long term coastal resilience. Coastal ecosystems such as mangroves, salt marshes, seagrasses, and tidal wetlands store carbon and support biodiversity while reducing exposure to hazards. Protecting and restoring these ecosystems supports emission reduction and strengthens resilience at the same time. Coastal policies can also encourage low carbon infrastructure and align coastal development with national climate goals. Linking local action with broader climate commitments improves consistency across policy levels. Strong coordination across sectors is needed to integrate adaptation and mitigation effectively. Climate change affects many areas including environment disaster management urban development fisheries energy and finance. Coastal policy frameworks therefore require coordination mechanisms that connect coastal planning with national climate strategies and development plans. Better coordination helps avoid conflicts and ensures that adaptation and mitigation measures support one another.

Time horizons are also important in climate responsive coastal policies. Climate risks develop over long periods and policies must be able to adjust as conditions change. Flexible tools such as scenario-based planning periodic review and adaptive zoning help governance systems respond to uncertainty and new information. Social equity plays a key role in integrating climate change into coastal policy. Climate impacts often affect poorer and marginalized coastal communities more severely. Policies that support participation address unequal risks and protect livelihoods help ensure that climate action contributes to fair and sustainable development. Integrating adaptation and mitigation into coastal policy frameworks allows governance systems to manage current risks while supporting long term climate and development goals. Treating these approaches as connected priorities helps coastal regions respond to climate uncertainty protect ecosystems and build resilient and inclusive futures (de Freitas et al., 2019; Le et al., 2020).

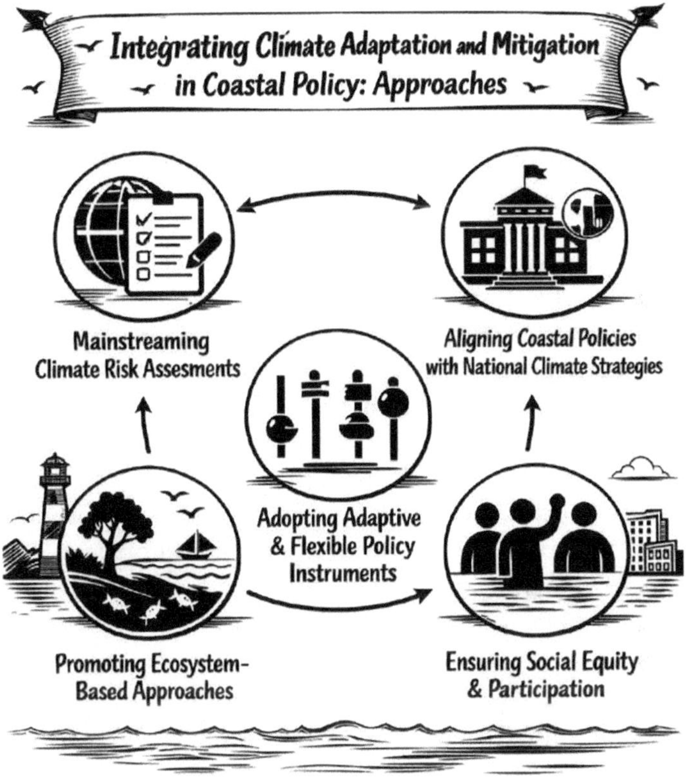

Fig. 8.1 Approaches of integrating climate adaptation and mitigation in coastal policy (Prepared by the authors)

8.7 Community Engagement and Stakeholder Participation

Community engagement and stakeholder participation are central to effective coastal resilience governance. Coastal areas are shaped by close interactions between natural processes and human activities and the impacts of climate change and coastal hazards are felt most strongly at the local level. Involving communities and stakeholders in governance helps ensure that resilience policies reflect local conditions knowledge and social realities. This improves their relevance credibility and practical effectiveness. Community engagement supports better risk awareness adaptive capacity and social cohesion. Coastal communities often hold valuable experience and traditional knowledge related to environmental change hazard patterns and ecosystem conditions. This knowledge adds depth to scientific information and improves risk assessment and adaptation planning. Participation also builds a sense of shared responsibility which supports long-term commitment to resilience actions. Stakeholder participation is important for addressing equity in coastal resilience. Climate-related risks often affect marginalized groups such as small-scale fishers' informal settlers' women and indigenous communities more severely. Inclusive governance processes help ensure that their needs and priorities are reflected in decision making. Participation also improves transparency and accountability and helps build trust between public institutions and coastal populations.

Participatory approaches in coastal governance extend beyond consultation. Stakeholders can be involved in planning implementation monitoring and evaluation. Community-based coastal management allows local groups to work with government agencies to manage resources and reduce risks. These approaches strengthen local capacity and support practical actions such as ecosystem restoration disaster preparedness and livelihood support. Multi stakeholder platforms bring together government agencies civil society researchers' private actors and community representatives. These platforms support dialogue coordination and conflict resolution in coastal areas where different interests often compete. Participatory vulnerability and risk assessments also strengthen governance by involving communities in identifying hazards exposure and coping capacities. Challenges remain including power imbalances limited capacity and symbolic participation. Effective governance supports meaningful participation through access to information capacity building and clear institutional roles. When participation is inclusive and sustained it helps connect policy goals with local action and shared responsibility and strengthens long-term coastal resilience (Mattera et al., 2025; Malakar et al., 2023; Sebos et al., 2023).

8.8 International Cooperation and Policy Networks

International cooperation and policy networks play an important role in strengthening coastal resilience by addressing climate risks and environmental processes that cross national boundaries. Coastal hazards such as sea level rise extreme weather events and ecosystem degradation affect many countries at the same time. No single country can manage these challenges alone. International cooperation helps countries align policies, share knowledge, and mobilize resources to respond to common coastal risks. Global climate and disaster governance frameworks strongly influence coastal resilience policies. International climate agreements encourage countries to include climate risks in development planning and to strengthen adaptation in vulnerable areas such as coastal zones. National climate strategies increasingly recognize the importance of coastal protection ecosystem-based approaches and resilient infrastructure. These commitments link coastal resilience to wider climate goals and accountability systems. International disaster risk reduction frameworks also support coastal resilience by promoting prevention preparedness and risk informed planning. These frameworks focus on understanding risks strengthening institutions investing in resilience and improving readiness for extreme events. Their multi-hazard approach fits well with the range of threats faced by coastal regions including floods storms and erosion. They support the integration of coastal risk reduction into national and local governance. Global development agendas further strengthen cooperation for coastal resilience. International development goals highlight the close links between healthy oceans climate action and human well-being. These goals encourage balanced approaches that combine ecosystem protection disaster risk reduction and economic development. Coastal resilience is therefore treated as a shared development priority rather than a narrow sectoral issue.

Regional cooperation has a practical role in managing shared coastal challenges. Regional agreements and coastal management initiatives allow neighboring countries to work together on marine pollution, fisheries governance, sediment movement, and coastal erosion. These platforms support joint planning common standards and the exchange of good practices. Policy networks also support learning capacity building and collaboration among governments researchers and civil society. International cooperation provides financial and technical support for coastal resilience. Climate funds, development banks, and donor agencies finance adaptation ecosystem restoration and resilient infrastructure projects in coastal areas. Access to funding and technology is especially important for developing countries and small island states. International frameworks can face challenges due to differences in national priorities and capacity. Strong national ownership and alignment with domestic policies are needed for effective implementation. When well connected to national governance systems international cooperation strengthens the shared ability of coastal regions to manage climate risks and pursue long-term resilient development (Haque et al., 2025).

And the knowledge sharing and capacity building are essential to international cooperation for strengthening coastal resilience. Coastal regions face fast changing and uncertain risks linked to climate change environmental damage and social and economic pressures. Responding to these challenges requires more than funding and regulations. It also depends on access to knowledge, technical skills, strong institutions, and the ability to learn and adapt over time. International cooperation and policy networks help reduce gaps in knowledge and capacity between countries and support more effective coastal governance. Knowledge sharing in coastal resilience includes the creation exchange and use of scientific technical and practical knowledge. International organizations research networks and cooperative programs support the sharing of climate data hazard assessments, ecosystem monitoring tools, and good practices in adaptation and disaster risk reduction. Global scientific assessments provide a common evidence base for policy decisions at national and coastal levels. Translating scientific findings into clear guidance helps decision makers apply this knowledge in practice.

International cooperation also supports practical learning through regional platforms and thematic networks. Global and regional institutions promote exchange through training programs technical guidance and joint research. These platforms allow countries to learn from different coastal settings and adapt approaches such as ecosystem-based solutions early warning systems and integrated coastal planning to local needs. Capacity building works alongside knowledge sharing by strengthening the skills of individuals institutions and governance systems. Many coastal regions especially in developing countries and small island states face limits in technical expertise data access and institutional resources. International initiatives respond through training exchange programs and institutional support. Strengthening skills in coastal planning climate risk assessment mapping and monitoring helps authorities put resilience policies into action. Institutional capacity is important for long-term coastal resilience. International cooperation supports better coordination governance structures and links between science and policy. These efforts help embed resilience into routine decision making rather than treating it as a short-term activity. Cooperation among countries facing similar conditions has become more important. Regional and South to South partnerships support learning that is practical and relevant to local contexts. These networks encourage shared problem-solving joint training and coordinated responses to common coastal risks. Capacity building efforts increasingly involve communities' civil society and local practitioners. Inclusive approaches that value local and indigenous knowledge and address social differences strengthen trust and improve outcomes. These efforts build local capacity and increase acceptance of resilience measures. Challenges remain including weak coordination short project timeframes and limited links to national policies. Aligning capacity building with national priorities long-term partnerships and stronger feedback between knowledge providers and local users can improve results. Knowledge sharing and capacity building form a key part of international cooperation for coastal resilience. By improving access to knowledge, strengthening institutions, and supporting shared learning these efforts help coastal regions respond to climate risks and support sustainable development over time (Stojanov et al., 2021; Talukdar et al., 2025).

8.9 Monitoring, Evaluation, and Adaptive Governance

Monitoring, evaluation, and adaptive governance are essential for effective coastal resilience policy frameworks. Coastal systems are constantly changing due to climate conditions, ecological processes, and human activities. Because of this, policies that remain fixed over time often fail to respond to new risks. Governance systems need ways to observe, change, assess results, learn from experience, and adjust actions when necessary. Monitoring and evaluation provide the information needed to support these processes and help maintain long-term coastal resilience. Indicators form the basis of monitoring and evaluation systems. They help convert complex environmental social and institutional conditions into measurable information that can guide decisions. Coastal resilience indicators often cover physical hazards ecosystem health social vulnerability institutional capacity and governance performance. Using multiple indicators allows decision makers to better understand how natural systems and human actions interact and how resilience evolves over time.

Selecting indicators is both a technical and a governance decision. Indicator choices reflect policy priorities, available data, and institutional capacity. Effective indicator systems balance scientific accuracy with practical use so that responsible agencies can collect interpret and apply the data. Quantitative indicators such as shoreline change flooding frequency or ecosystem cover are often combined with qualitative indicators related to participation coordination and learning capacity. This combination supports a broader understanding of resilience. Monitoring systems turn indicators into action by producing timely and reliable data. Tools such as satellite imagery mapping systems and digital platforms have improved the ability to track coastal change and hazard trends. Early warning systems, climate observation networks, and ecosystem monitoring programs support both short-term risk management and long-term planning. Community-based monitoring also plays an important role by capturing local changes validating scientific data and strengthening community involvement (Djalante et al., 2011; Sparkes & Werners, 2023).

Evaluation links monitoring results with policy improvement. Evaluation helps assess whether resilience actions meet their goals identify unintended effects and understand trade-offs between different objectives. Rather than focusing only on compliance evaluation supports learning by examining what works under specific conditions and for different groups. Adaptive governance uses monitoring and evaluation to guide flexible decision making. This approach accepts uncertainty as part of coastal systems and allows policies to be adjusted based on evidence. Repeated cycles of planning action observation and revision help governance systems respond to changing risks social conditions and new knowledge. Feedback mechanisms ensure that information from monitoring and evaluation reaches decision makers and leads to adjustment. These mechanisms include policy reviews, stakeholder consultations, and learning platforms. Without strong feedback data collection loses its value and fails to improve outcomes. Learning is the key result of effective monitoring, evaluation, and adaptive governance. Institutions improve coordination procedures and skills through organizational learning. Social learning develops through participation and shared understanding of risks. Over time learning helps governance systems become more integrated inclusive and responsive. Challenges remain such as limited data capacity weak coordination and short planning horizons. Addressing these issues requires investment clear responsibilities and a governance culture that values learning and flexibility. Monitoring, evaluation, and adaptive governance are core elements of coastal resilience policy frameworks. They support informed decision-making, continuous improvement, and long-term adaptation. When well-integrated into governance systems they help coastal regions manage uncertainty and move toward more sustainable and resilient futures (Abdillah et al., 2025; Lin & Lee, 2023).

8.10 Conclusion and Arguments

This chapter has shown that building coastal resilience depends on a broad and integrated approach to policy and governance. Coastal resilience is more than the ability to cope with climate-related hazards.

It is a continuing governance goal that links climate adaptation, disaster risk reduction, ecosystem protection, and inclusive development. Achieving this requires policies that support long-term coordinated action across sectors and levels of decision making. Policy frameworks play an important role in turning resilience ideas into practical action. National strategies integrated coastal zone management land use planning and regulatory tools such as environmental assessment and adaptive management help reduce risk and guide sustainable development. These policies need strong institutional arrangements to work in practice. Clear roles coordination across governance levels and cooperation between public and private actors help ensure effective implementation. Financing mechanisms and economic tools are also essential for sustaining coastal resilience. Long-term investment and risk-sharing arrangements make resilience efforts financially viable. Community engagement and stakeholder participation strengthen governance by improving fairness trust and local capacity to adapt. When local knowledge and priorities are included resilience strategies become more relevant and effective. International cooperation adds further support through knowledge exchange, capacity building, and shared learning. Partnerships between countries help address common coastal challenges and reduce gaps in resources and expertise. Monitoring, evaluation, and adaptive governance connect all these elements by supporting learning flexibility and timely policy adjustment as conditions change. Coastal resilience should be seen as an ongoing process rather than a fixed outcome. Its success depends on coordinated policies, capable institutions, meaningful participation, and continuous learning. When these elements work together, they support sustainable and resilient coastal development in the face of climate change. The chapter discussions lead to the following arguments:

(a) Coastal resilience is an ongoing socio-ecological and governance process, not a fixed risk reduction outcome.
(b) Coastal governance requires alignment between climate adaptation, disaster risk reduction, and sustainable development.
(c) Legal and regulatory tools such as land-use planning and EIA help prevent unsafe coastal development.

(d) Strong institutions and multi-level governance support coordinated and context-specific resilience action.
(e) Long-term coastal resilience depends on stable financing, insurance, and economic instruments.
(f) Community and stakeholder participation improve equity, trust, and effectiveness of resilience measures.
(g) International cooperation strengthens national and local capacity through shared knowledge and resources.
(h) Monitoring, evaluation, and adaptive governance support learning and long-term coastal resilience.

References

Abdillah, A., Widianingsih, I., & Buchari, R.A. et al. Adapting to climate change and multi-risk governance: Toward sustainable adaptation and enhancing urban resilience—Indonesia. Discover Applied Sciences 7, 81 (2025). https://doi.org/10.1007/s42452-025-06491-7

Almutairi, A., Mourshed, M., & Ameen, R. F. M. (2020). Coastal community resilience frameworks for disaster risk management. *Natural Hazards, 101*, 595–630. https://doi.org/10.1007/s11069-020-03875-3

Anderies, J. M., & Barreteau, O. (2019). Governance principles for robust and resilient coastal systems in the face of global change. *Regional Environmental Change, 19*, 1831–1833. https://doi.org/10.1007/s10113-019-01542-3

Chaudhuri, S. K. (2025). Socioecological resilience among coastal communities for the sustainable blue economy in India. In W. Leal Filho, A. L. Salvia, J. P. P. Eustachio, & M. A. P. Dinis (Eds.), *Handbook of sustainable blue economy*. Springer. https://doi.org/10.1007/978-3-031-32671-4_19-1

de Freitas, D. M., de Carvalho, D. A., & Hosokawa, E. K. (2019). Adapting to a changing climate: An operational space for local adaptation Committee in Santos Coastal Area. In L. Nunes, R. Greco, & J. Marengo (Eds.), *Climate change in Santos Brazil: Projections, impacts and adaptation options*. Springer. https://doi.org/10.1007/978-3-319-96535-2_15

Djalante, R., Holley, C., & Thomalla, F. (2011). Adaptive governance and managing resilience to natural hazards. *International Journal of Disaster Risk Science, 2*, 1–14. https://doi.org/10.1007/s13753-011-0015-6

Haque, M., et al. (2025). *Disasters, climate change and sustainability in South Asia: Challenges and the way forward*. Springer Cham. https://doi.org/10.1007/978-3-032-04750-2. isbn:978-3-032-04749-6.

Krause, G., & Stead, S. M. (2017). Governance and offshore aquaculture in multi-resource use settings. In B. Buck & R. Langan (Eds.), *Aquaculture perspective of multi-use sites in the Open Ocean*. Springer. https://doi.org/10.1007/978-3-319-51159-7_7

Le, T. D. N. (2020). Climate change adaptation in coastal cities of developing countries: Characterizing types of vulnerability and adaptation options. *Mitigation and Adaptation Strategies for Global Change, 25*, 739–761. https://doi.org/10.1007/s11027-019-09888-z

Lin, B. C., & Lee, C. H. (2023). Conducting an adaptive evaluation framework of importance and performance for community-based earthquake disaster management. *Natural Hazards, 115*, 1255–1274. https://doi.org/10.1007/s11069-022-05594-3

Malakar, K. D., Kumar, M., Anand, S., & Kuzur, G. (2023). Disaster, Policy Instruments, and Sustainability. In *Climate Change and Socio-Ecological Transformation. Advances in Geographical and Environmental Sciences*. Springer, Singapore. https://doi.org/10.1007/978-981-99-4390-6_6

Mattera, M., Erokhin, D., Baldelli, M., et al. (2025). Climate experiences to adaptation participation: A survey study in Catalonia. *Theoretical and Applied Climatology, 156*, 568. https://doi.org/10.1007/s00704-025-05706-6

Moaddine, H. J., & Fosse, J. (2025). Shaping a sustainable blue economy for inclusive and resilient coastal futures. In W. Leal Filho, A. L. Salvia, J. P. P. Eustachio, & M. A. P. Dinis (Eds.), *Handbook of sustainable blue economy*. Springer. https://doi.org/10.1007/978-3-031-32671-4_125-1

Nichols, C. R., Wright, L. D., & Zarillo, G. (2025). *Integrated coastal resilience*. Springer Cham. https://doi.org/10.1007/978-3-031-68153-0. isbn:978-3-031-68152-3.

Owusu, A. B., Adu-Boahen, K., & Dadson, I. Y. (2024). Institutional arrangement for mitigating and adapting to climate change-related flood risk in Greater Accra metropolitan area (GAMA). *City and Environment Interactions, 21*, 100129. https://doi.org/10.1016/j.cacint.2023.100129

Ramieri, E., Bocci, M., & Markovic, M. (2019). Linking integrated coastal zone management to maritime spatial planning: The Mediterranean experience. In J. Zaucha & K. Gee (Eds.), *Maritime spatial planning*. Palgrave Macmillan. https://doi.org/10.1007/978-3-319-98696-8_12

Robinson, S., Carlson, D., Bouton, E., et al. (2023). The dynamics of institutional arrangements for climate change adaptation in small Island developing states in the Atlantic and Indian oceans. *Sustainability Science, 18*, 251–264. https://doi.org/10.1007/s11625-022-01186-z

Rogge, E. (2025). Blue bonds: Sustainable financing for the blue economy around the Indian Ocean. *Jindal Global Law Review, 16*, 155–177. https://doi.org/10.1007/s41020-025-00263-5

Roux, D. J., Kingsford, R. T., & Cockburn, J. (2025). Strategic adaptive Management for Transparency, accountability and learning: Insights from more than a quarter century of practice. *Environmental Management, 75*, 2521–2539. https://doi.org/10.1007/s00267-025-02272-5

Sebos, I., Nydrioti, I., Katsiardi, P., et al. (2023). Stakeholder perceptions on climate change impacts and adaptation actions in Greece. *Euro-Mediterranean Journal for Environmental Integration, 8*, 777–793. https://doi.org/10.1007/s41207-023-00396-w

Singh, A. (2020). Building resilience in coastal ecosystems: Problems and prospects. In A. Singh, R. L. S. Fernando, & N. P. Haran (Eds.), *Development in coastal zones and disaster management. Disaster research and management series on the global south*. Palgrave Macmillan. https://doi.org/10.1007/978-981-15-4294-7_10

Sparkes, E., & Werners, S. E. (2023). Monitoring, evaluation and learning requirements for climate-resilient development pathways. *Current Opinion in Environmental Sustainability, 64*, 101329. https://doi.org/10.1016/j.cosust.2023.101329

Stojanov, R., Rosengaertner, S., de Sherbinin, A., et al. (2021). Climate mobility and development cooperation. *Population and Environment, 43*, 209–231. https://doi.org/10.1007/s11111-021-00387-5

Sundar, V. (2022). Sustainable hard and soft measures for coastal protection. In C. N. S. Reddy & S. Sassa (Eds.), *Scour-and erosion-related issues. Lecture notes in civil engineering* (Vol. 177). Springer. https://doi.org/10.1007/978-981-16-4783-3_4

Talukdar, S., Chatterjee, R., Bera, S., Prashar, A., & Shaw, R. (2025). Disaster risk Management in South Asia through Innovations, communication, and technological advances. In S. Talukdar, R. Chatterjee, S. Bera, A. Prashar, & R. Shaw (Eds.), *Communication, science, technology, and innovation in disaster risk management. GIScience and geo-environmental modelling*. Springer. https://doi.org/10.1007/978-3-031-77189-7_1

Valente, S., & Pinho, P. (2025). Adaptive planning approaches for coastal climate adaptation: Process and key-elements. *Environmental Management, 75*, 1013–1038. https://doi.org/10.1007/s00267-025-02117-1
Wu, T., & Barrett, J. (2022). Coastal land use management methodologies under pressure from climate change and population growth. *Environmental Management, 70*, 827–839. https://doi.org/10.1007/s00267-022-01705-9

9

Sustainable Coastal Development and Blue Economy

This chapter introduces sustainable coastal development as an important approach for addressing the growing environmental, economic, and social pressures faced by coastal regions. Coastal areas are living spaces where ecosystems, communities, and activities such as fisheries, tourism, and infrastructure development are closely linked. Rapid urban growth, climate change impacts, and increasing use of coastal and marine resources are reshaping these regions and creating new challenges for long-term sustainability. The chapter brings together ideas related to sustainable development and the Blue Economy, showing how economic progress can take place while protecting coastal ecosystems and supporting community well-being. It highlights practical approaches such as integrated coastal planning, ecosystem-based management, climate adaptation, disaster risk reduction, and nature-based solutions. The role of governance, policy support, innovation, and local participation is also emphasized, with special attention to fairness, inclusion, and the use of traditional knowledge. Coastal development is presented as a continuous process that requires learning, cooperation, and flexibility rather than a fixed goal. The chapter encourages readers to engage with the ideas and practices discussed, offering guidance

K. D. Malakar et al., *Coastal Ecologies at Risk*, Palgrave Studies in Environmental Sustainability, https://doi.org/10.1007/978-3-032-26081-9_9

for informed decision-making and for building resilient, inclusive, and sustainable coastal futures.

Key Points of the Chapter

- Describes sustainable coastal development as a balance between environmental protection, economic growth, and social equity across land and sea areas.
- Explains the blue economy as effective only when it supports ecosystem health, climate resilience, and inclusive livelihoods.
- Highlights the role of nature-based solutions and sustainable infrastructure in reducing coastal risks and supporting long-term development.
- Examines how community participation, social inclusion, and livelihood diversity shape fair and accepted coastal development.
- Explores integrated, adaptive, and participatory governance as essential for turning sustainability goals into coastal action.

9.1 Sustainable Coastal Development

Sustainable coastal development refers to an approach that plans, manages, and uses coastal and marine areas in ways that protect the environment, support economic activities, and promote social well-being over the long term. Coastal zones are among the most active and productive regions in the world. They contain important ecosystems such as mangroves, coral reefs, estuaries, and beaches, while also supporting large populations and a wide range of economic activities. These areas play a major role in food security, employment, trade, tourism, and cultural identity. At the same time, rapid urban growth, poorly planned infrastructure, excessive resource extraction, and climate-related pressures have placed coastal systems under serious stress. In this context, sustainable coastal development offers a practical way to balance development needs with environmental protection and fairness across generations. Sustainable coastal development is based on the idea that land and sea systems are closely connected and that human actions directly affect natural processes. Instead of treating sectors separately, this approach encourages coordination across fisheries, tourism, ports, energy, conservation, and settlement planning. It supports ecosystem-based management, careful decision-making, and inclusive governance that values scientific knowledge as well as local and traditional experience. By recognizing environmental limits within economic planning, sustainable coastal development seeks to avoid long-term ecological damage while creating shared economic benefits (Kotowska et al., 2020; Stylios et al., 2015).

The importance of sustainable coastal development has increased in the context of climate change. Coastal regions are highly exposed to sea level rise, erosion, saline intrusion, stronger storms, and rising ocean temperatures. These impacts often affect vulnerable communities whose livelihoods depend on fisheries, coastal farming, and tourism. Sustainable coastal development responds to these risks by linking climate adaptation and disaster risk reduction with development planning. Actions such as protecting natural coastal buffers, promoting climate resilient infrastructure, and supporting diverse livelihood options help reduce risks while strengthening the ability of communities to adapt. The idea of the Blue Economy has gained growing attention within this sustainability context. The Blue Economy promotes the use of ocean and coastal resources in ways that generate income and employment while maintaining healthy marine ecosystems. It includes traditional activities such as fisheries and shipping, as well as emerging sectors like offshore renewable energy, marine biotechnology, and nature-based tourism. Sustainable coastal development provides the planning and governance foundation needed for the Blue Economy to succeed. Without careful coastal planning, Blue Economy initiatives may lead to environmental damage or social inequality. International policy agendas have also highlighted the importance of sustainable coastal development. Global commitments emphasize the conservation and responsible use of marine and coastal resources, along with the need for climate action. These efforts encourage integrated planning, nature-based solutions, and cooperation across different levels of governance. Putting these ideas into practice remains challenging due to competing interests, weak coordination among institutions, limited data, and unequal power relations. Addressing these issues requires strong policy frameworks, inclusive decision-making, and adaptive management that can respond to change and uncertainty. Sustainable coastal development stands as a key foundation for a fair and resilient Blue Economy. It offers a clear way to manage coastal spaces so that environmental health, economic progress, and social well-being are advanced together. As pressures on coastal systems continue to grow, moving toward sustainable development pathways has become essential for securing the future of coastal regions.

9.2 Blue Economy and its Significance

The Blue Economy refers to an approach that supports the careful and sustainable use of ocean, sea, and coastal resources to strengthen economic growth, improve livelihoods, and create employment, while protecting the long-term health of marine and coastal ecosystems. It has developed from the growing recognition that oceans are essential for economic development but are also under increasing environmental stress. Instead of relying on activities that focus only on resource extraction, the Blue Economy promotes development that respects ecological limits and supports social inclusion. Within the wider context of sustainable coastal development, it offers a shared framework that connects environmental protection with economic and social progress. The Blue Economy is built on the principles of sustainability, ecosystem-based management, and fairness across generations. It views oceans as complex systems where activities such as fisheries, shipping, tourism, energy production, and coastal infrastructure are closely linked with marine biodiversity and ecosystem services. Healthy oceans contribute to food security, climate regulation, and protection from coastal hazards, while damaged ecosystems weaken local economies and community well-being. For this reason, the Blue Economy places the protection of marine ecosystems at the center of economic planning rather than treating it as an afterthought. This reflects a shift away from development paths that exhaust natural resources toward approaches that support renewal and long-term stability.

The relevance of the Blue Economy has increased as the world faces growing environmental change and economic uncertainty. Oceans play an important role in addressing climate change, food insecurity, and employment challenges. Marine and coastal sectors support the livelihoods of millions of people, especially in developing countries and small island regions where economic options are limited. Oceans also absorb large amounts of heat and carbon dioxide, making them critical for climate regulation. Promoting responsible and sustainable ocean use therefore supports both climate adaptation and economic resilience. Global development agendas have increasingly highlighted the importance of the Blue Economy. International organizations have emphasized its potential to align ocean governance with sustainability and equity goals, while also supporting job creation and economic diversification. These perspectives show that the Blue Economy is not a single sector, but a broad approach that requires coordination across economic planning, environmental management, and social development policies. Strong links exist between the Blue Economy and sustainable coastal development. Coastal areas connect land and sea and host ports, cities, tourism centers, and sensitive ecosystems. Poorly planned coastal development can weaken Blue Economy goals through pollution, habitat loss, and increased disaster risks. In contrast, well-planned coastal development supports ocean-based activities by protecting ecosystems, reducing conflicts among users, and strengthening resilience to climate hazards. This close relationship makes integration between coastal planning and Blue Economy strategies essential. Key sectors commonly linked with the Blue Economy include sustainable fisheries and aquaculture, environmentally responsible maritime transport and ports, coastal and marine tourism, offshore renewable energy, and marine biotechnology. Each of these sectors depends on healthy ecosystems and effective governance. Despite its potential, the Blue Economy also raises concerns related to governance and social equity. Without strong safeguards, development efforts may favor large-scale investments and marginalize small-scale users, especially fishing communities. Effective governance is therefore needed to ensure fair access to resources, shared benefits, and meaningful participation in decision-making. Inclusive policies, transparent institutions, and science-based management remain essential for balancing economic development with social justice and environmental protection (Fig. 9.1) (Hazra & Bhukta, 2022; Sardá et al., 2025).

The Blue Economy represents a shift in how oceans and coasts are valued and managed. Its strength lies not only in economic opportunity, but also in its ability to bring together conservation, climate resilience, and social inclusion within a single approach. When aligned with sustainable coastal development, the Blue Economy can support long-term prosperity while safeguarding marine and coastal ecosystems for present and future generations.

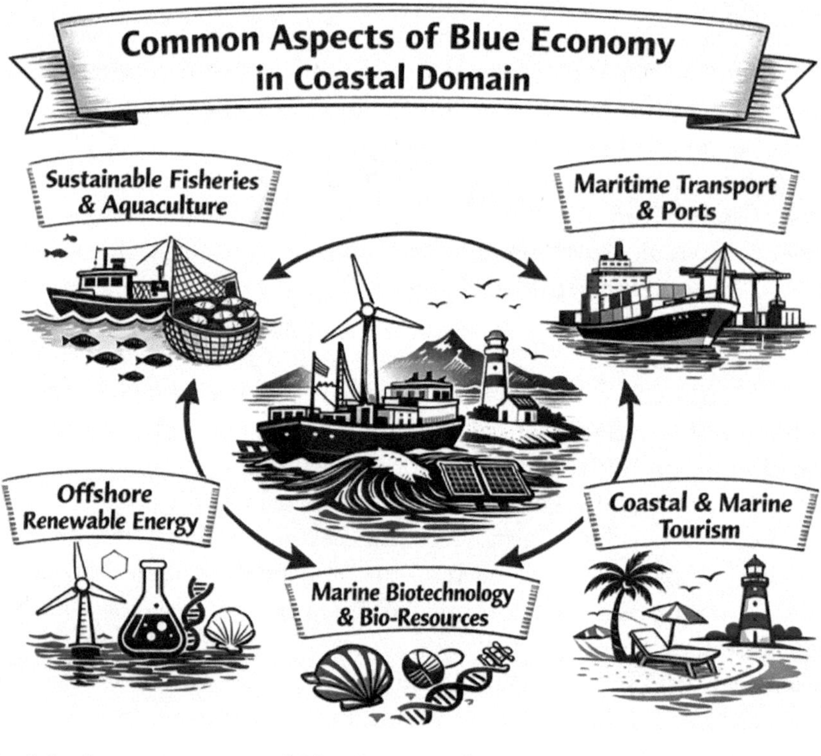

Fig. 9.1 Common aspects of Blue Economy (Prepared by the authors)

9.3 Sustainable Fisheries and Aquaculture

Sustainable fisheries and aquaculture form a central part of the Blue Economy and play an important role in supporting sustainable coastal development. Fisheries and aquaculture provide food, nutrition, employment, and cultural value to millions of people living in coastal and island regions. These activities rely directly on healthy marine and coastal ecosystems. Overfishing, habitat loss, pollution, and climate-related changes have placed increasing pressure on fish stocks and aquaculture systems. Sustainability in fisheries and aquaculture has therefore become a key development concern that is closely linked to food security, poverty reduction, and the resilience of coastal communities. Sustainable fisheries focus on keeping fish stocks healthy and productive while supporting economic viability and social fairness. This involves managing fishing activities in ways that prevent overfishing, reduce unwanted catch, protect important habitats, and respect the natural limits of marine ecosystems. The emphasis is on long-term resource care rather than short-term gains. Sustainable aquaculture involves farming aquatic species in ways that reduce environmental impacts, limit dependence on wild fish for feed, and support local livelihoods. Together, sustainable fisheries and aquaculture help meet growing demand for aquatic food while protecting ecosystem health.

In many developing countries, small-scale and artisanal fisheries remain the backbone of coastal livelihoods. They provide income, employment, and nutrition, especially for vulnerable and marginalized groups. These communities are often the most affected by declining fish stocks, climate variability, and competition from large commercial fleets. Sustainable fisheries approaches address these challenges by encouraging community participation, secure access to resources, and recognition of traditional fishing practices. Strengthening local governance and shared management systems helps promote social inclusion and stability in coastal areas. Aquaculture has expanded rapidly worldwide and is often seen as a way to reduce pressure on wild fish stocks. Within the Blue Economy, sustainable aquaculture offers opportunities for economic growth, innovation, and job creation. Poorly managed aquaculture, however, can lead to water pollution, habitat damage, disease outbreaks, and conflicts over land and water use. Sustainable aquaculture focuses on careful site selection, efficient feed use, waste control, and low-impact farming systems. When integrated with coastal planning, aquaculture can support capture fisheries and improve overall coastal productivity.

Climate change further highlights the importance of sustainability in fisheries and aquaculture. Rising sea temperatures, ocean acidification, shifting currents, and extreme weather events are changing fish distribution and ecosystem conditions. These changes create uncertainty for fishers and producers and place pressure on food systems. Sustainable management supports adaptation through flexible fishing rules, protection of resilient habitats, and livelihood diversification. Climate resilient aquaculture promotes species diversity, improved biosecurity, and adaptable farming practices. Effective governance remains essential for sustainable fisheries and aquaculture. International guidance promotes ecosystem-based management, precaution in decision-making, and stakeholder participation. At national and regional levels, aligning fisheries and aquaculture policies with broader coastal development and Blue Economy strategies helps reduce conflicts and improve coordination. Strong monitoring, enforcement, and data systems are needed to translate sustainability principles into practice (Arkhipov, 2022; Rahimi-Midani, 2023).

Sustainable fisheries and aquaculture depend on responsible harvesting, low-impact aquaculture systems, community involvement, and climate adaptation. In coastal areas where fisheries share space with tourism, ports, energy infrastructure, and conservation areas, integrated planning becomes especially important. Marine spatial planning and coastal management help balance competing uses while protecting ecosystem services. Despite growing attention, several challenges remain. Illegal fishing, limited financial resources, weak institutions, and restricted access to technology continue to slow progress, particularly for small-scale producers. Addressing these issues requires investment in skills, innovation, inclusive financing, and cooperation across regions. Sustainable fisheries and aquaculture are essential for achieving sustainable coastal development and a strong Blue Economy. They support food security, livelihoods, and the ecological foundations of coastal and marine systems. Strengthening sustainability, resilience, and fairness in fisheries and aquaculture governance represents a long-term investment in the future of coastal communities (Fig. 9.2).

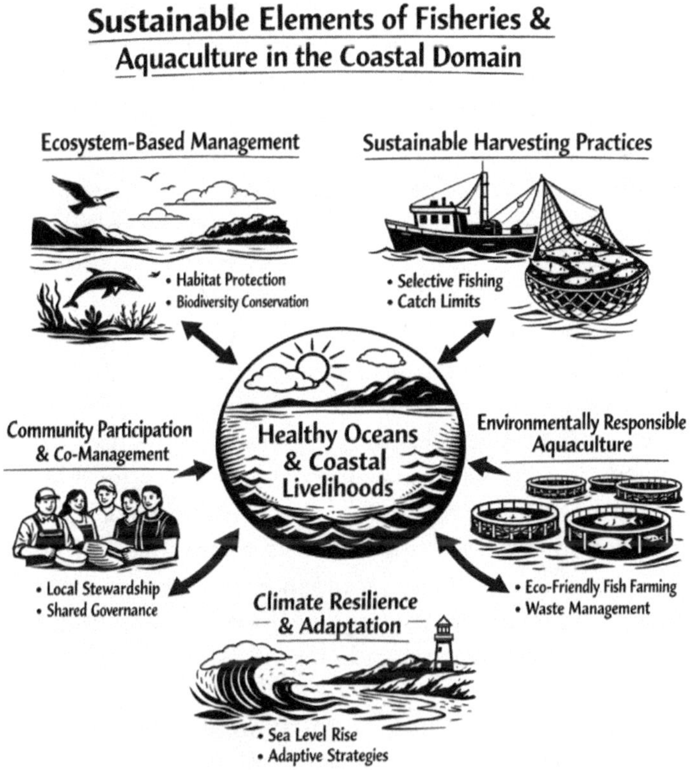

Fig. 9.2 Sustainable elements of fisheries and aquaculture (Prepared by the authors)

9.4 Sustainable Tourism and Coastal Recreation

Sustainable tourism and coastal recreation are an important part of the Blue Economy and play a key role in supporting sustainable coastal development. Coastal and marine tourism is among the fastest growing sectors worldwide and provides income, employment, and infrastructure for many coastal regions. Beaches, coral reefs, mangroves, coastal landscapes, and cultural heritage sites attract large numbers of visitors each year and serve as major economic resources for coastal and island communities. At the same time, poorly planned or mass tourism can place heavy pressure on fragile coastal environments through habitat damage, pollution, overuse of natural resources, and social disruption. Sustainable tourism and coastal recreation aim to balance economic benefits with environmental protection and community well-being. Sustainable coastal tourism is based on respect for the environment, local culture, and inclusive economic growth. It focuses on reducing environmental impacts while increasing benefits for local people. The long-term success of tourism depends on the health of coastal and marine ecosystems that support activities such as swimming, diving, boating, and wildlife viewing. Sustainable tourism encourages responsible visitor behaviors, environmentally friendly facilities, and low-impact recreational practices so that natural and cultural attractions remain valuable over time.

The role of sustainable tourism in the Blue Economy is closely linked to livelihoods and local development. Coastal tourism creates jobs across many activities, including accommodation, transport, guiding services, crafts, and food supply chains. When tourism is well planned, it can support community-based enterprises and help fishing and farming households diversify their incomes. Poorly managed tourism often leads to unequal benefit sharing, short-term employment, and financial losses for local areas. Sustainable tourism approaches therefore emphasize local participation, support for small-scale initiatives, and fair distribution of economic gains. Environmental protection is a central concern in coastal tourism and recreation. Ecosystems such as coral reefs, dunes, mangroves, and seagrass beds are especially sensitive to physical disturbance, waste, and unplanned construction linked to tourism. Sustainable tourism applies ecosystem-based management through visitor limits, environmental standards, and nature-based solutions such as habitat restoration and green infrastructure. These actions help conserve biodiversity while improving the quality of recreational experiences.

Climate change has increased the need for sustainability in coastal tourism. Rising sea levels, coastal erosion, coral bleaching, and extreme weather events threaten tourism facilities and destinations. Sustainable tourism responds by promoting resilient infrastructure, protecting natural coastal buffers, and expanding tourism activities beyond climate-sensitive areas. Planning that accounts for future climate risks helps tourism remain viable while reducing damage from natural hazards. Governance and policy play an important role in shaping sustainable coastal tourism. Effective coastal management, land use planning, and regulatory enforcement are needed to prevent overdevelopment and environmental harm. Coordination among tourism authorities, environmental agencies, local governments, and communities helps manage competing interests and align tourism development with Blue Economy goals. Sustainable tourism and coastal recreation offer strong opportunities for economic development but require careful management. When guided by sustainability principles, tourism can support livelihoods and encourage the protection of coastal and marine ecosystems. Integrating environmental care, community participation, and climate resilience into tourism planning is essential for maintaining healthy and inclusive coastal systems for present and future generations (Gössling et al., 2018; Suresh et al., 2025).

9.5 Sustainable Coastal Infrastructure Development

Sustainable coastal infrastructure development is a key part of sustainable coastal development and the Blue Economy because it provides the physical base for economic activities, settlements, and environmental management in coastal areas. Coastal infrastructure includes ports and harbors, roads, housing, tourism facilities, energy systems, water and sanitation services, and protective structures such as seawalls and breakwaters. These facilities support trade, movement, livelihoods, and access to essential services, making them vital for coastal economies. Traditional approaches to infrastructure development, however, have often led to habitat loss, higher disaster risk, and long-term environmental damage. Sustainable coastal infrastructure responds to these challenges by integrating environmental care, climate resilience, and social inclusion into planning and design. Coastal areas are dynamic spaces shaped by tides, waves, sediment movement, and changing sea levels. Infrastructure that ignores these natural processes can increase erosion, flooding, and ecosystem damage. Sustainable approaches aim to work with natural systems rather than against them. This includes the use of ecosystem-based and nature-based solutions such as restoring mangroves, dunes, and wetlands alongside or instead of hard engineering structures. These solutions can reduce maintenance costs over time while strengthening natural protection and ecological functions.

Climate change has increased the need for sustainable coastal infrastructure. Rising sea levels, storm surges, coastal flooding, and extreme weather events place growing pressure on existing and planned infrastructure. Ports, transport networks, tourism facilities, and coastal cities are especially exposed, with direct impacts on safety and economic stability. Sustainable infrastructure planning addresses these risks through elevated designs, durable materials, flexible land use planning, and the protection of natural coastal buffers. These measures help reduce damage, protect investments, and support long-term coastal stability. Within the Blue Economy, sustainable coastal infrastructure supports the growth of ocean-based sectors. Well-planned ports improve fisheries and maritime trade while limiting pollution and damage to habitats. Sustainable tourism infrastructure allows visitors to enjoy coastal areas without degrading beaches, reefs, and landscapes. Renewable energy projects such as offshore wind and tidal systems require careful design and location to avoid environmental harm. When infrastructure follows sustainability principles, it supports economic diversification, innovation, and employment while protecting ecosystem health.

Governance and policy play an important role in guiding sustainable coastal infrastructure. Tools such as integrated coastal zone management and marine spatial planning help coordinate development across sectors and levels of government. Environmental assessments and climate risk assessments support better decision-making and reduce long-term risks. Strong coordination among institutions is especially important in coastal regions, where unclear responsibilities often lead to poor planning. Social equity is central to sustainable coastal infrastructure development. Infrastructure projects can affect local communities through land use changes, loss of livelihoods, and reduced access to coastal resources. Sustainable approaches prioritize inclusive planning, community participation, and fair sharing of benefits. Infrastructure that improves access to markets, services, and safe housing can strengthen the ability of vulnerable groups to adapt to change. Poorly planned projects can deepen inequality and create conflict. Technological advances also support sustainable coastal infrastructure. Improved materials, monitoring systems, and planning tools allow for more efficient and environmentally sensitive designs. Green buildings, smart ports, and low-carbon transport systems reduce emissions and environmental impacts while improving performance. Combining modern technologies with local knowledge and traditional practices can further improve outcomes, especially in resource-limited coastal areas. Sustainable coastal infrastructure development forms the backbone of sustainable coastal development and a resilient Blue Economy. It supports livelihoods, economic growth, and environmental protection in coastal regions. By integrating ecological awareness, climate resilience, and social inclusion into infrastructure planning and design, coastal societies can reduce risks and create lasting benefits. As climate change and development pressures increase, adopting sustainable approaches to coastal infrastructure has become essential for building resilient, productive, and fair coastal futures (Suresh et al., 2025; Steven et al., 2023).

9.6 Conservation and Restoration of Coastal Ecosystems

Conservation and restoration of coastal ecosystems are central to sustainable coastal development and the Blue Economy because these ecosystems support ecological balance, economic activity, and social resilience in coastal regions. Mangroves, coral reefs, seagrass meadows, salt marshes, and coastal wetlands provide essential services such as shoreline protection, fisheries support, carbon storage, and biodiversity conservation. Despite their importance, many coastal ecosystems have been heavily degraded by urban growth, pollution, overuse of natural resources, and poorly planned infrastructure. Protecting and restoring these ecosystems is therefore critical for sustaining the natural base of coastal livelihoods and economies. Conservation focuses on protecting existing ecosystems from further damage, while restoration aims to rebuild ecosystems that have been degraded or lost. Sustainable coastal development brings these approaches together through ecosystem-based management. This approach recognizes that healthy ecosystems are closely linked to economic and social well-being. Conservation tools such as protected areas, regulated resource use, and habitat protection help maintain key ecological functions. Restoration efforts help recover ecosystem services that support fisheries, tourism, and climate regulation. Together, these actions strengthen the capacity of coastal systems to cope with human pressures and environmental change.

The importance of conserving and restoring coastal ecosystems has increased as climate-related risks grow. Coastal ecosystems act as natural defenses against storms, flooding, and erosion. Mangroves and wetlands reduce wave energy, coral reefs weaken storm impacts, and seagrass beds help stabilize sediments. These nature-based solutions often provide cost-effective and flexible alternatives to hard engineering structures. Integrating ecosystem conservation and restoration into coastal planning allows development efforts to address climate adaptation, disaster risk reduction, and biodiversity protection at the same time. Coastal ecosystem conservation is closely linked to economic sustainability within the Blue Economy. Healthy ecosystems support productive fisheries, attract tourism, and reduce long-term management costs.

Degraded ecosystems, in contrast, lead to declining fish stocks, loss of tourism value, and greater exposure to coastal hazards. Experience shows that ecosystem restoration can deliver environmental, economic, and social benefits. Conservation in this sense supports development rather than limiting it. Governance and community involvement are critical to successful conservation and restoration. Approaches that exclude local communities or overlook traditional knowledge often fail to produce lasting outcomes. Sustainable strategies encourage shared management, local participation, and fair benefit sharing. Involving coastal communities in restoration activities can improve ecological results while creating jobs and strengthening local stewardship. Clear legal frameworks, secure access rights, and effective monitoring systems are necessary to support long-term conservation efforts. Conservation and restoration of coastal ecosystems are essential for sustainable coastal development and a resilient Blue Economy. Protecting and rebuilding natural systems helps maintain ecosystem services, reduce climate risks, and support sustainable livelihoods. Integrating ecosystem conservation into development planning aligns environmental protection with economic progress and social well-being. As pressure on coastal zones continues to increase, investment in conservation and restoration has become both an environmental need and a strategic priority for long-term coastal development (Dimeyeva, 2001; Rudianto et al., 2022; Saunders et al., 2024).

9.7 Blue Technologies and Innovation for Sustainable Development

Blue technologies and innovation play an important role in sustainable coastal development and the Blue Economy by offering practical ways to connect economic growth with environmental protection and social well-being. Blue technologies refer to the use of scientific, engineering, and digital tools in ocean and coastal sectors such as energy, fisheries, transport, conservation, and climate adaptation. Coastal and marine systems are facing growing pressure from population growth, intensive resource use, and climate change. Innovation has therefore become essential for improving efficiency, reducing environmental damage, and strengthening the management of ocean resources. Blue innovation is guided by ideas of sustainability, resilience, and long-term thinking. Traditional technologies often focus mainly on increasing production, while blue technologies aim to use resources carefully and protect ecosystem health. These technologies include ocean observation systems, tools for sustainable harvesting, low carbon energy solutions, and methods for ecosystem restoration. Better understanding of complex marine systems supports informed decision-making and flexible management in changing coastal environments. Innovation thus helps convert scientific knowledge into practical solutions for coastal development.

A key contribution of blue technologies is their role in climate action and environmental protection. Offshore renewable energy such as wind, tidal, and wave power supports low-carbon development while creating new opportunities for coastal economies. Digital monitoring tools improve tracking of coastal erosion, water quality, and biodiversity, and strengthen early warning systems for storms and flooding. These advances support disaster risk reduction, climate adaptation, and nature-based approaches in coastal planning. Blue technologies also improve the sustainability of established Blue Economy sectors. In fisheries and aquaculture, innovations such as vessel tracking systems, improved fishing gear, and efficient feeding practices help reduce overfishing, unwanted catch, and waste. Marine biotechnology supports the development of products such as medicines, bio-based materials, and alternative energy sources while reducing pressure on natural stocks. In maritime transport, cleaner fuels, energy-efficient vessels, and smart port systems help lower emissions and improve environmental performance. These changes support both economic efficiency and environmental responsibility. Innovation also supports economic development in coastal regions by encouraging diversification, entrepreneurship, and skilled employment. New businesses and research centers focused on ocean technologies help build knowledge-based industries. Access to technology, skills, and finance, however, remains uneven across regions. Supportive policies, training programs, and partnerships are needed to ensure that innovation benefits coastal communities more widely. Governance plays an important role in guiding blue technologies toward sustainable outcomes.

Clear regulations, environmental safeguards, and ethical standards help prevent new risks linked to technological change. Aligning innovation policies with coastal planning and marine management ensures that technologies support broader sustainability goals. Cooperation among governments, researchers, industry, and coastal communities strengthens the relevance and acceptance of innovation. Bluc technologies and innovation are central to advancing sustainable coastal development and a resilient Blue Economy. They provide useful tools to address climate change, resource pressure, and environmental decline while creating new economic opportunities. When guided by sustainability principles and inclusive governance, blue innovation can support coastal development that is environmentally responsible, economically sound, and socially fair (Dimeyeva, 2001; Rudianto et al., 2022; Saunders et al., 2024).

9.8 Socio-Economic Considerations in Sustainable Coastal Development

Socio-economic considerations are central to sustainable coastal development and the effective realization of the Blue Economy because coastal areas are not only ecological spaces but also places where people live, work, and maintain social and cultural connections. Coastal regions support a wide range of economic activities such as fisheries, aquaculture, tourism, ports, and emerging blue industries, while also hosting large populations that are often vulnerable to economic and environmental change. Sustainable coastal development therefore requires a people-focused approach that balances economic growth with social equity, poverty reduction, and respect for cultural values. Ignoring these dimensions can lead to development outcomes that are environmentally sound but socially unfair and difficult to sustain. Livelihood security is a major concern in coastal development. Millions of people depend directly on coastal and marine resources for income and employment, especially through small-scale fisheries, aquaculture, and informal tourism. These livelihoods are highly sensitive to environmental degradation, climate variability, and market changes. Sustainable coastal development aims to strengthen livelihood resilience by promoting diversification, value addition, and skills development. Reducing dependence on a single resource or activity helps improve economic stability and lowers vulnerability in coastal communities. Equity and inclusion are essential to sustainable coastal development. Coastal development projects often benefit large commercial interests more than small-scale resource users, women, and indigenous communities. Addressing these imbalances requires governance systems that ensure fair access to resources, meaningful participation in decision-making, and equitable sharing of benefits. Community-based management, recognition of traditional knowledge, and secure access rights help strengthen local stewardship and promote social justice in coastal areas.

The Blue Economy offers opportunities for socio-economic transformation, but outcomes depend on how development is planned and governed. Blue economy activities can create jobs, attract investment, and support innovation, but they can also increase social exclusion if social impacts are not carefully considered. Large infrastructure projects, tourism expansion, and offshore energy development may disrupt livelihoods and displace communities. Sustainable coastal development frameworks address these risks through social impact assessments, stakeholder engagement, and benefit sharing arrangements. Climate risks further shape socio-economic conditions in coastal regions. Coastal populations face frequent exposure to cyclones, storm surges, flooding, and sea level rise. Poor households and marginalized groups are often the most affected due to limited assets, insecure housing, and weak access to basic services. Sustainable coastal development integrates disaster risk reduction and climate adaptation measures that prioritize vulnerable groups, including early warning systems, social protection, and resilient housing. Strong institutions and effective governance are critical for achieving positive socio-economic outcomes. Weak coordination among agencies, limited enforcement, and fragmented responsibilities often reduce the effectiveness of coastal development efforts. Strengthening institutional capacity, improving policy coordination, and promoting participatory planning help align social goals with environmental and economic objectives. Socio-economic considerations are therefore inseparable from sustainable coastal development and the Blue Economy.

Giving priority to livelihoods, equity, inclusion, and resilience allows coastal development to support human well-being while protecting the natural systems on which coastal societies depend. Integrating these dimensions into coastal planning and governance is essential for building development pathways that are socially fair, environmentally responsible, and economically viable (Burbridge, 1999; Dronkers & Stojanovic, 2016; Neumann & Kenchington, 2017).

9.9 Sustainable Coastal Governance and Policy

Sustainable coastal governance and policy provide the institutional foundation for sustainable coastal development and the Blue Economy. Coastal zones are complex spaces where many activities take place within a limited land and sea area. Fisheries, tourism, ports, energy facilities, conservation areas, and coastal settlements operate side by side, often under different authorities and rules. This complexity can lead to conflicts over resource use, weak coordination, and environmental damage. Sustainable coastal governance addresses these challenges through clear policies, coordinated institutions, and participatory decision-making that balance economic development, social equity, and environmental protection. Sustainable coastal governance is guided by principles such as integration, transparency, accountability, and flexibility. Coastal and marine systems function as connected social and ecological systems, and governance approaches must reflect these linkages. Fragmented management, where sectors such as fisheries, environment, transport, and urban development are handled separately, has often failed to address long-term impacts and cumulative risks. Integrated governance frameworks promote coordination across sectors, the use of scientific knowledge in policymaking, and planning that considers future needs. These principles help align coastal development with the broader goals of the Blue Economy.

Policy tools such as Integrated Coastal Zone Management and Marine Spatial Planning are important for putting sustainable governance into practice. Integrated Coastal Zone Management supports coordinated management of land and sea activities while accounting for ecological processes and social priorities. Marine Spatial Planning helps organize coastal and marine space in ways that reduce conflicts, improve efficiency, and protect important ecosystems. Together, these tools support balanced decisions between development and conservation. The growth of the Blue Economy has increased the need for strong coastal governance. Blue Economy strategies aim to expand ocean-based economic activities while maintaining ecosystem health and social inclusion. Without effective governance, these strategies can lead to environmental harm or social inequality. Sustainable coastal governance ensures that Blue Economy initiatives follow clear rules, respect environmental limits, and remain accountable. Governance reform is especially important in developing coastal and island regions where institutional capacity is often limited.

Participation and coordination across different levels of governance are essential for effective coastal management. Coastal governance involves national and regional authorities, local governments, communities, private actors, and civil society groups. Inclusive approaches recognize that local communities have valuable knowledge and a direct interest in sustainable resource use. Participation improves trust, compliance, and long-term success of policies. Shared management in fisheries, community-based conservation, and local coastal planning are examples of such approaches. Climate change presents a major challenge for coastal governance. Sea level rise, erosion, saline intrusion, and extreme weather events are increasing risks for coastal populations and infrastructure. Sustainable coastal governance integrates climate adaptation and disaster risk reduction into policies and planning. This includes climate informed land use planning, protection of natural coastal buffers, resilient infrastructure standards, and early warning systems. Flexible governance that allows learning and adjustment over time is important in responding to uncertainty.

Legal and regulatory frameworks also play a key role in sustainable coastal governance. Clear laws that define coastal boundaries, resource rights, environmental standards, and institutional responsibilities are necessary for effective implementation. Weak enforcement and overlapping mandates often reduce policy effectiveness. Strengthening legal clarity and enforcement helps reduce illegal activities and supports responsible investment. Equity and social justice are important elements of sustainable coastal governance. Coastal development policies can affect groups differently, often placing greater pressure on small-scale fishers, informal settlers, and indigenous communities. Sustainable governance addresses these concerns through rights-based policies, social safeguards, and fair sharing of benefits. Social impact assessments and grievance mechanisms help reduce conflict and improve acceptance of policies. Knowledge, data, and institutional capacity strongly influence governance outcomes. Reliable information on ecosystems, livelihoods, and climate risks is essential for informed decision-making. Data gaps and limited technical capacity remain challenges in many coastal regions. Strengthening monitoring systems, research efforts, and links between science and policy improves governance quality. Capacity building for policymakers, local authorities, and communities supports more effective and inclusive planning. Sustainable coastal governance and policy are essential for achieving sustainable coastal development and a resilient Blue Economy. They provide the structures through which competing interests are balanced, environmental limits are respected, and long-term development choices are shaped. Promoting integration, participation, flexibility, and equity help coastal societies pursue economic opportunities while protecting ecosystems and supporting social well-being. Strengthening governance and policy frameworks has become a strategic priority for building sustainable coastal futures (Gonçalves et al., 2024; Mimura et al., 2008).

9.10 Conclusion and Arguments

Sustainable coastal development brings together the themes of the Blue Economy, sustainable fisheries and aquaculture, tourism and recreation, coastal infrastructure, ecosystem conservation, blue technologies, socio-economic inclusion, and governance into a single and future focused development approach. Coastal areas function as living systems where the environment, economic activities, and human well-being are closely linked. Experience shows that sector-based or short-term solutions are not enough to address the combined pressures created by climate change, rapid urban growth, overuse of resources, and growing social inequality. Sustainable coastal development therefore calls for an integrated and people focused framework that strengthens resilience while respecting ecological limits. The Blue Economy provides a shared perspective for shaping this approach by redefining growth in ways that support environmental protection, innovation, and social fairness. Healthy coastal ecosystems support fisheries, tourism, climate regulation, and protection from natural hazards, while sustainable infrastructure and appropriate technologies enable economic activities with lower environmental impacts. Social factors such as secure livelihoods, inclusion, and fairness influence whether sustainability efforts gain long-term support. Governance and policy frameworks play a central role by linking these elements, improving coordination across sectors, encouraging participation, and turning sustainability principles into practical action. Sustainable coastal development should be viewed as an ongoing process that manages risks, balances competing needs, and builds long-term resilience so that coastal and marine resources continue to support present and future generations. The chapter discussions lead to the following arguments:

(a) Sustainable coastal development is an integrated socio-ecological process rather than a sector-based activity.
(b) The Blue Economy is sustainable only when ecosystem health and social equity guide development.
(c) Ecosystem conservation and restoration strengthen economic resilience instead of limiting growth.
(d) Climate adaptation and disaster risk reduction must be part of all coastal development actions.
(e) Inclusive and adaptive governance is essential for achieving long-term sustainable coastal development.

References

Arkhipov, A. G. (Ed.). (2022). *Sustainable fisheries and aquaculture: Challenges and prospects for the blue bioeconomy*. Springer Cham. https://doi.org/10.1007/978-3-031-08284-9. isbn:978-3-031-08283-2.

Burbridge, P. R. (1999). The challenge of demonstrating the socio-economic benefits of integrated coastal management. In W. Salomons, R. K. Turner, L. D. de Lacerda, & S. Ramachandran (Eds.), *Perspectives on integrated coastal zone management*. Environmental Science. Springer. https://doi.org/10.1007/978-3-642-60103-3_3

Dimeyeva, L. A. (2001). Methods of conservation and restoration of vegetation cover on the Aral Sea coast. In S. W. Breckle, M. Veste, & W. Wucherer (Eds.), *Sustainable land use in deserts*. Springer. https://doi.org/10.1007/978-3-642-59560-8_6

Dronkers, J., & Stojanovic, T. (2016). Socio-economic impacts—Coastal management and governance. In M. Quante & F. Colijn (Eds.), *North Sea region climate change assessment. Regional climate studies*. Springer. https://doi.org/10.1007/978-3-319-39745-0_19

Gonçalves, C., & Pinho, P. (2024). A manifesto for coastal landscape governance: Reframing the relationship between coastal and landscape governance. *Ambio, 53*, 1454–1465. https://doi.org/10.1007/s13280-024-02040-5

Gössling, S., Hall, C. M., & Scott, D. (2018). Coastal and ocean tourism. In M. Salomon & T. Markus (Eds.), *Handbook on marine environment protection*. Springer. https://doi.org/10.1007/978-3-319-60156-4_40

Hazra, S., & Bhukta, A. (Eds.). (2022). *The blue economy: An Asian perspective*. Springer Cham. https://doi.org/10.1007/978-3-030-96519-8. isbn:978-3-030-96518-1.

Kotowska, I., Mańkowska, M., & Pluciński, M. (2020). Socio-economic costs and benefits of seaport infrastructure development for a local environment. The case of the port and the City of Świnoujście. In A. Carpenter & R. Lozano (Eds.), *European port cities in transition. Strategies for sustainability*. Springer. https://doi.org/10.1007/978-3-030-36464-9_18

Mimura, N., Shikida, A., & Yamao, M. (2008). Policy for conservation and sustainable development of the coastal zone. In N. Mimura (Ed.), *Asia-Pacific coasts and their management. Coastal systems and continental margins* (Vol. 11). Springer. https://doi.org/10.1007/978-1-4020-3625-5_4

Neumann, B., Ott, K., & Kenchington, R. (2017). Strong sustainability in coastal areas: A conceptual interpretation of SDG 14. *Sustainability Science, 12*, 1019–1035. https://doi.org/10.1007/s11625-017-0472-y

Rahimi-Midani, A. (2023). *Deep Technology for Sustainable Fisheries and Aquaculture*. Springer Singapore. https://doi.org/10.1007/978-981-99-4917-5. isbn:978-981-99-4916-8.

Rudianto, R., Darmawan, V., Isdianto, A., et al. (2022). Restoration of coastal ecosystems as an approach to the integrated mangrove ecosystem management and mitigation and adaptation to climate changes in north coast of East Java. *Journal of Coastal Conservation, 26*, 37. https://doi.org/10.1007/s11852-022-00865-4

Sardá, R., Pogutz, S., Theodorou, N. A., & Ramakrishna, K. (2025). The blue economy: What it is, what is not, and how to get there. In W. Leal Filho, A. L. Salvia, J. P. P. Eustachio, & M. A. P. Dinis (Eds.), *Handbook of sustainable blue economy*. Springer. https://doi.org/10.1007/978-3-031-32671-4_48-1

Saunders, M. I., Cannard, T., Fischer, M., Sheppard, M., Twomey, A., Morris, R., & Waltham, N. J. (2024). A roadmap to coastal and marine ecological restoration in Australia. *Environmental Science & Policy, 159*, 103808. https://doi.org/10.1016/j.envsci.2024.103808

Steven, A., et al. (2023). Coastal development: Resilience, restoration and infrastructure requirements. In J. Lubchenco & P. M. Haugan (Eds.), *The blue compendium*. Springer. https://doi.org/10.1007/978-3-031-16277-0_7

Stylios, S., Floqi, T., Marinski, J., & Damiani, L. (Eds.). (2015). *Sustainable development of sea-corridors and coastal waters: The TEN ECOPORT project in south East Europe*. Springer Cham. https://doi.org/10.1007/978-3-319-11385-2. isbn:978-3-319-11384-5.

Suresh, A., Wartman, M., Rasheed, A. R., & Macreadie, P. I. (2025). Tourism and recreation in blue carbon ecosystems: Exploring synergies, trade-offs and pathways to sustainability. *Ocean & Coastal Management, 266*, 107697. https://doi.org/10.1016/j.ocecoaman.2025.107697

10

Coastal Disaster Risk Reduction Practices: Future Challenges and Opportunities

Coastal regions are among the most dynamic and vulnerable environments on Earth, shaped by the close interaction between natural systems and human activities. In recent decades, climate change has intensified coastal disasters such as storms, flooding, erosion, and sea-level rise, placing increasing pressure on coastal ecologies and the communities that depend on them. This chapter provides an introduction to climate-related disasters in coastal areas and explores how these events interact with fragile coastal ecosystems. The chapter begins by outlining the basic concepts of climate disasters and coastal ecologies, helping readers understand how environmental processes, human development, and climate variability are closely linked. It highlights how changes in temperature, rainfall patterns, and ocean systems are altering coastal landscapes and increasing disaster risks. Attention is also given to the role of mangroves, coral reefs, wetlands, and other ecosystems in reducing disaster impacts and supporting coastal resilience. The chapter emphasizes the need to view coastal disasters not only as natural events but as outcomes influenced by social, economic, and ecological factors. By setting this foundation, the chapter prepares readers for deeper discussions on coastal risk reduction, adaptation, and sustainable management

K. D. Malakar et al., *Coastal Ecologies at Risk*, Palgrave Studies in Environmental Sustainability, https://doi.org/10.1007/978-3-032-26081-9_10

in later sections. It aims to encourage a holistic understanding of coastal systems in an era of accelerating climate change.

Key Points of the Chapter

- Describes how coastal disaster risks emerge from the combined effects of climate change, development, ecosystems, and governance
- Explains why prevention-focused and adaptation-driven approaches work better than response-based disaster management in coastal areas
- Highlights the role of healthy coastal ecosystems and nature-based solutions in building long-term resilience
- Examines the importance of community empowerment and inclusive governance in effective coastal disaster risk reduction
- Identifies how innovation in technology planning and policy can strengthen future coastal resilience.

10.1 Coastal Disaster Risk Reduction

Coastal disaster risk reduction (CDRR) has become an important field of study and practice as coastal hazards continue to grow in frequency, intensity, and complexity across the world. Coastal zones are places where land and sea meet and where large populations, vital infrastructure, valuable ecosystems, and major economic activities are concentrated. Fisheries tourism ports energy facilities and coastal settlements depend heavily on these areas for their survival and growth. At the same time coastal regions face increasing risks from both natural processes and human activities including cyclones, storm surges, flooding, erosion, sea-level rise, tsunamis, and saltwater intrusion. In this context CDRR offers a structured way to understand, anticipate, and reduce disaster risks while protecting livelihoods, ecosystems, and long-term coastal sustainability.

Coastal disaster risk reduction is closely linked to the wider disaster risk reduction approach which focuses on reducing exposure vulnerability and potential losses rather than concentrating only on emergency response after disasters occur. In coastal areas this approach is especially important because disaster risk is shaped by the interaction of physical coastal processes, ecological conditions, social and economic factors, and systems of governance. Rapid and unplanned urban growth, degradation of natural protective ecosystems, climate variability, and weak coordination among institutions often increase the scale of disaster impacts. CDRR therefore promotes an integrated and preventive approach that combines scientific risk assessment, ecosystem-based management, resilient infrastructure, and community preparedness. Climate change has further increased the importance of coastal disaster risk reduction. Rising sea levels, changing storm patterns, and warming coastal waters are altering hazard conditions and intensifying existing risks. Traditional strategies that depend mainly on hard engineering structures are no longer sufficient on their own. Current CDRR practice supports a balanced combination of structural and non-structural measures including early warning systems, land-use planning, ecosystem-based adaptation, climate-resilient livelihoods, and inclusive governance arrangements. Collaboration across different levels and actors is a central element of coastal disaster risk reduction. Coastal hazards often extend beyond administrative boundaries and require coordination at local, regional, national, and sometimes international levels. Active involvement of local communities' indigenous groups and vulnerable populations is equally important since their knowledge and experience provide valuable insights for risk reduction. Bringing scientific understanding together with local practices helps create solutions that are more effective, realistic, and sustainable (Ferreira, 2022; Setiawati et al., 2024).

This chapter presents coastal disaster risk reduction as a field that continues to evolve in response to environmental change, social transformation, and emerging risks. The ideas discussed here provide a foundation for linking disaster risk reduction with sustainable coastal development and for building coastal futures that are safer, more equitable, and resilient.

10.2 Coastal Disaster Risk Reduction Practices

Present coastal disaster risk reduction (CDRR) practices reflect a steady movement away from response-focused disaster management toward more proactive, integrated, and resilience-focused approaches. Coastal regions are widely recognized as high-risk areas because of the combined influence of natural hazards, climate change, rapid urban growth, and declining coastal ecosystems. Present-day CDRR practices therefore aim not only to reduce loss of life and property but also to address the deeper causes of vulnerability and exposure. These approaches are shaped by advances in scientific understanding, technological development, evolving policy frameworks, and growing awareness of the close links between social systems and coastal environments. Hazard and risk assessment plays a central role in present CDRR practice. Coastal hazard mapping, vulnerability assessment, and risk analysis are commonly used to identify areas exposed to cyclones, storm surges, flooding, erosion, and sea-level rise. Improvements in remote sensing, geographic information systems, and modeling techniques have increased the accuracy and usefulness of these assessments. Such tools support informed planning by helping authorities identify priority areas, guide land-use decisions, and design appropriate risk reduction measures. Climate change projections are now more often included to reflect future changes in hazard patterns. Structural measures remain important in coastal disaster risk reduction especially in densely populated and economically important coastal zones. Seawalls, embankments, breakwaters, storm surge barriers, and cyclone shelters are widely used to reduce direct exposure to hazards. Design standards have improved in many regions to better account for extreme events and long-term sea-level rise. At the same time reliance on hard infrastructure alone is increasingly questioned because of high costs, environmental impacts, and limited long-term effectiveness. As a result, structural measures are now more commonly combined with non-structural and ecosystem-based approaches.

Ecosystem-based approaches have gained strong attention in recent CDRR practice. Protection and restoration of mangroves, coral reefs, salt marshes, and sand dunes help reduce wave energy, limit storm surge impacts, stabilize shorelines, and support livelihoods. These approaches also provide additional benefits such as biodiversity conservation, carbon storage, and income opportunities. Many coastal risk reduction efforts now place ecosystem-based measures within wider coastal management and climate adaptation strategies. Early warning systems are another important element of current CDRR practice. Advances in weather forecasting, ocean monitoring, and communication systems have improved the accuracy and timing of warnings for coastal hazards. Present systems focus on linking forecasting with risk communication, evacuation planning, and local response capacity. Attention is also given to ensuring warnings reach vulnerable and remote coastal communities in forms that are clear, practical, and locally appropriate. Community-based disaster risk reduction has become a key part of coastal risk management. Coastal communities are often the first to face disasters and play a major role in response and recovery. Current practices therefore emphasize awareness programs, training activities, and participatory planning. Local knowledge, social networks, and community institutions are increasingly recognized for their role in strengthening preparedness and adaptive capacity. Inclusive participation is encouraged with attention to women, children, older people, and marginalized groups who often face greater risks (Busayo & Kalumba, 2020; Setiawati et al., 2024).

Policy and institutional arrangements strongly influence the success of CDRR practices. Many countries now follow national disaster risk reduction strategies that emphasize prevention, preparedness, and resilience. In coastal areas, these policies are more closely linked with coastal zone management, climate adaptation, and sustainable development planning. Coordination across sectors such as environment, urban planning, fisheries, and disaster management is essential to address complex coastal risks. Decentralized governance has also supported more flexible and locally relevant implementation. Capacity building and knowledge sharing remain essential elements of present-day CDRR practice. Training programs for professionals, officials, and community leaders help strengthen technical, organizational, and social skills. Efforts to connect research, policy, and practice support the use of scientific knowledge in real-world decision-making. Regional cooperation also enables the exchange of data, experience, and early warning information across coastal regions. Present coastal disaster risk reduction practices show that reducing coastal risk requires integrated, adaptive, and people-focused approaches. Progress has been made in preparedness, protection, and resilience, but challenges remain related to governance capacity, resource constraints, and uneven implementation. Reviewing existing practices provides a strong foundation for identifying future challenges and opportunities and for advancing safer, more resilient, and sustainable coastal development pathways (Malakar et al., 2023).

10.3 Climate Change Adaptation in Coastal Disaster Risk Reduction

Climate change adaptation has become an essential part of coastal disaster risk reduction CDRR, as coastal regions face increasing and interconnected risks linked to climate change. Rising sea levels, stronger cyclones, changing rainfall patterns, coastal erosion, saltwater intrusion, and warming oceans are reshaping coastal hazard conditions. These changes are increasing existing risks and creating new and uncertain threats for coastal communities' ecosystems and economies. Integrating climate change adaptation into CDRR is therefore necessary to reduce long-term vulnerability and to strengthen the capacity of coastal systems to respond to change. Climate change adaptation in coastal disaster risk reduction involves planned adjustments in natural, social, economic, and institutional systems to reduce harm and manage climate-related impacts. Unlike short-term disaster response, this approach focuses on anticipation, long-range planning, and flexible strategies that can adjust to uncertainty. In coastal settings, adaptation-oriented CDRR seeks to reduce exposure to climate-driven hazards, strengthen adaptive capacity, and address deeper causes of vulnerability, such as poverty, environmental degradation, and weak governance. This approach supports broader goals of sustainable development and climate resilience. Climate science plays an important role in adaptation-focused CDRR through its use in coastal risk assessment and planning. Present practices increasingly include climate projections, sea-level rise scenarios, and extreme-event analysis in hazard mapping and vulnerability assessments. These forward-looking assessments help decision-makers move beyond past hazard patterns and design measures that remain effective under future climate conditions. This is especially important for long lasting infrastructure, settlements, and livelihood systems, where poor planning can increase future risk.

Nature-based solutions are an important element of climate change adaptation in coastal disaster risk reduction. Protecting, restoring, and managing coastal ecosystems such as mangroves, coral reefs, seagrasses, dunes, and wetlands strengthen natural protection against storm surges, waves, and erosion, while allowing coastlines to adjust to rising sea levels. These approaches are valued for their flexibility, affordability, and additional benefits, including biodiversity protection, carbon storage, and livelihood support. Stronger ecosystems directly reduce climate-related disaster risks. Climate-resilient infrastructure and spatial planning also support adaptation-focused CDRR. Actions include designing and upgrading coastal infrastructure to withstand future climate extremes, elevating critical facilities, and applying resilient building standards. Non-structural measures, such as land-use zoning, coastal setback lines, and managed relocation from highly exposed areas, are equally important. These strategies aim to limit exposure while allowing gradual adjustment to changing conditions. In many regions, risk-informed development planning and relocation are increasingly seen as necessary responses to growing climate threats, despite social challenges.

Community-based adaptation is a central part of climate change adaptation in coastal disaster risk reduction. Coastal communities often hold valuable knowledge of environmental change and long-standing coping practices. Adaptation-focused initiatives, therefore, emphasize participatory planning, livelihood diversification, skills development, and social support. Enabling communities to anticipate risks and take part in decision-making strengthens social resilience and reduces disaster impacts, especially for vulnerable groups. Institutional and governance arrangements strongly influence the effectiveness of climate change adaptation in CDRR. Successful action depends on coordination across sectors such as disaster management, environment, urban development, fisheries, and climate policy. Strong links between adaptation plans, disaster risk reduction strategies, and coastal management frameworks support consistent implementation. Governance systems that encourage learning, flexibility, and gradual adjustment are particularly important for managing uncertainty in coastal climate risks. Climate change adaptation is reshaping coastal disaster risk reduction by shifting attention from short-term protection toward long-term resilience and transformation. Integrating climate knowledge, ecosystem-based approaches, resilient infrastructure, community participation, and adaptive governance allows CDRR to respond more effectively to rising and uncertain climate risks. This approach supports the protection of lives, livelihoods, and ecosystems while advancing sustainable and climate-resilient coastal development.

10.4 Community Engagement and Empowerment

Community engagement and empowerment play a vital role in coastal disaster risk reduction (CDRR) because coastal communities are directly exposed to risk and actively contribute to reducing vulnerability and building resilience. People living in coastal areas interact closely with changing marine and land environments and, over time, develop local knowledge, social relationships, and coping practices that are essential during disaster events. Present day CDRR approaches recognize that lasting risk reduction depends on the meaningful involvement of communities across planning, implementation, and evaluation stages. Community engagement in coastal disaster risk reduction emphasizes participation and shared responsibility rather than simple consultation. Through participatory risk assessment, hazard mapping, and vulnerability analysis, community members share insights on local hazard behavior, evacuation routes, livelihood risks, and social conditions. These contributions strengthen technical assessments and improve the relevance of risk reduction actions. Engagement also helps build trust between communities, local authorities, and technical agencies. Clear understanding of risks and planned actions encourages long-term support for CDRR initiatives.

Empowerment in coastal disaster risk reduction is closely linked to building local capacity and adaptive ability. Training in early warning interpretation, first aid, evacuation procedures, and emergency response enables communities to act quickly and effectively during hazard events. Strengthening local organizations such as community groups, cooperatives, and disaster management committees supports collective action and coordination. Well organized communities are better prepared to mobilize resources, support vulnerable households, and work with external agencies during emergencies. Inclusive participation is an essential element of community-based coastal disaster risk reduction. Women, children, older people, persons with disabilities, and marginalized livelihood groups such as small-scale fishers often experience greater disaster impacts due to social and economic conditions. Empowerment-focused approaches seek to ensure fair representation in decision-making and address diverse needs. Inclusive practices contribute to more equitable and effective risk reduction outcomes. Livelihood security is another important aspect of community empowerment in coastal areas. Many households rely on livelihoods that are sensitive to climate and hazard impacts. CDRR initiatives increasingly support skill development, alternative income options, and access to social support programs. Strengthening livelihoods reduces economic vulnerability and improves the ability of families to prepare for, recover from, and adapt to disasters. Local and indigenous knowledge also strengthens community empowerment in coastal disaster risk reduction. Traditional practices related to settlement location, resource use, and hazard awareness reflect long experience with coastal environments. Combining this knowledge with scientific tools and planning approaches supports solutions that are locally appropriate and culturally meaningful. Collaboration between communities and experts encourages shared learning and strengthens ownership of risk reduction actions. Community engagement and empowerment help move coastal disaster risk reduction away from top-down, response-driven approaches toward participatory and resilience-focused practice. Valuing local knowledge, strengthening capacity, promoting inclusion, and supporting livelihoods place communities at the center of risk reduction efforts. This people centered approach supports effective disaster risk reduction while contributing to social equity, sustainable development, and long-term coastal resilience (Malakar and Roy, 2024; Ferreira, 2022; Mall et al., 2019; Temali, 2026).

10.5 Governance and Policy Considerations

Governance and policy frameworks play a strong role in shaping how effective coastal disaster risk reduction (CDRR) can be because disaster risk in coastal areas is complex and crosses many sectors. Coastal regions are managed through overlapping responsibilities and competing interests related to urban development fisheries tourism conservation infrastructure and disaster management. In such settings, clear governance arrangements and well-coordinated policies are essential to reduce risk, avoid inappropriate development, and support long-term coastal resilience. Effective CDRR governance therefore relies on integrated, inclusive, and flexible policy approaches rather than isolated, sector-focused actions. Institutional coordination is a major challenge in coastal disaster risk reduction. Coastal risks are influenced by environmental processes, social and economic change, and development decisions that extend across multiple sectors and levels of government. Policy approaches increasingly stress coordination among sectors such as environment, climate policy, urban planning, water management, and disaster management, as well as cooperation between national, regional, and local authorities. Integrated governance helps ensure that risk reduction is embedded within development planning, infrastructure investment, and resource management, instead of being treated only as an emergency concern. Policy mainstreaming is an important feature of present-day CDRR governance. Disaster risk reduction and climate adaptation goals are increasingly incorporated into development policies, coastal management plans, and regulatory systems. Land-use planning, building regulations, environmental assessment processes, and coastal setback rules are used to limit exposure and reduce vulnerability. When applied effectively, these tools help prevent unsafe development in hazard-prone coastal areas and promote safer growth patterns. In practice, weak enforcement and political or economic pressures often limit their effectiveness, which points to the need for stronger institutional capacity and accountability.

Local governance plays a critical role in coastal disaster risk reduction because local authorities are often responsible for preparedness, response, and recovery. Many local institutions, however, face limitations related to authority, funding, and technical expertise. Strengthening CDRR governance requires clear roles, adequate resources, and access to reliable risk information at the local level. Participation by communities' civil society organizations and private actors further improves transparency, trust, and local ownership of risk reduction efforts. Inclusive governance arrangements are particularly important for addressing the needs of vulnerable and marginalized coastal populations. Adaptive governance has become increasingly important in managing coastal disaster risks. Climate change is altering hazard conditions and increasing uncertainty, which reduces the effectiveness of fixed policies over time. Governance systems that encourage learning, regular review, and scenario-based planning are better suited to respond to new information and emerging risks. Flexible decision-making supports timely adjustment of strategies and more effective long-term planning. Governance and policy frameworks form the foundation of successful coastal disaster risk reduction. Coordination across institutions, policy integration, decentralization, participation, and adaptive decision-making strongly influence how well risk reduction efforts perform. Strong governance improves current outcomes and creates conditions for innovation and resilience building. As coastal regions face growing and interconnected risks, effective forward-looking governance remains essential for sustainable and equitable coastal disaster risk reduction.

10.6 Emerging Challenges in Coastal Disaster Risk Reduction

Coastal disaster risk reduction (CDRR) is entering a period of growing complexity as environmental, social, economic, and governance-related challenges increasingly interact and intensify risks in coastal regions. While progress has been made in managing coastal hazards, the nature of risk is changing due to climate change, rapid development, population pressure, and long-standing structural weaknesses. These changes are reshaping coastal risk conditions and placing limits on traditional approaches to disaster risk reduction. New circumstances require strategies that are more adaptive, integrated, and forward-looking. Climate-related hazards represent one of the most serious emerging challenges for coastal disaster risk reduction. Rising sea levels, stronger cyclones, shifting storm patterns, and more frequent extreme rainfall events are altering both the scale and location of coastal risks. Historical records that once guided hazard assessment and infrastructure planning are becoming less reliable. Decision-making now takes place under conditions of uncertainty, where future hazards are difficult to predict and planning must consider longer time horizons. This creates challenges for infrastructure and settlements designed to last for many decades. Rapid urban growth and coastal development add further pressure to CDRR efforts. Coastal regions continue to attract people and investment because of economic opportunity, trade, and tourism. Poorly planned development often expands into flood-prone, low-lying, and erosion-sensitive areas, which increases exposure to hazards. Informal settlements in high-risk locations face greater vulnerability due to limited access to safe housing, basic services, and risk reduction support. Managing development while reducing risk remains a persistent challenge. Compound and cascading risks are becoming more common in coastal areas. Hazards such as storm surge, heavy rainfall, river flooding, and high tides increasingly occur together, leading to severe flooding events. These combined hazards can overwhelm protection systems and emergency response capacity. Disasters may also disrupt infrastructure, transport networks, supply chains, and livelihoods, spreading impacts far beyond the immediate hazard area. Addressing these linked risks requires coordination across sectors, which remains limited in many contexts.

Ecosystem degradation presents another growing challenge for coastal disaster risk reduction. Coastal ecosystems such as mangroves, coral reefs, dunes, and wetlands once provided natural protection against hazards, but are under pressure from pollution, land-use change, resource overuse, and climate stress. Their decline reduces natural protection and increases dependence on engineered solutions that may be costly and less flexible over time. Governance and institutional constraints continue to limit effective responses to emerging coastal risks. Fragmented responsibilities, weak coordination, and separate policy systems for disaster management, climate adaptation, and coastal planning create gaps and conflicting priorities. Limited resources, technical capacity, and short political timeframes further restrict long-term planning and action. Social inequality is becoming more visible within coastal disaster risk reduction. Marginalized groups, including small-scale fishers indigenous communities' women, and low-income households, often face higher exposure and fewer recovery options. Limited access to information, resources, and decision-making processes reduces their ability to prepare for and recover from disasters.

Gaps in data and knowledge also remain a challenge. Despite improvements in monitoring and analysis, many regions lack long-term and locally specific data on social, economic, and environmental conditions. Limited integration of scientific knowledge with local experience can reduce the relevance and acceptance of risk reduction measures. Emerging challenges in coastal disaster risk reduction reflect increasing complexity, uncertainty, and interconnected risks. Climate change, development pressures, ecosystem loss, governance limitations, and social inequality together redefine coastal risk conditions. Addressing these challenges requires approaches that support adaptive planning, integrated thinking, and inclusive governance to promote resilient, equitable, and sustainable coastal futures (Aziz et al., 2024; Busayo & Kalumba, 2020; Tonmoy et al., 2020).

10.7 Lessons Learned from Past Coastal Disasters

Past coastal disasters have strongly influenced current thinking and practice in coastal disaster risk reduction (CDRR.) Events such as cyclones, tsunamis, storm surges, and extensive coastal flooding revealed not only physical exposure but also deep social, institutional, and governance weaknesses. These disasters acted as important learning moments by showing the limits of response-focused approaches and highlighting the need for preventive, integrated, and resilience-focused strategies. Lessons from past events continue to shape policies, planning processes, and community-based risk reduction efforts in coastal regions. Experience from past disasters showed the importance of prioritizing risk reduction rather than relying mainly on emergency response. Earlier coastal disaster management placed heavy emphasis on relief and reconstruction after disasters. Repeated losses demonstrated that this approach alone is costly and ineffective. Many disasters showed that weak preparedness limited warning reach, and poor evacuation planning increased loss of life and damage. These outcomes led to greater focus on early warning systems, preparedness planning, and risk-informed development in coastal areas.

Past disasters also revealed the protective value of coastal ecosystems. Locations with healthy mangroves, coral reefs, dunes, and wetlands often experienced less damage than areas where these systems were degraded or removed. Severe impacts in places where natural buffers were replaced by unplanned development highlighted the consequences of ecosystem loss. These experiences reinforced the importance of ecosystem protection and restoration as part of coastal disaster risk reduction. Land-use planning and regulation emerged as another key lesson. Post-disaster assessments frequently showed that the worst impacts occurred in low-lying, flood-prone, and erosion-sensitive areas with dense settlement. Weak enforcement of zoning rules, informal housing, and economic pressure contributed to unsafe coastal occupation. These patterns emphasized the need for proactive spatial planning, coastal setback zones, and risk-sensitive development controls to reduce exposure.

Early warning systems and communication were shown to save lives when warnings were timely, clear, and trusted. Where warnings were delayed, misunderstood, or poorly communicated, avoidable casualties occurred. Past events demonstrated that warning systems must link scientific monitoring with local communication, evacuation planning, and public trust. Community preparedness proved to be a decisive factor in disaster outcomes. Communities with prior training, disaster awareness, strong social networks, and local leadership were better able to respond and recover. Indigenous knowledge and informal support systems often played an important role during evacuation, rescue, and early recovery. Social inequality became clearly visible during coastal disasters. Poor households, small-scale fishers' women, children, older people, and marginalized groups often faced greater losses and slower recovery due to limited access to resources, information, and decision-making. These experiences highlighted the need to address social vulnerability alongside physical hazards.

Governance weaknesses were also revealed. Fragmented responsibilities, poor coordination, and separation between disaster management, coastal planning, and climate adaptation reduced effectiveness. Integrated governance and clear institutional roles emerged as essential requirements. Recovery and reconstruction provided further lessons. Rapid rebuilding without addressing risk often recreated vulnerability, while recovery efforts that applied risk reduction principles improved long-term resilience. Past disasters also highlighted the importance of learning and institutional memory. Each event generated valuable knowledge, but this knowledge was not always carried forward into future policy and practice. Lessons from past coastal disasters reshaped coastal disaster risk reduction by showing that disasters result from the interaction of hazards, development choices, governance systems, and social conditions. These lessons emphasize prevention, ecosystem protection, community empowerment, inclusive governance, and risk informed recovery as foundations for resilient, equitable, and sustainable coastal futures.

10.8 Future Opportunities and Innovations

Future opportunities and innovations in coastal disaster risk reduction (CDRR) are shaped by rapid technological progress, changing governance practices, and a growing need to strengthen long-term resilience in coastal regions. Climate change is intensifying hazards while social and economic pressures continue to reshape coastal environments. In response, CDRR is moving beyond traditional protection-focused measures toward approaches that are more adaptive, forward-looking, and integrated. These opportunities not only help reduce disaster losses but also support sustainable coastal development and resilience over time. Technological progress offers important opportunities to improve coastal disaster risk reduction. Advances in satellite observation, real-time ocean monitoring, and weather forecasting have strengthened coastal risk assessment and early warning systems. Improved forecasting tools now allow better anticipation of storm surges, coastal flooding, and erosion under changing climate conditions. Digital communication systems and mobile networks help ensure that warnings and risk information reach coastal communities quickly and clearly. These developments support stronger decision-making and allow risk reduction actions to be planned earlier and more effectively.

Nature-based solutions provide another major opportunity for future CDRR practice. Growing attention is being given to ecosystem-based approaches such as mangrove restoration, coral reef recovery, dune stabilization, and wetland conservation. These measures offer flexible and cost-effective protection while allowing natural systems to respond to changing conditions. New approaches to ecosystem monitoring, management, and financing, including carbon-related initiatives and incentive-based conservation programs, help connect environmental protection with disaster risk reduction goals. Innovations in coastal planning and infrastructure design also offer promising opportunities. Climate-resilient infrastructure approaches, such as adaptable coastal defenses, elevated structures, and multipurpose protective landscapes, are gaining wider use. These designs accept uncertainty and aim to function under a range of future conditions rather than a single fixed standard. Risk informed spatial planning, including flexible zoning and planned relocation from highly exposed areas, can reduce long-term exposure while supporting fair and orderly transitions.

Community-centered innovation plays a key role in advancing coastal disaster risk reduction. Participatory planning, local innovation efforts, and shared knowledge creation help strengthen social resilience and encourage solutions that reflect local needs and conditions. Combining indigenous and local knowledge with scientific tools supports approaches that are practical, culturally appropriate, and effective. Community-based insurance schemes, local warning networks, and livelihood diversification initiatives further support local adaptive capacity. Institutional and governance innovation is equally important for realizing future opportunities in CDRR. Governance systems that support learning, coordination, and flexibility are better able to respond to changing coastal risks. Stronger alignment between disaster risk reduction, climate adaptation, coastal management, and development policies reduces fragmentation and improves effectiveness. New financing approaches, including climate-related funds, resilience-focused investments, and partnerships with private actors, help mobilize resources for long-term coastal resilience. Future opportunities in coastal disaster risk reduction point toward a shift to more transformative and resilience-focused practice. Using new technologies, expanding nature-based solutions, rethinking infrastructure, empowering communities, and improving governance can strengthen responses to future risks. These innovations provide a solid foundation for building coastal futures that are safer, more inclusive, and environmentally sustainable.

10.9 Conclusion and Arguments

Coastal disaster risk reduction (CDRR) has shifted from a largely reactive and hazard focused practice toward a broader approach that looks ahead and emphasizes resilience. This chapter presented CDRR as a necessary response to increasing coastal risks shaped by climate change, environmental degradation, and rapid social and economic change. The discussion showed that present practices, such as risk assessment, early warning systems, ecosystem-based approaches, and community participation, have improved preparedness and reduced disaster losses in many places. At the same time, the level of implementation and effectiveness remains uneven across regions. Climate change adaptation emerged as a central part of coastal disaster risk reduction. Planning for future risks now requires anticipation, flexibility, and long-term thinking rather than reliance on past experience alone. Infrastructure development, risk governance, and coastal planning must be able to function under uncertainty and adjust as hazard conditions continue to change. The chapter also highlighted the importance of community engagement and empowerment. Approaches that place people at the center, value local knowledge, support inclusion, and strengthen livelihoods are essential for sustainable risk reduction in coastal settings.

Governance and policy considerations were shown to strongly influence outcomes. Effective coordination among institutions, clear policy alignment, and governance systems that allow learning and adjustment are necessary to manage complex coastal risks. Discussion of emerging challenges drew attention to compound hazards, rapid urban growth, ecosystem loss, and increasing social inequality. Lessons from past coastal disasters demonstrated that inaction leads to repeated losses, while prevention, preparedness, and learning reduce long-term risk. Future opportunities and innovations pointed to practical pathways for strengthening coastal disaster risk reduction. Advances in technology, nature-based solutions, adaptable infrastructure, and inclusive governance approaches offer new ways to address future risks and support resilience building. Coastal disaster risk reduction should be understood as a continuous process rather than a one-time intervention. Its success depends on how well it is integrated into sustainable coastal development. Stronger integration, equity, and innovation will be essential for responding to uncertainty and for shaping coastal futures that are safer, fairer, and more resilient over time. The chapter discussions lead to the following arguments:

(a) Coastal disasters result from the interaction of hazards, development choices, governance, and social vulnerability.
(b) Risk reduction is more effective and less costly than response and relief-focused disaster management.
(c) Climate change demands adaptive, flexible, and forward-looking coastal disaster risk reduction strategies.
(d) Healthy coastal ecosystems provide essential long-term protection and multiple resilience benefits.
(e) Community empowerment and inclusive governance are key to resilient coastal disaster risk reduction.

References

Aziz, F., Wang, X., Mahmood, M. Q., Awais, M., & Trenouth, B. (2024). Coastal urban flood risk management: Challenges and opportunities—a systematic review. *Journal of Hydrology, 645*, 132271. https://doi.org/10.1016/j.jhydrol.2024.132271

Busayo, E. T., & Kalumba, A. M. (2020). Coastal climate change adaptation and disaster risk reduction: A review of policy, programme and practice for sustainable planning outcomes. *Sustainability, 12*(16), 6450. https://doi.org/10.3390/su12166450

Ferreira, Ó. (2022). Modelling risk reduction measures to minimise future impacts of storms at coastal areas. In C. Kondrup et al. (Eds.), *Climate Adaptation Modelling*. Springer Climate. Springer. https://doi.org/10.1007/978-3-030-86211-4_8

Malakar, K. D., Kumar, M., Anand, S., & Kuzur, G. (2023). Disaster, policy instruments, and sustainability. In *Climate change and socio-ecological transformation. Advances in geographical and environmental sciences*. Springer. https://doi.org/10.1007/978-981-99-4390-6_6

Malakar, K. D., & Roy, S. (2024). Community Cartography and Participatory GIS. In *Mapping Geospatial Citizenship*. Springer Cham. https://doi.org/10.1007/978-3-031-63107-8_5

Mall, R. K., Srivastava, R. K., Banerjee, T., et al. (2019). Disaster risk reduction including climate change adaptation over South Asia: Challenges and ways forward. *International Journal of Disaster Risk Science, 10*, 14–27. https://doi.org/10.1007/s13753-018-0210-9

Setiawati, M. D., Nandika, M. R., Hernawan, U. E., et al. (2024). Application of coastal hazard index to advance nature based protection for coastal communities in the small islands. *Discover Applied Sciences, 6*, 462. https://doi.org/10.1007/s42452-024-06164-x

Temali, I. K. (2026). Climate change adaptation green Technologies for Coastal Risk Prevention: Case studies summary from coastal countries. In E. Berisha, J. Moodie, L. Allkja, & M. Jeftić (Eds.), *The Palgrave handbook of just green transitions in the Western Balkans and beyond*. Palgrave Macmillan. https://doi.org/10.1007/978-3-031-95075-9_19

Tonmoy, F. N., Hasan, S., & Tomlinson, R. (2020). Increasing coastal disaster resilience using smart city frameworks: Current state, challenges, and opportunities. *Frontiers in Water, 2*, 3. https://doi.org/10.3389/frwa.2020.00003

Part IV

Participatory Approaches in Coastal Disaster Risk Reduction and Sustainability: Risk Perceptions and Field Narratives

11

Risk Perceptions in Coastal Disaster Risk Reduction: Participatory Insights

This chapter brings together voices from the Indian Sundarbans and coastal Medinipur to reflect on how coastal risk is understood and lived in everyday life. It grows out of sustained field engagement and participatory interactions with communities who face multiple hazards as part of their normal surroundings. Rather than approaching disaster risk through technical categories alone, the chapter begins from the ways people speak about loss, change, and uncertainty in their own terms. The purpose of this chapter is to shift attention from events to experiences. Coastal hazards are not encountered as isolated moments, but as ongoing conditions that shape livelihoods, homes, and future choices. By organizing the discussion around seven narratives, the chapter traces how different risks intersect and how responses are shaped by memory, work, environment, and generational change. Cyclones, floods, erosion, sea-level rise, and environmental degradation are presented not as separate problems, but as connected pressures that influence everyday decision-making. The chapter also pays close attention to differences in perspective. Older residents draw on memory and experience, while younger people look ahead with concern about opportunity, dignity, and belonging. Together, these voices reveal that disaster risk is not only

K. D. Malakar et al., *Coastal Ecologies at Risk*, Palgrave Studies in Environmental Sustainability, https://doi.org/10.1007/978-3-032-26081-9_11

about safety, but about the possibility of sustaining life on the coast. This preface sets the stage for a participatory reading of coastal disaster risk reduction, grounded in lived realities rather than abstract frameworks.

Key Points of the Chapters

- Describes coastal risk as a continuous lived condition shaped by memory, livelihoods, and daily adaptation
- Explains the gap between technical risk assessments and community understanding of hazards and warnings
- Highlights slow environmental changes as deeper long-term risks than sudden extreme events
- Examines the close link between mangrove livelihoods and ecosystem-based disaster risk reduction
- Identifies how youth perspectives reshape resilience through aspirations, mobility, and decisions to remain on the coast
- Highlight narratives (Field narratives on the following):
  - Cyclone risk grounded in shared memory and everyday preparedness
  - Erosion lived as gradual loss and repeated adjustment
  - Flooding experienced as seasonal disruption and daily strain
  - Tsunami risk understood through warnings rather than experience
  - Sea-level rise felt through livelihood insecurity and rising costs
  - Mangroves valued for protection but limited by survival needs
  - Youth facing coastal futures through aspiration and belonging.

11.1 Introduction

Coastal regions are among the most dynamic and vulnerable social and ecological spaces in the world where natural processes, climate forces, and human livelihoods interact in complex ways (Malakar et al., 2025a; Ramteke et al., 2026). Along the eastern coast of India, especially in the Indian Sundarbans and coastal Medinipur, disaster risk is not an abstract idea limited to reports or projections. It is part of everyday life shaped by repeated cyclones, frequent flooding, gradual erosion, saline intrusion, ecological decline, and constant uncertainty about what lies ahead (Ghosh & Mistri, 2023; Hasnine & Nagdeve, 2025). This chapter places disaster risk reduction DRR within these lived realities and argues that understanding how people perceive, interpret, and respond to coastal risk is central to building meaningful and lasting resilience. Risk perception strongly influences preparedness, response, and long-term adaptation. Technical assessments often describe risk through measurements of hazards exposure and vulnerability. Coastal communities however understand risk through memory, experience, and daily interaction with their environment. In the Sundarbans and coastal Medinipur risk is rarely seen as a single event (Dhar et al., 2023). Cyclones are remembered through broken embankments, disrupted livelihoods, and long recovery periods. Floods are recalled through damaged homes, interrupted schooling, and repeated rebuilding. Erosion is experienced through the loss of land, homes, and emotional ties to place. These experiences shape how people judge threats, trust warnings, and decide whether to remain, adapt, or migrate (Ghosh & Mistri, 2023).

This chapter adopts a participatory approach that places community voices at the center of analysis. The narratives presented here are connected accounts that show how different hazards, both sudden and slow, are experienced together. Repeated exposure to cyclones has created strong collective memories that influence preparedness and trust in warning systems. While cyclones are destructive, they are also familiar leading to both awareness and at times acceptance of risk (Malakar & Roy, 2024; Malakar et al., 2025b). Erosion presents a quieter but persistent threat. It unfolds gradually and often falls outside formal disaster timelines but it reshapes settlements, livelihoods, and identities. Communities describe erosion as an ongoing disaster that demands constant adjustment rather than emergency response. This challenges event-focused disaster frameworks and highlights the importance of recognizing slow environmental change. Flooding occupies a space between sudden shock and ongoing stress. In many coastal areas floods are expected seasonal events rather than rare crises. Local narratives speak of fatigue, disruption, and uncertainty, showing gaps between warning messages and practical preparedness. Tsunami risk is shaped mainly by imagination rather than direct experience. Although rare along this coast tsunamis remain part of public awareness through media, memories, and drills. Reliance on warning systems highlights the importance of trust when lived experience is limited. Confusion between tsunami and cyclone warnings also points to communication challenges in multi-hazard settings. Sea-level rise is often understood through its impact on livelihoods rather than visible shoreline change. Salinized farmland, declining fish catches, damaged infrastructure, and rising daily costs shape how communities experience this risk. These perceptions show that disaster risk reduction must address economic security alongside physical safety.

Mangroves hold a complex place in community perceptions. They are valued as protection from storms and erosion but also as sources of fuel, food, and income. This dual role creates tension between conservation needs and everyday survival, highlighting the importance of participatory ecosystem-based approaches that respect both ecological and social realities. Youth perspectives add an important generational dimension. Young people often view coastal risk through questions of education, mobility, and future opportunity. While they play key roles in communication and response, limited livelihood options encourage migration, raising concerns about the long-term sustainability of coastal resilience. These narratives show that coastal disaster risk reduction is not only a technical task but a social, economic, and cultural process shaped by memory, livelihoods, institutions, and generational change. Communities possess detailed and context-specific knowledge of risk that is often overlooked. Recognizing these perspectives helps shift disaster risk reduction toward approaches grounded in lived experience (Patra & Shukla, 2017; Pattanaik, 2021).

This chapter lays the foundation for a people-centered approach to coastal disaster risk reduction. By linking memory adaptation and anticipation it shows that resilience grows from everyday experience as much as from planning. Participatory knowledge strengthens scientific understanding by grounding policy and practice in real lives. In rapidly changing coastal regions listening to how people perceive risk is essential for building resilient fair and lasting coastal futures. In this context, the chapter focuses on the coastal region of West Bengal (especially Indian

Sundarbans and coastal Medinipur) (Fig. 11.1) and draws on diverse community narratives across multiple, interconnected stages (Fig. 11.2).

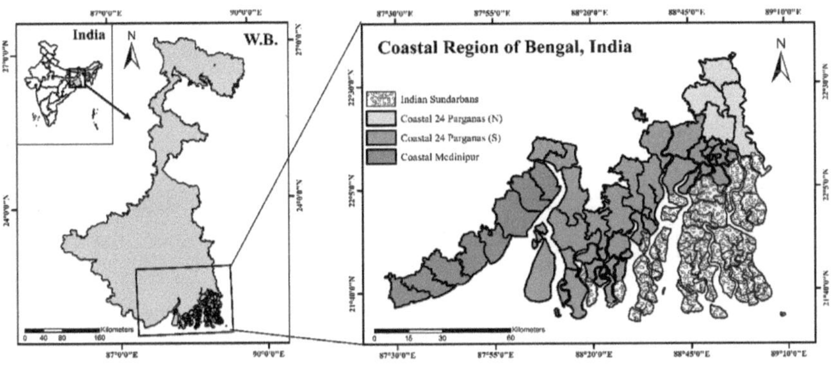

Fig. 11.1 Coastal region of West Bengal (especially Indian Sundarbans and coastal Medinipur) (prepared by the authors)

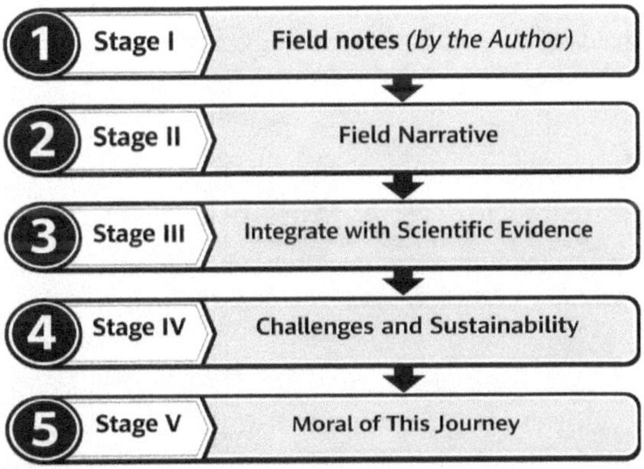

Fig. 11.2 Stages of field narrative collection and processing (prepared by the authors)

11.2 Exploring Risk Perceptions Through Field Investigation

11.2.1 Echoes of Cyclone Risks: Collective Memory and Preparedness

(a) ***Field Notes (By the Authors)***

Fieldwork in the Indian Sundarbans and coastal Medinipur was not limited to documenting physical landscapes. It involved listening to memories that remain closely tied to place. Villages in this region are shaped by tides and embankments, but they are also shaped by accounts of past cyclones that continue to circulate long after the events themselves. These stories influence how people think about risk in their everyday lives. Disaster risk is not understood as a distant possibility or a future calculation. It is experienced as something ongoing, present in decisions about housing, livelihoods, and mobility. This narrative shows how shared memories of earlier cyclones guide local practices of preparedness. In some cases, these memories support adaptive responses and strengthen resilience. In other cases, they encourage acceptance and fatalism, limiting the scope for change in the face of recurring environmental threats (**Source:** Authors' field diary, August 2025).

(b) ***Field Narratives (Interaction with Locals)***

Interactions with local residents in the Sundarbans showed that memories of cyclones are expressed through lived experiences rather than through fixed timelines. Elderly fishers often spoke of storms as events that altered the course of rivers, recalling how channels shifted after cyclones such as Aila and Amphan. Their accounts were linked to concrete moments of disruption, including the loss of boats, breaches in embankments, and long periods spent in relief shelters. Women described practical responses at the household level, such as storing dry food, identifying safer and higher places, and reading signs from the sky and nearby mangroves to anticipate weather changes. In coastal Medinipur, farmers narrated cyclone impacts through the slow spread of salinity. They spoke of fields turning white, declining crop yields, and gradual shifts from paddy cultivation to aquaculture. Younger residents described a different relationship to risk. For them, mobile phone warnings and routine evacuation procedures had made cyclones familiar events. Preparedness was treated less as an emotional reaction and more as a regular part of everyday administration (**Source:** Authors' field diary, August 2025).

(c) ***Integrate with Scientific Evidence***

These local narratives reflect patterns identified in disaster research on risk perception in areas exposed to repeated hazards. Studies in this field show that frequent exposure to disasters can increase awareness, while at the same time reducing the sense of urgency, depending on levels of trust in institutions and past experiences of recovery. Research from the Bay of Bengal points to a rise in cyclone intensity alongside notable improvements in forecasting and early warning systems. Field observations from this study support this mixed picture. Communities express greater confidence in warnings than in the past, but long-standing structural weaknesses continue to shape the scale of damage. Fragile embankments, informal housing, and insecure livelihoods remain key factors that intensify cyclone impacts. Local discussions about mangroves as natural forms of protection also echo findings from ecological research, which show that mangroves can reduce storm surges. At the same time, residents are aware that such protection has limits during particularly severe events (Malakar, 2026; Alam et al., 2023; Saini et al., 2025).

(d) ***Challenges and Sustainability***

A key issue that emerges from these narratives is the uneven movement from knowledge to long-term action. Although preparedness drills and cyclone shelters are in place, their effectiveness is often limited by poor maintenance, social exclusion, and ongoing pressures on livelihoods. In both study areas, seasonal migration makes evacuation planning more difficult, as many households are not consistently present during warning periods. Repeated exposure to cyclones also appears to reduce concern among younger residents, for whom such events have become routine. Long-term safety depends not only on physical infrastructure, but also on keeping risk awareness alive. This requires sustained efforts through education, exchanges between generations, and forms of planning that treat lived experience as equal in value to technical expertise (Khondoker et al., 2024; Masud-All-Kamal et al., 2025; Paul & Dutt, 2010).

(e) ***Moral of the Narrative Journey***

This narrative argues that reducing disaster risk in coastal areas cannot be treated as a technical task alone. It is also shaped by cultural practices and shared ways of remembering past events. Collective memory works as an informal record of risk, influencing how people prepare in ways that numbers and models cannot fully explain. When scientific approaches take these local memories seriously, responses to risk become more rooted in everyday realities and better able to adjust over time.

11.2.2 Living on the Edge: Erosional Risks and Coastal Community Resilience

(a) ***Field Notes (By the Authors)***

This narrative is based on sustained field engagement along the fragile edges of the Indian Sundarbans and coastal Medinipur, where land, water, and livelihoods are in constant negotiation. A straightforward but unsettling observation guides the discussion. In these places, erosion is not experienced as a single incident, but as a slow and continuing process. Its effects build up over time and are closely woven into everyday life. Households track time not only through seasons or years, but also through the gradual loss of land to rivers and the sea. The narrative explores how coastal communities perceive, describe, and respond to the risks of erosion. It shows that resilience emerges as a practical and adaptive response, shaped by living with ongoing uncertainty rather than by reacting to sudden events (**Source:** Authors' field diary, August 2025).

(b) ***Field Narratives (Interaction with Locals)***

In the Sundarbans, conversations with residents often began with gestures toward the riverbank rather than references to maps. Elderly men described homesteads that no longer exist, tracing their outlines in the air and recalling how entire hamlets slowly disappeared into tidal creeks over the course of a decade. Women spoke about the repeated work of rebuilding their homes, including moving kitchens further inland, relocating tube wells, and using nearby trees as markers to judge how close the water had come. In these accounts, erosion was explained through the loss of place, as shrines were submerged, burial grounds were washed away, and ancestral land survived mainly in memory. In coastal Medinipur, the tone of these conversations was different, though the sense of unease remained strong. Fishers and small traders described the sea's advance as more sudden and seasonal, often following cyclones or monsoon storms when beaches narrowed within a short period. Youth groups spoke about roads breaking apart and tourism stalls being moved each year. Across both regions, erosion was described as a *"silent danger."* It was seen as less dramatic than cyclones, but more persistent, shaping everyday decisions related to schooling, marriage, and migration (**Source:** Authors' field diary, August 2025).

(c) ***Integrate with Scientific Evidence***

These everyday accounts closely match scientific research on coastal and delta change in eastern India. Studies of landforms describe the Sundarbans as one of the most erosion-prone mangrove deltas in the world, shaped by rising sea levels, reduced sediment flow, and strong tidal forces. Research from coastal Medinipur shows a more uneven pattern, with areas of land gain and land loss existing side by side, although the overall trend points to continuing loss. Storm surges and human activities such as seawalls and port development have played an important role in accelerating this process. Local descriptions of erosion as slow but unavoidable reflect findings from disaster research, which show that gradual hazards often lead to a sense of familiarity rather than panic. Community references to mangroves as forms of protection also align with ecological studies that show their ability to reduce wave energy. At the same time, both scientific evidence and local experience point to clear limits, as mangroves alone cannot counter the effects of rapid sea-level rise (Carobene & Carobene, 2022; Mahato, 2024).

(d) ***Challenges and Sustainability***

One key concern that emerges from these accounts is the gap between the risks created by erosion and the responses offered by institutions. Disaster management systems continue to focus mainly on sudden events such as cyclones and floods, while erosion develops slowly and often falls outside standard emergency planning. Short-term embankment repairs, temporary seawalls, and relocation efforts struggle to keep pace with the steady retreat of the shoreline. Livelihood pressures further limit relocation options, as many households depend on coastal and river-based resources for their survival. At the same time, the narratives point to existing forms of adaptation. These include community-led monitoring of embankments, informal warning practices based on tidal patterns, and efforts to diversify sources of income. Long-term responses will depend on linking these local practices with wider planning approaches, including managed retreat, ecosystem-based adaptation, and forms of governance that involve affected communities and recognize erosion as a serious risk in its own right (Ellis & Bajracharya, 2023; Devi, 2025; Sahay, 2025).

(e) ***Moral of the Narrative Journey***

This narrative argues that resilience along the coast grows out of everyday experiences of loss and adjustment. Living with erosion encourages communities to adapt gradually, to draw on shared memory, and to think of home as something that can move rather than remain fixed. Coastal disaster risk reduction needs to pay close attention to these local perspectives. In this context, resilience is not about resisting change at all costs, but about the ability to continue living, adapting, and planning as the shoreline continues to shift.

11.2.3 Flood Risk Perceptions Along the Coast: Rising Waters, Rising Concerns

(a) ***Field Notes (By the Authors)***

This narrative draws on repeated field visits to flood-affected settlements in the Indian Sundarbans and coastal Medinipur, where flooding has become part of everyday life rather than an occasional hazard. Floods in these areas are experienced less as sudden disasters and more as expected seasonal events. This experience shapes how communities anticipate risk, live with loss, and adjust their preparedness over time. The narrative examines how memories of past floods, local ways of interpreting warnings, and everyday coping practices shape coastal disaster risk reduction at the local level (**Source:** Authors' field diary, August 2025).

(b) ***Field Narratives (Interaction with Locals)***

In the Sundarbans, accounts of flooding were closely tied to the rhythms of the tide. Residents explained that even what were once considered normal high tides now enter courtyards and homes. A middle-aged farmer observed, *"Aage jol ashto nodite, ekhon jol ashe barite"* (Earlier the water stayed in the river, now the water comes into the house). Women spoke at length about the everyday strain caused by saline floodwater, including spoiled stored rice, contaminated ponds, and rising illness among children. An elderly respondent recalled, *"Prottek bochhor bandh bhange, amra shudhu jinish guchhiye uthchi"* (Every year the embankment breaks, and we only keep lifting our belongings). In these narratives, flood risk was described less as a moment of shock and more as a condition of ongoing exhaustion.

In coastal Medinipur, residents framed their experiences around monsoon flooding and storm surges. Fishers described sudden inundation following heavy cyclonic rainfall, while shopkeepers spoke of knee-deep water that disrupted markets for weeks at a time. A young respondent noted, *"Alert ashe phone-e, kintu jol koto uthbe seta keu bole na"* (Alerts come on the phone, but no one tells us how high the water will rise). Unlike the Sundarbans, where floodwater advances gradually with the tide, floods in Medinipur were described as abrupt and unpredictable. Despite these differences, residents across both regions pointed to similar ways of coping. These included raising house plinths, keeping essential items ready, and sending children to relatives' homes during the peak flood months.

(c) ***Integrate with Scientific Evidence***

These lived experiences closely match scientific findings on coastal flooding in eastern India. Studies of climate and water systems show that rising sea levels, heavier rainfall, and stronger tidal effects are increasing both the frequency and duration of floods in the Sundarbans. Research from coastal Medinipur points to a growing overlap between cyclonic rainfall and river flooding, which has made flood events more intense and harder to predict. The widespread feeling that floods have become part of everyday life reflects insights from disaster research, which note that repeated low to moderate flooding often leads to familiarity rather than panic. Local attention to embankment failures also reflects engineering studies, which identify aging and poorly maintained embankments as major weak points during extreme flood events (Bala & Boza, 2023; Das & Das, 2023; Need et al., 2024).

(d) ***Challenges and Sustainability***

One major concern that emerges from these accounts is the gap between receiving flood warnings and being able to act on them in practical ways. Although early warning systems have improved, many respondents spoke about uncertainty related to how deep the water would be, how long flooding would last, and whether safe shelters would be accessible. Structural measures such as embankments, sluice gates, and drainage canals often provide short-term relief, but they are difficult to sustain as sea levels continue to rise. Social conditions further shape people's ability to respond. Landlessness, dependence on daily wage work, and care responsibilities, particularly for women, limit options for evacuation and recovery. At the same time, local forms of adaptation are clearly visible. These include community-managed shelters, informal flood calendars based on past experience, and collective efforts to repair embankments. Such practices point to the importance of incorporating local flood knowledge into formal disaster planning and decision-making (Kreibich et al., 2016; Perera et al., 2020).

(e) ***Moral of the Narrative Journey***

This narrative argues that flood risk along the coast is shaped less by fear of rare events and more by long experience with repeated flooding. Rising waters have taught communities to live with uncertainty, adjust their practices over time, and rely on shared memory alongside official warnings. Coastal disaster risk reduction needs to move beyond responses that focus only on single events. It must recognize floods as part of seasonal life and treat community knowledge as a vital part of building resilient coastal futures.

11.2.4 Risks of Tsunami: Stories of Warning and Response

(a) ***Field Notes (By the Authors)***

This narrative draws on participatory interactions in the Indian Sundarbans and coastal Medinipur, where tsunami risk holds a particular place in local thinking. Unlike floods or cyclones, tsunamis are not part of yearly experience in these areas. Even so, they remain present as remembered threats, shaped by accounts of distant disasters, media coverage, and official drills. In this context, tsunami risk is commonly seen as unlikely but potentially severe. Preparedness therefore depends more on trust in warning systems and official guidance than on personal experience. The narrative explores how coastal communities understand tsunami risk, interpret warnings, and prepare for an event that is rare but carries serious consequences (**Source:** Authors' field diary, August 2025).

(b) ***Field Narratives (Interaction with Locals)***

In the Sundarbans, discussions about tsunami risk often began with a clear distancing from direct experience. Many respondents emphasized that they had never witnessed a tsunami themselves. An elderly fisherman explained, *"Amra tsunami dekhi ni, kintu shunechi 2004-e onno deshe ki holo"* (We have not seen a tsunami, but we heard what happened in other countries in 2004). The 2004 Indian Ocean tsunami appeared repeatedly in these conversations as a shared point of reference, carried through television coverage and later reinforced through disaster awareness programs. Younger residents referred mainly to school-based mock drills rather than lived events. A schoolteacher noted, *"Chhatroder bolano hoy, jodi hothat jol shore jay, tahole upore uthbe"* (Students are told that if the sea suddenly withdraws, they must move to higher ground).

In coastal Medinipur, local accounts placed greater emphasis on warning systems. Fishers described sirens and mobile phone alerts as the main indicators of tsunami risk. A middle-aged fisher stated, *"Siren bajle bujhi boro bipod, tokhon shob kaj chere douraite hoy"* (When the siren sounds, we know it is a big danger; then we must leave all work and run). At the same time, some uncertainty and doubt were also expressed. A shopkeeper remarked, *"Alert to anek kichur jonno ashe, tsunami aar jhor ek hoye jay"* (Alerts come for many things; tsunami and cyclones get mixed up). Across both regions, tsunami risk was described as distant but serious. It was spoken of as a threat that demands respect, even though it has not been personally experienced by most residents.

(c) ***Integrate with Scientific Evidence***

These local perceptions are consistent with scientific assessments of tsunami risk along the eastern coast of India. Geophysical research describes the Bay of Bengal as an area where tsunamis occur less frequently than in major Indian Ocean subduction zones, but where risk cannot be ruled out. Modeling studies show that tsunamis generated in the Andaman and Sumatra region can travel toward eastern India and affect low-lying coastal areas such as the Sundarbans and Medinipur. Field observations of strong reliance on warning systems reflect a broader scientific view that tsunami preparedness in regions with little direct experience depends largely on timely warnings, clear evacuation planning, and public education rather than on past memory. The confusion reported by community members between cyclone and tsunami alerts also reflects known challenges in communicating risk in areas exposed to multiple hazards (Prerna et al., 2015; Singh et al., 2012).

(d) ***Challenges and Sustainability***

One important issue that emerges from these accounts is the difficulty of keeping attention focused on a hazard that occurs rarely. Many respondents noted that tsunami awareness efforts tend to be occasional. They are often linked to drills or anniversaries of major past events, and then gradually lose priority as more immediate concerns such as floods and erosion take over. Physical and social constraints further limit preparedness. These include the lack of vertical evacuation structures, unclear evacuation routes, and mobility challenges faced by older residents. Long-term preparedness depends on making tsunami risk education part of everyday disaster awareness, rather than treating it as a special or distant threat. Trust in sirens and warning alerts remains strong within communities, but clear messages, consistent use, and regular engagement are necessary to avoid confusion and fatigue over time (Lindell, 2023; Paton, 2017; Rafliana, 2022).

(e) ***Moral of the Narrative Journey***

This narrative suggests that perceptions of tsunami risk along the coast are shaped less by direct experience and more by learned attention and preparedness. Communities prepare not because they have lived through local devastation, but because they have come to imagine it through shared stories, drills, and warning messages. Coastal disaster risk reduction therefore needs to support this form of awareness over time. Rare risks must remain visible, understandable, and practical, so that when the sea behaves in unfamiliar ways, communities are able to respond quickly and with confidence.

11.2.5 Rising Concern About Sea Level: Economic Risk and Coastal Livelihoods

(a) ***Field Notes (By the Authors)***

This narrative draws on long-term observations and participatory interactions in the Indian Sundarbans and coastal Medinipur, where sea-level rise is now seen as an everyday economic concern rather than a distant climate issue. In these regions, rising sea levels are understood mainly through their effects on livelihoods. People speak of declining fish catch, increasing salinity in agricultural fields, damage to ponds, and growing uncertainty in local markets. The narrative examines how coastal communities experience sea-level rise as a slow and ongoing economic risk. It shows how this process reshapes livelihood security, influences decisions about work and investment, and affects plans for the future (**Source:** Authors' field diary, August 2025).

(b) ***Field Narratives (Interaction with Locals)***

In the Sundarbans, residents often described the sea and rivers as slowly moving inward. Farmers spoke less about visible changes in water levels and more about the steady spread of salinity. One smallholder explained, *"Jol ekhono chokhe dekha moto bareni, kintu mati aar aager moto nei"* (The water hasn't risen visibly, but the soil is no longer what it used to be). Paddy fields that once produced a reliable crop now fail without warning, pushing many households toward brackish aquaculture or daily wage work. Women responsible for managing household water highlighted the added financial burden caused by salinity. As one woman noted, *"Pond-er jol kharap hole taka diye jol kinte hoy"* (When Pond water becomes unusable, we have to buy water). For them, sea-level rise is felt through daily expenses and extra labor rather than through dramatic changes along the shoreline.

In coastal Medinipur, livelihood concerns were closely tied to fishing and small-scale coastal trade. Fishers spoke about changing nearshore conditions and repeated damage to basic infrastructure. A middle-aged fisher explained, *"Samudro ta shob shomoy bhitore dhukche, ghat aar rasta bachte parchhe na*" (The sea keeps moving inward; jetties and roads cannot survive). Small tourism vendors shared similar concerns, describing how beach erosion and frequent flooding reduce seasonal earnings. A young respondent observed, *"Aaj dokan tule niye gele, kal abar notun jaygay dite hoy"* (If we move the shop today, tomorrow we have to set it up somewhere else). Across both regions, sea-level rise was described as a source of economic instability. It was understood as a force that demands constant adjustment, rather than as a sudden event leading to immediate displacement (**Source:** Authors' field diary, August 2025).

(c) ***Integrate with Scientific Evidence***

These everyday experiences closely reflect scientific research on sea-level rise and coastal vulnerability in eastern India. Measurements from tide gauges and satellite records show that sea levels in the northern Bay of Bengal are rising, with the effects becoming stronger due to land subsidence in delta regions such as the Sundarbans. Research also shows increasing salinity in soils and freshwater sources, which directly affects agriculture and access to drinking water. These changes mirror the concerns raised by local residents. In coastal Medinipur, studies of shoreline processes indicate that even small rises in sea level can increase erosion and damage roads, jetties, and other infrastructure during storm surges. The way communities describe sea-level rise mainly as a threat to livelihoods aligns with findings from disaster research. Slow and gradual hazards tend to weaken income security and long-term economic stability rather than creating immediate risks to life (Carobene & Carobene, 2022; Pramanik et al., 2015).

(d) ***Challenges and Sustainability***

One major challenge that emerges from these accounts is that sea-level rise often remains difficult to address within standard disaster management systems. Sea-level rise does not appear as a single, clearly defined event, which makes compensation, insurance, and policy responses uneven. Changes in livelihoods, such as shifts toward aquaculture or informal tourism, can help households cope in the short term, but they also introduce new risks. These include unstable markets and added pressure on local environments. The repeated cost of adjusting to these conditions places a heavy burden on households with limited land and resources. At the same time, the narratives point to growing forms of resilience. These include diversifying income sources, community-based water sharing, and increased awareness of economic risks linked to climate change. Long-term sustainability will depend on stronger coordination between livelihood planning, ecosystem restoration, and social protection within coastal disaster risk reduction efforts (Hecht, 2009; Schwoerer et al., 2023; Wulandari et al., 2022).

(e) ***Moral of the Narrative Journey***

This narrative argues that sea-level rise is felt most strongly through everyday economic pressure rather than through visible changes along the shoreline. Coastal communities may not speak in technical terms, but they understand rising seas through declining harvests, unstable work, and growing daily expenses. Disaster risk reduction therefore needs to treat sea-level rise first as an economic risk and then as a physical hazard. Placing livelihoods at the center of adaptation is essential, so that resilience is not only about staying on the coast, but about sustaining life and work there.

11.2.6 Mangroves at Risk: Understanding Perception Dynamics

(a) ***Field Notes (By the Authors)***

This narrative draws on extended field engagement in the Indian Sundarbans and coastal Medinipur, where mangroves hold a complex place in everyday life. Local residents speak of mangroves as sources of protection, as resources that support livelihoods, and as fragile systems facing increasing pressure. In these settings, mangroves are not viewed only as natural ecosystems. They are understood as living forms of infrastructure that shape how people think about risk and safety along the coast. The narrative explores how communities value the protective role of mangroves, how they recognize emerging threats to these landscapes, and how they manage ongoing tensions between conservation efforts, livelihood needs, and coastal security (**Source:** Authors' field diary, August 2025).

(b) ***Field Narratives (Interaction with Locals)***

In the Sundarbans, mangroves are spoken of with familiarity and respect, often as part of everyday life. Elderly residents recalled earlier times when dense mangrove belts extended farther toward the sea. One respondent reflected, *"Bon ta aage chhilo shobcheye boro pahara"* (Earlier the forest was our biggest guard). Fishers described how mangroves help reduce wave force and support fish breeding, while also noting that these forests have become thinner over time. Women who collect fuelwood spoke openly about the difficult choices involved. As one woman explained, *"Jodi gachh na kati, chulha jole na; abar beshi kati, bipod bare"* (If we don't cut trees, the stove won't burn; if we cut too much, danger increases). In this context, perceptions of mangrove risk are closely tied to daily survival as well as long-term safety.

In coastal Medinipur, where mangroves are more scattered and often established through plantation programs, local views took a different form. Younger residents tended to associate mangroves with government initiatives rather than inherited landscapes. A youth volunteer remarked, *"Amra jani mangrove jhor thamay, tai lagano hoy"* (We know mangroves stop storms, that's why they are planted). At the same time, doubts were also expressed. A fisherman stated, *"Choto gachh diye boro jhor ke rokhha kora jay na"* (Small trees cannot protect against big storms). Across both regions, residents clearly recognized mangroves as protective features of the coast. However, many questioned whether they are sufficient on their own in the face of increasing climatic extremes (**Source:** Authors' field diary, August 2025).

(c) ***Integrate with Scientific Evidence***

These local views closely reflect scientific research on the role of mangroves in coastal risk reduction. Ecological and disaster studies show that mangroves help reduce wave force, lower storm surge levels, and limit erosion, especially during moderate cyclones. Research from the Sundarbans indicates that areas with healthy mangrove cover suffered less damage during recent storms than stretches of coast protected only by embankments. At the same time, scientific findings also support the doubts expressed by residents. Mangroves on their own cannot fully protect coastal areas from severe cyclones, particularly when these events occur alongside rising sea levels. Studies also point to ongoing mangrove degradation linked to changes in land use, the spread of aquaculture, and increasing salinity. These pressures were often mentioned indirectly in local accounts. Seeing mangroves as both a form of protection and a fragile resource reflects this shared understanding between scientific research and everyday experience (McIvor et al., 2012; Paul et al., 2023).

(d) ***Challenges and Sustainability***

A key issue that emerges from these accounts is the challenge of balancing mangrove conservation with everyday livelihood needs. Strict conservation rules can sometimes distance local users from decision-making, while uncontrolled use can speed up environmental damage. Plantation efforts, especially in Medinipur, often face practical problems. These include unsuitable species selection, low survival rates, and limited involvement of local communities. Long-term sustainability is also affected by slow and ongoing pressures such as rising salinity and changing sediment patterns, which make mangrove regeneration difficult even in protected areas. At the same time, the narratives point to signs of resilience. Community-led protection efforts, shared rules to limit excessive cutting, and growing awareness of mangroves as long-term sources of safety all suggest possible paths forward. Sustainable outcomes will depend on forms of management that actively involve local communities and recognize mangroves not only as ecological buffers, but also as systems closely tied to social and economic life (Biswas et al., 2024; Lugo et al., 2014).

(e) ***Moral of the Narrative Journey***

This narrative suggests that mangroves offer the greatest protection when they are treated as a shared responsibility rather than as an external solution imposed from outside. Coastal communities understand the value of mangroves through everyday experience, storms that feel less destructive, fish stocks that continue to support livelihoods, and risks that are partly reduced. Disaster risk reduction efforts therefore need to move beyond counting plantations or repeating protection messages. Greater attention must be given to strengthening everyday relationships between people and mangroves. When communities care for mangroves, they are also safeguarding their own futures, grounding resilience in connections between ecology, livelihoods, and shared memory.

11.2.7 Youth and Future: Next Generation Perspectives on Coastal Risks

(a) ***Field Notes (By the Authors)***

This narrative draws on interactions with young people in the Indian Sundarbans and coastal Medinipur, where questions about the future of coastal life are increasingly shaped by the next generation. Many young residents describe coastal risk in ways that differ from older generations. Their views are shaped less by memories of past loss and more by uncertainty about what lies ahead. Disaster risk is closely tied to decisions about education, work, and whether it is possible to build a stable life on the coast. The narrative examines how young people speak about coastal hazards, adaptation, and future plans. Their accounts reflect both concern and a sense of agency, showing how youth are actively thinking about what resilience might look like in their own lives (**Source:** Authors' field diary, August 2025).

(b) ***Field Narratives (Interaction with Locals)***

In the Sundarbans, young people often spoke about coastal risk with a mix of awareness and restlessness. School and college students showed familiarity with cyclones, floods, and erosion through disaster education programs. A college-going youth explained, *"Amra jani jhor, bonna shob barbe, kintu ekhane theke bhobishshot kora khub chap-er"* (We know storms and floods will increase, but building a future while staying here is very stressful). While older residents often described land loss in emotional terms, younger respondents framed risk through the loss of opportunity. They spoke about disrupted schooling during disasters, weak internet connections, and limited options for stable employment.

Young women highlighted layered forms of vulnerability. One respondent shared, *"Jhor-er shomoy porashona bondho hoye jay, tarpor abar ghore kaj beshi"* (During storms, studies stop, and household work increases). At the same time, many young people spoke with pride about their involvement in local response efforts. Youth volunteers described assisting with evacuation and relief work. A teenage volunteer said, *"Amra phone diye lokjon-ke alert kori, eta amader kaj"* (We alert people using phones; this is our role). For these youth, digital tools were not only practical resources but also part of how they understood their place in the community.

In coastal Medinipur, youth perspectives were shaped by wider exposure beyond the village. Many had migrated seasonally or worked in nearby towns and cities. A young fisherman noted, *"Baire giye dekhi, shob jaygay jol-er problem nei, tokhon mone hoy ekhane thaka thik kina"* (When we go outside and see places without water problems, we wonder if staying here makes sense). Even so, strong ties to the coast remained. A tourism worker reflected, *"Samudro amar kaaj-o, bipod-o"* (The sea is both my work and my danger). Young people in Medinipur described coastal risk as a constant negotiation between economic security and cultural belonging, rather than a simple choice between staying and leaving (**Source:** Authors' field diary, August 2025).

(c) ***Integrate with Scientific Evidence***

These views reflect findings from research on how different generations understand climate and disaster risk. Studies show that younger people are often more aware of climate-related threats, but less confident about the long-term ability of coastal areas to remain safe and stable. Research from delta regions also points to youth migration as a common response to risk. While moving away can help individuals cope, it can also create challenges for community resilience. The strong use of digital tools noted during fieldwork matches wider evidence that young people often play a key role in sharing warnings and supporting disaster response. At the same time, scientific assessments caution that the loss of educated youth can weaken local capacity to adapt. This concern was also raised indirectly by older residents during community discussions (Goodale, 2025; Haynes & Tanner, 2015; Jennath & Paul, 2022).

(d) ***Challenges and Sustainability***

One major issue that emerges from these accounts is the tension young people feel between ambition and belonging. Many youths have growing knowledge, practical skills, and comfort with digital tools. At the same time, limited opportunities in coastal areas push them toward migration, which can weaken the transfer of knowledge and practices needed for long-term resilience. Disaster policies often treat young people mainly as recipients of support rather than as active partners in planning and decision-making. This limits the use of their skills and energy. Sustainable futures on the coast will depend on creating real opportunities for young people to stay and contribute. These include education that connects with local livelihoods, work in environmentally sustainable sectors, roles in community-based disaster management, and wider access to digital resources. Without such pathways, resilience may remain fragile, sustained only by those whose capacity continues to decline (Seddiky et al., 2024; Zaremohzzabieh et al., 2014).

(e) ***Moral of the Narrative Journey***

This narrative argues that the future of coastal resilience depends on how young people are engaged in the present. Youth do not deny coastal risks. They understand them clearly, but they think about risk through questions of opportunity, dignity, and choice. Disaster risk reduction therefore needs to look beyond protecting existing conditions and focus on enabling viable futures. Young people should not be placed in a position where they must choose between safety and a sense of belonging. Instead, they need space and support to shape resilient coastal lives in ways that reflect their own goals and aspirations.

11.3 Conclusion and Arguments

The narratives presented in this book, ranging from cyclone risks to youth perspectives on the future, reveal a shared understanding of coastal risk shaped by everyday life in the Indian Sundarbans and coastal Medinipur. Risk is not viewed as an abstract calculation or a distant possibility. It is experienced as something that builds over time, shaped by memory, disrupted livelihoods, environmental change, and ongoing conversations between generations. Sudden hazards such as cyclones and tsunamis exist alongside slower processes like erosion, flooding, sea-level rise, and mangrove loss. Together, these forces create a persistent sense of uncertainty that influences how people live and plan for the future. Communities respond to these risks through daily practices rather than formal measures alone. They adjust their livelihoods, modify homes, rely on local knowledge, and decide carefully when to trust official warnings. These responses show that disaster risk reduction is closely tied to social and economic life. Flooding brings exhaustion, erosion leads to displacement, rising seas unsettle incomes, and environmental decline weakens natural forms of protection. Mangroves are valued as sources of safety, but their care is limited by everyday survival needs. Youth perspectives add another important dimension. Young people express strong ties to place alongside concern about whether a secure future on the coast is possible. Their views suggest that resilience is no longer only about recovery after disaster, but also about the ability to remain and build a life. These accounts show that effective coastal disaster risk reduction must connect shared memory, scientific knowledge, livelihood security, care for ecosystems, and the inclusion of younger generations. In this context, resilience does not mean holding on to the past unchanged. It refers to the ability to adapt in meaningful ways while sustaining dignity, identity, and hope in uncertain coastal settings. The chapter discussions lead to the following arguments:

(a) Coastal risks are lived as ongoing conditions rather than isolated disasters.
(b) Collective memory strongly influences preparedness, trust, and adaptation.
(c) Slow-onset hazards create deeper long-term economic stress than sudden events.
(d) Mangroves act as protective systems that depend on shared local governance.
(e) Early warnings work only when they are clear and usable at the local level.
(f) Youth are vital to future resilience but face few options to stay on the coast.
(g) Disaster risk reduction must focus on livelihoods and futures, not only events.

References

Alam, E., Collins, A., Islam, A. R. M. T., Paul, A., & Islam, M. K. (2023). Change in cyclone disaster vulnerability and response in coastal Bangladesh. *Disasters*. https://doi.org/10.1111/disa.12608

Bala, G., & Boza, A. (2023). *Trends in the frequency and intensity of compound coastal flooding events along the Indian coastline during 1980–202*. https://doi.org/10.5194/egusphere-egu23-17187

Biswas, C., Channarayapatna, S., & Pandey, C. (2024). *Livelihood and beyond relations of local communities with Sundarban mangroves*. https://doi.org/10.21203/rs.3.rs-4212772/v1

Carobene, G., & Carobene, G. (2022). Climate change and coastal hazards in Sunderbans. *Springer Water*, 175–198. https://doi.org/10.1007/978-3-031-18846-6_9

Das, K., & Das, K. (2023). *Perils of premature reclamation: Case studies from the Indian Sundarbans* (pp. 241–250). https://doi.org/10.1007/978-3-031-42231-7_18

Devi, I. B. (2025). Adaptive strategies and livelihood resilience among riverine communities facing flood and erosion hazards in Assam, India. *Indian Journal of Science and Technology, 18*(29), 2396–2406. https://doi.org/10.17485/ijst/v18i29.1087

Dhar, T., Bornstein, L., Lizarralde, G., & Nazimuddin, S. M. (2023). Risk perception—A lens for understanding adaptive behaviour in the age of climate change? Narratives from the global south. *International Journal of Disaster Risk Reduction, 95*, 103886. https://doi.org/10.1016/j.ijdrr.2023.103886

Ellis, M., & Bajracharya, B. (2023, October). *Reducing community risk to coastal erosion with managed relocation.* https://doi.org/10.47389/38.4.52

Ghosh, S., & Mistri, B. (2023). Cyclone-induced coastal vulnerability, livelihood challenges and mitigation measures of Matla–Bidya inter-estuarine area, Indian Sundarban. *Natural Hazards, 116*(3), 3857–3878. https://doi.org/10.1007/s11069-023-05840-2

Goodale, T. A. (2025). *Building perceptional progression among adolescents concerning climate impacted coastal resilience in economically divergent communities* (pp. 59–70). https://doi.org/10.4324/9781003508793-7

Hasnine, M., & Nagdeve, D. A. (2025). Natural hazard induced coastal vulnerability in Indian Sundarbans: A village-level study by using geospatial and statistical techniques. *Natural Hazards Research, 5*(2), 262–275. https://doi.org/10.1016/j.nhres.2024.10.004

Haynes, K., & Tanner, T. (2015). Empowering young people and strengthening resilience: Youth-centred participatory video as a tool for climate change adaptation and disaster risk reduction. *Children's Geographies, 13*(3), 357–371. https://doi.org/10.1080/14733285.2013.848599

Hecht, A. D. (2009). The tripping points of sea level rise. *Environmental Research Letters, 4*(4), 041001. https://doi.org/10.1088/1748-9326/4/4/041001

Jennath, A., & Paul, S. K. (2022). *Migration as a response to climate disasters in coastal areas: Insights from Kerala, India* (pp. 1–10). https://doi.org/10.1109/OCEANSChennai45887.2022.9775396

Khondoker, N. I., Haque, H. A., & Faysal, S. A. (2024). *Structural mitigation approach for evacuation of cyclone on the lens of intersectionality at Ranipur, Charkhali, Deuli, Kakrabunia in Mirzaganj Upazilla, Patuakhali.* https://doi.org/10.70771/jocw.pp.11

Kreibich, H., Pech, I., Schröter, K., Müller, M., & Thieken, A. H. (2016). New insights into flood warning and emergency response from the perspective of affected parties. *Natural Hazards and Earth System Sciences, 1–13*. https://doi.org/10.5194/NHESS-2016-133

Lindell, M. K. (2023). US Pacific coast communities' past preparedness and preparedness intentions for Cascadia subduction zone tsunamis. *International Journal of Disaster Risk Reduction, 84*, 103466. https://doi.org/10.1016/j.ijdrr.2022.103466

Lugo, A. E., Medina, E., & McGinley, K. (2014). Issues and challenges of mangrove conservation in the Anthropocene. *Desafíos de la conservación del mangle en el antropoceno., 20*, 11–38. https://data.fs.usda.gov/research/pubs/iitf/ja_iitf_2014_Lugo004.pdf

Mahato, P. S. (2024). *A study of morphological changes in the coastal areas and Offshore Islands of Sudarban coastline using remote sensing*. IntechOpen. https://doi.org/10.5772/intechopen.112243

Malakar, K. D., & Roy, S. (2024). Voice mapping of coastal communities: Field narratives from the coastal Medinipur and Sundarbans. In *Mapping geospatial citizenship. SpringerBriefs in GIS*. Springer. https://doi.org/10.1007/978-3-031-63107-8_8

Malakar, K. D., Roy, S., & Kumar, M. (2025a). *Geospatial Technologies in Coastal Ecologies Monitoring and Management*. Springer Cham. ISBN:978-3-031-92016-5. https://doi.org/10.1007/978-3-031-92017-2

Malakar, K. D., Roy, S., & Kumar, M. (2025b). Monitoring and Management of Coastal Ecologies in India: Community narratives through PGIS. In *Geospatial Technologies in Coastal Ecologies Monitoring and Management. Advances in geographic information science*. Springer. https://doi.org/10.1007/978-3-031-92017-2_15

Malakar, K. D. (2026). Geographies of Disaster Risk Reduction Practices: New Approaches, Progress, and Contemporary Challenges. Emerald Publishing Limited. ISBN: 978-1-83662-011-2. https://doi.org/10.1108/978-1-83662-010-5

Masud-All-Kamal, M., Monaem, M. A., & Malak, M. A. (2025). *Sociocultural and governance barriers to cyclone evacuation in coastal Bangladesh., 1*(1). https://doi.org/10.1007/s44367-025-00016-0

McIvor, A. L., Spencer, T. J., Möller, I., & Spalding, M. (2012). Storm surge reduction by mangroves. https://www.nacionmulticultural.unam.mx/mezinal/docs/3151.pdf

Need, S. F., Metcalfe, A. V., Sen, D. J., Bhaskaran, P. K., & Lambert, M. F. (2024). *Typhoons and tigers—Flood risk in the Sundarbans*. https://doi.org/10.21914/anziamj.v63.17196

Paton, D., Johnston, D., Johnston, D., Rossiter, K., Buergelt, P., Richards, A., & Anderson, S. (2017). Community understanding of tsunami risk and warnings in Australia. *The Australian Journal of Emergency Management, 32*(1), 54–59. https://researchers.cdu.edu.au/en/publications/community-understanding-of-tsunami-risk-and-warnings-in-australia

Patra, S., & Shukla, J. (2017). Specific coastal issues of Digha coast, West Bengal. *International Journal of Current Research, 9*(7), 53714–53718. https://www.journalcra.com/sites/default/files/issue-pdf/24041.pdf

Pattanaik, A. (2021). *Mangrove-dependent small-scale fisher (SSF) communities in the Sundarbans-vulnerable yet viable*. Thesis submitted to University of Waterloo.

Paul, A. K., Paul, A., Sardar, J., Ray, R., Khatun, K., Chand, S., Maji, R., & Ali, S. (2023). *Mangroves in cyclone-battered Sundarbans, India: A geoinformatics-based multi-temporal study* (pp. 327–341). https://doi.org/10.1007/978-3-031-42231-7_24

Paul, B. K., & Dutt, S. (2010). Hazard warnings and responses to evacuation orders: The case of Bangladesh's cyclone sidr. *Geographical Review, 100*(3), 336–355. https://doi.org/10.1111/J.1931-0846.2010.00040.X

Perera, D., Perera, D., Perera, D., Agnihotri, J., Seidou, O., Seidou, O., & Djalante, R. (2020). Identifying societal challenges in flood early warning systems. *International Journal of Disaster Risk Reduction, 51*, 101794. https://doi.org/10.1016/J.IJDRR.2020.101794

Pramanik, M., Biswas, S. S., Mukherjee, T., Roy, A. K., Pal, R., & Mondal, B. (2015). Sea level rise and coastal vulnerability along the eastern coast of India through geo-spatial technologies. *Journal of Remote Sensing & GIS, 4*(2), 1–8. https://doi.org/10.4172/2469-4134.1000145

Prerna, R., Kumar, T. S., Mahendra, R. S., & Mohanty, P. C. (2015). Assessment of tsunami Hazard vulnerability along the coastal environs of Andaman Islands. *Natural Hazards, 75*(1), 701–726. https://doi.org/10.1007/S11069-014-1336-8

Rafliana, I. (2022). Tsunami risk communication and management: Contemporary gaps and challenges. *International Journal of Disaster Risk Reduction, 70*, 102771. https://doi.org/10.1016/j.ijdrr.2021.102771

Ramteke, K. K., Akter, S., Vineeth, P., Ronad, S., Barman, M. J., & Thanga Anusya, S. (2026). Integrating remote sensing and geographic information science for sustainable coastal and marine ecosystem management. In K. K.

Basheer Ahammed, A. C. Pandey, B. R. Parida, & I. W. G. A. Karang (Eds.), *Monitoring coastal marine ecosystems under climate change*. Springer. https://doi.org/10.1007/978-3-032-09137-6_14

Sahay, A. (2025). *Bridging the gap: The need for science-based policies to address slow-onset hazards in India.* https://doi.org/10.5194/egusphere-egu25-557

Saini, A., Singh, A., & Jothiprakash, V. (2025). *Socioeconomic impacts and preparedness for intensifying cyclones in the bay of Bengal*. https://doi.org/10.5194/egusphere-egu25-16644

Schwoerer, T., Berry, K., Rasmus, S., Joe, J., & Ravens, T. M. (2023). Climate policy must account for community-specific socio-economic, health, and biophysical conditions—Evidence from coastal Alaska. *Regional Environmental Change, 23*(3). https://doi.org/10.1007/s10113-023-02080-9

Seddiky, M. A., Ara, E., & Karim, A. (2024). Climate change-induced hazard risks and migration in Bangladesh: A case study. *Current Research in Environmental Sustainability, 7*, 100253. https://doi.org/10.1016/j.crsust.2024.100253

Singh, A. P., Murty, T., Rastogi, B. K., & Yadav, R. B. S. (2012). Earthquake generated tsunami in the Indian Ocean and probable vulnerability assessment for the East Coast of India. *Marine Geodesy, 35*(1), 49–65. https://doi.org/10.1080/01490419.2011.637849

Wulandari, A., Shohibuddin, M., & Satria, A. (2022). Strategi Adaptasi Rumah Tangga Nelayan dalam Menghadapi Dampak Abrasi. *Jurnal Sosial Ekonomi Kelautan Dan Perikanan, 17*(2), 269. https://doi.org/10.15578/jsekp.v17i2.10364

Zaremohzzabieh, Z., Abu Samah, B., Omar, S. Z., Bolong, J., & Mohamad Shaffril, H. A. (2014). Youths' sustainable livelihood with information and communication technologies: Toward an ICT for development quality model. *American Journal of Applied Sciences, 11*(6), 947–958. https://doi.org/10.3844/AJASSP.2014.947.958

12

Voices from the Coast: Field Narratives on Participatory Coastal Disaster Risk Reduction and Sustainability

This chapter brings together a series of field narratives shaped through long-term engagement with coastal communities in the Indian Sundarbans and Coastal Medinipur. The writing draws from repeated visits, conversations, and shared time with people who live with tides, storms, erosion, and uncertainty as part of everyday life. Instead of treating disaster risk reduction only as a technical concern, the chapter focuses on how coastal residents understand risk, safety, and sustainability through their daily practices and experiences. The chapter aims to bring forward voices that are often placed at the margins of policy and planning. Fishers, women, youth, children, and community volunteers appear not as passive recipients of support, but as active participants who shape coastal resilience in practical ways. Activities such as monitoring the shoreline, repairing embankments, adjusting livelihoods, caring for mangroves, sharing warnings, and spreading climate awareness are presented as forms of knowledge expressed through action. Each narrative highlights a different aspect of participatory resilience. Together, they show how disaster risk reduction is woven into everyday life along the coast. The chapter does not offer ideal models or quick solutions. It documents ongoing efforts, challenges, and learning processes that reflect

K. D. Malakar et al., *Coastal Ecologies at Risk*, Palgrave Studies in Environmental Sustainability, https://doi.org/10.1007/978-3-032-26081-9_12

the realities of living in fragile coastal environments. By focusing on lived experience and local practice, the chapter invites readers to see resilience as something built gradually through collective effort and from within coastal communities themselves.

Key Points of the Chapter

- Describes coastal resilience as an everyday, community-driven process.
- Explains disaster risk reduction as lived practice rooted in local context.
- Examines how monitoring, defenses, and livelihoods shape coastal adaptation.
- Highlights the central roles of women, youth, and children in resilience.
- Explains the need to connect community action with institutional support.
- Highlight narratives (Field narratives on the following):
  - Community vigilance and local observation form the first line of coastal safety.
  - Collective effort supports the building and upkeep of coastal protection structures.
  - Sustainable fishing helps reduce livelihood risk and pressure on coastal ecosystems.
  - Women provide everyday leadership in preparedness, care, and recovery.
  - Eco-tourism supports livelihoods while encouraging care for coastal environments.
  - Youth contribute through communication, environmental action, and disaster response.
  - Climate education helps children share risk awareness and ideas of sustainability.

12.1 Introduction

Coastal regions are among the most dynamic and fragile social and ecological spaces in the world. Human settlements along the coast exist alongside shifting shorelines, tidal rhythms, and increasing climate extremes (Roy, 2025). Along the eastern coast of India, especially in the Indian Sundarbans and Coastal Medinipur, everyday life is closely tied to risk (Ghosh et al., 2015; Sánchez-Triana et al., 2018). Cyclones, storm surges, saline intrusion, erosion, and uncertainty in livelihoods are not occasional disruptions. They are recurring conditions that shape social relations, memory, and daily decision-making (Abijith et al., 2023; Thambidurai & Dikshit, 2025). In such settings, disaster risk reduction cannot be understood only through infrastructure, policies, or external interventions. It needs to be approached as a lived and participatory process that grows from local context. This chapter, *Voices from the Coast*, introduces that lived reality through field narratives that place community agency, local knowledge, and sustainability-oriented practices at the center.

The main argument of this chapter is that coastal resilience is built from within communities through everyday acts of observation, cooperation, care, and learning. Formal disaster management systems remain important, but their effectiveness depends on how they connect with local practices. Early warnings are meaningful when communities trust and interpret them. Embankments last when they are collectively maintained. Livelihoods remain viable when ecological limits are respected. Adaptation becomes sustainable when younger generations are actively involved. The narratives presented here move away from abstract ideas of vulnerability and instead show how people deal with risk in daily life (Malakar et al., 2023). The Sundarbans and Coastal Medinipur offer different but related coastal settings. The Sundarbans, shaped by tidal rivers, mangrove forests, and embankment-protected islands, present a landscape where nature acts as both protection and threat. Coastal Medinipur, marked by open beaches, fishing harbors, and erosion-prone shores, reflects a more exposed meeting point between land and sea. Despite these differences, communities in both regions share the experience of living with uncertainty and responding through collective action.

This chapter places these shared experiences within a broader discussion of participatory disaster risk reduction and sustainability.

The first narrative, *Community-based Coastal Monitoring: Watchful Shores*, presents vigilance as the basis of coastal safety. Close attention to tides, river behavior, embankments, and weather signals forms the first line of protection. Community members act as local observers who turn environmental signs into timely action. This narrative shows how local monitoring connects scientific warning systems with household preparedness through trust and shared responsibility. *Building Barriers Together: Community-led Coastal Defenses* focuses on collective efforts around physical protection. Embankments, sand bunds, bamboo fencing, and mangrove belts appear not only as engineering structures, but as outcomes of shared labor, negotiation, and memory. This narrative also points to the limits of community effort and raises questions about long-term sustainability and the need for external support. *Nets of Tomorrow: Sustainable Fishing Initiatives* explores how fishing communities respond to declining resources, uncertain weather, and growing risk. Decisions about fishing methods, timing, and diversification reflect attempts to balance immediate survival with long-term sustainability. The narrative also brings attention to women's roles in post-harvest work and food security, linking household resilience with ecological care.

Gendered dimensions of resilience are examined further in *Leading the Wave: Women's Roles in Coastal Resilience*. This narrative documents women's involvement in preparedness, shelter management, document protection, and social support. It challenges portrayals of women as passive victims and shows how their everyday work sustains community resilience, while also pointing to gaps in recognition and decision-making power. As livelihoods change, *Green Shores: Developing Eco-Tourism for Sustainability* introduces eco-tourism as an emerging response to climate risk. Community-led initiatives connect income generation with environmental protection and disaster awareness. This narrative shows how sustainability is shaped through everyday choices such as managing waste, limiting visitor pressure, and sharing ecological knowledge. Intergenerational engagement is central to long-term resilience, as shown in *Youth Rising: Empowering New Generations for Coastal Stewardship*. Young people act as links between local knowledge and new tools, especially in communication, environmental action, and disaster response. Their involvement reshapes how communities prepare for hazards while navigating education, work, and migration.

The final narrative, *Climate Kids and Climate Education: Promoting Sustainable Thinking in Disaster Risk Reduction*, focuses on children as active contributors to resilience. Climate education moves from classrooms into homes, shaping preparedness and environmental attitudes. Children become carriers of risk awareness and sustainability thinking within families and communities.

Therefore, these narratives offer an integrated understanding of participatory coastal disaster risk reduction. Resilience appears as layered and relational, shaped through monitoring, infrastructure, livelihoods, gender relations, education, and intergenerational learning. The chapter does not present ideal solutions. It recognizes tensions, limits, and inequalities that influence community action and highlights the need for sustained institutional support. By cantering voices from the coast, this chapter challenges top-down approaches to disaster risk reduction. It argues for approaches that respect lived experience, local knowledge, and collective responsibility. The introduction sets the stage for the narratives that follow and invites readers to engage with coastal resilience as a shared and deeply human process shaped along the water's edge. In this context, the chapter focuses on the coastal region of West Bengal (especially Indian Sundarbans and coastal Medinipur) (Fig. 12.1) and draws on diverse community narratives across multiple, interconnected stages (Fig. 12.2).

12.2 Unveiling Field Interactions: A Compilation of Field Narratives

12.2.1 Community-Based Coastal Monitoring: Watchful Shores

(a) ***Field Notes (By the Authors)***

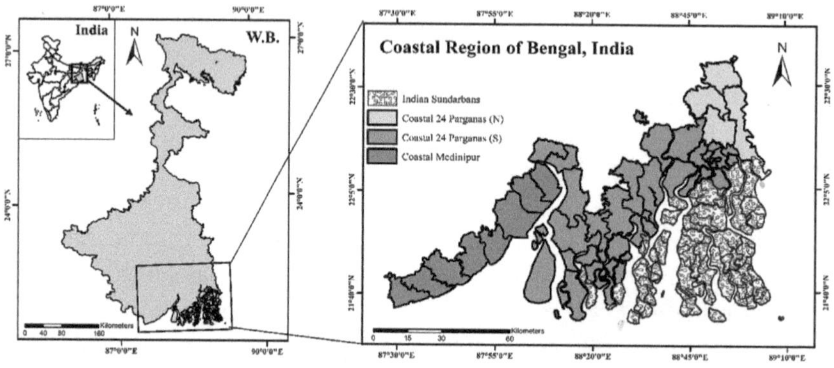

Fig. 12.1 Coastal region of West Bengal (especially Indian Sundarbans and coastal Medinipur) (Prepared by the authors)

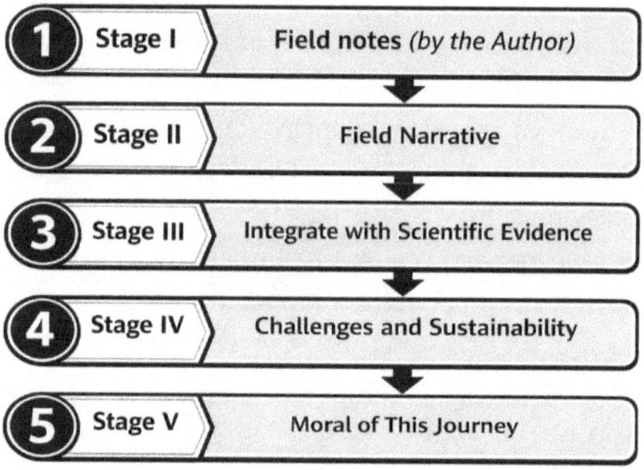

Fig. 12.2 Stages of field narrative collection and processing (prepared by the authors)

This narrative is based on extended field engagements across the low-lying islands of the Indian Sundarbans and the exposed shoreline villages of Coastal Medinipur. It begins with a simple but important observation drawn from fieldwork: coastal safety in these areas is not shaped only through embankments or cyclone shelters. It is also produced through everyday forms of collective vigilance. Community-based coastal monitoring has become a living practice in these coastal settings. Informal watch groups, regular tide observation, care of mangroves, and the sharing of early warnings are part of daily life. These practices are rooted in memory, experience, and a shared sense of responsibility, developed through long-term engagement with changing coastal conditions. The narrative describes how such "*watchful shores*" are sustained through ordinary actions, local knowledge, and ongoing partnerships between communities and institutions. By focusing on these everyday practices, the narrative highlights how coastal safety is maintained not only through physical structures but through continuous attention and collective effort (Source: Authors' field diary, August 2025).

(b) ***Field Narratives (Interaction with Locals)***

Field interactions in the Indian Sundarbans often began early in the morning, when fishers and crab collectors paused to read the river before stepping onto their boats. These moments revealed how close attention to water, tide, and weather forms part of everyday life. A middle-aged fisherman from Satjelia Island explained this attentiveness clearly: "*The water speaks before the storm comes. When the tide behaves strangely, we warn each other even before the radio says anything.*" Such observations reflect a shared practice of watching and interpreting environmental signs before formal warnings arrive. Women's self-help groups also described how routine monitoring has become part of their daily responsibilities. After high tides, checking embankments for cracks and signs of weakness is treated as necessary work. One woman noted, "*If we wait for officials, it may be too late. We mark weak points and inform the panchayat the same day.*" These actions show how responsibility for safety is locally organized and carried out through timely communication and collective effort.

In Coastal Medinipur, attention shifted toward the open shoreline and beach-based vigilance. Youth volunteers spoke about informal night patrols during the cyclone season, especially when tides run high. As one young volunteer explained, "*We are not experts, but we know our beach. When the waves start cutting deeper, we know danger is close.*" Elder fishers connected present actions to past experience, recalling how memory of earlier cyclones continues to guide decisions. One recalled, "*Aila taught us that silence before the storm is dangerous. Now, we never ignore it.*" These field interactions show that coastal monitoring is not understood locally as a project or formal program. Instead, it functions as a social habit, shaped by livelihood practices, shared memory, and everyday relationships with land and water (Source: Authors' field diary, August 2025).

(c) ***Integrate with Scientific Evidence***

The observations shared by coastal communities closely reflect scientific work on disaster risk reduction. Research shows that paying early attention to unusual tides, embankment erosion, and changes in mangrove cover can reduce losses during cyclones. Community-based monitoring supports formal early warning systems by filling local gaps, especially in places where official alerts arrive late or do not reach all households. In the Sundarbans, daily attention to mangrove health matches ecological studies that describe mangroves as effective natural barriers. These studies show that mangroves help reduce wave force and limit the impact of storm surges. Local practices of watching and caring for mangrove areas reflect this knowledge through long experience with changing coastal conditions. In Coastal Medinipur, fishers' observations of beach erosion correspond with scientific evidence of shoreline retreat linked to rising sea levels and more frequent storms. What fishers notice through daily movement along the coast reflects patterns identified through measurement and mapping. Local accounts therefore do not stand apart from scientific knowledge. They add depth to it by connecting broader data to lived experience and specific coastal places (Das & Das, 2025; McIvor et al., 2012).

(d) ***Challenges and Sustainability***

Community-based coastal monitoring, while effective in many ways, faces ongoing challenges. Monitoring work is largely voluntary, and over time this leads to fatigue, especially when there is little formal recognition or support. Younger generations, who are increasingly moving toward non-coastal livelihoods, sometimes see these practices as demanding and difficult to sustain. Women play a central role in monitoring embankments and shelters, but they often have limited influence in formal decision-making spaces. Trust between communities and authorities also shapes the long-term viability of monitoring efforts. When responses are delayed or infrastructure is poorly maintained, confidence in institutions can weaken. Such experiences affect willingness to continue collective monitoring and reporting. At the same time, both study sites show strong capacity for adaptation. Communities have adjusted by rotating volunteer responsibilities, linking monitoring activities with everyday livelihood practices, and building informal connections with local disaster management offices. In this context, sustainability is not defined by fixed arrangements, but by the ability to respond, adapt, and continue under changing conditions (Corbau et al., 2025; McKenna & Hanrahan, 2024).

(e) ***Moral of the Narrative Journey***

This narrative points to a clear lesson. Coastal disaster risk reduction grows from careful listening. Watchful shores depend not only on technology, but on communities who observe their surroundings with care and draw on memory and experience shaped over time. Such attention guides how risks are noticed and addressed in everyday settings. Support from scientific and institutional systems strengthens these local practices and helps them continue. When local monitoring is taken seriously, it becomes more inclusive and easier to maintain. Voices from coastal areas show that resilience develops through daily effort. It is built through shared responsibility, steady observation, and collective care for fragile coastal environments.

12.2.2 Building Barriers Together: Community-Led Coastal Defences

(a) ***Field Notes (By the Authors)***

This field narrative is grounded in prolonged and immersive engagements across the fragile deltaic islands of the Indian Sundarbans and the wave-battered coastal belt of Coastal Medinipur. It begins with an observation drawn from everyday encounters in these regions: barriers against coastal hazards are not only physical structures, but collective processes shaped through shared effort. Embankments, mangrove belts, sand bunds, and temporary bamboo fencing are built, repaired, and watched over through shared labor and local agreement. These activities rely on cooperation, trust, and ongoing discussion within the community. Defense-building in this context is closely tied to how people live with risk and respond to repeated exposure to coastal change. The narrative shows how community-led coastal defenses take shape at the meeting point of lived experience, historical memory, and participatory action. It highlights defense-building as a practice that is technical in form, but deeply social in process, rooted in everyday work and collective responsibility (Source: Authors' field diary, August 2025).

(b) ***Field Narratives (Interaction with Locals)***

In the Sundarbans, embankments carry meanings that go beyond their physical form. During field visits to river-facing villages, residents often described embankments as the line that separates survival from loss. Many pointed to stretches that had been rebuilt after recent cyclones. An elderly farmer explained this sense of responsibility clearly: "*The river takes a little every year. If we do not put it back with our own hands, no one else will.*" Groups of men spoke about working through the night to close breaches using mud-filled sacks during storm surges. Women described their role in organizing food and drinking water for those involved in repair work, stressing that such efforts depend on collective support. As one woman noted, "*When the embankment breaks, it breaks for everyone. So, the repair is also for everyone.*"

In Coastal Medinipur, community-led defenses take a different shape. Villagers build seasonal sand bunds and bamboo barriers to reduce wave impact on homes and fish drying areas. A young fisher explained their purpose in practical terms: "We know these barriers will not stop the sea forever, but they slow it down. That time helps us move, save nets, save children." Elder residents reflected on changes over time, recalling earlier reliance on distant authorities and contrasting it with present practices. One resident shared, "*Before, we waited for help. Now, after every storm, we first gather ourselves.*" These accounts show defense-building as an adaptive process shaped by repeated encounters with coastal hazards. Learning, adjustment, and collective action emerge through experience, making defense-building both a practical response and a shared social practice (Source: Authors' field diary, August 2025).

(c) ***Integrate with Scientific Evidence***

Community practices described in this narrative align closely with scientific work on coastal disaster risk reduction. Research shows that embankments, when maintained regularly, help reduce the duration of flooding and limit saline intrusion in deltaic regions such as the Sundarbans. Local efforts to strengthen weak sections of embankments reflect engineering studies that identify small breaches as major points of failure during cyclones. Attention to mangrove planting and care also reflects ecological research. Studies show that mangrove belts reduce wave force and help stabilize coastal sediments. In Coastal Medinipur, the use of temporary sand bunds and bamboo fencing reflects approaches discussed in coastal geomorphology that aim to reduce erosion and slow the impact of extreme events. Community-led defenses do not stand apart from scientific knowledge. They put scientific ideas into practice through everyday work shaped by local experience and repeated exposure to coastal risk (Dhar et al., 2023; Islam & Islam, 2025; Malakar et al., 2023).

(d) ***Challenges and Sustainability***

Community-built coastal defenses often work well, but they also face serious challenges. Physical fatigue, limited financial resources, and uneven access to materials can slow down repair work. In the Sundarbans, repeated saline flooding weakens embankments faster than communities can rebuild them. Migration of younger residents for work further reduces the labor available for maintenance and emergency response. In Coastal Medinipur, villagers explained that temporary barriers are frequently damaged or washed away during stronger storms. A local volunteer described this situation by saying, "*Sometimes it feels like we are fighting a losing battle.*" Such experiences point to the limits of community effort as coastal hazards become more intense. Long-term sustainability depends on support from outside that builds on local initiative. Access to technical guidance, stronger materials, and formal recognition from institutions can reduce the burden on communities. Without such support, resilience may become an unsustainable form of self-reliance that communities cannot continue on their own (Ferreira et al., 2024; Puri et al., 2024).

(e) ***Moral of the Narrative Journey***

This narrative points to a clear lesson. Coastal protection is strongest when it is built together. Community-led defenses show that resilience does not rest only in concrete or stone, but in cooperation, trust, and shared responsibility for risk. These efforts depend on people working together and taking collective ownership of their surroundings. Voices from the coast remind us that while the sea continues to pose constant challenges, collective action can reduce vulnerability and create a sense of agency. Building barriers together is therefore also about building solidarity. Each repaired embankment and each planted mangrove reflect a community's determination to protect its shore and remain in place despite ongoing uncertainty.

12.2.3 Nets of Tomorrow: Sustainable Fishing Initiatives

(a) ***Field Notes (By the Authors)***

This narrative is drawn from repeated field visits to fishing-dependent settlements along the tidal creeks of the Indian Sundarbans and the open coastal villages of Coastal Medinipur. It begins with an observation shaped by everyday interactions along these shores. Fishing is no longer understood only as an extractive livelihood. It is increasingly seen as a practice that involves care and responsibility toward coastal environments. The idea of "*nets of tomorrow*" refers not only to fishing gear or technology, but also to social networks, shared knowledge, and collective decisions. These elements link sustainable fishing practices with efforts to reduce disaster risk. Fishers and fishing families are adjusting their practices in response to declining catches, frequent cyclones, and growing ecological stress. The narrative outlines how coastal communities are rethinking fishing practices while working to secure both livelihoods and long-term coastal resilience (Source: Authors' field diary, August 2025).

(b) ***Field Narratives (Interaction with Locals)***

In the Sundarbans, fishing takes place within a network of rivers, creeks, and mangrove channels. Fishers spoke with concern about changing seasons and declining catches, while also reflecting on the growing risks involved. A senior fisher described this change by saying, "*Earlier, the net would come up heavy even near the village. Now we must go farther, and the risk is more.*" Younger fishers spoke about shared decisions to avoid fine-mesh nets that trap juvenile fish. One of them explained, "*If we catch the small ones today, what will remain tomorrow?*" Women involved in crab sorting and fish drying described how sustainable fishing practices help protect household security. They spoke about stored dry fish as an important support when fishing activity stops after a cyclone. One woman noted, "*When fishing stops after a cyclone, stored dry fish becomes our support.*" These accounts show how everyday practices link sustainability with family needs and survival.

In Coastal Medinipur, field interactions pointed to ongoing learning and adjustment. Fishers described reducing risky deep-sea trips during unstable weather and shifting toward nearshore fishing combined with other forms of work. A middle-aged boat owner explained, "*Cyclones come faster now. We listen to warnings and stay back, even if it means less income that day.*" Women's groups spoke about collective savings and cooperative marketing as ways to reduce pressure on daily catches. One participant said, "*If the price is fair, we don't need to overfish.*" These voices show sustainability as a balance between making a living and practicing restraint. Fishing decisions are shaped by shared responsibility, household needs, and growing awareness of risk along the coast (Source: Authors' field diary, August 2025).

(c) ***Integrate with Scientific Evidence***

Field narratives from both regions closely reflect scientific work on sustainable fisheries and disaster risk reduction. Marine and estuarine studies show that overfishing, especially through the use of fine-mesh nets, speeds up the decline of fish stocks and weakens long-term livelihoods. Community decisions to limit certain types of fishing gear reflect widely accepted ideas about managing fisheries in ways that protect ecosystems. In the Sundarbans, fishers described changes in breeding seasons and fish availability. These observations match research that links climate variability and increasing salinity to shifts in fish populations. Decisions to follow early warning systems and reduce time at sea during cyclone alerts also reflect evidence showing lower loss of life and fishing assets when risks are anticipated and avoided. Scientific research has also highlighted the role of women in post-harvest work and household food security. This role is clearly visible in field accounts from both regions, where women's labor supports families during periods when fishing is disrupted. Local practices therefore connect closely with scientific knowledge, placing ecological understanding into everyday decisions shaped by experience and shared responsibility (Du et al., 2021; Ghosh et al., 2023; Milne, 2010).

(d) ***Challenges and Sustainability***

Sustainable fishing initiatives face many challenges, even as awareness increases. Economic pressure remains the most persistent concern. Many fishers explained that restraint becomes difficult during lean seasons. A fisher from the Sundarbans expressed this clearly: "*We know it is wrong, but hunger pushes us.*" Limited access to affordable and suitable fishing gear adds to the problem, and the enforcement of regulations is uneven. In Coastal Medinipur, small-scale fishers spoke about competition from mechanized boats, which weakens their efforts to fish in sustainable ways. Climate uncertainty adds further difficulty, as sudden storms disrupt even careful plans and increase daily risk. Long-term sustainability depends on governance that supports local efforts. Access to credit, fair market prices, inclusive regulations, and recognition of community-based monitoring systems all play an important role. Without such support, the commitment of fishing communities may weaken over time (Gedam & Avhad, 2024; Malakar & Roy, 2024).

(e) ***Moral of the Narrative Journey***

This narrative offers a clear message. Sustainable fishing depends as much on foresight as it does on nets and gear. Voices from the coast show that when fishers choose restraint, cooperation, and awareness of risk, they strengthen both ecological balance and the ability to face disasters. These choices reflect an effort to think beyond immediate needs and consider long-term security. The idea of "*nets of tomorrow*" represents a shared sense of responsibility. Caring for the sea in the present helps protect livelihoods in the face of climate uncertainty and changing coastal conditions. These narratives suggest that coastal sustainability grows through the steady weaving of knowledge, care, and community into everyday fishing practices, where each cast of the net carries attention to both the present and the future.

12.2.4 Leading the Wave: Women's Roles in Coastal Resilience

(a) ***Field Notes (By the Authors)***

This field narrative is rooted in long-term engagement with women-led practices across the estuarine islands of the Indian Sundarbans and the hazard-prone shoreline villages of Coastal Medinipur. Fieldwork across these areas repeatedly pointed to a reality that is often overlooked. Women are not only affected by coastal disasters; they are also central to building and sustaining resilience. Women play active roles in managing cyclone shelters, ensuring food security, monitoring embankments, and supporting household livelihoods. Much of this work takes place as part of everyday routines and often remains unrecognized. These actions shape how communities prepare for hazards, respond during emergencies, and recover in the aftermath. Through steady involvement and shared responsibility, women guide local efforts to adapt to changing coastal risks. The narrative describes how women's daily actions, social networks, and experience-based knowledge support participatory disaster risk reduction and long-term sustainability. These practices anchor resilience in everyday life and highlight the importance of women's contributions to shaping safer and more sustainable coastal landscapes (Source: Authors' field diary, August 2025).

(b) ***Field Narratives (Interaction with Locals)***

In the Sundarbans, women's roles in building resilience become visible well before a cyclone reaches land. During field interactions, women from self-help groups described how they follow weather updates, keep important documents safe, and prepare dry food together. A woman from a river-facing village explained, "*When the warning comes, men are often outside. We gather the children, elders, and important papers. This is our first duty.*" Others spoke about regular checks along embankments after high tides. One respondent said, "*We walk the embankment every morning, because if there is a crack, water will enter at night.*" In Coastal Medinipur, women's leadership was most visible in cyclone shelters and during recovery after storms. Local women volunteers described how they manage shelter kitchens, maintain cleanliness, and ensure safety for elderly women and girls. A middle-aged volunteer shared, "*During the cyclone, fear spreads fast. Women calm people, because we stay together.*" Women involved in fish drying and small-scale trade also spoke about their role in supporting household stability. One explained, "*If fishing stops for days, our dried fish and savings help the family survive.*" These accounts present resilience as a practice shaped through everyday care, coordination, and continuity. Women's actions connect preparation, response, and recovery, making resilience part of daily life rather than a separate task (Source: Authors' field diary, August 2025).

(c) ***Integrate with Scientific Evidence***

Scientific research points to the importance of women's participation in disaster preparedness and recovery. Studies on coastal risk reduction show that households where women take part in decision-making are often better prepared for evacuation, food security, and recovery after disasters. Accounts from the Sundarbans, where women monitor embankments and protect important documents, reflect disaster management approaches that focus on preparedness and clear communication at the household level. In Coastal Medinipur, women's involvement in managing cyclone shelters reflects findings from research that link inclusive shelter management with reduced vulnerability, especially for children and older people. Research on climate adaptation also highlights women's roles in sustaining livelihoods through post-harvest activities, savings groups, and small-scale trade. These practices were repeatedly observed across both study sites. Women's everyday experiences connect policy goals with practical outcomes on the ground. Their actions show how preparedness and recovery are carried out in daily life, turning broad ideas about resilience into effective local practice (Azad & Pritchard, 2023; Cvetković et al., 2018).

(d) ***Challenges and Sustainability***

Women play a central role in building coastal resilience, but they face ongoing challenges in sustaining leadership. Heavy workloads, unpaid care responsibilities, and limited recognition often restrict their participation in formal decision-making spaces. Several women shared their frustration that their labor is expected but rarely acknowledged. One respondent expressed this clearly: "*When things go well, no one remembers us. When things go wrong, they call us first.*" Access to training, credit, and institutional support remains uneven in both regions. These gaps make it difficult for women to move from informal responsibility to recognized leadership roles. Long-term sustainability of women-led resilience depends on turning participation into empowerment through skill development, shared decision-making, and policy recognition. Without such support, women's contributions risk becoming an added burden rather than a supported role. Strengthening coastal resilience therefore requires valuing women's time, knowledge, and leadership, not only relying on their labor (Azad & Pritchard, 2023; Chatterjee & Malakar, 2023; Malakar & Roy, 2024).

(e) ***Moral of the Narrative Journey***

This narrative points to a clear understanding of how coastal resilience is sustained. Care is as important as construction in shaping safer coastal lives. Voices from the coast show that when women take leading roles, resilience becomes more inclusive, forward-looking, and lasting. These forms of leadership are often quiet, rooted in everyday work, and carried out without formal recognition. The idea of *"leading the wave"* refers not to visibility alone, but to the steady labor through which women support families, coordinate communities, and reduce disaster risk. These narratives show that the future of sustainable coastal regions depends on recognizing and strengthening the leadership that already exists on the ground. Valuing this everyday work is essential for building coasts that can endure ongoing environmental change.

12.2.5 Green Shores: Developing Eco-tourism for Sustainability

(a) ***Field Notes (By the Authors)***

This narrative is based on grounded fieldwork across emerging eco-tourism areas in the Indian Sundarbans and selected coastal villages of Coastal Medinipur. In these locations, communities are exploring tourism as an alternative livelihood that places less pressure on fragile coastal environments. Field observations pointed to a gradual change in local aspirations. Many households are moving away from strong dependence on climate-sensitive livelihoods such as fishing and daily wage labor, and toward community-managed eco-tourism that values nature, culture, and safety. The idea of "*Green Shores*" reflects this shift. Sustainability in these settings is understood not only in ecological terms, but also in social terms. Eco-tourism links income diversification with environmental care and growing awareness of disaster risk. These efforts draw on local knowledge, shared responsibility, and collective decision-making. The narrative describes how locally led eco-tourism initiatives contribute to resilience building and risk reduction in fragile coastal landscapes. It shows how tourism, when shaped by community priorities, can support livelihoods while strengthening care for coastal environments and preparedness for future risks (Source: Authors' field diary, August 2025).

(b) ***Field Narratives (Interaction with Locals)***

In the Sundarbans, eco-tourism has become part of everyday life shaped by mangroves and tidal rhythms. During field interactions with homestay owners and boat guides, residents described hosting small groups of visitors interested in birds, mangroves, and village life rather than large-scale tourism. A woman who manages a homestay explained this change by saying, "*Tourists now come to see how we live with the forest, not to disturb it. That makes us careful too.*" Local youth trained as nature guides described their role in sharing information about tides, cyclones, and mangrove protection with visitors. One young guide noted, "*When we explain cyclone stories to visitors, we also remember why these forests matter.*" These exchanges show how eco-tourism creates space for learning within the community. In Coastal Medinipur, eco-tourism initiatives are more recent but continue to expand. Villagers spoke about efforts to keep beaches clean, reduce plastic use, and promote local food and crafts. A community volunteer involved in beach walks shared, "*Earlier, outsiders came, left garbage, and went away. Now we tell them this beach is our home.*" Women running small food stalls and handicraft activities emphasized the value of steady income. One explained, "*Fishing stops when storms come, but tourists return after. That money helps us repair our houses.*" These accounts present eco-tourism as a shared learning process that is shaping how communities relate to their environment and think about their future (Source: Authors' field diary, August 2025).

(c) ***Integrate with Scientific Evidence***

Scientific research supports community-based eco-tourism as a path toward sustainable coastal development. Studies show that having more than one source of income reduces vulnerability to climate shocks by lowering dependence on a single and often risky livelihood. Accounts from the Sundarbans, which focus on mangrove-centered tourism, reflect ecological research showing that tourism linked to conservation encourages forest protection and strengthens ecosystem services such as buffering storm surges. In Coastal Medinipur, efforts to keep beaches clean and manage visitor activities reflect approaches used in sustainable coastal zone management. These efforts aim to reduce pressure on shorelines that are already vulnerable to erosion. Research also shows that eco-tourism can increase awareness of disaster risk. Through interaction with visitors, local hosts share information about hazards, evacuation routes, and seasonal changes in risk. Experiences from coastal communities therefore connect closely with scientific thinking on sustainability and resilience. Eco-tourism brings together livelihood security, environmental care, and disaster awareness through everyday practice (Arabadzhyan et al., 2021; van Onselen et al., 2025).

(d) ***Challenges and Sustainability***

Eco-tourism development holds promise, but it also faces several challenges. In the Sundarbans, limits in transport, seasonal access, and fear of wildlife often reduce the number of visitors. Community members also raised concerns that tourism could harm mangroves if it expands without careful control. One resident expressed this concern clearly: "*If too many boats come, the forest will suffer, and then we will suffer too.*" Such views show a strong awareness of the need to protect the environment while seeking new sources of income. In Coastal Medinipur, uneven access to markets and a lack of formal training limit the quality and reach of eco-tourism services. Many initiatives remain small and informal, which makes it difficult to attract visitors on a regular basis. Climate-related risks add further uncertainty, as cyclones and coastal erosion can damage infrastructure and reduce visitor confidence. Long-term sustainability depends on balanced growth and clear support systems. Guidelines, skill development, and institutional backing are needed to strengthen community control over tourism activities. Without this support, eco-tourism may become another fragile livelihood exposed to the same risks faced by existing coastal occupations (Das & Chatterjee, 2023; Mallick, 2025).

(e) ***Moral of the Narrative Journey***

This narrative conveys a clear message about how green shores are created. Coastal sustainability is built through shared care rather than exploitation. Voices from the coast show that eco-tourism can support both livelihoods and landscapes when it is rooted in local knowledge and guided by collective governance. The idea of "*Green Shores*" refers to a form of development where welcoming visitors goes together with protecting ecosystems and preparing for coastal hazards. Field experiences show that sustainability along the coast is more lasting when communities themselves decide how nature is valued, shared, and protected. Such locally shaped approaches help ensure that coastal environments remain secure and productive for future generations.

12.2.6 Youth Rising: Empowering New Generations for Coastal Stewardship

(a) ***Field Notes (By the Authors)***

This narrative is shaped through sustained field engagement with young people across the island villages of the Indian Sundarbans and the erosion-prone shoreline settlements of Coastal Medinipur. Field interactions point to a clear generational change in these coastal areas. Young people are no longer seen only as passive recipients of risk or as migrants looking for opportunities elsewhere. They are increasingly taking active roles as stewards of their coastal environments. Youth involvement is visible in disaster preparedness, digital communication, environmental protection, and local volunteer activities. Young people bring new skills and ideas while also drawing on knowledge shared by elders and community members. Their participation reflects a growing sense of responsibility toward their surroundings. The narrative explains how youth engagement is reshaping local resilience. By combining traditional knowledge with new skills and aspirations, young people are contributing to participatory approaches to coastal disaster risk reduction and long-term sustainability (Source: Authors' field diary, August 2025).

(b) ***Field Narratives (Interaction with Locals)***

In the Sundarbans, youth involvement is often most visible in disaster preparedness activities. During field interactions, young men and women working as village volunteers described how they share cyclone warnings through mobile phones, loudspeakers, and door-to-door communication. A college student volunteer explained this clearly: "*Elders understand the river, but we understand the phone. Together, we make the warning reach faster.*" Others spoke about helping with evacuations and assisting in the management of shelters during storms. A young woman noted, "*When people see familiar faces, they trust us and move quickly.*" Youth groups in the Sundarbans are also closely involved in environmental care. Several respondents described mangrove planting activities organized by local clubs. One young participant explained, "*We plant mangroves because we know they protect our future homes.*" These efforts show how environmental protection and safety are closely connected in youth-led actions. In Coastal Medinipur, youth narratives focused on beach clean-up drives, awareness campaigns against plastic waste, and first-response support after cyclones. A school leaver involved in community work said, "*The beach is our playground and our shield. If we don't care for it, who will?*" These interactions show young people acting as links between scientific knowledge, everyday technology, and shared community experience (Source: Authors' field diary, August 2025).

(c) ***Integrate with Scientific Evidence***

Scientific research supports the role of youth participation in disaster risk reduction. Studies on community-based disaster management show that youth-led communication networks help spread early warnings more quickly and reach more households, which reduces loss of life during extreme events. Accounts from the Sundarbans reflect research findings that digital skills among younger people strengthen last-mile communication in hazard-prone areas. Environmental studies also show that mangrove restoration, often carried out by youth volunteers, helps stabilize shorelines and reduce the impact of storm surges. In Coastal Medinipur, youth-led beach clean-up activities reflect coastal management research that highlights waste reduction as an important step in protecting coastal ecosystems and supporting local tourism. Practices observed among young people in both regions closely match evidence-based approaches to resilience and sustainability. Youth involvement brings together communication, environmental care, and local action, strengthening community responses to coastal risk (Marchezini et al., 2017; Vasileiou et al., 2022).

(d) ***Challenges and Sustainability***

Young people contribute energy and commitment to coastal stewardship, but they face several challenges in sustaining their involvement. Many respondents spoke about the difficulty of balancing community work with education, employment, and pressures to migrate. A youth volunteer from the Sundarbans shared this concern clearly: "*Sometimes we want to help, but there is no support or recognition.*" Limited access to training, lack of formal roles in decision-making, and insufficient resources often weaken long-term engagement. In Coastal Medinipur, seasonal migration further disrupts the continuity of youth groups. These interruptions make it difficult to maintain collective efforts and reduce the long-term impact of youth-led activities. Sustained youth participation depends on stronger institutional support. Skill-building opportunities, recognition of volunteer work, and inclusion of young people in local governance and disaster planning can strengthen their role. Without such support, youth engagement is likely to remain irregular and short-term rather than leading to lasting change in coastal resilience (Hohenhaus et al., 2023; Kelly et al., 2022).

(e) ***Moral of the Narrative Journey***

This narrative points to a clear message about the future of coastal resilience. Strength lies in supporting those who will inherit the shore. Voices from coastal communities show that when young people are trusted, trained, and included, disaster risk reduction becomes more creative, inclusive, and oriented toward the future. The idea of "*Youth Rising*" refers not only to numbers or energy, but to stewardship. Young people carry local knowledge forward while learning to respond to new and changing risks. Their involvement reflects a balance between tradition and adaptation. These narratives show that sustainable coasts are shaped when younger generations are given both responsibility and space to speak. Coastal futures become stronger when youth are active participants in decisions that affect their lives, their environments, and the resilience of their communities.

12.2.7 Climate Kids and Climate Education: Promoting Sustainable Thinking in Disaster Risk Reduction

(a) ***Field Notes (By the Authors)***

This field narrative draws on immersive engagement with schools, informal learning spaces, and community interactions across hazard-prone villages in the Indian Sundarbans and the exposed coastal settlements of Coastal Medinipur. Field observations pointed to a clear shift in how children are viewed within these communities. Children are not only learners but also active carriers of climate knowledge within their households. In areas where cyclones, flooding, and coastal erosion are part of everyday life, climate education shapes how children understand risk, responsibility, and care for their surroundings. Lessons from classrooms and informal learning spaces influence how children talk about safety at home and respond to environmental change. The narrative outlines how "*climate kids*" are becoming important contributors to disaster risk reduction. By sharing what they learn with family members and neighbors, children help turn education into everyday preparedness and wider community awareness (Source: Authors' field diary, August 2025).

(b) ***Field Narratives (Interaction with Locals)***

In the Sundarbans, climate education often extends beyond formal textbooks and classrooms. During interactions with primary school students, children spoke with confidence about cyclones, high tides, and mangroves. One student explained, "*Our teacher says mangroves are like walls. If we cut them, the storm will come inside.*" Teachers described using drawings, songs, and mock drills to explain evacuation routes and safe shelters in ways children could easily understand. Parents also spoke about how these lessons reach the home. A mother shared, "*My son reminds me to keep documents in plastic before the rains. Earlier, I never thought of that.*" Such accounts show how children carry climate knowledge from school into everyday household practices. In Coastal Medinipur, schools and local clubs actively involve children in disaster awareness activities. Students described participating in cyclone mock drills and beach cleanliness campaigns. A middle-school girl explained, "*If the siren sounds, we know where to go. I also tell my grandmother not to panic.*" Informal conversations suggested that children often act as links between schools and households, passing on early warnings to elders who may not have access to digital media. One teacher observed, "*Children listen, remember, and repeat. That repetition saves lives.*" These interactions show climate education as an ongoing process that moves between school, home, and community. Learning does not remain confined to the classroom, but becomes part of daily communication and shared preparedness along the coast (Source: Authors' field diary, August 2025).

(c) ***Integrate with Scientific Evidence***

Scientific research highlights the importance of education in disaster risk reduction and climate adaptation. Studies show that early climate education helps improve understanding of risk, encourages preparedness, and supports the ability to adapt over time. Accounts from the Sundarbans reflect findings that learning led by children can strengthen disaster readiness at the household level, particularly in settings with limited resources. Research also shows that participatory methods such as drills, storytelling, and visual learning work better than abstract instruction in hazard-prone areas. In Coastal Medinipur, children often act as messengers between schools and homes. This role reflects communication studies that identify schools as important points for spreading warnings to households that may not have direct access to digital information. Environmental education that focuses on mangroves, coastal ecosystems, and waste reduction also aligns with work in sustainability studies. Such research links ecological understanding with everyday actions that support environmental care. Experiences shared by these "*climate kids*" show how education moves beyond classrooms and contributes to stronger and more resilient coastal communities (Torani et al., 2019; Tyas et al., 2025).

(d) ***Challenges and Sustainability***

Climate education holds promise, but it also faces several challenges. In both regions, schools work with limited resources, crowded classrooms, and heavy curriculum demands that leave little space for local climate content. Teachers spoke about concerns related to continuity. One teacher explained, "*Training happens once, but children need regular practice.*" Floods and cyclones also interrupt schooling, which weakens learning over time. Social attitudes within households create further barriers. Not all families give equal value to children's knowledge about climate and risk. A student shared this experience by saying, "Sometimes elders say we are too young to understand." Such responses can limit how children's learning is shared and used at home. Long-term impact depends on stronger institutional support. Climate education needs to be part of regular school curricula, supported by ongoing teacher training and recognition of children's voices within families and community planning. Without this support, climate education may remain uneven and informal rather than becoming a stable part of coastal resilience efforts (Singh & Ahmad, 2025; Winter et al., 2022).

(e) ***Moral of the Narrative Journey***

This narrative points to a clear lesson about coastal resilience. It begins early in life. Voices from the coast show that when children learn about climate risks and sustainability, this understanding moves beyond the classroom and into homes and communities. Children often carry this knowledge into everyday family decisions and conversations. The term "*climate kids*" refers not only to a future generation, but to children who are already contributing to disaster risk reduction. Through what they learn and share, they help shape household practices and community attitudes toward safety and environmental care. These field narratives show that investing in climate education is not something to delay. It is an immediate need in places facing regular climate stress. Along vulnerable coastlines, sustainable futures are being shaped through learning and dialogue. Schoolyards, classrooms, and family conversations become spaces where resilience grows, built through knowledge shared one child at a time.

12.3 Conclusion and Arguments

This chapter brings together insights from community-based coastal monitoring, livelihood practices, eco-tourism initiatives, and climate education across the Indian Sundarbans and Coastal Medinipur. The accounts presented here show that coastal resilience in these regions does not emerge by chance, nor is it shaped only through external interventions. It is built through everyday practices, social relationships, and learning that passes across generations and remains closely tied to place. Across the coast, communities monitor tides, repair embankments, manage fisheries, and explore alternative livelihoods such as eco-tourism. Women, youth, and children play active roles in these efforts, contributing labor, care, and knowledge in ways that often remain outside formal systems. These practices show disaster risk reduction as part of daily life, shaped through cooperation, attention, and shared responsibility rather than through technical measures alone. Sustainability appears here as a process rather than a final outcome. People adjust their actions over time, balancing the need to survive with care for coastal ecosystems and drawing on memory to guide present decisions. Scientific knowledge supports these efforts when it connects with local experience and everyday realities. When such links are missing, its influence remains limited. The chapter concludes with a clear message. Resilient coasts are built from within communities through participation, shared responsibility, and continued learning. Coastal futures depend not only on infrastructure and policy, but also on social relationships, everyday ethics, and the collective work of living with changing coastal environments. The chapter discussions lead to the following arguments:

(a) Coastal resilience is built through collective vigilance, cooperation, and shared memory, along with physical infrastructure.
(b) Knowledge gained from everyday local observation works alongside scientific understanding of early warning, ecosystem care, and changing livelihoods.
(c) Women, youth, and children play essential roles in preparedness, response, and long-term sustainability across coastal communities.
(d) Livelihood diversification through sustainable fishing and ecotourism helps reduce risk and improves the ability to recover after disasters.
(e) Climate education strengthens resilience by enabling children to share risk awareness within families and support future environmental care.
(f) Community-led resilience efforts require institutional recognition, access to resources, and inclusion in formal planning to remain effective over time.

References

Abijith, D., Saravanan, S., & Sundar, P. K. S. (2023). Coastal vulnerability assessment for the coast of Tamil Nadu, India—A geospatial approach. *Environmental Science and Pollution Research, 30*, 75610–75628. https://doi.org/10.1007/s11356-023-27686-8

Arabadzhyan, A., Figini, P., García, C., González, M. M., Lam-González, Y. E., & León, C. J. (2021). Climate change, coastal tourism, and impact chains–A literature review. *Current Issues in Tourism, 24*(16), 2233–2268. https://doi.org/10.1080/13683500.2020.1825351

Azad, M. J., & Pritchard, B. (2023). The importance of women's roles in adaptive capacity and resilience to flooding in rural Bangladesh. *International Journal of Disaster Risk Reduction, 90*, Article 103660. https://doi.org/10.1016/j.ijdrr.2023.103660

Chatterjee, B., & Malakar, K. D. (2023). Women empowerment for sustainable social development: Interlinkages and dynamics. *South Asian Journal of Social Studies and Economics, 20*(3), 219–226.

Corbau, C., Lazarou, A., Bajt, O., Filipović Marijić, V., Simčič, T., Coltorti, M., Pignoni, E., & Simeoni, U. (2025). Citizen science for monitoring plastic pollution from source to sea: A systematic review of methodologies, best practices, and challenges. https://doi.org/10.3390/w17182668

Cvetković, V. M., Roder, G., Öcal, A., Tarolli, P., & Dragićević, S. (2018). The role of gender in preparedness and response behaviors towards flood risk in Serbia. *International Journal of Environmental Research and Public Health, 15*(12), 2761. https://doi.org/10.3390/ijerph15122761

Das, K., & Das, K. (2025). *Climate resilience or impeding vulnerabilities in the light of mangroves of Indian Sundarban* (pp. 169–185). https://doi.org/10.9734/bpi/mono/978-93-48859-73-0/ch10

Das, M., & Chatterjee, B. (2023). Ecotourism a solution or deception for conservation: A case of Bhitarkanika Wildlife Sanctuary, Odisha, India. *Journal of Hospitality and Tourism Insights, 6*(3), 1380–1399. https://doi.org/10.1108/JHTI-12-2021-0336

Dhar, T., Bornstein, L., Lizarralde, G., & Nazimuddin, S. M. (2023). Risk perception—A lens for understanding adaptive behaviour in the age of climate change? Narratives from the Global South. *International Journal of Disaster Risk Reduction, 95*, Article 103886. https://doi.org/10.1016/j.ijdrr.2023.103886

Du, Y., Sun, J., & Zhang, G. (2021, October). The impact of overfishing on environmental resources and the evaluation of current policies and future guideline. In *2021 International conference on public relations and social sciences (ICPRSS 2021)* (pp. 1120–1124). Atlantis Press. https://doi.org/10.2991/assehr.k.211020.316

Ferreira, R., Davidson, T., Buttell, F., Contillo, C. M., Leddie, C., Leahy, C., Friedman, R., et al. (2024). Barriers to equitable disaster recovery: A scoping literature review. *International Journal of Disaster Risk Reduction, 110*, 104628. https://doi.org/10.1016/j.ijdrr.2024.104628

Gedam, A., & Avhad, S. B. (2024). Sustainable fisheries management in India: challenges and opportunities. In *Frontiers in animal science: Research and development* (Vol II). Bhumi Publishing. ISBN: 978-93-95847-13-1

Ghosh, A., Schmidt, S., Fickert, T., & Nüsser, M. (2015). The Indian Sundarban mangrove forests: History, utilization, conservation strategies and local perception. *Diversity, 7*(2), 149–169. https://doi.org/10.3390/d7020149

Ghosh, R., Mukherjee, J., Bandyopadhyay, A., Chatterjee, S., Choudry, A., Ghosh, P., Pathak, S., Sen, A., & Sinha, P. (2023). Analyzing scenarios and designing initiatives toward just transitions: coproducing knowledge with (in) the dried fish sector in the Indian Sundarbans. *Frontiers in Water, 5*, 1043628. https://doi.org/10.3389/frwa.2023.1043628

Hohenhaus, M., Boddy, J., Rutherford, S., Roiko, A., & Hennessey, N. (2023). Engaging young people in climate change action: A scoping review of sustainability programs. *Sustainability, 15*(5), 4259. https://doi.org/10.3390/su15054259

Islam, M. M., & Islam, A. R. M. T. (2025). Strengthening climate resilience in coastal Bangladesh: Analyzing the role of NGOs in adaptation governance from an intersectionality lens. *Progress in Disaster Science,* 100498. https://doi.org/10.1016/j.pdisas.2025.100498

Kelly, R., Elsler, L. G., Polejack, A., van der Linden, S., Tönnesson, K., Schoedinger, S. E., Wisz, M. S., et al. (2022). Empowering young people with climate and ocean science: Five strategies for adults to consider. *One Earth, 5*(8), 861–874. https://doi.org/10.1016/j.oneear.2022.07.007

Malakar, K. D., Kumar, M., Anand, S., & Kuzur, G. (2023). Climate change and socio-ecological transformation: Applications and field narratives. In *Climate change and socio-ecological transformation. Advances in geographical and environmental sciences.* Springer. https://doi.org/10.1007/978-981-99-4390-6_7

Malakar, K. D., Roy, S. (2024). Voice mapping of coastal communities: Field narratives from the coastal Medinipur and Sundarbans. In *Mapping geospatial citizenship. SpringerBriefs in GIS*. Springer. https://doi.org/10.1007/978-3-031-63107-8_8

Mallick, N. (2025). Eco-tourism development in Sundarbans: Challenges and opportunities. *Management Journal for Advanced Research, 5*(2), 70–76. https://doi.org/10.5281/zenodo.15455120

Marchezini, V., Trajber, R., Olivato, D., et al. (2017). Participatory early warning systems: Youth, citizen science, and intergenerational dialogues on disaster risk reduction in Brazil. *International Journal of Disaster Risk Science, 8*, 390–401. https://doi.org/10.1007/s13753-017-0150-9

McIvor, A. L., Spencer, T. J., Möller, I., & Spalding, M. (2012). *Storm surge reduction by mangroves*. https://www.nacionmulticultural.unam.mx/mezinal/docs/3151.pdf

McKenna, F., & Hanrahan, J. (2024). Meaningful community engagement through advanced indicator systems for sustainable destination planning. *Environmental and Sustainability Indicators, 22*, Article 100392. https://doi.org/10.1016/j.indic.2024.100392

Milne, G. (2010). *India marine fisheries: Issues, opportunities and transitions for sustainable development*. The World Bank. https://doi.org/10.11588/xarep.00003772

Puri, A., Elkharboutly, M., & Ali, N. A. (2024). Identifying major challenges in managing post-disaster reconstruction projects: A critical analysis. *International Journal of Disaster Risk Reduction, 107*, Article 104491. https://doi.org/10.1016/j.ijdrr.2024.104491

Roy, S. D. (2025). Coastal ecosystem of India: Recent development and perspective planning. In S. Chattopadhyay (Ed.), *Marine and coastal resources of India. Sustainable Development Goals Series*. Springer. https://doi.org/10.1007/978-3-031-81276-7_22

Sánchez-Triana, E., Ortolano, L., & Paul, T. (2018). Managing water-related risks in the West Bengal Sundarbans: Policy alternatives and institutions. *International Journal of Water Resources Development, 34*(1), 78–96. https://doi.org/10.1080/07900627.2016.1202099

Singh, M., & Ahmad, A. (2025). Climate change and sustainability: A critical analysis of Indian curriculum. *Discov Educ, 4*, 502. https://doi.org/10.1007/s44217-025-00947-z

Thambidurai, P., & Dikshit, A. K. (Eds.) (2025). *Coastal environments of India: A comprehensive approach*. Springer. ISBN: 978-3-031-97340-6. https://doi.org/10.1007/978-3-031-97341-3

Torani, S., Majd, P. M., Maroufi, S. S., Dowlati, M., & Sheikhi, R. A. (2019). The importance of education on disasters and emergencies: A review article. *Journal of Education and Health Promotion, 8*(1), 85. https://doi.org/10.4103/jehp.jehp_262_18

Tyas, R. A., Wilujeng, I., Rosana, D., Kuswanto, H., & Purwasih, D. (2025). A review of disaster risk reduction education implementation: Integration, trends, and trajectories. *Social Sciences & Humanities Open, 12*, Article 102015. https://doi.org/10.1016/j.ssaho.2025.102015

van Onselen, V., Bayrak, M. M., Gladfelter, S., & Lin, T. Y. (2025). Community insights on tourism development and nature-based solutions for increased resilience to coastal hazards at Caota Sand Dunes Geopark in Taiwan. Available at SSRN 5184284. https://doi.org/10.1016/j.pdisas.2025.100472

Vasileiou, K., Barnett, J., & Fraser, D. S. (2022). Integrating local and scientific knowledge in disaster risk reduction: A systematic review of motivations, processes, and outcomes. *International Journal of Disaster Risk Reduction, 81*, Article 103255. https://doi.org/10.1016/j.ijdrr.2022.103255

Winter, V., Kranz, J., & Möller, A. (2022). Climate change education challenges from two different perspectives of change agents: Perceptions of school students and pre-service teachers. *Sustainability, 14*(10), 6081. https://doi.org/10.3390/su14106081

Part V

Applications of GIScience and Participatory Approaches in Coastal Disaster Risk Reduction and Sustainability: Case Studies

13

Geospatial Science for Disaster Risk Monitoring, Management, and Sustainability: Insights from the Indian Coast

India's coastal regions are facing increasing pressure from natural hazards, climate variability, and rapid socio-economic change. These pressures have intensified disaster risks and raised concerns about the long-term sustainability of coastal environments and the livelihoods that depend on them. Addressing these challenges requires approaches that can capture spatial patterns, track change over time, and examine the links between environmental processes and human activities. In this context, geospatial science plays an important role in improving understanding of coastal dynamics and supporting informed decision-making. This chapter presents case studies from different parts of the Indian coast that demonstrate the application of geospatial techniques in disaster risk monitoring, management, and sustainability planning. The studies focus on a range of coastal issues, including exposure to cyclones and sea-level rise, flood inundation, urban expansion, environmental degradation, and population vulnerability. Each case shows how spatial data and analytical methods can be used to assess risk, identify vulnerable areas, and support planning and adaptation at local and regional scales. The chapter seeks

K. D. Malakar et al., *Coastal Ecologies at Risk*, Palgrave Studies in Environmental Sustainability, https://doi.org/10.1007/978-3-032-26081-9_13

to enhance understanding of how geospatial approaches can strengthen coastal risk management and support more resilient and sustainable coastal development. It is intended to be useful for researchers, planners, and practitioners working in coastal studies and disaster risk reduction.

Key Points of the Chapter

- Describes the role of geospatial science in coastal disaster risk monitoring.
- Explains spatial patterns of vulnerability across India's coastal regions.
- Examines coastal hazards using GIS and remote sensing techniques.
- Highlights case-based applications for risk-informed coastal planning.
- Supports sustainable and resilient coastal management approaches.
- A collection of casebooks for monitoring and management of coastal ecological risks:
  - GIS-based coastal vulnerability assessment (casebook 1)
  - Coastal wetland vulnerability to urban expansion and pollution (casebook 2)
  - Coastal dynamics and its impact on population (casebook 3)
  - Multi-temporal SAR-based flood inundation mapping coastal region (casebook 4)
  - Impact of sea-level rise on the coastal active delta (casebook 5)
  - Climate change adaptation and sustainable coastal planning (casebook 6)

13.1 Introduction

India has a coastline of more than 7,500 km, marked by wide variation in landforms, ecosystems, and human use. Coastal regions support key economic activities such as ports, fisheries, tourism, industry, and large urban settlements. At the same time, they are among the most exposed and sensitive environments in the country. Natural hazards including cyclones, storm surges, coastal flooding, shoreline erosion, sea-level rise, and salinity intrusion occur frequently along the coast and are becoming more intense under changing climatic conditions. These risks are further increased by rapid urbanization, industrial growth, and long-term environmental degradation. As a result, many coastal hazards have shifted from being occasional events to persistent risks that affect livelihoods, infrastructure, ecosystems, and regional development (Das, Shahfahad et al., 2024; Roy et al., 2023).

Managing these complex challenges requires approaches that can account for spatial variation, environmental change, and human pressures at multiple scales. Geospatial science has become increasingly important in this context. Tools such as Geographic Information Systems (GIS), remote sensing, Global Navigation Satellite Systems (GNSS), and radar-based techniques including Synthetic Aperture Radar (SAR) and Interferometric SAR (InSAR) enable systematic observation and analysis of coastal processes over space and time. These technologies support the integration of diverse datasets, help track environmental change, and allow relationships between physical processes and human activities to be examined in a spatially explicit manner. Along the Indian coast, their use has expanded steadily over the past two decades, supporting vulnerability assessment, hazard mapping, impact analysis, and climate adaptation planning. This shift reflects a growing emphasis on prevention, preparedness, and long-term sustainability rather than short-term disaster response (Chinmayi et al., 2024; Kara & Hastaoğlu, 2026; Malakar et al., 2025a).

The case studies presented in this chapter highlight how geospatial approaches are applied across different coastal settings in India. They represent a range of physical environments, including deltaic regions, urban coastlines, industrial zones, and cyclone-prone areas. The studies also address multiple sources of risk, such as sea-level rise, flooding, urban expansion, environmental pollution, and population exposure. Together, they show how geospatial methods can connect environmental processes with patterns of human vulnerability and support informed planning and management decisions. One case study examines a GIS-based Coastal Vulnerability Index assessment for the Odisha coast, a region that frequently experiences cyclones, storm surges, and coastal erosion. The assessment combines physical indicators such as shoreline change, coastal slope, geomorphology, sea-level trends, and wave conditions to produce a spatial representation of relative vulnerability along the coast. This approach helps identify areas that are more exposed to coastal hazards and provides useful input for disaster preparedness, mitigation planning, and coastal zone management. Urban coastal environments present a different set of challenges, which are illustrated through the assessment of coastal wetland vulnerability in the Chennai metropolitan region. Coastal wetlands play a critical role in flood regulation, shoreline protection, and ecological functioning, but they are under increasing pressure from land-use change, infrastructure development, and pollution. Using remote sensing and GIS, this study examines changes in land cover and environmental stress affecting wetlands. The findings highlight how geospatial tools can support the monitoring of fragile ecosystems and inform sustainable urban and environmental planning in rapidly growing coastal cities. Another case study focuses on coastal dynamics and their influence on population distribution in the Bharuch district of Gujarat. This region is shaped by energetic coastal processes, estuarine systems, and industrial development. By analyzing shoreline change, erosion and accretion patterns, and population data, the study explores the spatial relationship between changing coastal conditions and human settlement patterns. The analysis demonstrates how geospatial methods can be used to assess population exposure and support risk-sensitive land-use planning and adaptation strategies.

Flooding is one of the most common and damaging hazards in India's coastal and deltaic regions, particularly during the monsoon season. This issue is addressed in a case study from Thiruvarur district in Tamil Nadu using multi-temporal SAR data. By integrating InSAR techniques with backscatter analysis, the study maps flood inundation patterns under different conditions. The use of SAR data allows reliable flood detection even during cloud-covered periods, which is a major limitation of optical imagery. This case highlights the value of radar-based geospatial analysis for flood monitoring, emergency response, and post-event assessment.

Long-term risks associated with sea-level rise are examined through a case study of the West Bengal delta. This low-lying and densely populated region is highly sensitive to inundation and salinity intrusion. The study combines elevation data, land-use information, and sea-level trends to identify areas at risk. The results demonstrate how geospatial analysis supports long-term risk assessment and adaptation planning, particularly in environmentally sensitive and socially vulnerable deltaic regions. The final case study focuses on climate change adaptation and sustainable coastal planning In Visakhapatnam, Andhra Pradesh. As a major port city with expanding urban and industrial infrastructure, Visakhapatnam faces multiple climate-related risks, including cyclones and coastal flooding. The study uses geospatial techniques to integrate hazard assessment, land-use analysis, and planning scenarios. This approach supports more coordinated and forward-looking coastal development strategies that balance economic growth with risk reduction and environmental protection.

These case studies illustrate the wide range of applications of geospatial science in coastal disaster risk monitoring and management across India. They show how spatial analysis can be adapted to different hazards, environments, and planning needs. More importantly, they demonstrate the contribution of geospatial approaches to sustainability by supporting informed decision-making, strengthening ecosystem protection, and improving the resilience of coastal communities and infrastructure. The collective insights from these studies highlight the value of geospatial science as a practical and analytical foundation for coastal risk management in India. They provide useful reference points for researchers, planners, and practitioners working toward safer, more resilient, and sustainable coastal regions in the face of ongoing environmental and climatic change.

13.2 A Collection of Casebooks on Monitoring and Management of Coastal Ecological Risks

13.2.1 GIS-Based Coastal Vulnerability Index (CVI) Assessment for the Odisha Coast (Casebook 1)

Working Summary

Odisha located eastern coast of India, which faces compound threats from deltaic subsidence, cyclone, and shoreline changes. The main objective of this research is to develop a Coastal Vulnerability Index by integrating the most important geomorphic and ecological parameters. In this study most effective six parameters were selected (shoreline/soil erosion, Slope, Elevation, Wind/ Wave proximity, LULC, and NDVI). These all parameters were delineated by GEE platform and integrated through weighted overlay analysis using AHP method. The Coastal Vulnerability Index of this study area has been classified into four zones, Very High vulnerability zone which falls under 21.3% area, 15.7% is high, 44.1% is Medium, and 18.9% is low vulnerability zones. The results indicate that the northern part of this study area need spatial ecosystem restoration to rebuild geomorphological resilience, while the southern part needs managed retreat planning.

Keywords: Coastal Vulnerability Index; Shoreline Erosion Assessment; Geospatial multi-criteria analysis; Mangrove ecosystem protection; Deltaic morpho-dynamics; Coastal Odisha

Introduction

Coastal zones act as dynamic ecological defender between terrestrial and marine systems. This zone's highly sensitive to both natural and anthropogenic point of views. The erosional nature and inundation along the coastline due to the cyclone or storm surges a critical concern for coastal management, mostly in regions where geomorphological instability intersects with high exposure to tropical cyclones (Mohanty & Kumar, 2025). The Odisha coast approximately 574.7 km long, where 35% of the Indian eastern coast's cyclones were landfalls, with over a quarter of the coastline experiencing erosion rates of 20–35 meter/year (Mishra et al., 2022; Rao et al., 2020; Sahoo & Bhaskaran, 2018). This coastal region exhibits significant spatial heterogeneity. The northern sectors of Odisha coast (e.g., (Das, Dutta et al., 2025; Das, Talukdar, Shahfahad, Baig, et al., 2024; Das, Talukdar, Shahfahad, Naikoo, et al., 2024; Parida et al., 2018), Bhadrak) show accretion trends, that sustained by dense mangroves forest, on the other side, southern parts of this state's (e.g., Ganjam) face active erosion exacerbated by port construction and river mouth dynamics (Das, Talukdar, Shahfahad, Naikoo, et al., 2024; Parida et al., 2018; Rajakumari et al., 2022). The Coastal Vulnerability Index (CVI) is a standard tool for assessing these coastal hazards. Existing studies on the Odisha coast often classify physical and socio-economic factors and also frequently overlook vegetation mediated protection (mangrove) and apply uniform parameter weights across diverse landscapes. This approach fails to excuse for the differences between mangrove-dominated area and erosion-prone regions. The protective mechanisms generated by spectral indices such as NDVI, along with terrestrial sediment loss indicators, remain underrepresented in standard frameworks. To overcome these limitations, in this study employs an integrated geospatial methodology by synthesizing six complementary datasets (Malakar et al., 2025a, 2025b; Kumar, Dharanirajan et al., 2021; Yacouba et al., 2010) wave and wind directions, topography (slope and elevation), and soil types. By considering these all parameters as interconnected components of a socio-ecological-physical system rather than independent variables, the method captures the dynamic complexity of fundamental coastal vulnerability. Soil erosion serves as a composite metric reflecting sediment system stability, which is critical for designing targeted interventions and prioritizing ecosystem restoration investments (Dey & Mazumder, 2023; Hastuti et al., 2022; Mohapatra & Sharma, 2019; Sahoo & Bhaskaran, 2018; Saravanan et al., 2023). Empirical validation against observed shoreline change establishes this integrated CVI framework as a climate adaptation planning, and strategic resource allocation in vulnerable regions.

In this study's main objective is to delineate and identify a GIS-based coastal vulnerability zones, which will help to the local people and governing bodies to develop sustainable plan.

Methods and Methodology

Study Area

The Odisha coast extends 480 km along the Bay of Bengal (17°49′N–22°34′N, 81°27′E–87°29′E), spanning six administrative districts (Balasore, Bhadrak, Kendrapara, Jagatsinghpur, Puri, Ganjam) with territorial width of 10–100 km, bordered by West Bengal (northeast) and Andhra Pradesh (southeast) (Behera et al., 2019; Parida et al., 2025). The coast exemplifies hexadeltaic morphology formed by six major river systems (Subarnarekha, Budhabalanga, Baitarani, Brahmani, Mahanadi, Rushikulya), creating three geomorphic sections: northern deltaic plain (Balasore), central Mahanadi-Brahmani-Baitarani compound delta system, and southern Chilika Lagoon (1,165 km^2, India's largest coastal wetland with >200 fauna and flora species) (Kantamaneni et al., 2019). Quaternary-Holocene unconsolidated sediments (sands, clays, silts) overlie Paleogene-Neogene formations, reflecting post-Cretaceous coastal evolution (Narendr et al., 2022). The coast exhibits pronounced morpho-dynamic variability: 25.6% actively erodes (20–34 m/year), 51% accretes, and 23.4% remains stable (Fig. 13.1). The eastern coast of India considers as a cyclonic corridor, among all Odisha coast experiences 29% landfall, predominantly during pre-monsoon (May) and post-monsoon (October–November) seasons (Bera & Maiti, 2021; Hastuti et al., 2022; Parida et al., 2025).

Methodology

This study employs Google Earth Engine (GEE) as the primary platform for all multi-spectral remote sensing analysis and geospatial processing,

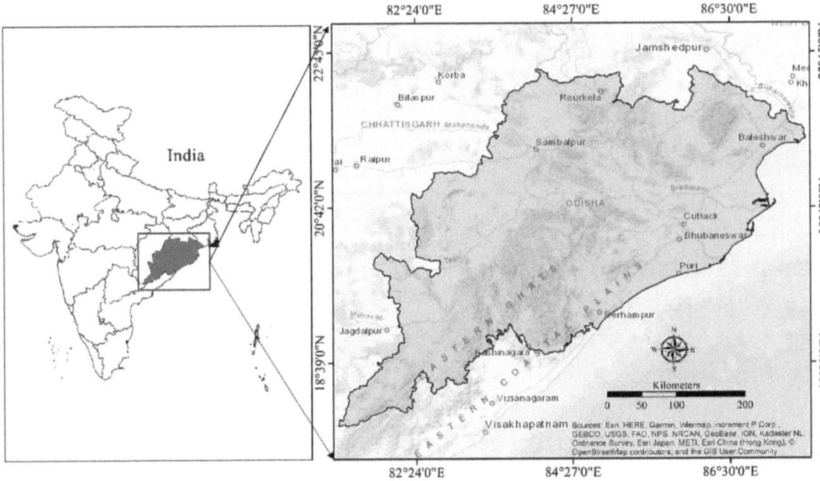

Fig. 13.1 Mapping study area: coastal Odisha (prepared by the authors)

leveraging its cloud-computing infrastructure and extensive satellite data archive (Behera et al., 2022; Moharrami et al., 2021; Pandey et al., 2022; Tiwari et al., 2020). ArcGIS 10.4 was utilized exclusively for cartographic output generation and final vulnerability map production. All geospatial data were standardized to WGS84 UTM Zone 45N projection at 30-meter cell resolution. Landsat 7 and 9 multispectral time series (2000 and 2023) were accessed and processed within GEE's environment (Behera et al., 2019; Bera & Maiti, 2021; Mohanty & Kumar, 2025; Parida et al., 2025). The Modified Normalized Difference Water Index (MNDWI) was computed to delineate water-land boundaries with sub-pixel precision (Eq. 13.1), and Normalized Difference Vegetation Index (NDVI) assessment (Eq. 13.2). Automated image filtering removed cloud-contaminated scenes; temporal image differencing quantified shoreline displacement as erosion/accretion rates (m/year) for each 30-meter coastal segment. SRTM DEM (30 m resolution) was imported into GEE and processed using focal neighborhood algorithms to generate slope derivatives reflecting coastal steepness gradients (Kantamaneni et al., 2019; Narendr et al., 2022). Concurrent elevation raster extraction provided the topographic baseline for inundation vulnerability assessment. Both products were exported as GeoTIFF files for ArcGIS reclassification into standardized vulnerability scales. Normalized Difference Vegetation Index was calculated from Landsat bands and reclassified into vegetation condition classes (high, moderate, degraded, non-vegetated). ESA based Sentinels-2B data for 2025, land cover product was queried within GEE to distinguish water, vegetation, built-up, and bare-land categories, enabling mangrove extent delineation through spectral threshold analysis (Parida et al., 2025). ERA5 Climate reanalysis data (significant mean wave height, wind speed) were extracted from GEE for 2010–2023 decadal normals, with wind magnitude serving as a proxy for wave energy generation in this fetch-limited coastal system (Sahana et al., 2019). DEM-derived slope and flow-accumulation layers were processed in GEE and exported for soil loss estimation, integrating rainfall erosivity, soil erodibility, topographic factors, vegetation cover, and management practices. Erosion severity was classified into seven categories (Lalrindika et al., 2024). All six parameter rasters were imported into ArcGIS 10.4 and standardized to a common 1–5 vulnerability scale (Fernandez et al., 2016; Hoque et al., 2017; Thenmozhi et al., 2024) (Eq. 13.3). Analytical Hierarchy Process weighted linear combination generated the final Coastal Vulnerability Index (Eq. 13.4), classified into four categories (Very High, High, Moderate, and Low) and visualized as publication-quality vulnerability maps (Biswas et al., 2023; Hoque et al., 2017, 2018, 2019; Rautela et al., 2023; Varikkodan et al., 2023) (Fig. 13.2).

$$\text{MNDWI} = \frac{Green - SWIR}{Green + SWIR} \tag{13.1}$$

$$NDVI = \frac{NIR - RED}{NIR + RED} \tag{13.2}$$

$$\text{A} = \text{R} \times \text{K} \times \text{LS} \times \text{C} \times \text{P} \tag{13.3}$$

$$CVI = \sum\nolimits_{i=1}^{n} (Wi \times Pi) \tag{13.4}$$

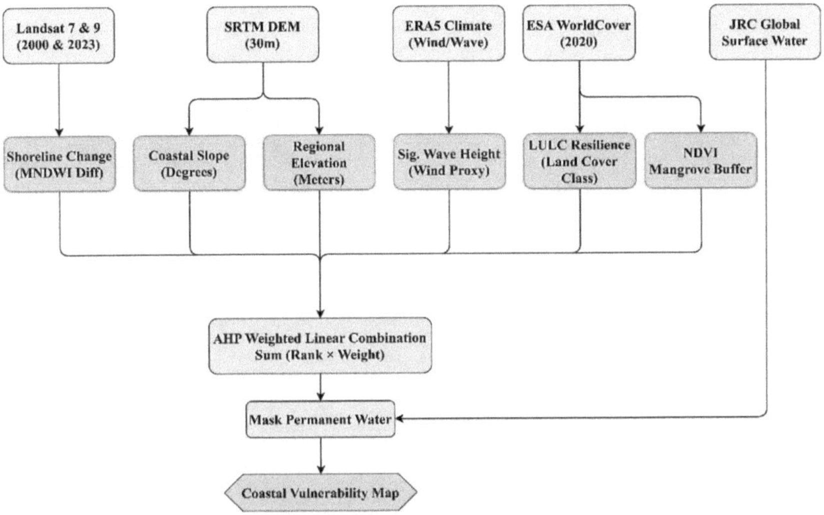

Fig. 13.2 Methodological flowchart: Casebook 1 (prepared by the authors)

Result and Discussion

Shoreline erosion emerged as the dominant vulnerability concern, with the Shoreline Change/Soil Erosion Index (AHP weight 0.25) revealing that 92.3% of the coast faces high-to-very-high vulnerability (Mohanty & Kumar, 2025; Parida et al., 2025) (Fig. 13.3). The Bay of Bengal's dynamic sediment transport system creates persistent zones of net erosion, especially where wave refraction concentrates energy and where river sediment supply has been disrupted by coastal structures (Parthasarathy & Deka, 2021; Sahana et al., 2019).

Historical shoreline mapping confirmed that areas now classified as very high vulnerability (52.8%) have experienced sustained retreat over decades, reflecting ongoing sediment imbalance (Parida et al., 2025). Protected embayments and river mouth deltas comprise the modest 1.4% low-vulnerability class. Slope conditions (AHP weight 0.15) showed that 53.3% of the coastal zone is very highly vulnerable, concentrated across the extensive deltaic plains where slopes rarely exceed 1 degree. However, northern sections with Quaternary terraces and laterite ridges maintained low vulnerability (23.0%), demonstrating

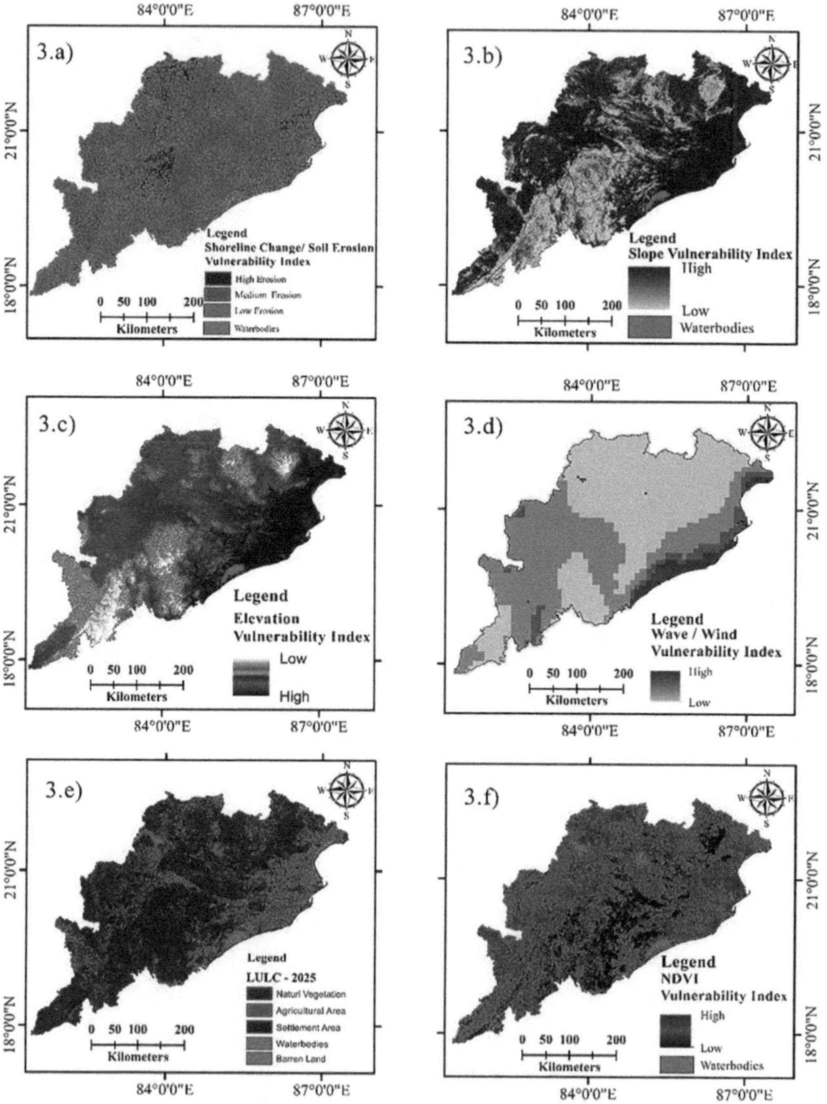

Fig. 13.3 (a) soil erosion vulnerability index; (b) slope vulnerability index; (c) elevation vulnerability index; (d) wave/wind vulnerability index; (e) LULC-2025; (f) NDVI vulnerability index (prepared by the authors)

how subtle topographic relief- often just 13 meters can dramatically alter inundation behavior (Abijith et al., 2023; Bera & Maiti, 2021). Slopes steeper than 23 degrees provided natural resistance to storm surge penetration, whereas the flat deltaic landscape offers virtually no topographic defines. Elevation patterns (AHP weight 0.20) underscore the region's fundamental exposure problem (Table 13.1).

Nearly half the coast (49.2%) lies in very high vulnerability classes, with mean elevations between 0 2 meters above spring tide levels. The critical threshold emerges at 3 meters elevation 57.3% of the coast falls below this height, meaning most settlements and agricultural land face regular or episodic saltwater intrusion. The nonlinear increase in vulnerability below 2 4 meters marks the zone separating occasional flooding from chronic hazard exposure. Wave and wind conditions (AHP weight 0.10) surprisingly show moderate vulnerability despite the Bay of Bengal's notorious cyclone activity. Some 50.9% ranks low vulnerability because the extended continental shelf dissipates wave energy

Table 13.1 AHP-based score of Coastal Vulnerability Index

Parameters	Classes	Rank	AHP-based score
Shoreline Change/ Soil Erosion Vulnerability Index	High	4	0.25
	Medium	3	
	Low	2	
Slope Vulnerability Index	High	2	0.15
	Medium	3	
	Low	4	
Elevation Vulnerability Index	High	1	0.20
	Medium	3	
	Low	4	
Wave/ Wind Vulnerability Index	High	4	0.10
	Medium	3	
	Low	2	
LIULC-2025	Natural vegetation	1	0.15
	Agricultural area	3	
	Settlement area	4	
	Waterbodies	1	
	Barren land	1	
NDVI Vulnerability Index	High	1	0.15
	Medium	3	
	Low	4	

before reaching shore (Sahana et al., 2019). Annual mean significant wave heights (1.2–1.8 meters) against storm peaks exceeding 4 meters, but shoreline erosion responds weakly to waves alone. Vulnerability only in exposed (5.2% high vulnerability) where bathymetric protection fails. Land-use patterns (AHP weight 0.15) reveal an inconsistency. Natural vegetation covers 73.8% of the coastal zone, but 37.2% of the area still rates high vulnerability because uneven mangroves, degraded vegetation, and agricultural lands lack effective protective capacity (Dey & Mazumder, 2023). Settlement expansion (3.2% of coast) concentrates population precisely in the most erosion-prone and flood-prone segments, increasing socio-economic risk. Barren and sparsely vegetated zones (0.7%) show extreme vulnerability, typically sites of recent catastrophic erosion. Vegetation strength (NDVI, AHP weight 0.15) mirrors these patterns. Only 35.3% of coast maintains strong vegetation (NDVI > 0.6), primarily in northern mangrove complexes near river mouths where freshwater maintains ecosystem health. Conversely, 20.7% shows degraded vegetation (NDVI < 0.4), concentrated in the southern part of this study area, where saltwater intrusion and anthropogenic pressure have damaged protective plant populations. The 42.1% with intermediate values reflects areas undergoing vegetation recovery (Narendr et al., 2022; Parida et al., 2025).

The combination of these 6 parameters generates an indication of a coastline where erosion, low elevation, and gentle slopes lead to the significant vulnerability, which neither moderate wave energy nor partially grown vegetation can adequately mitigate (Bera & Maiti, 2021). To significantly reduce threat, especially the most vulnerable populations can have their vegetation restored and strategic coastal protection.

Coastal Vulnerability Index (CVI)

Identification of coastal vulnerability is an essential for the development of a planning strategy for coastal area's population (Mohapatra & Sharma, 2019; Saravanan et al., 2023). The final Coastal Vulnerability Index of this state's was generated through a statistical based systematic Analytical Hierarchy Process (AHP) and weighted overlay method (Fig. 13.4). The weighted overlay analysis conducted in ArcGIS 10.4. All six most vulnerability parameter's Shoreline Change/

Soil Erosion (Fig. 13.3a), Slope (Fig. 13.3b), Elevation (Fig. 13.3c), LULC (Fig. 13.3e), NDVI ((Fig. 13.3f), and Wave/Wind ((Fig. 13.3d) were preliminarily standardized to a common 1–5 vulnerability scale using linear reclassification methods, that ensure dimensional consistency among parameters (Parida et al., 2025). The AHP methodology based different assigned weights to each parameter's has been decided based on expert judgment and literature-derived influence coefficients (Dey & Mazumder, 2023). In this regard, the weights for Shoreline Change/ Soil Erosion were decided 0.25 (3, for the Elevation 0.20, for Slope 0.15, for LULC 0.15, for NDVI 0.15, and lastly weights for the Wave/ Wind 0.10, these all reflecting their relative importance in driving coastal vulnerability (Eq. 13.5 and Table 13.1).

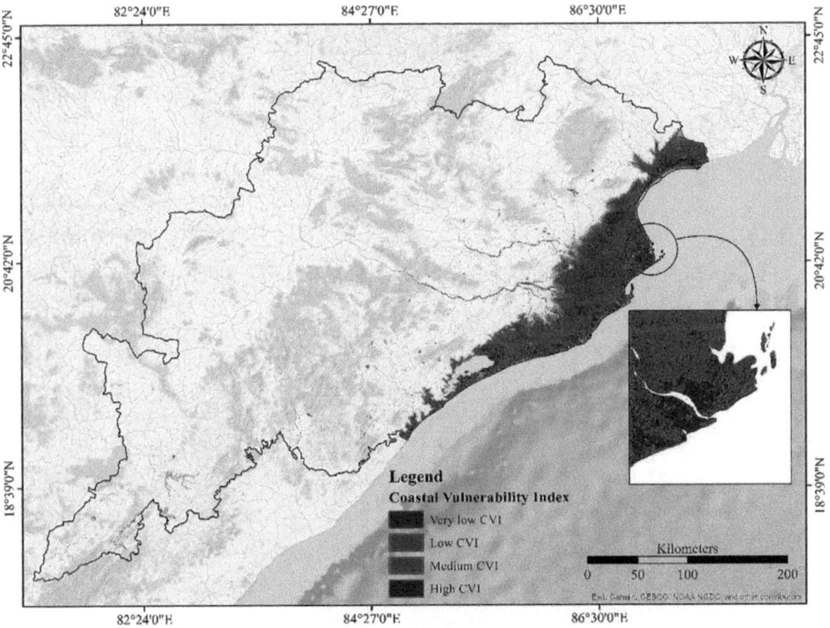

Fig. 13.4 Coastal vulnerability index in the study area (prepared by the authors)

The Weighted Overlay tool in ArcGIS 10.4 executed combination of all weighted parameters using the formula (Eq. 13.5):

$$CVI = \sum_{i=1}^{n} (Wi \times Pi) \tag{13.5}$$

This generated a continuous CVI into discrete vulnerability classes, which classified into: Low (CVI < 1.5), Medium (1.5–2.0), High (2.0–2.5), and Very High (>2.5) zones (Mandal et al., 2022; Sahoo & Bhaskaran, 2018; Saravanan et al., 2023). The Spatial clustering analysis identified vulnerability hotspots. These hotspots were concentrated in southern deltaic plains of this study area (particularly: Ganjam, Puri) where coastal erosion (52.8% very high), low elevation (49.2% very high), gentle slopes (53.3% very high), and degraded vegetation (20.7% high NDVI vulnerability) converge (Table 13.1). The northern part of Odisha (Kendrapara, Bhadrak) exhibited moderate-to-low CVI values due to protective mangrove ecosystems (73.8% vegetation coverage), elevated terrain, and coastline's stability. The final CVI map delineates priority intervention zones where approximately 45–55% of coastline requires urgent management. That management plan provide evidence based on decision support for climate adaptation, ecosystem restoration, and coastal land protection planning under alarming sea-level rise and cyclonic scenarios.

Conclusion

Coastal Vulnerability across the Odisha coast vary significant. This vulnerability requires serious management strategies run to local conditions rather than uniform coastal policies. This assessment identifies critical CVI zones where deltas were subsidence. The low elevation, and declining nature of vegetation combining to increase possibilities for flood and erosion risk along the coast. Southern districts (Ganjam-Puri) of this state's face severe hazard (coastal erosion) and that would benefit from integrated shoreline protection and mangrove restoration, while northern areas should focus on preserving existing mangrove ecosystems. Geospatial monitoring tools now allow regular tracking of coastal changes; however, converting these findings into effective adaptation measures that requires coordination between government agencies, local communities, and planners across state and district boundaries. The framework presented can be adapted for other deltaic coasts worldwide facing similar river-related, coastal erosion and cyclone pressures along the coastlines, particularly under projected sea-level rise and changing monsoon patterns.

13.2.2 Assessment of Coastal Wetland Vulnerability (CWV) to Urban Expansion and Pollution in Chennai Metropolitan (Casebook 2)

Working Summary

This research investigates coastal wetland vulnerability in the Chennai metropolitan, which is under rapid industrial and infrastructure development pressure. The objective of this research is to find coastal wetland vulnerability in reference to the expansions of Chennai metropolitan area. In this study best suitable cloud-based GIS methodology adopt. The Sentinel-2B satellite imagery data has been used from 2019 to 2025 were processed in Google Earth Engine cloud platform, to develop an Enhanced coastal Vulnerability Index which evaluate wetland degradation. That index integrates with human-driven parameters such as urbanization, soil, ecological indicators like vegetation condition and surface water. The findings reveal that significance land erosion over the last six-year period, with about 75.31 km^2 of previously vulnerable land converted into urban infrastructure expansions. The vulnerability assessment indicates that 577.31 km^2 of the area falls within the high to very high vulnerability zones, which particularly connected to industrial, settlement expansion and widespread fly ash deposition near to the coast. It was found that, only 253.39 km^2 of wetlands remain in comparatively good ecological condition. The findings suggest that many wetlands are approaching critical thresholds beyond which natural recovery will be difficult. In this situation need an urgent and stronger industrial regulations implication to safe ecologically sensitive areas, and active restoration of shrinking tidal mudflats to support long-term coastal resilience.

Keywords: Coastal wetland vulnerability; Enhanced vulnerability index; Encroachment; Vulnerability; Sentinel-2 imagery; Chennai Coast

Introduction

Coastal wetlands act as critical protective barriers for any cities, by absorbing flooded waters, recharging groundwater table, and regulating tidal and storm surges. But, in the rapidly expanding and population pressure on the Chennai metropolitan area, these ecosystems are often shift as wasteland ripe for development (Abijith et al., 2023; Vasudevan et al., 2022) The Ennore Creek, situated in the northern of this city's, plays a vital role to maintain hydrological cycle, linking the Kosasthalaiyar River and Pulicat Lake to the Bay of Bengal (Hastuti et al., 2022). Though this estuarine creek essential for controlling flood and life line for fisheries and fishermen communities, become a flashpoint of ecological conflict (Pednekar & Siva Raju, 2020). The rapid industrial expansion identified northward by thermal power plant, petrochemical refineries industries, and port facilities has steadily decayed the wetland's ecological health. Two primary underlie parameters identified for the Ennore Creek's decline that are the physical encroachment of the land and industrial waste (Johnson et al., 2020). The North Chennai Thermal Power Station and the expanding Kamarajar Port have replaced vast tidal mudflats with developing the concrete infrastructure and fly ash deposits (Vincent & Owens, 2021). These artificial barriers prevent the natural tidal flow, which trapping tidal water and allowing sediment deposition at the river mouth (Pednekar & Siva Raju, 2020). Simultaneously, this estuarine creek has become a repository for industrial discharge, like hot water, fly ash, and slurry leakage from thermal power, have degraded sediment composition and water quality of this creek's, which create hostile condition for the aquatic organisms, and weakening the ecosystem's capacity to filter contaminants. Traditional research methodological approaches are identified as slow and spatially limited in capturing such extensive degradation and contamination (Kantamaneni et al., 2019; Malakar et al., 2025c; Narendr et al., 2022; Parida et al., 2025). This research employs remote sensing and GIS technique to overcome these constraints, delivering a landscape-scale assessment.

The main objective of this research to evaluate Coastal Wetland Vulnerability by conceptualizing vulnerability as the gap between environmental stressors and ecosystem resilience. This GIS-based method transcends simple observation, offering a systematic way for identifying which areas of Chennai metropolitan have transitioned from functional wetland to degraded industrial landscape.

Materials and Methods

Study Area

The research emphasizes on the Chennai metropolitan area, specifically the estuarine zone along the Coromandel Coast of Tamil Nadu in India. The area is located between the latitudes of 12°45′ N and 13°45′ N, and the longitudes of 80°01′ E and 80°22′ E (Saleem Khan et al., 2020). This wetland system serves as an essential hydrological junction where freshwater from the Kosasthalaiyar River mixes with the brackish waters of Pulicat Lake before it flows into the Bay of Bengal (Abijith et al., 2023). The wetlands provide natural protection against coastal erosion, but the area is largely influenced by industrial structures, particularly the North Chennai Thermal Power Station and Kamarajar Port. The conflict between delicate mangrove ecosystems and large-scale industrial development highlights the region as a crucial example for exploring environmental vulnerability and the degradation of ecosystems (Fig. 13.5).

Data Collection and Pre-processing

For this research satellite-based Sentinel-2B Surface Reflectance remote sensing data has been collected and pre-processed on Google Earth Engine cloud-based platform (Abijith et al., 2023; Rani et al., 2023).

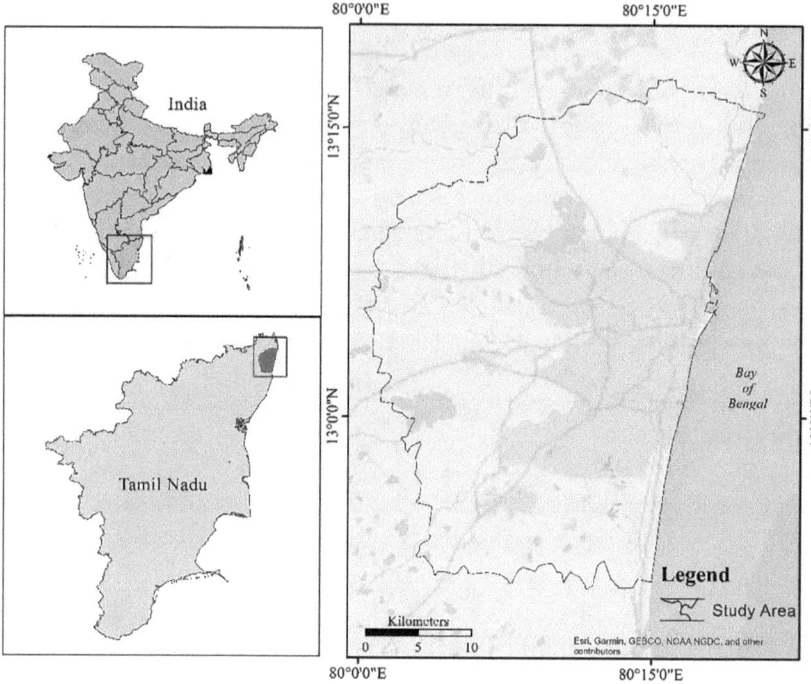

Fig. 13.5 Mapping study area: coastal Chennai metropolitan area (prepared by the authors)

In this study two six-month period's data were used:

- From January to June 2019
- From January to June 2025

Before start analysis, atmospheric error has been corrected. The QA60 quality band was applied to screen the imagery, retaining only scenes with cloud cover below 1% (Johnson et al., 2020; Saleem Khan et al., 2020). This filtering essential because cloud cover data obscure the underlying land cover and would have compromised the accuracy of spectral measurements.

Geospatial Indices

In this research the assessing the wetland vulnerability employs five spectral indices required, which were generated from Sentinel-2B data (bands are Blue, Green, Red, Near-Infrared (NIR), and Short-Wave Infrared (SWIR) (Fig. 13.6). These all indices were generated using established mathematical formulas on cloud-based platform GEE (Abijith et al., 2023; Pednekar & Siva Raju, 2020; Rani et al., 2023). For stressor indicators, the NDBI (Eq. 13.6) was applied to map the resistant surfaces, while the BSI (Eq. 13.7) was used to identify exposed fly ash and degraded land of this study area.

$$NDBI = \frac{SWIR - NIR}{SWIR + NIR} \tag{13.6}$$

$$NDVI = \frac{NIR - RED}{NIR + RED} \tag{13.7}$$

$$BSI = \frac{(SWIR + NIR) - (NIR + BLUE)}{(SWIR + NIR) + (NIR + BLUE)} \tag{13.8}$$

$$NMDWI = \frac{GREEN - SWIR}{GREEN + SWIR} \tag{13.9}$$

For resilience indicators, the Normalized Difference Vegetation Index (NDVI) was applied to assess the natural vegetation and mangrove density (Eq. 13.7), further for the better result Modified Normalized Difference Water Index (MNDWI) was used to delineate open water bodies (Kumar & Acharya, 2016; Mondal & Pal, 2018) (Eq. 13.9). Water quality was evaluating purpose applied the Normalized Difference Turbidity Index (NDTI) (Eq. 13.10), which identified the water clarity and suspended sediment concentrations across the study area basically industrial pollutant (Bid & Siddique, 2019) (Fig. 13.6).

$$NDTI = \frac{RED - GREEN}{RED + GREEN} \tag{13.10}$$

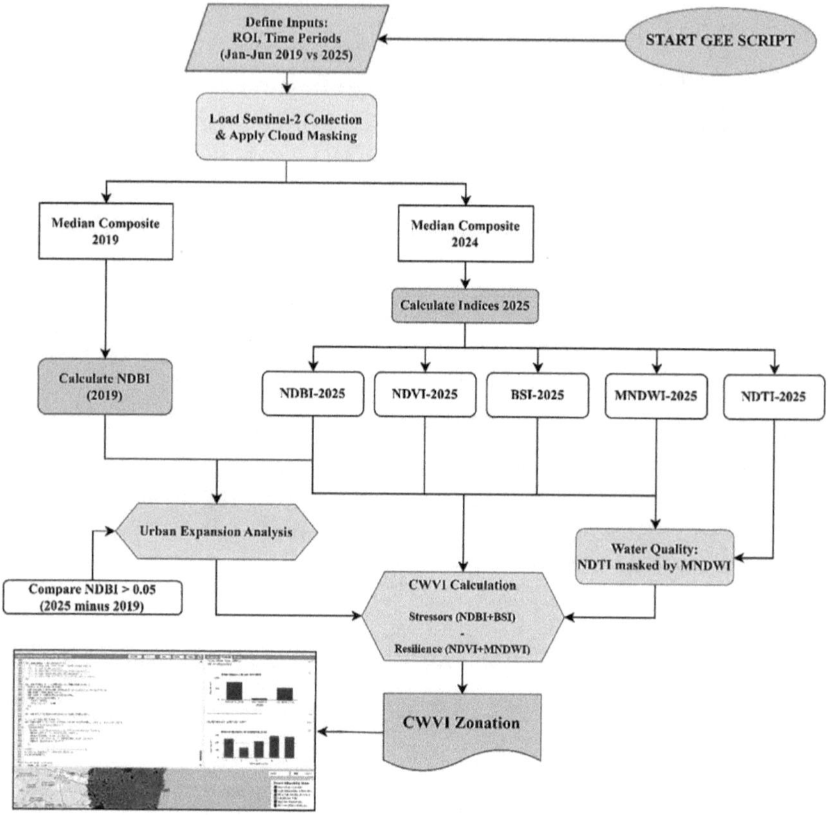

Fig. 13.6 Methodological flowchart: Casebook 2 (prepared by the authors)

Results and Discussion

The 6 years interval's (from 2019 to 2025) satellite data based built-up area expansion in this study areas indicated a substantial shift (Abijith et al., 2023; Vasudevan et al., 2022). From 2019 and 2025 reveals substantial shifts in land-use patterns across Chennai (Saleem Khan et al., 2020; Theocharidis et al., 2025). The NDVI was applied to identify and map impervious surfaces such as concrete, distinguishing them from natural wetland vegetation areas. Urban extent measured 786.55 km^2 in

2019, with an additional increase 75.31 km^2 of 2025. Which is almost rise 9.57% from 2019 to 2025 (Fig. 13.7).

This expansion basically happened permeable surfaces like mudflats, salt pans, and transitional wetlands. This conversion of land seriously harming the ecological balance. By 2025, total urban area had grown to 548.48 km^2 (Saleem Khan et al., 2020). The concentrated pattern of new development suggests that industrial growth is clustered in specific zones particularly in the northern part of these metropolitan cities that compromising the creek's capacity to attenuate floodwaters, and natural ecological balance potentiality (Vincent & Owens, 2021). The Bare Soil Index identified non-vegetated lands, which different from typical urban areas.

High index values (0.63) of BSI appeared in zones where fly ash had accumulated, and in areas cleared for future development (Fig. 13.8). These highly reflective surfaces indicate topsoil removal or burial beneath industrial deposits.

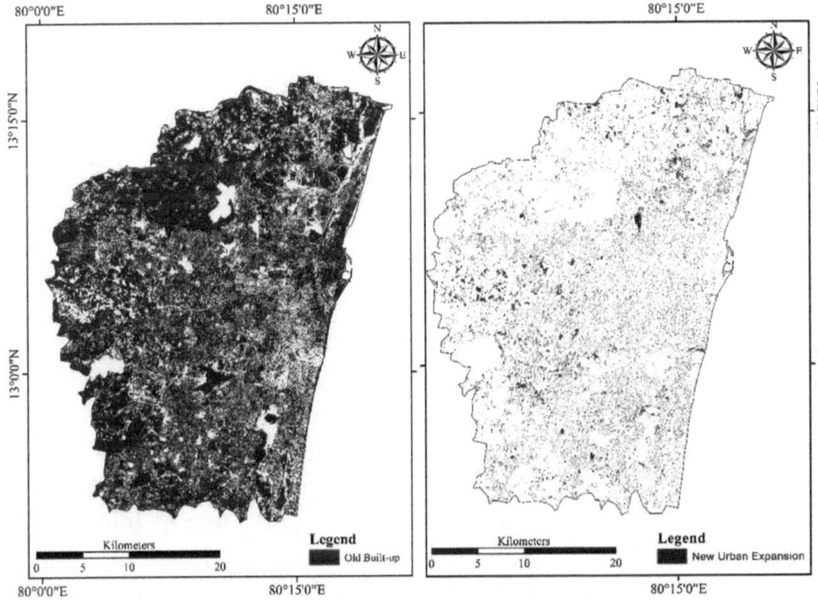

Fig. 13.7 Built-up area (2019 and 2025) (prepared by the authors)

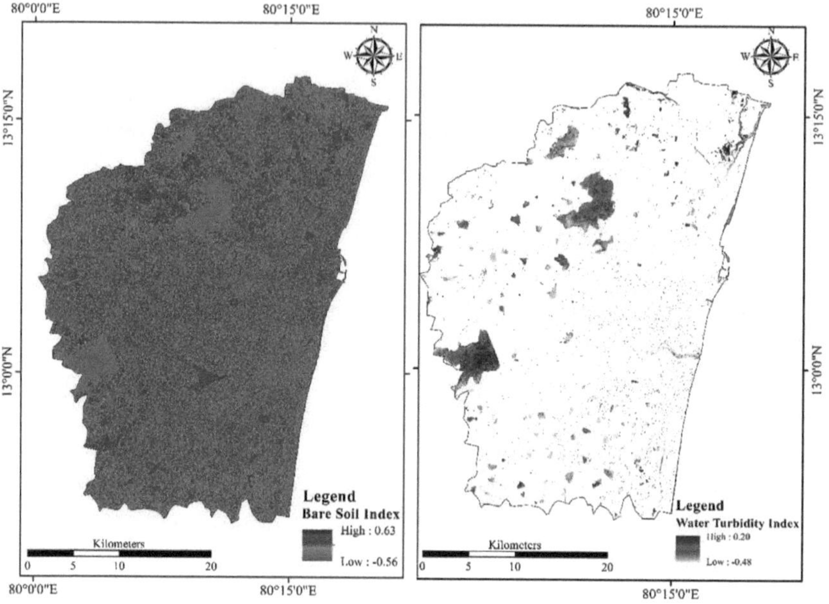

Fig. 13.8 Bare soil index and water turbidly index (prepared by the authors)

This index (BSI) identified the fly ash storage region, which is typically excluded from standard urban indices (NDBI). This is essential to differentiate fly ash and bare soil because they disturb and shift the mangrove growing soil near the coastal area. The Normalized Difference Turbidity Index (ranging from 0.20 to –0.48) measured water clarity using red and green wavelengths, revealing spatial variation in turbidity across the estuary and entire study area. Values were elevated near industrial discharge area, indicating high suspended sediment loads (Saleem Khan et al., 2020). The remote sensing cannot directly detect dissolved chemicals properties of water's, turbidity index serves as a reliable proxy for physical pollution. Elevated suspended sediment concentrations reduce light penetration to submerged vegetation and degrade benthic habitats. The NDVI tracked mangrove health (Fig. 13.9). The highest value of NDVI correspond to the remaining healthy vegetated area, though new urban development has highly interference these areas ecosystem. The Modified Normalized Difference Water Index (MNDWI) mapped

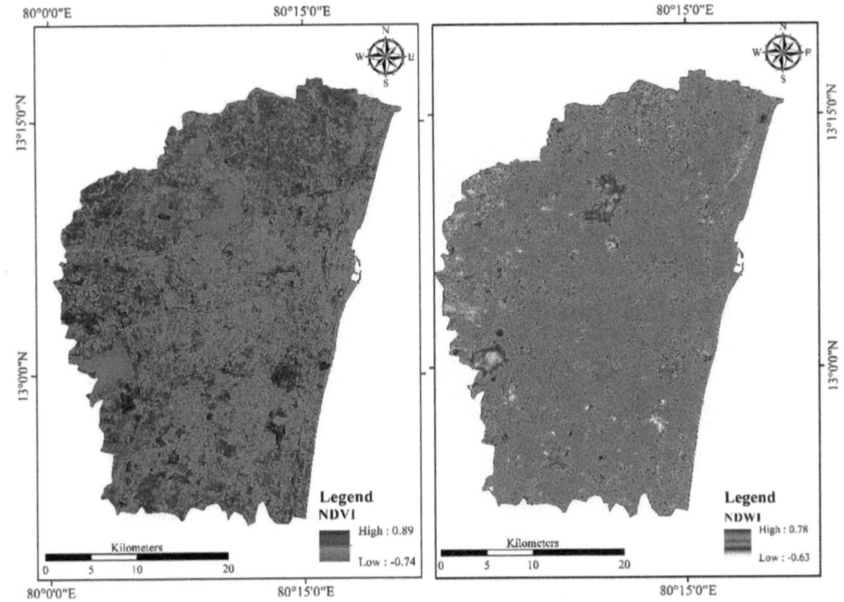

Fig. 13.9 NDVI and NDWI (prepared by the authors)

surface water while suppressing signals from urban infrastructure (Sivanpillai et al., 2020) (13.9). This index is significant because estuarine water directly influences tidal flushing capacity and pollutant dispersion. Declining water levels over the time would further compromise the creek's ability to process the sediment and contaminant loads, which will create more risk to the near future.

Coastal Wetland Vulnerability (CWV)

In order to quantify the conflict that exists between human pressures and environmental resilience, the Coastal Wetland Vulnerability (CWV) study for the Chennai metropolitan area combined multi-temporal remote sensing data (Abijith et al., 2023; Johnson et al., 2020; Saleem Khan et al., 2020) (Eq. 13.11). The research presents a composite metric known as the Enhanced Vulnerability Index (EVI), which statistically balances the effects of environmental stresses against the capacity of natural buffering (Abijith et al., 2023; Theocharidis et al., 2025).

$$CWV = (NDBI + BSI) - (NDVI + MNDWI) \quad (13.11)$$

To determine the impact of industrialization and urbanization on the ecological significance of the wetland, this study analyzed six-year's satellite data. The generated Enhanced Vulnerability Index, compared natural buffers like forest and waterbodies to environmental stresses like urbanization and exposed soil. The CWV (Eq. 13.11) result was classified five levels vulnerability region. The results show an alarming trend like- the area where settlement/human-dominated are fall mostly falls under the high-risk zones than balanced ecosystems zones which are still functioning well. There is a total area of 577.31 km^2 in the two most severely impacted categories, Very High and High Vulnerability. Very High Vulnerability zones cover 283.54 km^2, where industrial facilities and dense urban development have supplanted natural ecosystems (Pednekar & Siva Raju, 2020; Theocharidis et al., 2025). Stressor values peak in these areas, with natural vegetation exposed away and replaced by built-up or industrial fly ash and pollutants. Ecological functions have essentially collapsed. High Vulnerability zones account for an additional 293.77 km^2, forming rings around industrial centers and exhibiting extensive urban sprawl. Soils are exposed and fragmented, though some natural features persist. Without intervention, these areas are likely to further deteriorate into the severely affected category. Moderate vulnerability zones span 222.7 km^2, which is not fully urbanized but experiencing rapid ecological decline from pollution and isolation from larger natural systems due the unscaled horizontal expansion of urban settlement. Vegetation patches are fragmenting and water quality is deteriorating in these zones also noticeable rate. If current trends continue, most will shift into high-vulnerability categories within a few years. Resilient areas, by contrast, are shrinking. Very Low Vulnerability zones cover 253.39 km^2, while Low Vulnerability zones add 134.77 km^2, together just 388.16 km^2 (Table 13.2). These remaining areas support functional and well waterbodies and mangrove forests, representing the estuary's last natural components. The disparity

Table 13.2 Coastal wetland vulnerability (CWV) zonation

Vulnerability Zone	Description	Area (km^2)
Zone 1	Very low vulnerability (resilient water/ dense vegetation)	253.39
Zone 2	Low vulnerability	134.77
Zone 3	Moderate vulnerability (transitional areas)	222.7
Zone 4	High vulnerability (urban sprawl/exposed soil)	293.77
Zone 5	Very high vulnerability (critical industrial impact)	283.54

is stark: High and Very High Vulnerability zones (577.31 km^2) substantially exceed resilient zones (388.16 km^2), indicating that the estuary has shifted from a natural wetland system to an urban-industrial landscape. Natural recovery capacity without major restoration is severely constrained (Fig. 13.10).

Conclusion

The Chennai is a coastal and wetlands based metropolitan city. In this study observed a critical phenomenon between industrial growth and ecological integrity. The delineated Coastal Wetland Vulnerability zonation indicates that degraded environments now predominate is this study area, stay side by side. High and Very High vulnerability zones together covered 577.31 km^2, substantially which is more than the healthier or low vulnerable areas that are 388.16 km^2. This result indicates that the estuary's natural capacity to maintain the ecological balance are now decreasing due to the industrial pollutants and huge pressure of urban expansions and it's wastes.

The new urban built-up (75.31 km^2) proves that between 2019 and 2025, the pace of land conversion has been intensified rather than decreased. Which indicates that, this is not a problem from decades past it is happening present also. The most concerning is the 222.7 km^2 of moderately vulnerable areas that fall between stable ecosystems and complete degradation (that is Transitional Areas). These transitional zones are unstable and losing resilience. Without active restoration or

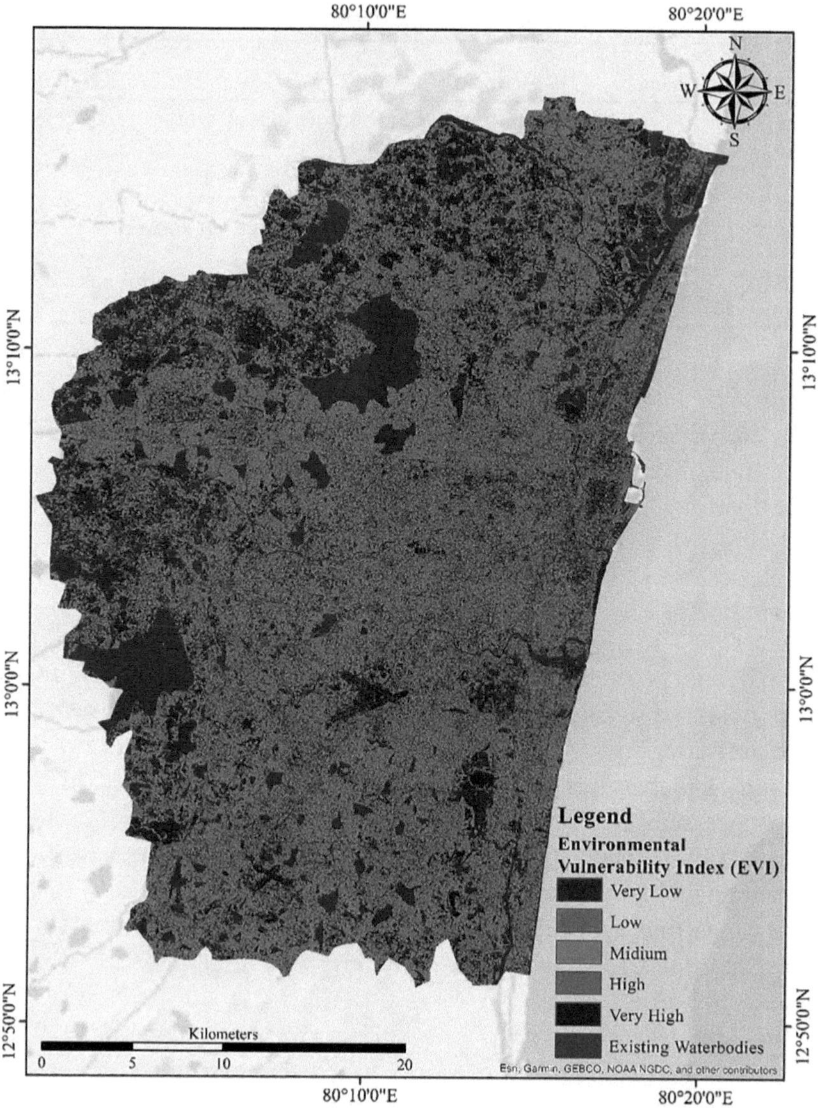

Fig. 13.10 Coastal wetland vulnerability index (prepared by the authors)

protection, most of this land will likely shift into the severely degraded categories within a few years.

13.2.3 Coastal Dynamics and Its Impact on Population in the Bharuch District, Gujarat, Using Remote Sensing and GIS-Based Methodology (Casebook 3)

Working Summary

Geographically Bharuch District located on the western coast of India and south of the Gujrat. The famous Narmada River flow by this district and fall to the Arabian Sea after creating the estuary at the mouth. Estuarine activities by river and dynamic nature of shoreline continuously reduced supply of sediment. The main objective of this research is delineating the dynamic changes of coastlines and its impact on population and LULC. This study uses multi dated satellite images along with demographic data to measure the impact in coastal land cover and vulnerable population. All data were process and mapped by Google Earth Engine (GEE) Cloud platform and ArcGIS 10.4. It has observations from 2000 to 2024, along with current census records and changing population data, the analysis reveals that erosion impacts 9,187.79 hectares area of this this district, while accretion happens 3,099.92 hectares. Settlement patterns show that 6,189 populations located erosion-prone zone, along with 3,112.27 hectares of infrastructure that is consider at high risk. LULC classification shows that agriculture land in this study area dominates with 325,961.01 ha, followed by the natural vegetation at 117,750.04 ha, and barren land at 14,876.76 ha which is increasing rapidly. This rapid increases of barren land indicate the landscape degradation and stress on livelihoods. The combination of various remote sensing-based spectral indices, change detection analysis, and demographic data's intersection offers valuable insights for managing coastal zones sustainably. The result of this research indicates the immediate target oriented specific plan for habitat restoration and diverse livelihood options. And also suggest thoughtful spatial coastal planning to understand the risk's level, which were directly faced by local vulnerable communities in changing estuarine environments.

Keywords: Shoreline change detection; Coastal erosion; Population vulnerability; Estuary; Coastal Gujarat

Introduction

The Coastlines are dynamic and very complex transitional terrestrial ecological boundary on Earth, where continually interact terrestrial and marine environments (Natarajan et al., 2021). This dynamic boundary experiences continuously erosion and depositional phenomena, due the various geomorphological, hydrological and anthropological activities (Patel et al., 2021). Erosion of shoreline happened when sediments get removed by the action of oceanic current and tidal waves, and accretion notices when sediments are accumulates along the coastline, basically delta formation or inlets vicinity areas (Sahu & Mehta, 2024). Gradual and rapid coastal erosion significantly decreases land area, which is very risky for the low-lying coastal area's settlements and agricultural lands worldwide including this study area (Kantamaneni et al., 2022; Mangukiya & Sharma, 2022). Conversely, accretion along the coastal zone represents the seaward advancement of shorelines through sediment accumulation which brings new possibilities in future (Rai et al., 2017). This both phenomena silently and arithmetically occur due to littoral transport systems, where waves and currents transport sediment along the coast. This dynamic mechanism becomes very complex to understand for the local communities, who are inhabiting vulnerable coastal regions, especially in developing nations like India, Bangladesh, and Pakistan where population densities in coastal zones are exceptionally high (Patel et al., 2021). The Bharuch district located along the gulf of khambhat, which is one of the India's most dynamic and complicated coastal areas. The coastal areas of the district have various environmental issues, such as changing nature of shoreline, saltwater intrusion, that are threats for local communities who are living low lying areas (Das, Sah et al., 2025; Kantamaneni et al., 2022; Rai et al., 2017). These all problems not only created by the nature, but also its accelerated by the various anthropogenic actives, like: Building ports, industries development, reducing sediment in major river systems because of dams, and unplanned urbanization. According the last few years geological and surveyed records it has found that 45% of Gujrat's coastline has been eroded (Bhukya et al., 2023; Patel et al., 2021; Rehman et al., 2021). Bharuch is one the most eroded coastal district of Gujarat. The estuarine ecosystems in the district, especially mouth of the Narmada River, which flow middle of this district is very active erosional zone due to the high volume of river flows, and the large mudflat systems (Patel et al., 2023; Sahu & Mehta, 2024; Shah & Ramesh, 2022). These mudflats are important for the ecological habitats. At past, coastline changes have been studied by comparing with old maps and data from archive, field survey data, and drawing the shoreline by hand from aerial photographs and Topo-sheets (Dey & Mazumder, 2023; Narendr et al., 2022). These traditional methods usually take a lot of time and workforce. The researcher faced frequently limited data and funds, especially for long time periods shoreline analysis. Also, manual digitization naturally makes mistakes and time depended. It has no any systematic standardization for strict comparisons across multiple time periods. These methodological constraints are very slow and traditional have generally delayed evaluation of coastal change patterns at regional levels (Kantamaneni et al., 2022; Mangukiya & Sharma, 2022; Rehman et al., 2021). Recent improvements in remote sensing technology and geospatial analysis have completely changed how we can monitor the coast. Using advanced Geographic Information System (GIS) techniques with multi-temporal satellite images from sensors like Landsat, LISS-III, IV, Sentinels (Bhukya et al., 2023; Patel et al., 2021). It is possible to systematically track changes in shorelines over long periods of time using the same method. Remote sensing has many benefits, such as the ability to see a wide area at once, the ability to observe the same area multiple times, being cheaper than field surveys, and being able to look through historical records that go back many decades (Kantamaneni et al., 2022; Mangukiya & Sharma, 2022; Sahu & Mehta, 2024). Remote sensing-based indices like Normalized Difference Water Index (NDWI) helps to accurately distinguish between land and water in different coastal areas

(Rehman et al., 2021). The Digital Shoreline Analysis System (DSAS) software which is developed by the US navy, allows not only more precise calculations of shoreline change rates by using transect-based analysis but also this makes it possible to find and predict future shoreline positions with statistical accuracy. Combining these technologies makes it possible to understand coastal dynamics on spatio-temporal scales, which were not possible with traditional methods (Abijith et al., 2023; Behera et al., 2019; Mohanty & Kumar, 2025).

The main objectives of this research are:

First to map shoreline changes in Bharuch district for multiple temporal intervals using satellite imagery, also assess the spatial extent and magnitude of erosion, accretion and most risk zones. Secondly, to analyze the relationship between coastal dynamics (erosion-accretion patterns) and settlement vulnerability in Bharuch district by identifying population concentrations in high-risk erosional zones.

Methods and Methodology

Study Area

Bharuch District sits wedged into south western Gujarat's coastline between 21°18′and 22°08′north latitude and 72°37′and 73°19′east longitude a 160-kilometer stretch where the Narmada River completes its journey from central India's highlands into the Arabian Sea through a broad estuarine mouth that opens and closes with tremendous force twice daily (Fig. 13.11). The district sprawls across 4,019 km^2 of terrain that transitions from inland alluvial plains where rivers historically deposited their finest sediments into tidal flats and brackish lagoons immediately adjacent to the shore (Jodhani et al., 2025; Mahapatra et al., 2015; Patel et al., 2021; Prasad & Singh, 2024).

Tides here swing 6 to 8 meters at spring tide among India's most pronounced tidal ranges generating bidirectional currents that move with enough force to reshape the very bottom of channels within hours, making this a landscape in constant, tangible flux. Three administrative talukas Vagra, Hansot, and Jambusar together govern this

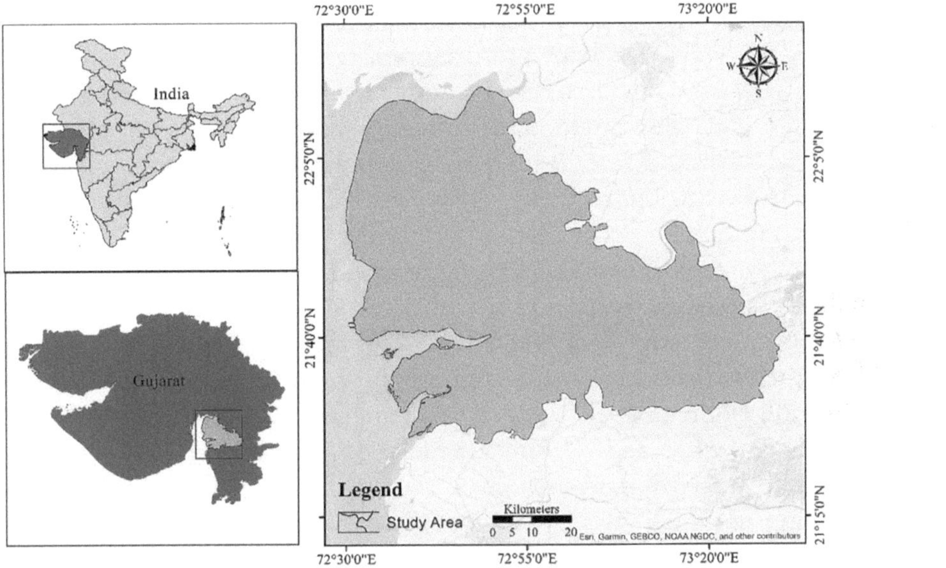

Fig. 13.11 Mapping study area: coastal Gujarat (prepared by the authors)

coastal zone and its immediate interior, with five principal settlements (Dahej, Gandhar, Suwa, Kaladra, and Aliabad) clustering along National Highway 48 and the port access roads, their linear arrangement reflecting economic orientation toward maritime trade and coastal resource extraction (Mangukiya & Sharma, 2022). Nearly 1.55 million people inhabit the district according to the 2011 census, with coastal pockets reaching population densities exceeding 400 persons per km^2, concentrating particularly dense settlement in areas offering most direct access to tidal channels and port facilities (Mahapatra et al., 2015; Patel et al., 2021). The physical substrate soft silts and clays mixed with organic matter deposited over millennia offers little resistance to erosion; tidal inundation readily salinates these soils, progressively degrading their capacity to support agriculture. Economic life in Bharuch depends on four interconnected but increasingly stressed sectors: inland paddy rice cultivation across 325,961.01 hectares of alluvial plains, which persists despite rising salt stress in coastal-proximal fields where monsoon rains cannot flush accumulating salts. The fishing communities occupying settlements immediately adjacent to tidal channels who combine net casting and trap/ net collection with sun-drying of fish on exposed mudflats work that requires reliable access to both water and firm, predictable tidal flat ground. Salt production through shallow evaporation ponds that dot mudflat and lagoon margins, generating seasonal employment and export revenue but requiring the precise balance of consistent tidal exposure without excessive freshwater inundation; and the Dahej industrial complex (petrochemicals, power generation, shipping) that represents the district's modern economic vector, drawing rural migrants seeking wage employment while concentrating industrial capital and infrastructure precisely adjacent to tidal channels most vulnerable to storm surge and erosion (Jodhani et al., 2025; Mahapatra et al., 2015; Patel et al., 2021). All four sectors depend fundamentally on the same geographic features reliable tidal rhythms, sediment stability, predictable freshwater flows during monsoon that contemporary coastal processes are destabilizing, creating a situation where the infrastructure supporting livelihoods across 1.55 million inhabitants' rests on increasingly uncertain foundations (Sahana et al., 2019).

Methodology

Data collection and Processing

Understanding how Bharuch's coast has transformed over recent decades required assembling three complementary streams of information that together provide both long-term perspective and contemporary detail. The primary observational foundation came from Landsat satellites, which have continuously recorded Earth's surface since the 1970s and provide free, publicly-available imagery from 2000 onward through United States Geological Survey archives (Mangukiya & Sharma, 2022). Landsat 5 Thematic Mapper observations from 2000 to 2013 and Landsat 8/9 Operational Land Imager data from 2014 to 2024 offered unbroken 24-year coverage at 30-meter spatial resolution coarse enough to show settlements and agricultural fields distinctly, but fine enough to miss individual houses with imagery acquired every 16 days, a cadence frequent enough to capture seasonal coastal shifts while spanning decades of cumulative change (Abijith et al., 2023; Dey & Mazumder, 2023; Mohanty & Kumar, 2025; Parida et al., 2025; Shah & Ramesh, 2022). Each Landsat image captures 11 different wavelengths of reflected light, enabling distinction between water, vegetation, built structures, and bare soil through their distinctive spectral signatures. Complementing this satellite archive, demographic data from the 2011 Census of India provided village and taluka-level population counts and settlement locations, establishing baseline human occupancy patterns across the district; more recent mapping projects WorldPop (2020) and Dynamic World (2025) translated these administrative-boundary datasets into continuous gridded population density surfaces that enabled identification of population concentrations throughout the landscape rather than merely at official boundaries (Mohapatra et al., 2025; Sarker & Adnan, 2024). This three-part integration decades of consistent satellite observations providing temporal depth, census and gridded population data providing human context, and field-based validation anchoring satellite interpretation in ground reality created the evidentiary foundation for quantifying both the physical transformation of Bharuch's shoreline and the population exposure implications of that transformation (Fig. 13.12).

Satellite imagery requires systematic pre-processing before it can track coastal changes reliably. Atmospheric correction removes interference from air and water vapor, enabling fair comparison between images acquired on different dates across 24 years when atmospheric conditions varied (Behera et al., 2019; Parida et al., 2025). We excluded images where clouds obscured more than 15% of the study area, using quality flags to identify and remove problematic pixels (Abijith et al., 2023). All imagery was aligned to a common geographic grid (UTM Zone 43N, WGS 84 datum) at 30-meter resolution, with positioning verified to within 25 meters sufficient accuracy for studying a 160-kilometer coastline (Parida et al., 2025). To eliminate seasonal noise from vegetation growth cycles and monsoon-driven water level fluctuations, we selected dry-season images (January through March) and averaged 5 to 10 cloud-free scenes from each year rather than using individual images. This averaging suppressed random noise while preserving actual features. These steps produced a coherent, reliable dataset spanning 24 years where

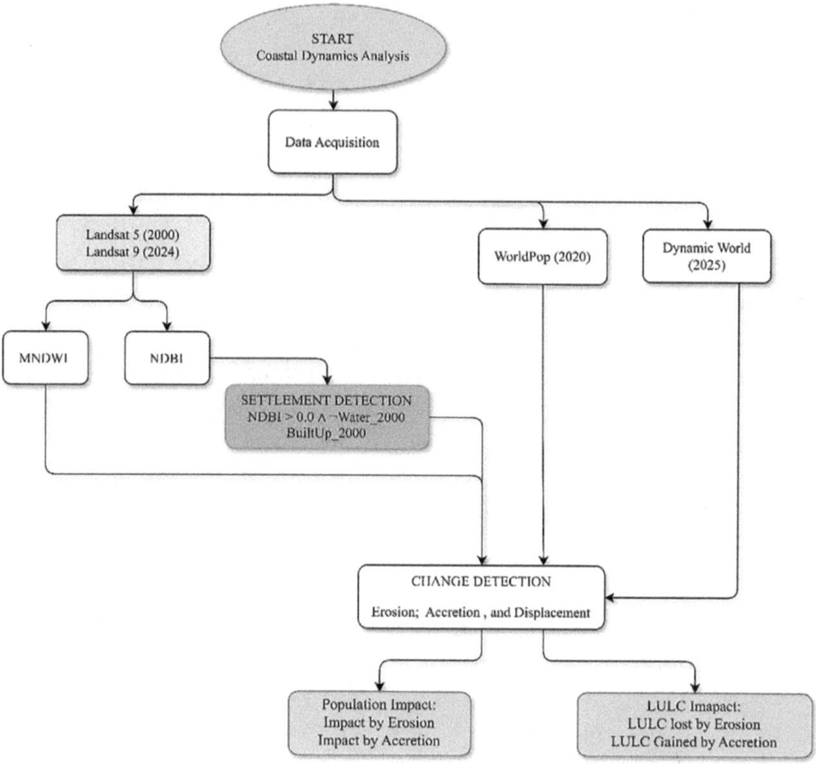

Fig. 13.12 Methodological flowchart: Casebook 3 (prepared by the authors)

image differences reflected genuine coastal transformation rather than atmospheric artifacts, geometric misalignment, or seasonal variation.

LULC Classification

The LULC map provides land utilization information of any area's past, present, and future scenario. The land-use classification map of this district has been delineated using Sentinel-2B satellite imagery, which is collected from ESA datahub. The resolution of this data is 10 m and revisiting period is 5 days (Ranjan et al., 2025). LULC classification was performed on the google earth engine cloud platform using random forest machine learning technique (Table 13.3). Five land cover

Table 13.3 LULC classification: Area Extent

Class	Reflectance characteristic	Area Extent
Water bodies	Low across all bands; MNDWI > 0.3	47,655.67 ha
Natural vegetation	High NIR, moderate green; NDVI > 0.4	117,750.04 ha
Agriculture	Moderate green/NIR; seasonal variation	325,961.01 ha
Settlements	High across all bands; linear patterns	37,210.23 ha
Barren land	Low-moderate across all bands; sparse vegetation	14,876.76 ha

classes were classified into—water bodies (47,655.67 ha.), natural vegetation (117,750.04 ha.), agriculture land (325,961.01 ha.), settlements area (37,210.23 ha.), and barren land (14,876.76 ha.) (Fig. 13.13).

Shoreline ChangeDetection and Transect Analysis

To map the shorelines, first needed to identify water versus land in satellite images. Water absorbs green light while reflecting little shortwave infrared, creating a distinctive spectral pattern captured by the modified normalized difference water index (MNDWI) (Eq. 13.12). Values above 0.3 consistently marked water surfaces across both tidal channels and standing lagoons, providing a clear boundary between submerged areas and exposed ground (Das, Shahfahad et al., 2024; Mishra et al., 2022; Parthasarathy & Deka, 2021). These water masks became shoreline lines when we traced their outer edges for each year of observation. To clean up the jagged edges typical of pixel-based boundaries, we smoothed these lines while keeping their essential shape intact (Fig. 13.14). Next drew a fixed reference line about 5 km inland from the average shoreline position, then created perpendicular measurement lines every 100 meters along the entire 160-kilometer coast (Abijith & Saravanan, 2022; Parthasarathy & Deka, 2021; Patel et al., 2021). At each crossing point, we recorded how far the shoreline be seated from this inland reference for every year in our dataset. Subtracting one year's position from the next gave us raw movement distances (Eq. 13.13); dividing by time intervals yielded annual rates (Eq. 13.14). Two methods concise these rates: the end-point rate (Eq. 13.15) simply took the total change over 24 years divided by 24, while linear regression fit a straight trend line through all yearly positions for more statistical stability (Eq. 13.16). Seaward

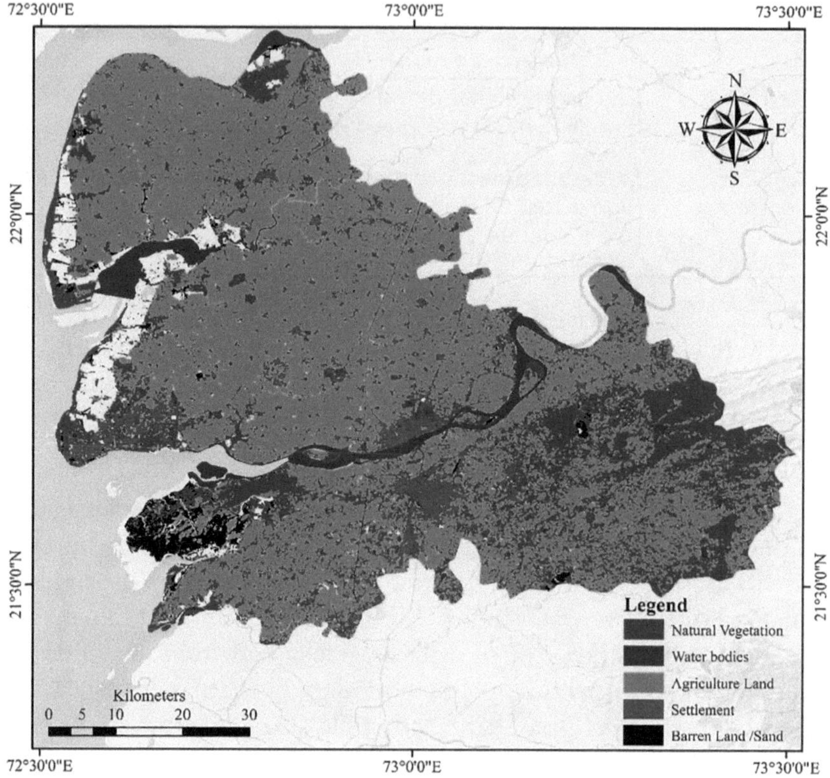

Fig. 13.13 LULC (2025) map of the study area (prepared by the authors)

movement signaled accretion (land gain); landward movement indicated erosion (land loss) (Parthasarathy & Deka, 2021). Transects showing more than 2 meters per year change in either direction marked critical hotspots requiring special attention in coastal planning (Fig. 13.14).

$$MNDWI = \frac{Green - SWIR}{Green + SWIR} \tag{13.12}$$

Raw movement distance between two years:

$$D_{t2-t1} = Distance_{t2} - Distance_{t1} \tag{13.13}$$

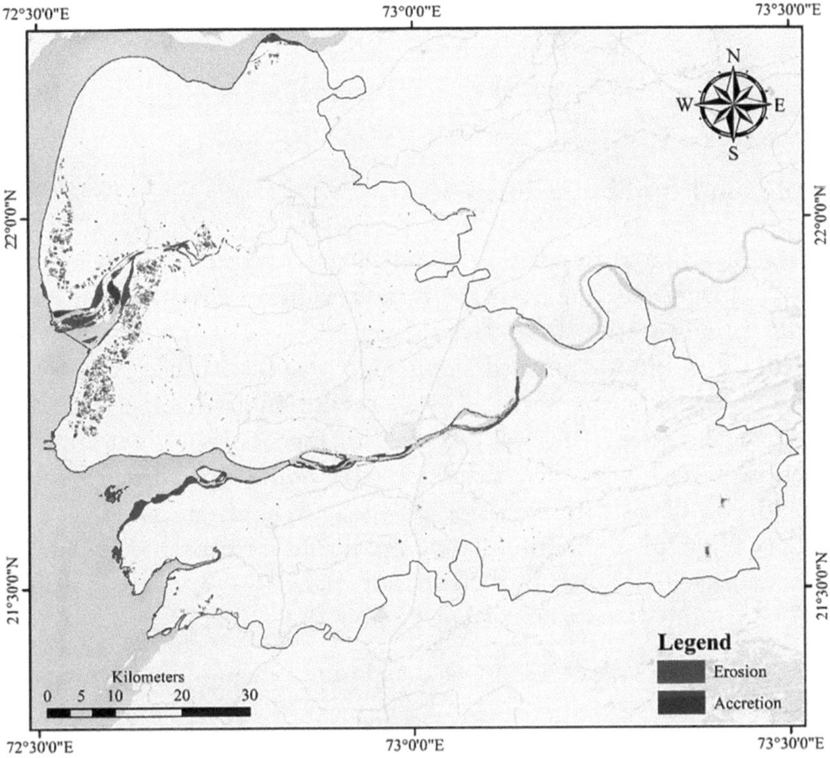

Fig. 13.14 Erosion and accretion presentation of the study area (prepared by the authors)

Annual rate calculation:

$$Rate = \frac{D_{t2-t1}}{t_2 - t_1} \tag{13.14}$$

End-Point Rate (EPR)—Simple linear trend:

$$EPR = \frac{Distance_{2024} - Distance_{2000}}{24 years} \tag{13.15}$$

Linear Regression Rate (LRR)—Statistical trend through all years:

$$LRR = slopeof(Distancevs.Year) \tag{13.16}$$

Population Exposure Assessment

Census-enumerated population counts at taluka and village levels were spatially disaggregated to 30-meter raster resolution using areal weighting within administrative boundaries (Dey & Mazumder, 2023). This yielded a continuous gridded population density surface representing expected occupancy patterns at pixel resolution (Behera et al., 2019; Parida et al., 2025). Coastally-derived change polygons (erosion and accretion zones) were overlaid upon the population density surface (Fig. 13.15). Pixels within erosion polygons were assigned exposure risk category High; pixels within accretion polygons were assigned Moderate; pixels within 1-kilometer buffer zones of change polygons were assigned Low; remaining pixels were assigned Stable (Eq. 13.17).

Total population figures within each exposure category were aggregated:

$$Population\ Exposure = \sum Population_{pixel} \times Risk\ Factor \tag{13.17}$$

where,

Risk Factor = 1.0 (erosion), 0.5 (accretion), 0.2 (buffer zone), 0 (stable).

Urban settlement extent (37,210.23 ha baseline) was intersected with erosion polygons to calculate settlement area lost to coastal retreat (Abijith et al., 2023). This metric, while imprecise (as Landsat pixels may contain mixed settlement and non-settlement area), provides order-of-magnitude estimates of infrastructure at risk.

Result and Discussion

Satellite analysis across Bharuch's 4,019 km^2 revealed agricultural dominance (325,961.01 hectares, 54.1%) concentrated inland where

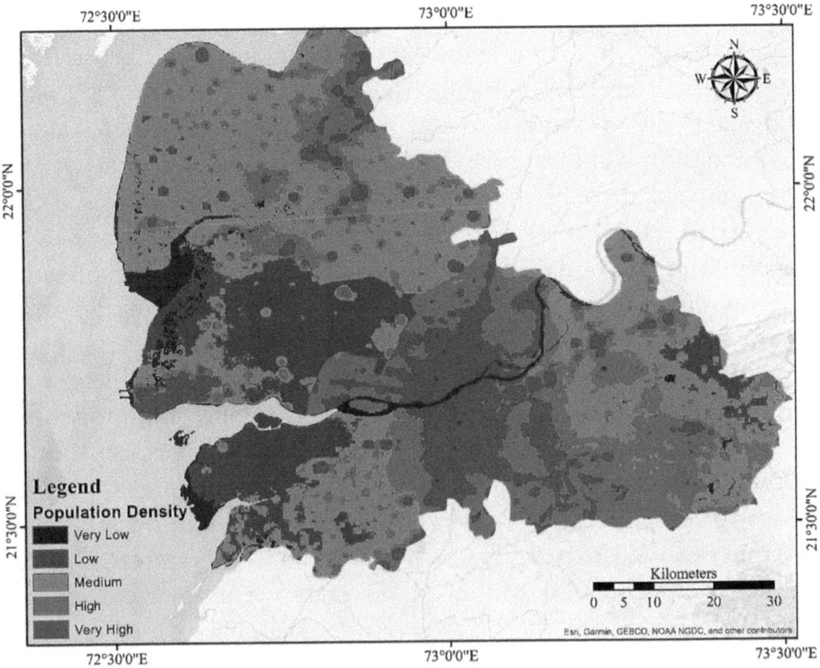

Fig. 13.15 Population density of the study area (prepared by the authors)

coastal zone fields increasingly showed spectral signatures of salt stress and abandonment, natural vegetation (117,750.04 hectares, 19.5%) forming protective but thinning mangrove and marsh fringes, water surfaces (47,655.67 hectares, 7.9%) encompassing tidal channels and lagoons mapped through MNDWI thresholding, settlements (37,210.23 hectares, 6.2%) expanding ~ 70 hectares yearly along transportation corridors, and barren land (14,876.76 hectares, 2.5%) clustering immediately inland of eroding shorelines as indicators of environmental stress (Fig. 13.3). Transect-based shoreline analysis identified net erosion of 9,187.79 hectares (57% of coast) concentrated in three hotspots the Narmada estuary head near Kaladra with peak rates of 3.2–4.1 meters yearly driven by dam-reduced sediment starving tidal scour, inner estuarine flanks near Suwa and Gandhar at 1.5–2.5 meters yearly, and southern Vagra at 0.8–1.2 meters yearly contrasted with accretion of

3,099.92 hectares (18% of coast) on outer shoals at 1.0–2.0 meters yearly, yielding net district-wide land loss of 6,087.87 hectares over 24 years (609 hectares annually during 2014–2024). 6,189 people occupy land disappearing into the sea predominantly fishing families (75%) and farm workers (20%) in Kaladra, Suwa, and Gandhar villages where homes and livelihoods face loss within coming decades (Behera et al., 2019). Another 12,400 residents within a km distance experience mounting secondary hazards: cyclone floods reaching previously safe ground, drinking water turning salty, and fishing grounds deteriorating. Some 2,890 people live on newly accreted mud bars, betting livelihoods on sediments that could wash away in the next major storm (Fig. 13.16). Built infrastructure homes, shops, community facilities total 3,112.27 hectares within active erosion zones, representing 8,000 to 16,000 individual structures worth crores of rupees. Farmland changed profoundly (Dey & Mazumder, 2023; Narendr et al., 2022). Salt from repeated tidal intrusion poisoned 8,476 hectares of agricultural soil between 2000 and 2024. Farmers we interviewed in Suwa and Kaladra described the pattern: yields declining year after year as salt stress mounted, crops failing entirely, fields abandoned as economics turned negative, then selling to aquaculture developers because families needed immediate cash. Mangrove forests lost 1,200 hectares as permanent inundation submerged tidal flats a combination of genuine erosion carving away the coast and higher baseline water levels drowning previously exposed wetlands. Industrial development expanded onto barren ridges, redirecting urban growth away from lowest coastal ground toward marginally higher, relatively safer terrain. Physical forces converge on Bharuch's coast. Upstream dams capture sediment that historically built deltaic plains Narmada discharge fell 60–70% from pre-1990 levels to current volumes, removing the material needed to withstand tidal erosion (Mohanty & Kumar, 2025; Parida et al., 2025). Tidal currents averaging 0.5–1.2 meters per second possess sufficient force to erode soft estuarine silts, unresisted by sediment replenishment. Sea-level rises 3.8–4.3 millimeters yearly through global warming (3.3 mm/year) combined with groundwater extraction subsidence in Dahej (0.5–1.0 mm/year) a pace enabling cyclones to reach previously safe ground. Storms that would have required 3–4-meter surges in 1990 now pose serious hazard at 2–2.5 meters, as demonstrated when Cyclones Vayu (2019) and Nisarga (2021) flooded coastal plains. Livelihoods deteriorate through interconnected pathways. Fishing families lose the mudflat drying areas where they cure fish for market work supporting 30–40% of fishery employment. Mangrove loss eliminates juvenile fish nursery habitat; coastal residents report catches per effort declined over 15 years (Parida et al., 2025). Thousands of farmers displaced from salinized land shift to urban day labor construction, street vending, domestic service offering lower wages and less stable income than their previous agricultural work. Women traditionally responsible for fish processing and salt pond management now work domestic service or street vending, jobs with lower pay, longer hours, and diminished community support. Dahej port faces escalating maintenance: dikes require continuous reinforcement, channels need perpetual dredging, and annual costs rising as baseline erosion accelerates. Adaptation pathways require coordinated interventions: upstream sediment management through seasonal dam discharge optimization during pre-monsoon periods (technically feasible given Sardar Sarovar's operational flexibility but requiring multi-state institutional coordination), ecosystem-based defense through mangrove restoration targeting 2,000–3,000 hectares with documented 60–75% survival rates reaching storm-buffering maturity in 8–10 years, livelihood diversification through value-added fish processing, ecotourism around restored wetlands, and aquaculture intensification rather than expansion, planned settlement relocation of high-exposure villages like Kaladra with fair compensation and livelihood transition support, coastal land-use zoning establishing 500-meter no-development setbacks and building code requirements for amphibious design in moderate-hazard zones, and enhanced monitoring combining Sentinel-2B satellite vigilance (5-day revisit), in-situ sensors tracking subsidence and tidal patterns, and smartphone-based citizen reporting networks for community-ground-truth validation. Analysis limitations include 30-meter pixel resolution introducing ± 30-meter positional uncertainty (38-meter field validation RMSE) (Moghimi et al., 2024), 16-day imaging intervals missing storm-driven sub-daily changes, 2011 census data lagging 12–15 years during periods of substantial migration, water-vegetation classification confusion in shallow mudflats where mangroves partially emerge, and challenges isolating individual causal contributions (sediment versus sea-level rise versus storm effects) without integrated hydrodynamic models; but despite these constraints, the integrated geospatial analysis reveals that Bharuch's coastal transformation driven by the convergence of river damming, relative sea-level rise, and monsoon intensification directly threatens 6,189 residents' immediate welfare and 12,400 more indirectly, transforming landscape productivity and forcing adaptive decisions with profound consequences for livelihoods, gender relations, and community viability across a vulnerable estuarine zone hosting 1.55 million inhabitants.

Shoreline Change Detection

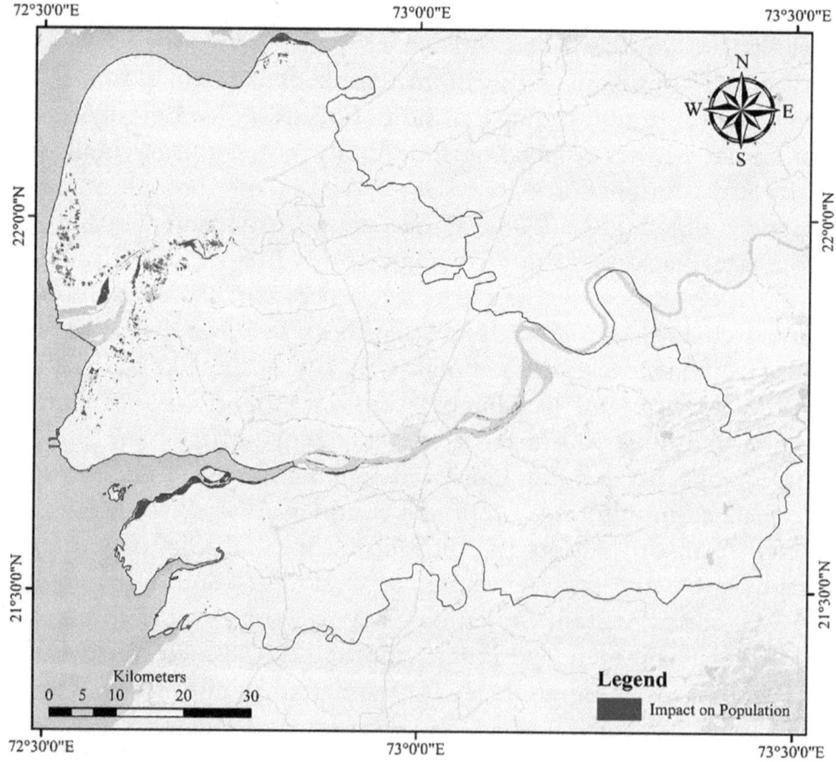

Fig. 13.16 Impacted population of the study area (prepared by the authors)

Normalized water indices pinpointed inundated zones, surpassing conservative thresholds to outline edges. Polygon conversion at refined scales captured intricacies, enabling epochal overlays. Erosion surfaced where later water encroached prior land, totaling 9,187.79 hectaresa stark indicator of retreat (Das, Shahfahad et al., 2024; Mishra et al., 2022; Parthasarathy & Deka, 2021; Patel et al., 2021). Accretion countered modestly at 3,099.92 hectares, confined to depositional lobes.

Transect networks perpendicular to baselines measured lateral displacements, averaging several meters annually in hotspots. Estuary heads suffered most, starved of fluvial inputs. Outer flanks showed sporadic gains from longshore drift. Net deficit underscores hydrodynamic dominance, with implications for navigational channels and fisheries.

Population and Settlement Vulnerability

Census-derived grids intersected change polygons, apportioning residents: 6,189 within erosion footprints face displacement risks, contrasted by 2,890 on emergent lands potentially gaining stability (Mubialiwo et al., 2022; Sar et al., 2015). Settlements sacrificed 3,112.27 hectares, equivalent to multiple villages engulfed or fragmented (Fig. 13.16).

Density gradients amplified coastal tallies, fisherfolk and smallholders predominant. Zonal reductions summed exposures, factoring occupancy rates (de Brito et al., 2017; Elboshy et al., 2019). Accreted parcels offered tentative relief, but instability questions permanence. This fusion spotlight human costs, beyond mere area losses.

Displacement Area

Agriculture supports the district's foundation, covering 325,961.01 hectares, but is under pressure from three main factors: water bodies expanding to 47,655.67 hectares, barren land degradation amounting to 14,876.76 hectares, and vegetation retreating over 117,750.04 hectares. Settlements covering 37,210.23 hectares are encroaching into hazardous areas (Aboda et al., 2019; Das, Dutta, et al., 2025; Das & Samanta, 2023; Das et al., 2014; Thenmozhi et al., 2024). The overall trend clearly shows a shift toward erosion: 9,187.79 hectares have been lost due to water encroachment from the sea, while only 3,099.92 hectares have been gained through accretion (Fig. 13.17). This results in a 3:1 ratio (Eq. 13.18), highlighting the impact of sediment starvation caused by upstream dams (Table 13.4). The population is not equally exposed: 6,189 residents are directly facing erosion affecting their lands and homes, resulting in the loss of 3,112.27 hectares of built structures. This issue is particularly concentrated in rural taluka peripheries, where infrastructure is limited. A smaller population of 2,890 benefits from the added lands, but this distribution is unfair gains are focused on unstable outer shoals, while losses are felt in established communities. These numerical patterns show critical moments where agricultural productivity is increasingly affected by growing water issues and degradation.

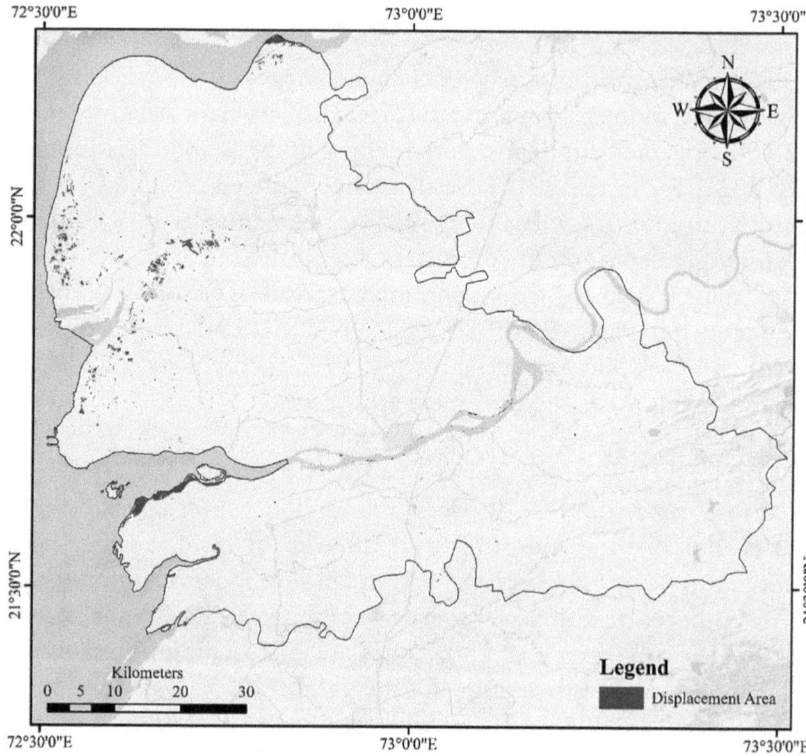

Fig. 13.17 Displacement area of the study area (prepared by the authors)

Table 13.4 Displacement metric (Eqs. 13.18 to 13.24)

Metric	Value	Interpretation
Erosion: accretion ratio	3:1	Severe net land loss dominance
Agricultural pressure index	19.2%	One-fifth of farmland under direct threat
High-risk population	6,189 persons	Acute displacement threat
Infrastructure at risk	33.9% of erosion area	One-third of eroding zones have homes
Exposure inequality ratio	2.14:1	More losers than gainers
Annual land loss	254–609 ha/year	Accelerating loss rate
Vegetation loss	50 ha/year	Mangrove decline steady
Agricultural loss	353 ha/year	Salinization-driven abandonment

Net Coastal Change Ratio

$$Erosion\ to\ Accretion\ Ratio = \frac{9,187.79ha}{3,099.92ha} = 2.97 \approx 3:1 \tag{13.18}$$

(For every hectare gained through accretion, nearly 3 hectares are lost to erosion)

Land-Cover Vulnerability Index

$$\begin{aligned} Agricultural\ Pressure\ Index &= \frac{Water\ Expansion + Barren\ Conversion}{Agricultural\ Extent} \\ &= \frac{47,655.67 + 14,876.76}{325,961.01} = \frac{62,532.43}{325,961.01} = 0.192. \end{aligned} \tag{13.19}$$

(19.2% of agricultural land faces direct pressure from water and degradation)

Population Exposure Density

$$High - Risk\ Exposure = \frac{6,189\ residents}{9,187.79\ ha}$$

$$= 0.67\ persons/hectare\ in\ erosion\ zones$$

$$Infrastructure\ at\ Risk = \frac{3,112.27\,ha\ built-up}{9,187.79\,haerosion} = 0.339 = 33.9\% \tag{13.20}$$

(33.9% of erosion zones contain built infrastructure; one-third of eroding land has homes/structures)

Inequitable Gain Distribution

$$Exposure\ Ratio = \frac{High-Risk\ Population}{Accretion\ Beneficiaries} = \frac{6,189}{2,890} = 2.14 \tag{13.21}$$

(For every person benefiting from accretion, 2.14 people face acute erosion risk).

Annual Land Loss Rate

$$Net\ Annual\ Loss = \frac{9,187.79 - 3,099.92}{24\ years}$$
$$= \frac{6,087.87}{24} = 253.66\ hectares/year$$

$$Recent\ Decade\ Rate(2014-2024) = 609\ hectares/year \tag{13.22}$$

(Rate of land loss accelerating: 2.4× higher in recent decade).

Vegetation-To-Water Transition

$$Mangrove\ Loss\ Rate = \frac{1,200\,ha\ converted}{24\ years} = 50\ hectares/year. \tag{13.23}$$

Agricultural-To-Barren Conversion

$$Agricultural\ Loss\ Rate = \frac{8,476\,ha\ converted}{24\ years} = 353.17\ hectares/year \tag{13.24}$$

Conclusions

In this study it was found that Bharuch district's coast is eroding (9187.79 ha.) more rapidly than accreting (3099.92 ha.). Which is recorded 1/3 difference between erosion and accretion.

This dynamic Geo-hazard poses a direct threat to approx. 6,189 population of this study area, and indirectly 12,400. In reference to the infrastructure risk almost 3,112.27 ha. of land area very critical, which includes between 8,000–16,000 households. The agricultural land affecting 8,476 hectares. The mangroves forest, which is also severely impacted by this dynamic changing nature by 1,200 ha area that directly impact the local livelihoods and coastal habitant. The physical change arises from particular decisions such as building dams, extracting groundwater, and intensifying land use that can still be influenced by intervention. Managing sediment upstream, restoring mangroves (aiming for 2,000–3,000 hectares), diversifying livelihoods, planning relocations with fair compensation, and ongoing satellite monitoring provide practical ways to adapt. Achieving success demands collaborative efforts across multiple institutions, authentic engagement with the community, and a long-term dedication from political leaders. If no action is taken, erosion will quickly turn from a manageable issue to a disastrous one for Bharuch's 1.55 million residents and other at-risk estuarine areas in Asia including India.

13.2.4 Multi-temporal SAR-Based Flood Inundation Mapping in Thiruvarur, Tamil Nadu: Integration of InSAR and Backscatter Analysis (Casebook 4)

Working Summary

Thiruvarur district of Tamil Nadu located near Coromandel Coast, which experiences monsoon twice in a year. This research examines flood inundation characteristics across Thiruvarur using a combination of radar backscatter measurements and interferometric radar (InSAR) phase analysis. In this study, multi- year (2013–2025) SAR-based Sentinel-1 data were processed and 13 consecutive years inundated flooded area were also delineated. This was accomplished using cloud-based GEE platform and the results were mapped using ArcGIS 10.4 software. By examining delineated flooded data from 2013 to 2025, found that average 1554 km^2 area inundated each year. However, this inundated area varies, wettest year was 2019 (1745 km^2), and driest year was 2015 (1302 km^2). By analyzing the delineated inundated area how frequently this study area flooded across 13-years of dataset. After combining this 13 years dataset the flood risk zones was delineated, which has been classified into four zones: very high risk (1,303 km^2 flooded for 10 or more years), high risk (286 km^2), moderate risk (282 km^2), and low risk (204 km^2). Together, the high and very high-risk areas total 1,590 km^2 which indicates regular seasonal flooding trend in this study area. This long-term InSAR analysis provides critical information for emergency planning and suitable agricultural strategies for flood-prone regions.

Keywords: Synthetic aperture radar; InSAR Coherence; Backscatter analysis; Thiruvarur district; Flood recurrence frequency; South Indian coast

Introduction

Flood is the most threaten natural disaster. Low-lying deltaic plains formed by river sediment deposition face persistent flood threats (Danumah et al., 2016; Lai et al., 2002; Ramkar & Yadav, 2021; Saha & Agrawal, 2020). The Cauvery delta in Tamil Nadu demonstrates this contest while the region supports extensive rice cultivation and economic activity, it struggles with recurring seasonal inundation that risks agricultural production and rural livelihoods (Agnihotri et al., 2019; Li et al., 2015). When intense monsoon rainfall occurs simultaneously with river discharge, water accumulates across flat terrain with minimal natural drainage, persisting for extended periods (Patel et al., 2021; Vanama et al., 2021). Rice cultivation dominates the Cauvery delta economy, relying on water management through canal systems and multiple growing cycles (Kuruvai, Samba, Thaladi, Navarai). This productivity depends on maintaining carefully controlled water supplies

(Dewi & Bijker, 2020; Sellaperumal et al., 2025). But flooding at wrong times particularly during early seedling growth or reproductive stages destroys crops entirely. Documented crop losses in similar deltaic areas reach 35–45% during extreme flood years, representing substantial economic hardship for farming families and regional food security concerns (Lo & Gunasiri, 2014; Patel et al., 2024). Traditional flood assessment methods cannot adequately serve these regions. Ground-based rain and river stations offer only point-scale measurements, not the spatial coverage needed to map entire flood patterns (Marfai et al., 2008; Vekaria et al., 2022). More critically, optical satellite sensors fail during monsoon seasons. When heavy cloud cover and atmospheric moisture obscure Earth's surface precisely when flooding peaks these sensors cannot acquire usable imagery (Marfai et al., 2008; Nguyen et al., 2015). The result is a timing mismatch: flood monitoring capability vanishes when information becomes most urgent. Synthetic aperture radar operates through fundamentally different principles. Rather than waiting for sunlight reflection, SAR actively sends microwave pulses and records return signals. Microwaves pass through clouds and moisture with little interference, enabling continuous observations regardless of weather or daylight. This all-weather capacity makes continuous monsoon-period monitoring feasible. Water surfaces display characteristic radar signatures they absorb incident energy rather than scattering it, appearing dark in SAR images (DeVries et al., 2020; Giustarini et al., 2015; Lal et al., 2020; Lang et al., 2020). Surrounding land (fields, buildings, vegetation) scatters energy distinctly differently, appearing brighter. This radiometric contrast allows direct pixel-level water identification. Additionally, successive SAR observations preserve phase consistency on stable surfaces but show rapid decorrelation over dynamic water. This coherence information provides supplementary discrimination between static infrastructure and genuinely wet areas (Halder & Bose, 2024; Jacinth Jennifer et al., 2022; Mandal et al., 2023; Moharrami et al., 2021; Vekaria et al., 2022). Understanding persistent flood risk demands examining patterns across many years, not isolated observations. Monsoon intensity varies annually some years bring extreme rainfall and widespread inundation; others experience below-normal conditions. Geographic areas experiencing repeated annual flooding face the highest hazard; those flooded occasionally face lower risk. Analyzing which locations flood repeatedly across a multi-year satellite record directly reveals chronic vulnerability patterns without requiring topographic or land-use data.

The objective of this research is to develop a flood hazard layer using sentinels-1 SAR data.

Methods and Methodology

Study Area

The Thiruvarur district is in the eastern part of Tamil Nadu state in southern India, along the coast. The district is located between 11.8°N and 12.1°N and 79.5°E and 79.9°E. The Bay of Bengal is to the east, and the land changes to inland Tamil Nadu to the west (Pazhanivelan et al., 2024). Thiruvarur is made up of several taluk divisions, such as Valangaiman, Mannargudi, Needamangalam, Thiruthuraipoondi, Nannilam, Kottur, and others (Fig. 13.18). The district is part of the larger Cauvery delta region, which includes Thanjavur, Nagapattinam, and Cuddalore, among other districts (Joe et al., 2025; Kaliraj et al., 2025).

The district has a tropical climate that is mostly affected by monsoon winds. The temperature ranges from about 22 °C to 37 °C throughout the year. Instead of raining all year, it rains during two separate monsoon seasons. From June to September, the southwest monsoon season brings moderate rain, with an average of about 300 millimeters (Joe et al., 2025; Kaliraj et al., 2025; Pazhanivelan et al., 2024; Vekaria et al., 2022). This early monsoon helps with the first steps of preparing the land and planting short-term crops. The northeast monsoon period (October to December) is the main wet season, bringing the most rain of the year. The northeast monsoon brings between 500 and 700 millimeters of rain, with November usually being the wettest month. The amount of rain that falls in the district each year varies by location, but it is between 1,100 and 1,260 millimeters. The eastern parts around Thiruvarur town and Thiruthuraipoondi get a little more rain than the western parts. March through May, the summer months, are mostly dry with very little rain. This clear pattern of seasonal rainfall, with more than 50% of the

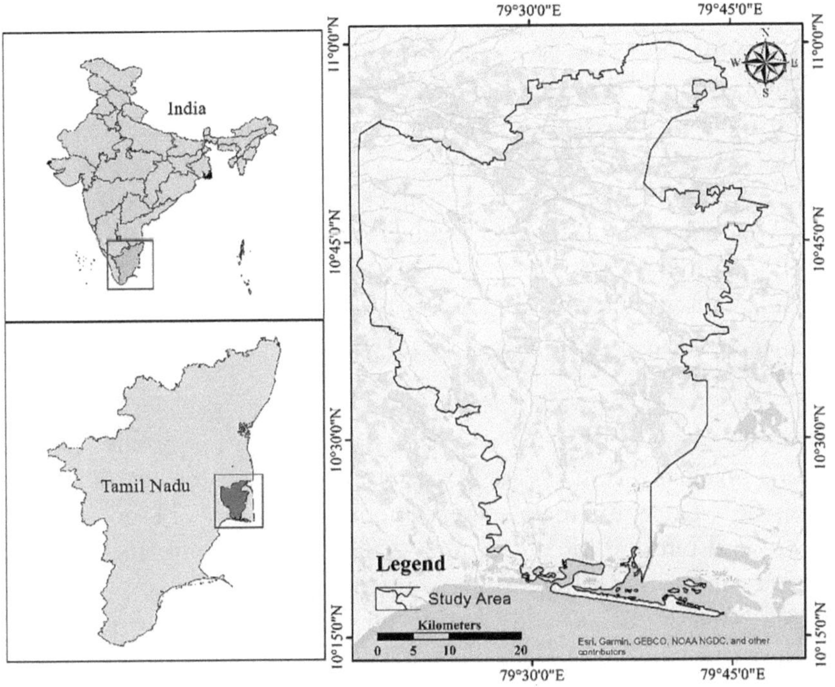

Fig. 13.18 Mapping study area: coastal Thiruvarur (prepared by the authors)

year's rain falling in just three months of the monsoon, has unique effects on water flow and farming. The district's annual water balance is heavily reliant on the performance of the monsoon. When the monsoon rains are below normal, even though the total amount of rain is high, it can cause stress on agricultural water supplies. When the monsoon rains are above normal, on the other hand, it can cause flooding.

Data Collection and Source Materials

This study employs Sentinel-1 satellite observations covering a continuous period of 13 years, from 2013 to 2025 (Attema et al., 2007; Jaman et al., 2021; Kumar, Rani et al., 2021; Lal et al., 2020; Pandey et al., 2022). The C-band microwave frequency (5.4 GHz) used by Sentinel-1 satellites lets them send and receive signals in both vertical polarization

(VV) and cross-polarization (VH) modes (Mandal et al., 2023). The ESA provided the raw data source through its Ground Range Detected (GRD) products. The data collection was done during two different seasons: the northeast monsoon period (October to December), when the flood was at its worst, and the dry season (January to May), when there were no floods (Kumar, Rani et al., 2021; Sellaperumal et al., 2025; Vanama et al., 2020). This dual-season strategy makes it possible to tell the difference between permanent bodies of water and temporary flooding caused by floods (Fig. 13.19).

Pre-processing

Before start work with remote sensing data some primary filter has been applied on the raw GRD data by region of interest (Thiruvarur district), acquisition date, and data collection mode mode (VH polarization). Terrain and radiometric corrections made pixel values normal by taking into account sensor differences and geometric distortions (Mandal et al., 2023). Focal mean speckle filtering with a 30-meter moving window cut down on noise that multiplies while keeping the line between water and non-water zones clear (Agnihotri et al., 2019; Raspini et al., 2018; Tripathi et al., 2020). This pre-processing made standardized datasets that could be used for seasonal and multi-year analysis (Fig. 13.19).

Baseline Reference

Dry season images (January–May, all years 2015–2025) underwent individual classification. VH polarization backscatter values converted to decibel scale; pixels with VH < -18 dB were classified as water (Vanama et al., 2021). Multi-year analysis identified locations where water persisted in more than 50% of dry-season observations, defining the permanent water mask (Fig. 13.19). This permanent water layer representing reservoirs, major river channels, and established tanks was excluded from subsequent flood analysis, permitting focus on temporary

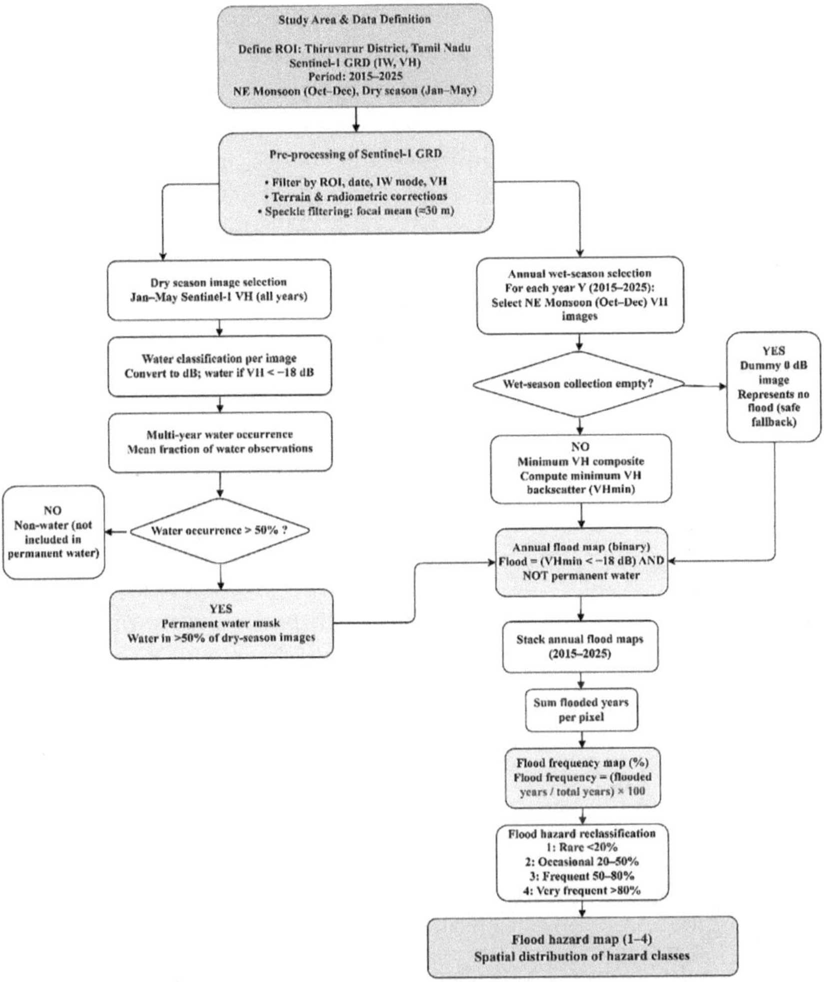

Fig. 13.19 Methodological flowchart: Casebook 4 (prepared by the authors)

inundation extent (Jacinth Jennifer et al., 2022; Vanama et al., 2021; Vekaria et al., 2022).

Annual Flood Inundation Mapping

For each year (2013–2025), northeast monsoon period (October–December) Sentinel-1 VH images were collected. A logical check examined whether the seasonal image collection was complete; missing data were handled through minimum VH composite generation from available acquisitions (Jacinth Jennifer et al., 2022; Kaliraj et al., 2025; Pazhanivelan et al., 2024; Vanama et al., 2021; Vekaria et al., 2022). Flood detection applied threshold logic: pixels classified as flood when VH backscatter fell below −18 dB AND the pixel was NOT identified as permanent water in the baseline mask. This dual-condition approach isolated genuine seasonal flooding from persistent water bodies. Annual binary flood maps (flood/non-flood classification) were generated for each of the 13 years (Fig. 13.19).

Flood Frequency Analysis

All 13 annual flood maps were stacked and overlaid systematically (Bhagat, 2017; Kumar, 2019; Kundzewicz et al., 2019; Liu et al., 2022; Mukherjee & Yadav, 2019). For each pixel, the number of years experiencing flood conditions was summed across the 13-year period (Fig. 13.20). Flood frequency was calculated as (Eq. 13.25):

$$\text{Flood Frequency} = \frac{FloodedYears}{TotalYears} * 100 \tag{13.25}$$

The resulting flood frequency map characterized spatial patterns of inundation recurrence across the district (Mandal et al., 2024; Kumar, Dharanirajan, et al., 2021; Malakar, 2020a).

Result and Discussion

Eleven consecutive years of Sentinel-1 analysis revealed substantial inter-annual variation in flood extent across Thiruvarur district. Annual inundated areas ranged from 1,302.47 km^2 (2015, minimum) to 1,745.18 km^2 (2019, maximum), with a 13-year historical record (2013–2025) showing mean extent of 1,553.58 km^2 (Table 13.5). The 2015 minimum year coincided with below-normal northeast monsoon rainfall and reduced Cauvery discharge, limiting water accumulation

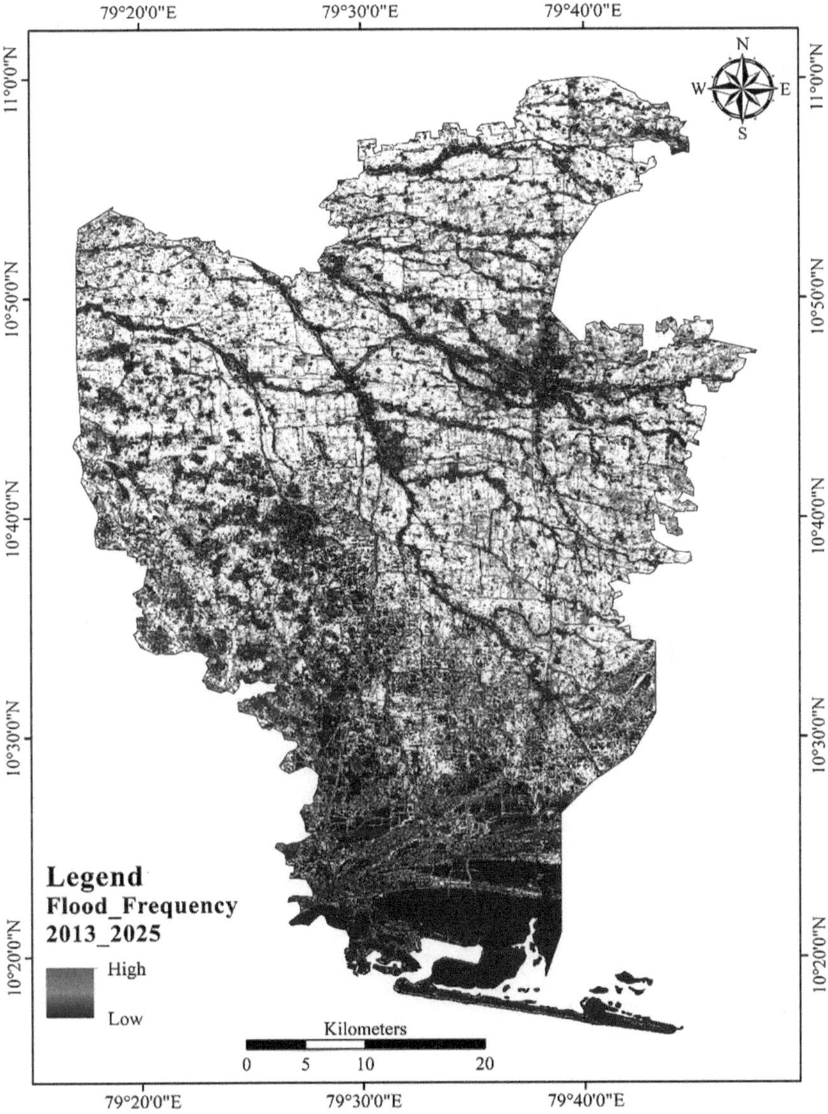

Fig. 13.20 Flood frequency analysis of the study area (prepared by the authors)

across the delta landscape (Fig. 13.21). Conversely, 2019's maximum extent reflected intense monsoon precipitation and coordinated river discharge surge (Jacinth Jennifer et al., 2022; Joe et al., 2025; Kaliraj et al., 2025). The 2022 inundation extent measured 1,503.6 km^2, positioning within the lower-middle range of historical values, indicating near-average monsoon intensity for that year. Year-to-year fluctuations reflected monsoon performance variability rather than systematic long-term trends (Fig. 13.22). Consecutive wet years (2016: 1,671.42 km^2, 2019: 1,745.18 km^2) alternated with relatively drier years (2015: 1,302.47 km^2, 2018: 1,419.9 km^2), demonstrating the region's dependence on seasonal precipitation timing and magnitude. Notably, the 442.71 km^2 difference between minimum and maximum extents (representing about 30% relative variation) underscores the substantial range in monsoon-driven flood behavior across the study period (Misra & Balaji, 2015; Sellaperumal et al., 2025).

Flood Hazard Zonation

Multi-year overlay produced four discrete flood hazard categories reflecting recurrence patterns. Very high-risk zones (1,303.15 km^2) encompassed areas experiencing flooding in 9 or more years within the 13-year analysis window essentially predictable annual inundation across

Table 13.5 Flooded inundated area

Year	Flooded area in km^2
2013	1500.16
2014	1567.94
2015	1302.47
2016	1671.41
2017	1552.92
2018	1419.9
2019	1745.18
2020	1701.93
2021	1478.02
2022	1503.8
2023	1564.16
2024	1597.94
2025	1590.66
Average flooded area	1553.58

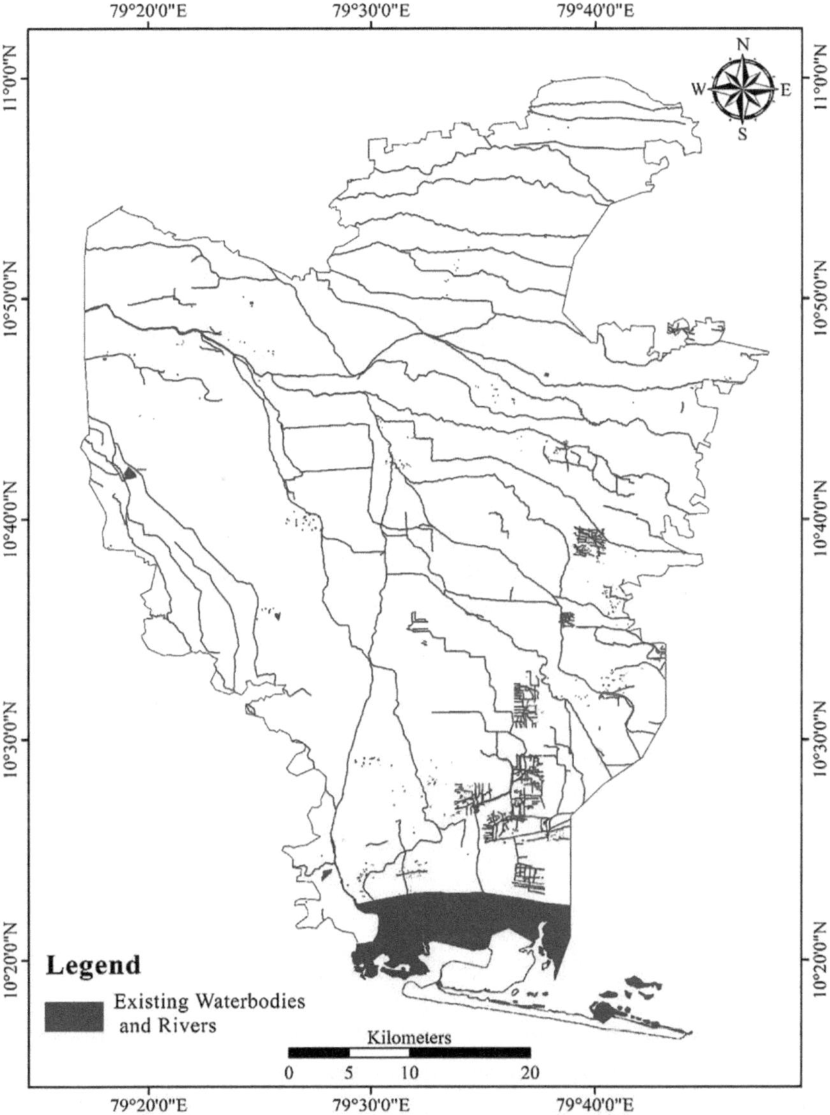

Fig. 13.21 Existing waterbodies of the study area (prepared by the authors)

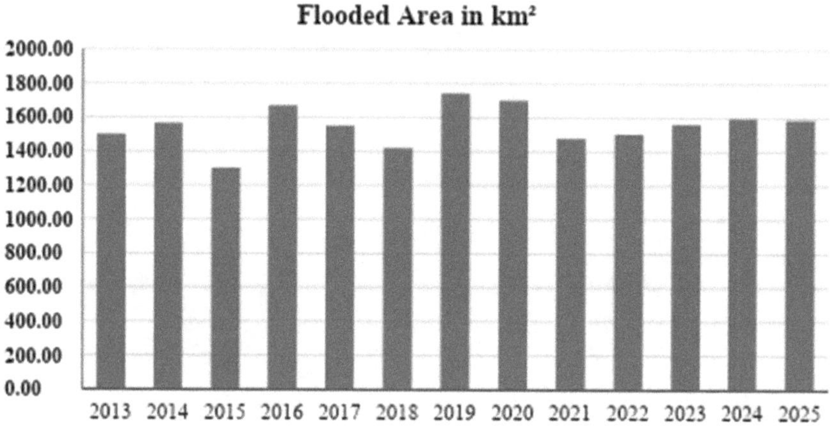

Fig. 13.22 Flooded area of Study area generated by InSAR Data of the study area (Prepared by the authors)

these lowland positions (Table 13.5). High risk zones (286.45 km^2) experienced flooding in 6–8 years, indicating frequent but not universal seasonal inundation. Moderate risk zones (282.18 km^2) flooded in 3–5 years, representing transitional landscapes subject to irregular monsoon variation (Eq. 13.26). Low risk zones (203.77 km^2) flooded rarely, in 1–2 years only, reflecting elevated terrain or effective drainage (Fig. 13.22). Combined very high- and high-risk zones totaled 1,589.6 km^2 (approximately 52% of district area), concentrating across the Cauvery River flood plains and adjacent deltaic lowlands where water naturally accumulates (Fig. 13.23). This spatial pattern aligned with known topographic lows and drainage convergence zones (Bhowmik et al., 2021; Dong et al., 2020; Iqbal, 2017; Popa et al., 2019; Suparyanto dan Rosad (2015, 2016).

The remaining 966.4 km^2 distributed across moderate and low risk categories occupied transitional and elevated positions where flooding occurred sporadically or never (Table 13.6).

$$\mathrm{F}_{\mathrm{total}}(\mathrm{x},\ \mathrm{y}) = \sum\nolimits_{2015}^{2025} Floodi(x, y), \tag{13.26}$$

where $\mathrm{F}_{\mathrm{total}}(\mathrm{x}, \mathrm{y})$ = cumulative of flood years at pixel (x,y).

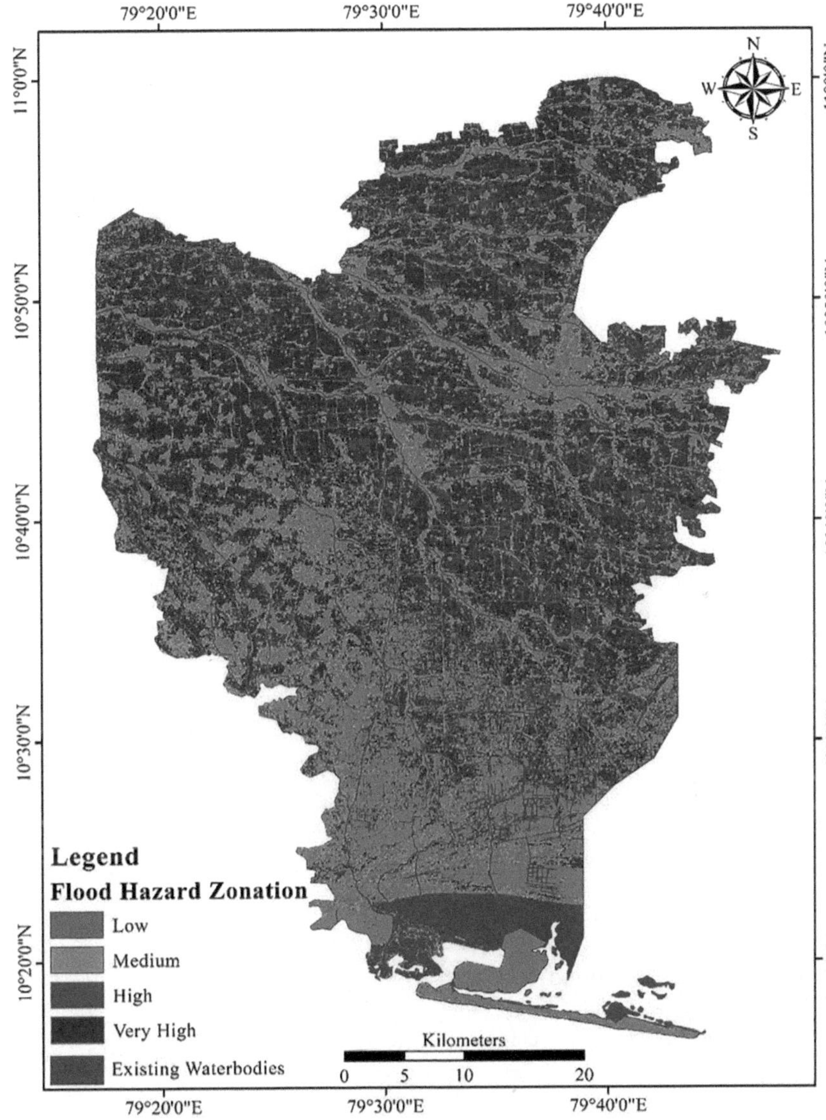

Fig. 13.23 Flood hazard zonation of the study area (prepared by the authors)

Table 13.6 Flood hazard zonation

Flood hazard zone	Area in km^2
Low risk	203.77
Moderate risk	282.18
High risk	286.45
Very high risk	1303.15

Floodi = binary flood classification (1 = flooded, 0 = non flooded) for the year i.

i = range from 2013 to 2025.

Conclusion

Satellite radar data from 2013 through 2025 mapped flood patterns across Thiruvarur district, revealing where water regularly accumulates. Zones experiencing flooding almost every year cover 1,303.15 km^2 these areas face predictable seasonal problems each monsoon. Another 286.45 km^2 flood frequently though not consistently. The total water coverage shifts by roughly 28.5% between wet and dry monsoon years, fluctuating between 1,302 and 1,745 km^2 (Table 13.6). By examining backscatter measurements alongside identifying permanent water bodies, the analysis distinguished actual flood water from permanent reservoirs and rivers. This approach required only satellite data spanning 13 years without needing additional maps of terrain or land use. The resulting flood risk map can help disaster response teams prepare contingency plans, guide agricultural extension services toward flood-resistant crop varieties suited to each zone, and inform long-term land-use planning in vulnerable deltaic landscapes. The direct methodology can be frequent in other flood-prone regions where satellite radar archives exist, offering practical flood monitoring capability for communities facing increasing water hazards from changing monsoon patterns.

13.2.5 Impact of Sea-Level Rise on the West Bengal Delta Using Remote Sensing and GIS-Based Methodology (Casebook 5)

Working Summary

The West Bengal deltaic state of India, developed after the depositional process of the Ganga and Brahmaputra Rivers. This region experiences accelerating relative sea-level rise of 5 mm/years due to combined marine transgression and terrestrial subsidence. The main objective of the study is finding the sea-level rising evidence and its impact on local community's habitants. This research integrates Sentinel-2B multispectral satellite imagery (10 m resolution, from ESA datahub), SRTM digital elevation models (30 m resolution), and WorldPop population data processed through Google Earth Engine (GEE) and ArcGIS 10.4 to quantify differential inundation impacts on the study area. In this study three sea-level rise scenarios were modeled based on the thresholding technique, there were: 1-meter low-lying (basically subsidence zone), 3-meter storm surge (cyclonic effect time), and 5-meter extreme (extreme climatic condition). This three-layer intersected with land-use and population distributions of this study area. Results demonstrate escalation nonlinear vulnerability of this study area. It has found that at 1-meter rise, 27.88 lakh people face exposure by coastal flooding risk with 430.08 km^2 mangrove loss and 1,558.80 km^2 agricultural land threatened; which highly increases to 152.96 lakh people, 3,968.97 km^2 mangrove forest loss, and 6,075.22 km^2 agricultural land impact under extreme scenarios. It was observed that Agricultural sectors emerge as most vulnerable, followed by natural vegetation and settlements of these states. Findings emphasize urgency for nature-based solutions including mangrove restoration and climate-responsive settlement strategies by implementing the integrated coastal area development plan. This spatially precise approach supports evidence based coastal management planning by providing replicable frameworks for deltaic monitoring globally.

Keywords: Sea-level rise; Inundation; Vulnerability; Google earth engine; Risk; Coastal West Bengal

Introduction

Coastal deltaic systems face unprecedented threats from rising sea levels, subsiding landscapes, and anthropogenic modifications (Halder et al., 2024; Hasan et al., 2024; Malakar, 2020b; Sorna et al., 2025). The West Bengal delta exemplifies these vulnerabilities acutely. Mean sea-level along the Bengal coast has risen 2–7.5 millimeters annually since the 1970s, while the landmass simultaneously subsides at approximately 2.9 millimeters per year (Parthasarathy & Deka, 2021; Sanyal & Lu, 2005). This convergence generates relative sea-level rise exceeding 5 millimeters annually accelerating inundation, saltwater intrusion, and erosion (Mahmood et al., 2025; Majumder et al., 2025). The Sundarbans region, a UNESCO World Heritage Site with elevations below 20 meters, remains particularly vulnerable to these compounding pressures. The West Bengal delta, comprising the Indian sector of the Ganges–Brahmaputra-Meghna system, exhibits ironic dynamics shaped by natural and anthropogenic forces. Between 1973 and 2022, approximately 478 square kilometers experienced erosional retreat while 408 square kilometers accreted through sediment deposition (Halder et al., 2024; Parthasarathy & Deka, 2021). However, this pattern masks critical vulnerabilities in the southern Sundarbans islands retreat at rates exceeding 63 meters annually, averaging 9 meters per year across the delta. Embankment construction, polder development, and upstream damming have disrupted sediment supply and natural flood rhythms, fundamentally destabilizing the morphodynamic equilibrium that historically sustained marginal land persistence (Dhanalakshmi et al., 2019; Halder et al., 2024; Naren & Maity, 2024; Parthasarathy & Deka, 2021). Traditional field-based coastal monitoring cannot capture the rapid, spatially heterogeneous patterns of inundation and geomorphological change characteristic of deltaic systems. Synthetic aperture radar (SAR) satellites overcome this limitation through all-weather imaging independent of cloud cover critical for monsoon-influenced environments. Optical remote sensing provides complementary vegetation and land-use information, while geographic information systems integrate multi-source datasets into spatially explicit vulnerability indices. Together, these technologies enable transformation from retrospective assessment to prospective risk quantification (Borrows & De Bruin, 2006; Mahapatra et al., 2015; Naren & Maity, 2024; Sahu & Mehta, 2024). Existing vulnerability studies adopt broad regional or national scales, insufficiently resolving local variations in subsidence, elevation, and land-use patterns that determine differential exposure. Few investigations quantify how embankment-induced subsidence acceleration compounds relative sea-level rise vulnerability in West Bengal specifically (Das et al., 2025; Halder et al., 2024; Mahmood et al., 2025; Parthasarathy & Deka, 2021). Climate projections employ global or pan-Indian scenarios without validation against local tide-gauge records and coastal-change trajectories. This study examines sea-level rise impacts on the West Bengal delta through integrated remote sensing and GIS-based vulnerability assessment.

Methods and Methodology

Study Area

The West Bengal delta occupies the vast Ganges–Brahmaputra-Meghna system, extending across three coastal districts South 24 Parganas, North 24 Parganas, and Purba Medinipur. The study area encompasses 220 kilometers of coastline, ranging between latitudes 21°29′00″ and 22°34′00″ N and longitudes 88°04′00″ and 89°05′00″ E (Fig. 13.24). Maximum elevation of West Bengal is 3636, which is located Darjeeling district, and minimum elevation approx. 0 to –7 m near the Sundarbans deltaic region (Majumder et al., 2025; Sanyal & Lu, 2005; Sorna et al., 2025). The rainfall is varied from south to north of this state, average rainfall recorded 175 220 cm, which is higher than the national average 125 cm (Das et al., 2025; Sanyal & Lu, 2005).

The region's coastal zone is legally demarcated from the High Tide Line extending 500 meters inland, delineating the area subject to tidal and coastal hazard influences (Sorna et al., 2025).

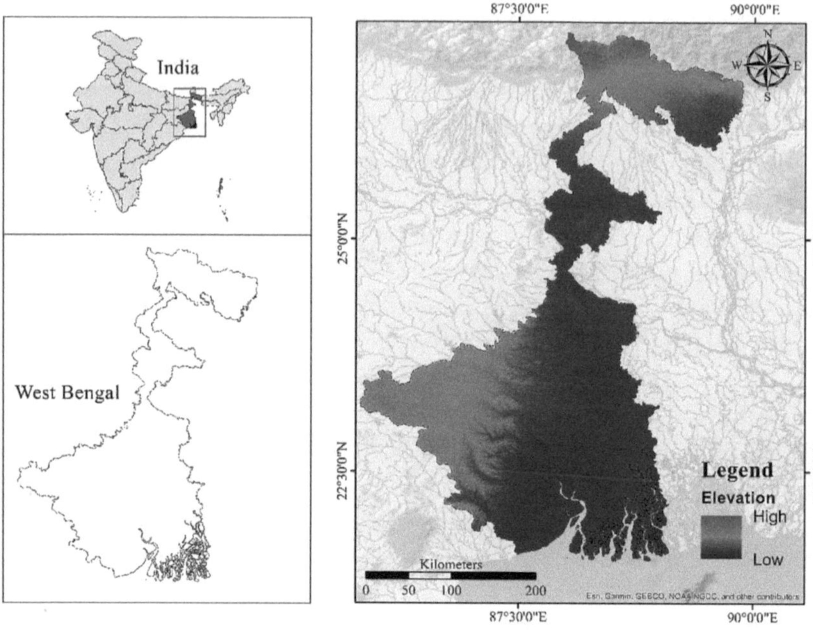

Fig. 13.24 Mapping study area: coastal West Bengal (prepared by the authors)

The Sundarbans forms the dominant geomorphological feature, covering approximately 5,600 square kilometers within West Bengal (India). This active deltaic region exhibits uniformly low relief, with elevations rarely exceeding 20 meters above mean sea level. The landscape consists of a complex network of tidal waterways, mangrove-vegetated islands, and estuarine channels. The Hooghly River, the westernmost Ganges distributary, traverses the study area for roughly 150 kilometers before entering the Bay of Bengal (Sanyal & Lu, 2005). The estuary experiences pronounced tidal amplification, with a tidal range of approximately 2.8 meters and flood-dominant asymmetric tidal propagation. Subsurface conditions reflect ongoing compaction and sediment loading, with documented subsidence rates between 1 and 3.65 millimeters annually. The delta's hydrology remains highly dynamic, influenced by seasonal monsoon discharge from upstream rivers and intricate salinity gradients shaped by freshwater inputs and marine tidal incursion.

Data collection and Analysis

The West Bengal delta encompasses the coastal zone extending from the Bengal Embankment to the seaward margins of the Sundarbans (Das et al., 2025). The study domain covers approximately 9,630 square kilometers of deltaic landscape subject to persistent relative sea level rise. Sentinel-2B multispectral satellite imagery was downloaded from Google Earth Engine for the year 2025. SRTM Digital Elevation Model (30 m resolution) provided baseline topographic data (Dhanalakshmi et al., 2019; Naren & Maity, 2024; Sorna et al., 2025). WorldPop population grid (100 m resolution, 2020) supplied distributed population estimates. All datasets were clipped to the study boundary within GEE using cloud masking algorithms to remove atmospheric contamination. Sentinel-2B imagery was processed using a random forest classifier in GEE to generate the 2025 LULC map (Table 13.7). Five classes were identified—water bodies, vegetation (mangroves), agricultural land, settlements, and bare soil (Fig. 13.25).

Classification accuracy was assessed using field reference data. SRTM DEM was processed in GEE to extract coastal elevation data (Anusha & Bharathi, 2020; Mudi et al., 2022; Samanta et al., 2018). Three relative sea-level rise scenarios were applied 1-meter (low-lying areas), 3-meter (storm surge conditions), and 5-meter (extreme scenarios). Raster-based threshold analysis delineated inundation-prone zones for each scenario. Inundation zones were intersected with LULC classes in GEE to quantify land-use vulnerability (Fig. 13.26). Three impact maps were produced showing LULC loss under 1 m, 3 m, and 5 m sea-level rise. WorldPop

Table 13.7 LULC classification: Area

LULC class	Area km^2	Area in %
Waterbodies	5689.94	5.28
Natural vegetation	35,225.01	39.69
Agricultural area	36,185.87	39.65
Settlement area	9800.03	10.48
Barren land	1859.55	0.95

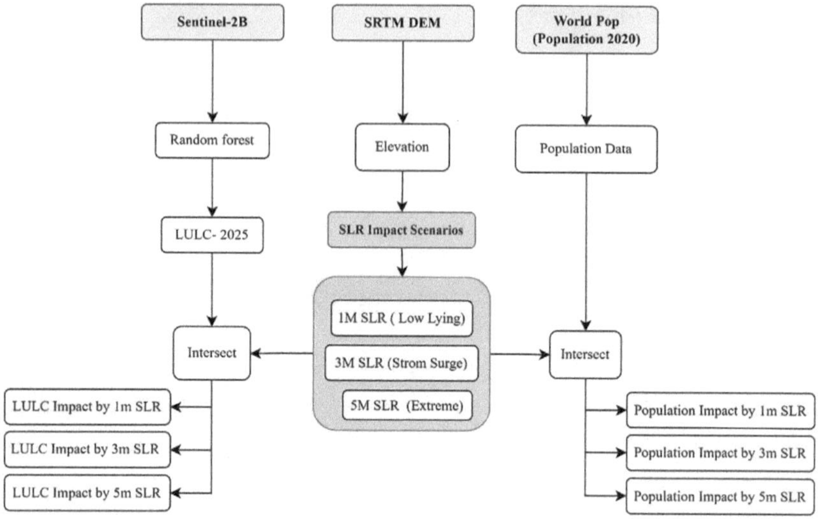

Fig. 13.25 Methodological flowchart: Casebook 5 (prepared by the authors)

data was masked by inundation zones in GEE to calculate exposed populations (Dhanalakshmi et al., 2019; Naren & Maity, 2024). Three population impact maps were generated corresponding to each SLR scenario. All processed data's outputs were exported from GEE as GeoTIFF files and imported into ArcGIS 10.4. Vulnerability indices combining elevation, LULC impact, and population exposure were computed in ArcGIS raster calculator. Thematic maps illustrating coastal vulnerability across multiple hazard scenarios were produced using ArcGIS tools.

Result and Discussion

The Study reveals severe escalation in inundation exposure across West Bengal delta with increasing sea-level rise situations. At 1-meter SLR, natural vegetation (basically mangroves) faces loss of 430.08 km^2, affecting 27.88 lakh people. (Table 13.9) This vulnerability increases dramatically at 3-meter SLR, where mangrove loss reaches 2,004.77 km^2 with 88.10 lakh people exposed (Mahapatra et al., 2015; Mahmood et al., 2025; Majumder et al., 2025). The extreme 5-meter scenario threatens 3,968.97 km^2 of mangrove and Natural Vegetation Forest

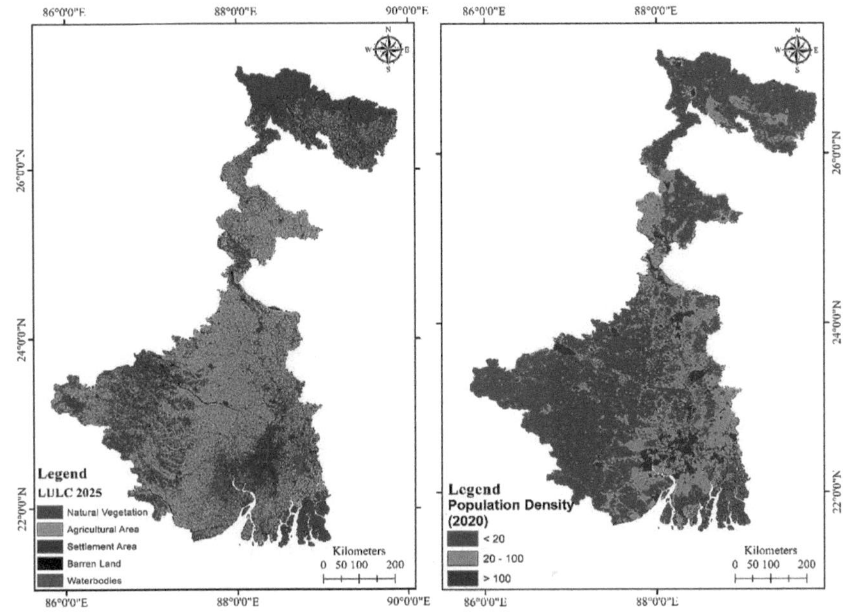

Fig. 13.26 Land use land cover and population map (prepared by the authors)

and 152.96 lakh people at direct risk a cumulative increase of 450% from baseline situations. Agricultural lands occur as the most affected sector, losing 1,558.80 km^2 under 1-meter rise (low lying), escalating to 6,075.22 km^2 under 5-meter SLR (Extreme). This exceeds mangrove losses equivalently, signaling serious risk to food security and rural livelihoods (Das et al., 2025; Halder et al., 2024; Mondal et al., 2025). Settlement areas, encompassing 10.48% of the delta baseline, experience progressive inundation ranging from 130.6 km^2 to 1,070.51 km^2 across situations. The compounding effect of aquaculture expansion amplifies agricultural vulnerability through saltwater intrusion and soil degradation (Fig. 13.27).

Mangrove forest degradation carries dual consequences. Loss of 430 km^2 at minimal SLR threatens critical storm-surge dissipation capacity, while simultaneously displacing fishing communities dependent on forest resources. Smaller mangrove patches in western delta

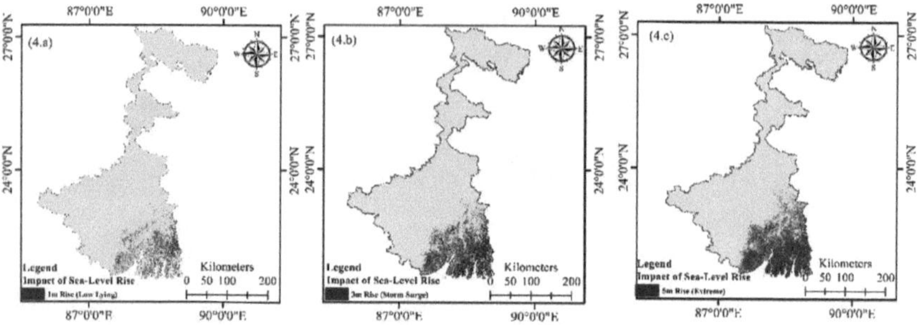

Fig. 13.27 Impact of sea-level rise (4.a.) 1 m rise; (4.b.) 3 m rise; (4.c.) 5 m rise (prepared by the authors)

regions show heightened erosion susceptibility. Current embankment-dependent infrastructure proves insufficient through predictable situations (Majumder et al., 2025). The 27.88 lakh people already at risk (Table 13.9) from storm surge now face amplified hazards through rising baseline water levels.

Sea-Level Rise (SLR) Vulnerability

Coastal vulnerability in the West Bengal delta was quantified using a three-component exposure-sensitivity-adaptive capacity framework (Fig. 13.27) (Dhanalakshmi et al., 2019; Mahapatra et al., 2015; Raaijmakers et al., 2008). Coastal elevation and subsidence rates, land-use patterns, and population density formed the core vulnerability indicators (Eq. 13.27).

$$V = \frac{E * S}{AC}, \tag{13.27}$$

where V = Vulnerability,

E = exposure reflects hazard intensity,

S = susceptibility, and

AC = Adaptive Capacity.

Elevation thresholds derived from SRTM DEM were applied to delineate inundation-prone zones under three SLR zones, 1-meter (low-lying), 3-meter (storm surge), and 5-meter (extreme) (Table 13.8). This "bathtub" approach overlapped reclassified elevation data with situation based

-specific water-level parameters in Google Earth Engine platform. Spatial intersection between LULC classes and inundation zones calculated sectoral vulnerability. Agricultural lands showed maximum exposure (6,075.22 km^2 at 5 m SLR), followed by natural vegetation/mangroves (3,968.97 km^2) and settlements (1,070.51 km^2) (Table 13.9). WorldPop raster data (100 m resolution) was masked by inundation layers to calculate exposed populations (Naren & Maity, 2024; Sanyal & Lu, 2005). Results ranged from 27.88 lakh people at 1-meter SLR to 152.96 lakh under extreme situations (Fig. 13.28).

Table 13.8 LULC impacted by the SLR

Scenario	Mangroves/Natural vegetation _ lost_ area km^2	Settlements _lost _ area km^2	Agriculture _lost _ area km^2
SLR 1 m	430.08	130.6	1,558.80
SLR 3 m	2,004.77	494.405	4,276.32
SLR 5 m	3,968.97	1,070.51	6,075.22

Table 13.9 Population influenced by the SLR

Scenario	Population _ affected (2020)
SLR 1 m	27,88,667
SLR 3 m	88,10,169
SLR 5 m	1,52,96,598

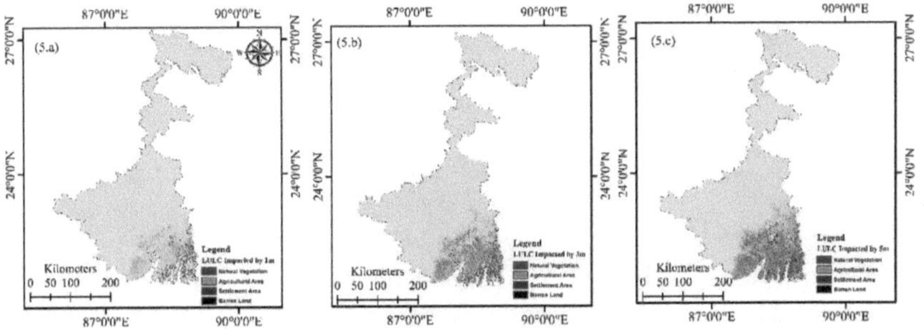

Fig. 13.28 Impact of sea-level rise (5.a.) LULC impacted by 1 m rise; (5.b.) LULC impacted by 3 m rise; (5.c.) LULC impacted by 5 m rise (prepared by the authors)

Conclusion

This study, which uses remote sensing and GIS, measures how the rising sea level (SLR) is becoming a bigger threat to the ecosystems and livelihoods in the West Bengal delta. The results show that vulnerability does not increase in a straight line. For example, at 1 meter of sea-level rise (SLR), 27.88 lakh people will have to move, but this number will rise to 152.96 lakh in extreme 5-meter scenarios, which is a 450% increase. Agricultural lands are the most vulnerable sector (6,075.22 km^2 loss at 5 m SLR), followed by mangrove ecosystems (3,968.97 km^2). This research identified food security and coastal protection at serious risk. Infrastructure that relies on embankments is not enough to protect against rising baseline water levels and subsidence rates that stay the same (2.9 mm/year). The adopted modern technology-based methodology provides more perfect result than the only field-based old methodology. Integration of both methodologies will provide best result. This integrated technique Facilitating spatially clear vulnerability mapping across diverse deltaic landscapes. This research suggests the nature-based solutions for risk calculation in this study areas by mangrove restoration, sustainable settlement relocation, and climate-resilient agricultural diversification.

13.2.6 Climate Change Adaptation and Sustainable Coastal Planning in Visakhapatnam, Andhra Pradesh, Using Geospatial Technique (Casebook 6)

Working Summary

Visakhapatnam faces severe coastal risks because serious facilities, growing cities, and failing climate impacts all crowd into the same small area. The main objective of this research is identifying the coastal risk and bio-shied zones. In this research adopted remote sensing-based methodology to identify the shoreline change analysis, LULC preparation, and Population density. The research found three distinct zones that need different plans to restore. 1st is port facilities and power plants (173 ha) sit dangerously close to rising seas, 2nd is the natural vegetation, mangrove and wetlands (5991.31 ha) need first restored to absorb storm surge, especially the mangrove (220 ha) reserves across Andhra Pradesh that deserve severer protection. Buildings, concrete walls everywhere in this study area are too expensive and damages nature, so by this research propose mixing strategies keeping walls where they problem is maximum, authorizing nature recover where it can, and changing how different government agencies work together. The research shows that when cities tie their adaptation plans to real evidence about where risks are greatest, and when they invest in both engineered defenses and ecosystem recovery, they build stronger resilience. If planning becomes truly integrated then, Visakhapatnam can protect its economy, keep communities safe, and preserve its coastal ecosystems all at the same time.

Keywords: Coastal vulnerability; Climate adaptation; Storm surge; Bioshield conservation; Shoreline erosion; Visakhapatnam Coast

Introduction

Coastal cities worldwide struggle with mounting climate pressures as sea levels rise and intense storms grow more frequent (Karnata & Kumari, 2025; Swarna Latha & Nageswara Rao, 2011). Visakhapatnam, India's major port hub, sits at the center of this challenge. The city brings together significant economic activity, large populations, and sensitive coastal ecosystems in a relatively small area (Baig et al., 2021; Sai Priya & Banerjee, 2024). The port itself moves millions of tonnes of cargo annually, while steel plants and power stations occupy the surrounding landscape (Dey & Mazumder, 2023; Sahana et al., 2019). Most of this critical infrastructure sits close to the shoreline just 2 to 5 meters above sea level making it vulnerable to flooding, storm surge, and the gradual retreat of the shoreline. Andhra Pradesh's coastline stretches nearly 1,000 kilometers along India's eastern seaboard. The region has always contended with cyclones during monsoon seasons and occasional severe flooding. In recent decades, however, the situation has worsened. Water-level measurements taken at Visakhapatnam's port show the sea rising steadily at 3 to 5 millimeters each year (Abijith et al., 2023; Mohanty &

Kumar, 2025). Satellite images collected over the past thirty years reveal that 319 kilometers of Andhra Pradesh's coast is actively eroding. In Visakhapatnam itself, the shoreline retreats by 0.5 to 1.0 meters annually in some areas (Das, Sah, et al., 2025; Sahu & Mehta, 2024). This ongoing erosion jeopardizes port facilities, fishing villages, farmland, and communities that depend on the coast. For decades, coastal protection relied almost entirely on engineering solutions concrete seawalls, breakwaters, and similar structures. These approaches have not worked well. They are expensive to build and maintain, they often fail to stop erosion, and they frequently damage nearby ecosystems (Baig et al., 2021; Karnata & Kumari, 2025; Malakar et al., 2022; Sai Priya & Banerjee, 2024). When a wall is built in one location, erosion often accelerates in the next area along the coast. More recent approaches combine these structural defenses with natural solutions. Mangrove forests, which cover 897,683 ha. across Andhra Pradesh, offer an example. These trees reduce storm surge heights by 1 to 2 meters and slow incoming waves by 20 to 70% (Karnata & Kumari, 2025; Swarna Latha & Nageswara Rao, 2011). They also support fishing communities by providing nurseries for young fish, capture carbon from the atmosphere more efficiently than land forests, and offer timber and other products that people depend on. Effective adaptation requires understanding where risks are greatest. Geographic information systems now allow researchers to combine satellite e data, terrain information, infrastructure locations, and community details to map vulnerable areas precisely. For Visakhapatnam, this mapping identified three types of zones: about 173 ha. where critical facilities are at severe risk and need strong protection; intermediate areas that require better drainage and stronger buildings; and nearly 6,000 ha. where restoring ecosystems would help protect the coast and support nature simultaneously. India has developed several policies to manage coastal zones (Das, Sah, et al., 2025; Kantamaneni et al., 2022; Sahu & Mehta, 2024). The coastal regulation zone rules restrict development in sensitive areas. The integrated coastal zone management program aims to coordinate planning across different government departments. However, these initiatives often work in isolation from one another. Different government agencies pursue their own agendas without coordinating (Mohanty & Kumar, 2025; Parida et al., 2025). Communities affected by coastal hazards rarely have a meaningful voice in planning decisions. Few systems exist to track whether adaptation efforts actually reduce vulnerability. Funding for nature-based solutions remains scarce compared to spending on traditional concrete structures (Kantamaneni et al., 2019; Sahana et al., 2019).

This study applies detailed geographic and scientific analysis to Visakhapatnam's coastal adaptation challenge. The research examines how physical hazards, ecological systems, and built infrastructure interact. It proposes pathways that combine physical barriers where essential with ecosystem restoration and improved urban planning. The aim is to build coastal resilience that endures and serves both the city's economic needs and community wellbeing.

Methods and Methodology

Study Area

The study encompasses Visakhapatnam's coastal district (17.69°N, 83.29°E; 79.79°E–79.88°E, 11.85°N–12.05°N), approximately 682 km along Andhra Pradesh's Bay of Bengal coastline, including rocky headlands near Rushikonda, harbor basins, sandy beaches with fishing settlements, and hinterland extending toward the Eastern Ghats with elevations of 5–50 meters (Baig et al., 2021; Sai Priya & Banerjee, 2024). The coastal zone definition spans from the high-tide line 500 meters inland, encompassing critical port facilities, steel plants, power infrastructure, and residential areas (Baig et al., 2021; Karnata & Kumari, 2025). Analysis integrated three complementary data streams processed through Google Earth Engine, a cloud-based platform enabling continental-scale geospatial computation: (1) multi-temporal Landsat surface reflectance imagery for shoreline dynamics, (2) digital elevation models for sea-level rise vulnerability, and (3) global land cover classification for infrastructure and ecosystem exposure (Fig. 13.29).

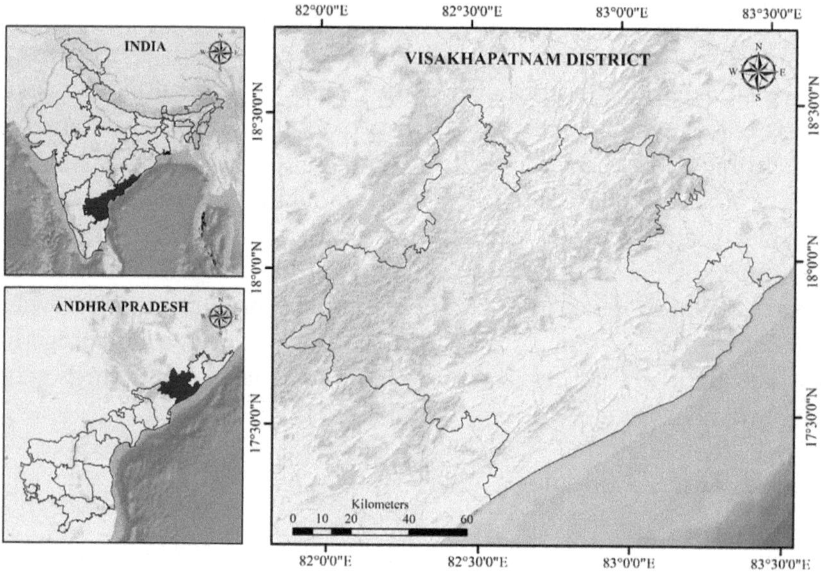

Fig. 13.29 Mapping study area: Visakhapatnam coast (prepared by the authors)

Shoreline ChangeDetection

Historical shoreline positions were extracted from Landsat 5 and 9 surface reflectance imagery spanning 1990–1993 (baseline) and 2021–2024 (current). Three-year periods were aggregated to minimize cloud effects and isolated monsoon-driven disturbances while capturing long-term erosion and accretion trends characteristic of Visakhapatnam's dynamic coastline (Behera et al., 2019; Das, Shahfahad, et al., 2024; Hastuti et al., 2022; Joe et al., 2025). The Modified Normalized Difference Water Index (MNDWI) was calculated as MNDWI = (Green − SWIR1)/(Green + SWIR1), where Green and SWIR bands derive from surface reflectance products. MNDWI effectively distinguishes water from land because green wavelengths reflect strongly from water while shortwave infrared reflects minimally (Mondal & Pal, 2018); SWIR additionally suppresses false water classification from urban areas and bare soil critical for developed port city coastlines (Saini et al., 2013). Median composite imagery aggregated pixel-wise MNDWI values across each three-year period, producing cloud-free water index surfaces (Sivanpillai et al., 2020). Water extent was extracted by thresholding MNDWI ≥ 0, and consecutive epochs were compared to quantify erosion (land-to-water transitions) or accretion in northern and southern Visakhapatnam coastal reaches (Fig. 13.30).

Inundation Vulnerability

Sea-level rise vulnerability was mapped using shuttle radar topography mission (SRTM) elevation data at 30-metre resolution. Terrain ≤ 2 meters above mean sea level appropriate for Visakhapatnam's microtidal Bay of Bengal setting combined with projected sea-level rise of 30–90 centimeters by 2100 was identified as vulnerable to permanent inundation (Bhaskaran et al., 2020; Haldar et al., 2016; Passeri et al., 2015;

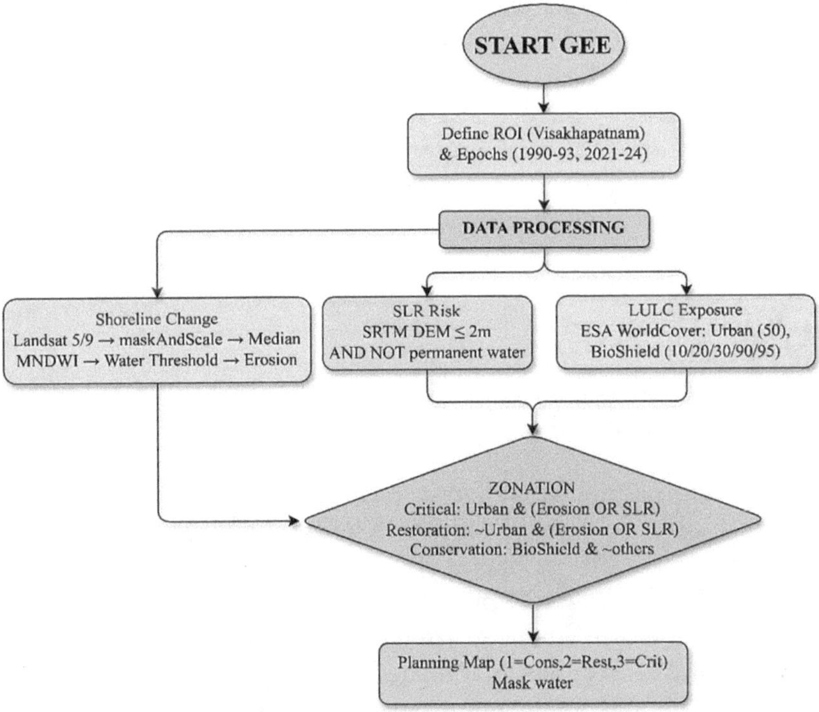

Fig. 13.30 Methodological flowchart: Casebook 6 (prepared by the authors)

Tripathy & Malladi, 2022). This layer was masked against permanent water (2021–2024 extent) to flag only terrestrial areas at risk, excluding lagoons and natural water bodies (ASF, n.d.; Attema et al., 2007; Nghia et al., 2022; Shen et al., 2019; Tiwari et al., 2020). Land cover classification derived from ESA WorldCover 2020 (10-metre resolution, Sentinel-1/2-based) was reclassified into three categories: built-up/urban (class 50) encompassing port zones and industrial areas, natural bioshield including vegetation and mangrove forests (classes 10, 20, 30, 90, 95), and other non-critical land. Urban exposure to coastal hazards was identified by intersecting built-up pixels with erosion or sea-level rise vulnerability, pinpointing critical infrastructure at direct risk (Fig. 13.30).

Land cover classification was derived from ESA WorldCover 2020, a 10-metre resolution global product based on Sentinel-1 and Sentinel-2 satellite observations. ESA WorldCover assigns each pixel into eleven categories; for this analysis, categories were reclassified into three management-relevant groups: (1) Built-up/Urban (WorldCover class 50), (2) Natural Bioshield including tree cover, shrubland, grassland, and mangrove forests (WorldCover classes 10, 20, 30, 90, 95), and (3) Other non-critical lands. Urban exposure to coastal hazards was identified by intersecting built-up land cover with both the erosion risk layer and the sea-level rise vulnerability layer, identifying infrastructure positioned in high-hazard zones.

Result and Discussion

Existing Bio-shield Conservation

Mangrove and natural vegetation forests, the bio shield, provide natural protection against coastal hazards while maintaining critical ecological functions (Feagin et al., 2010). Approx. total conservation area is 897683.52 ha. (Fig. 13.30). Visakhapatnam's existing bio shield comprises mangrove forests distributed across approximately 220 ha (Kamil et al., 2021). within broader Andhra Pradesh coastal zones, with significant concentrations in deltaic and estuarine sectors. The

Godavari and Krishna deltas, located 70–150 km south of Visakhapatnam, contain particularly extensive mangrove tracts (Athikalam & Karur Vaideeswaran, 2022; Feagin et al., 2010). Mangroves provide multifaceted coastal protection. Root systems and vegetative structure dissipate wave energy through friction and drag, reducing wave heights by 20–70% depending on density and hydraulic conditions. The intricate root architecture facilitates sediment deposition, reducing suspended sediment concentrations in mangrove-fringed creeks by 40–70% compared to open-water reaches and supporting coastal accretion (Karnata & Kumari, 2025; Naz & Chowdhury, 2022). Storm surge reduction through frictional water flow resistance can result in surge height reductions of 1–2 meters compared to unprotected shorelines, potentially the difference between catastrophic and manageable impacts (Fig. 13.31).

Mangrove forests support extraordinarily high biodiversity, providing critical habitat for fish and prawn larvae, seabirds, reptiles, and mammals. Nursery habitat functions support regional fishery productivity, with mangrove extent positively correlating with nearshore fishery yields (Das, Dutta, et al., 2025; Kamil et al., 2021; Nguyen et al., 2015). Carbon sequestration in mangrove biomass and soils contributes to climate change mitigation, with mangrove forests storing 3–4 times more carbon per unit area than terrestrial forests.

Andhra Pradesh's mangrove extent has declined significantly from its historical distribution due to the expansion of aquaculture, agricultural conversion, urban development, and infrastructure projects (Fig. 13.32). Despite legal protection under the Indian Forest Act and Coastal Regulation Zone notifications, enforcement challenges and economic pressures continue driving conversion. Contemporary conservation encompasses multiple strategies, including legal protection through CRZ and Forest Act provisions, community-based initiatives, eco-tourism that generates conservation incentives, and restoration in degraded areas. Blue carbon initiatives increasingly attract funding, emphasizing climate change mitigation co-benefits (Kamil et al., 2021). Restoration of historically mangrove-bearing areas within Visakhapatnam's coastal zone presents an opportunity to enhance coastal protection while restoring ecosystem functions. Successful restoration depends on adequate freshwater input,

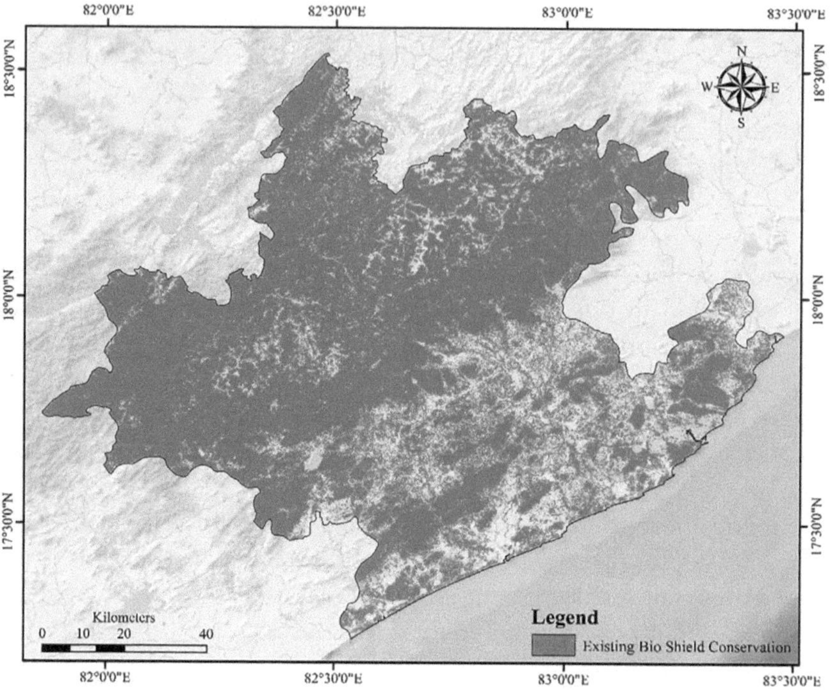

Fig. 13.31 Existing mio-shield conservation of the study area (prepared by the authors)

appropriate salinity levels, protection from disturbance during establishment, and community participation in management and benefit-sharing.

Ecological Restoration Potential

Visakhapatnam's coastal zone has about 5,991.31 ha. suitable for restoration (Fig. 13.32). These include degraded mangrove cover area with good soil and water conditions (Yang et al., 2024). Abandoned aquaculture ponds can convert to wetlands again. Riparian zones along coastal creeks need replanting where trees were vacant (Fig. 13.33). Some industrial and urban wastelands can also be restored through better land-use planning. These planning zones are identified through vegetation mapping, soil analysis, water assessment, and land tenure review. This creates a

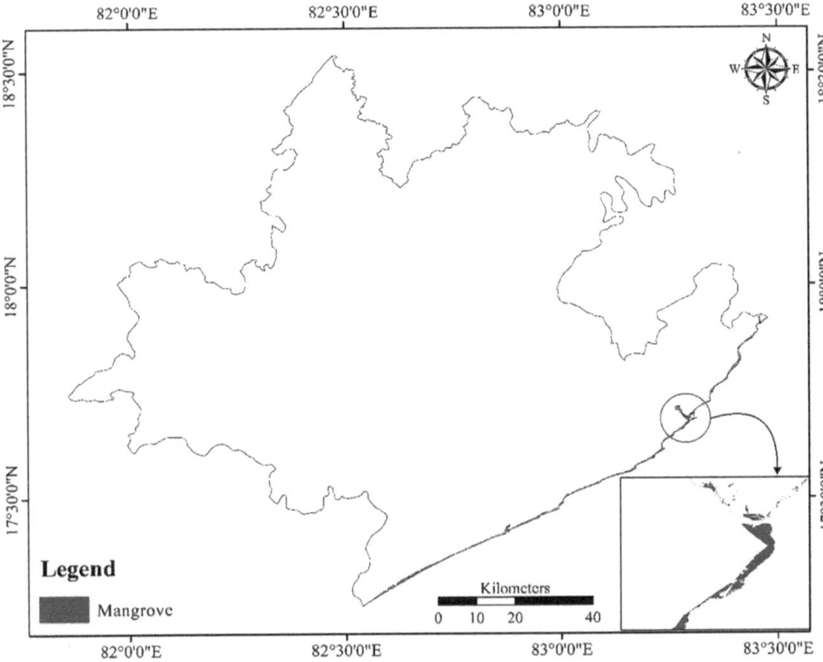

Fig. 13.32 Mangrove area of the study region (prepared by the authors)

basis for prioritizing restoration. Coastal planners now recognize that restoring ecosystems brings many benefits beyond traditional protection. Restoring mangroves, salt marshes, seagrass, and riparian vegetation reduce coastal hazards, conserves biodiversity, improves water quality, and helps address climate change. Mangrove restoration is the main ecological priority in Visakhapatnam because it provides many services (Bolliger et al., 2004; Yang et al., 2024). For successful pan implication, restoration needs the right water conditions a mix of salt and fresh water. Using local seeds and seedlings improves survival rates. Communities must participate as caretakers and benefit from timber, forest products, tourism, and also carbon credits. Regular monitoring of survival, growth, and ecosystem health allows for adaptive management (Fig. 13.33).

Planning Strategies

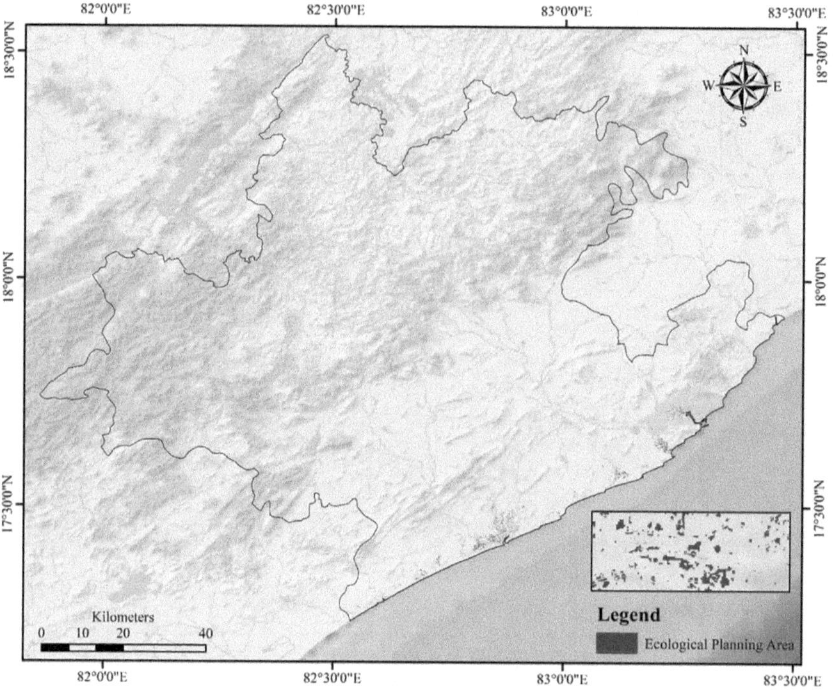

Fig. 13.33 Ecological planning area of the study region (prepared by the authors)

Visakhapatnam cannot solve its coastal crisis with uniform solutions. The city's hazards and opportunities vary dramatically across different neighborhoods and land types. The geospatial work has identified three distinct zones: high-risk urban areas totaling 173 hectares, places where nature can be restored across 5,991 ha (Baig et al., 2021; Bolliger et al., 2004; Yang et al., 2024). of Natural vegetation, and existing mangrove forests covering 220 ha. that need protection (Table 13.10). Each zone requires its own tactic.

Protecting Critical Infrastructure

Coastal wards 5–7, 17, 18, 25–26, 31, and 42–46 contain the port, steel mills, power plants, major roads, and homes for hundreds of thousands

Table 13.10 Area under risk, need to sustainable plan

Area in Ha	Zone
172.95	Critical (urban risk)
5991.31	Restoration (Eco-DRR)
897,683.52	Conservation (Bio-shield)

of people. These places cannot be abandoned, so they need protection. Building concrete walls along the entire eroding shoreline makes no economic sense it costs Rs. 15–20 crores per km and damages coastal ecosystems in the process. A smooth approach protects only the most critical infrastructure facilities (Fig. 13.34) like the port terminal, power plant intake systems, major transport routes. For these places, conventional seawalls, reinforced beaches, or elevated structures might all be options worth considering. The goal is choosing what works best in each location rather than applying the same solution everywhere. Elevating ground floors, using waterproof construction methods, and installing salt-resistant materials help existing buildings survive floods 0.6–1.5 meters deep. Better drainage systems drainage channels, retention ponds, permeable pavement help during monsoon storms, reducing flooding whether or not sea levels rise. Early warning systems that alert residents to approaching storms save lives. Training local emergency responders ensures rapid action when floods occur. The ward-level hazard mapping shows which areas are most vulnerable near Rushikonda Beach, in low-lying zones with poor drainage. Cities should restrict new building in these places or require buildings to meet higher flood-resistant standards. Communities living in informal settlements in the highest-risk areas should be offered the option to move, with support to find new livelihoods rather than returning to flood-prone work.

Restoring Ecosystems as Coastal Resistance

Nearly 220 ha. of degraded mangrove areas, abandoned shrimp ponds, dried-out stream valleys, and abandoned industrial sites can be converted to living, functioning ecosystems. Unlike concrete structures, restored mangroves do multiple jobs simultaneously—they break storm surge, provide fish habitat, improve water quality, and absorb carbon from the

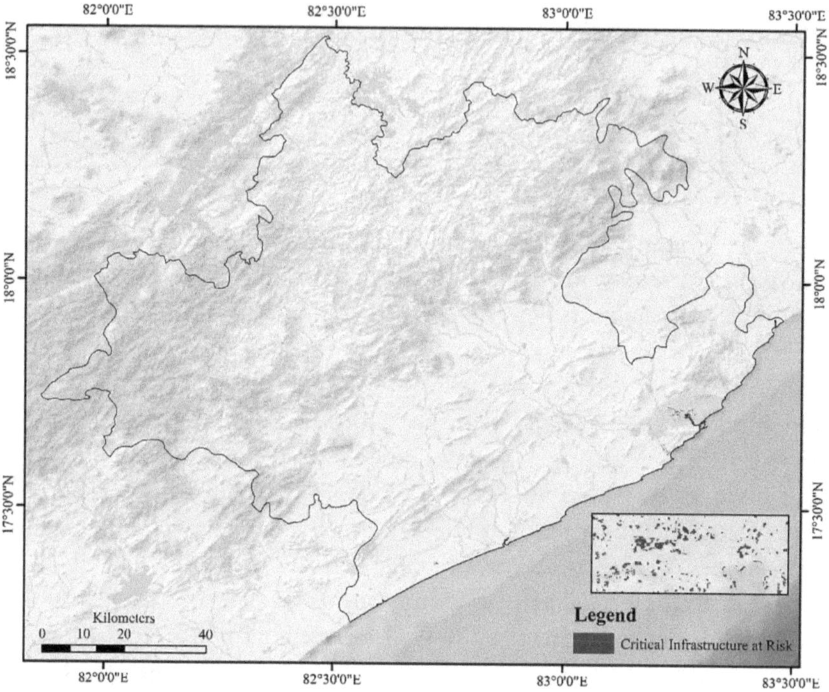

Fig. 13.34 Critical Infrastructure at risk of the study region (Prepared by the authors)

atmosphere (increases carbon credit). Mangrove restoration works when three things happen together (Naz & Chowdhury, 2022; Yang et al., 2024). First, the water conditions must be right mangroves need the correct salt balance and freshwater input. Second, the trees planted must come from local populations that evolved in Visakhapatnam's conditions, since they have higher survival rates than trees brought from elsewhere. Third, local people must participate and benefit. If fishermen, farmers, or laborers can harvest timber and other forest products, or find employment tending the forests, they become stewards rather than seeing restoration as an outside imposition. Even small mangrove patches as little as 5–10 ha. measurably reduce storm surge and wave damage in surrounding areas (Fig. 13.35). Multiple patches scattered across the coast create redundancy. If one patch gets damaged by a severe storm, others still provide protection (Yang et al., 2024). This differs from seawalls, which either work or fail completely. Mangrove restoration is just one option. Salt marshes can store water and slow inundation. Native forests along streams improve water quality and create wildlife corridors. Coastal scrub forests connect fragmented habitats. These different restoration types can be designed into urban areas mangrove corridors that also serve as parks, wetland retention systems that handle storm runoff while improving biodiversity. Local people should be involved in planning these areas, implementing restoration work, and benefiting from carbon finance programs that pay for mangrove conservation. Andhra Pradesh still has 220 ha. of mangrove forest, though this is less than half what existed historically. Every hectare lost to shrimp farming or urban sprawl weakens coastal protection (Feagin et al., 2010; Kamil et al., 2021). The state must enforce existing laws preventing mangrove conversion. Beyond protection, restoration should expand mangroves in places where they historically grew. Within Visakhapatnam district, perhaps 1,000–2,000 ha. could support new mangrove growth if conditions are managed properly. Priority should go to locations protecting the port, power plants, and residential areas where people are most exposed (Fig. 13.36).

Collaboration with Government Agencies to Work Together

Collaboration work is the key of development any areas. Not only the coastal adaptation plan fails but also most the development plan may fail when various governmental departments work individually (Clark, 1997; McKenna et al., 2008; Swarna Latha & Nageswara Rao, 2011). It is necessary to adopt this methodology for this study area. With the help of World Bank, India has adopted integrated coastal zone management (ICZM) frameworks by integrating all national development agencies or department (like—shipping development authority, industrial development authority, forest department, city development authority, and disaster management department). These frameworks start with baseline hazard mapping and vulnerability studies that should guide development planning. In India, this novel ICZM plan has been adopted by many

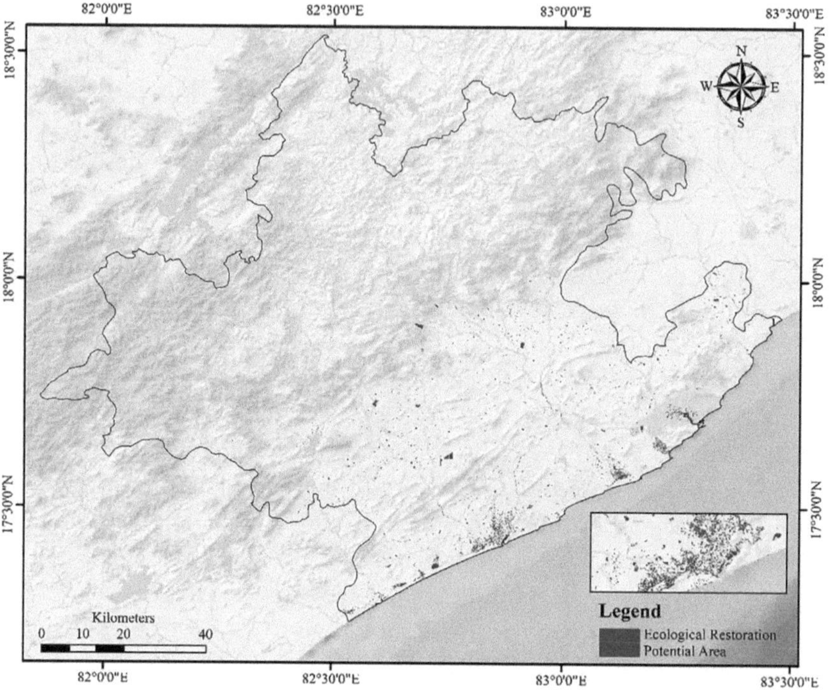

Fig. 13.35 Ecological restoration potential area of the study region (prepared by the authors)

states in paper but is rarely found in proper implication in reality (Alves et al., 2013; Bolliger et al., 2004; Clark, 1997; Yang et al., 2024).

Visakhapatnam needs serious institutional assurance for the implication of this plan, where this plan will arrange for faithful staff organizing across agencies, regular joint planning meetings, and clear accountability when strategies fail. The local communities, fishermen, informal residents, small business owners know their local level vulnerability, and risk. When, and how tides, coastal storm, cyclone impacted highly, where drainage fails, and what seasons are most dangerous. But these communities are rarely included when the planning strategies developed by the governing authorities. When communities help design their own protection measures, they use them better and maintain them longer than solutions designed by outsiders. Equal and better adaptation would result

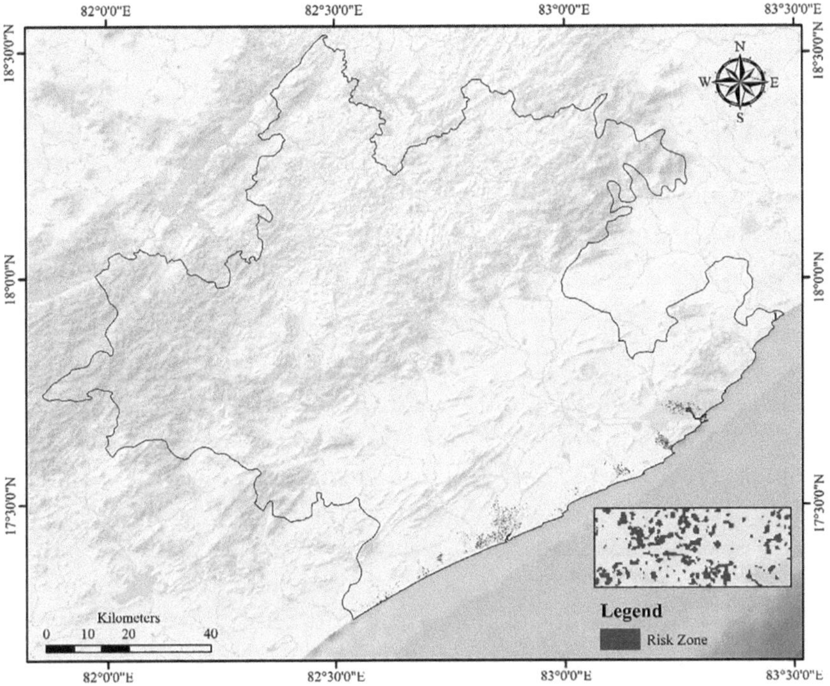

Fig. 13.36 Risk zone of the study area (prepared by the authors)

from formal mechanisms that involve coastal inhabitants in potential hazards mapping and strategy selection. Many states govt. build hazard mitigation plan but never check whether that plan work properly now or reducing damage. A seawall built in 1995 might be failing by 2020, but continues receiving maintenance funding. Mangrove restoration started in 2010 might have poor survival rates, but continued expansion without adjustment. Evidence-based management requires systematic tracking. In this regard satellite-based data and modern GIS technique help to track and monitor continuous shoreline position. When storms occur, level of storm surge, field surveys and satellite images can document actual inundation patterns (Alves et al., 2013; Clark, 1997).

The Adaptation techniques may fail for numerous undocumented reasons that appeared improbable during their formulation. Climate change might accelerate beyond existing projections. Unexpected problems could come up during restoration. New hazards might emerge in that area. Institutional coordination/governing authorities might falter. The adaptation plans should undergo a formal evaluation and revision every 5 years in to monitoring results. This means keeping alternative options open and being able to shift resources from failing approaches to ones that are more promising to work, than continuing with the first strategy choices that you made.

Conclusion

In this research is has found Visakhapatanam district facing various climate change and shoreline change related issues. The study found that most the infrastructure, which situated near to the coast are highly vulnerable, needs proper integrated proper adaption plans.

This analysis indicates that the successful coastal planning must be integrate the hazard mitigation, ecosystem restoration, development regulation, and social protection across different spatial and temporal scales.

Based on the remote sensing data it has calculated that total existing bio-shield area of this district are 897,683.60 ha, including natural vegetation and mangroves. Only mangroves is 220 ha. Which is very low. This bio-shied paly an essential role to protect the coastal region.

After analyzing data, it has observed critical infrastructure basically urban area in this region's falls under serious risk, which covering 172.95 ha. This urban risk prone area needs immediate ostentation for reduce vulnerability by enhancing coastal resilience plan. Visakhapatnam's coastal adaptation requires more than just the development of technologies; it also needs proper engagement of local communities. Then only possible to restore this district's coastal ecosystem sustainable way. That coastal sustainability depends on essential shift in development paradigms but prioritizing the ecosystem integrity alongside economic growth of this study area (Malakar et al., 2023). It provides an ecosystem-based adaptation plan, a solid evidence-based pathway to achieving sustainable future coastlines.

13.3 Conclusion and Arguments

The selected six case studies presented in this chapter show how geospatial science contributes to disaster risk assessment and management along the Indian coast. Covering different coastal settings, from the GIS-based Coastal Vulnerability Index assessment of Odisha to SAR-based flood inundation mapping in Thiruvarur, the studies demonstrate the practical use of remote sensing, GIS, and spatial analysis in identifying hazard-prone areas, sensitive coastal environments, and exposed populations. The findings indicate that coastal vulnerability in India is shaped by a combination of natural processes and human activities. Factors such as coastal landforms, climate-related changes, urban growth, population pressure, and development policies interact to influence risk levels across regions. The use of multi-temporal and multi-sensor datasets improves the quality of analysis by capturing both short-term changes and long-term trends, which supports more reliable risk assessment. The chapter highlights the value of geospatial science not only for analysis but also for planning and decision-making, offering useful inputs for coastal governance, climate change adaptation, and sustainable development along the diverse Indian coastline.

The case studies in this chapter show that geospatial techniques provide a practical and repeatable approach for assessing coastal risk across different physical and socio-economic conditions. Studies focusing on erosion, wetland degradation, flooding, and sea-level rise demonstrate how spatial analysis supports comparison of risk levels and helps identify priority areas for intervention. The findings also point to the need to combine physical coastal processes with human factors such as urban growth, land-use change, and population exposure. This combined perspective is necessary for effective disaster risk reduction and planning. The use of advanced geospatial methods, including SAR, InSAR, and multi-temporal satellite data, improves monitoring in coastal regions where field data are limited and cloud cover is common. These methods support both early warning and post-disaster assessment. The discussion shows that geospatial science plays an important role beyond technical analysis by contributing to sustainable coastal planning, climate change adaptation, and informed policy development. In the Indian coastal context, where environmental pressures continue to increase, geospatial science helps connect scientific evidence with governance and long-term coastal sustainability.

References

Abijith, D., & Saravanan, S. (2022). Assessment of land use and land cover change detection and prediction using remote sensing and CA Markov in the northern coastal districts of Tamil Nadu, India. *Environmental Science and Pollution Research, 29*(57), 86055–86067. https://doi.org/10.1007/s11356-021-15782-6

Abijith, D., Saravanan, S., & Sundar, P. K. S. (2023). Coastal vulnerability assessment for the coast of Tamil Nadu, India—A geospatial approach. *Environmental Science and Pollution Research, 30*(30), 75610–75628. https://doi.org/10.1007/s11356-023-27686-8

Aboda, C., Mugagga, F., Byakagaba, P., & Nabanoga, G. (2019). Development induced displacement; A review of risks faced by communities in developing countries. *Sociology and Anthropology*, *7*(2), 100–110. https://doi.org/10.13189/sa.2019.070205

Agnihotri, A. K., Ohri, A., Gaur, S., Shivam, Das, N., & Mishra, S. (2019). Flood inundation mapping and monitoring using SAR data and its impact on Ramganga River in Ganga basin. *Environmental Monitoring and Assessment*, *191*(12). https://doi.org/10.1007/s10661-019-7903-4

Alves, F. L., Sousa, L. P., Almodovar, M., & Phillips, M. R. (2013). Integrated Coastal Zone Management (ICZM): A review of progress in Portuguese implementation. *Regional Environmental Change, 13*(5), 1031–1042. https://doi.org/10.1007/s10113-012-0398-y

Anusha, N., & Bharathi, B. (2020). Flood detection and flood mapping using multi-temporal synthetic aperture radar and optical data. *Egyptian Journal of Remote Sensing and Space Science, 23*(2), 207–219. https://doi.org/10.1016/j.ejrs.2019.01.001

ASF, A. (n.d.). *How to map regional inundation with Sentinel-1 using Sentinel-1 Toolbox | ASF*. Retrieved February 7, 2021, from https://asf.alaska.edu/how-to/data-recipes/map-regional-inundation-with-sentinel-1-using-s1tbx/

Athikalam, P. T., & Karur Vaideeswaran, A. (2022). Vegetation bioshield for coastal protection in South Asia: Status and way forward. *Journal of Coastal Conservation, 26*(1), 3. https://doi.org/10.1007/s11852-022-00850-x

Attema, E., Bargellini, P., Edwards, P., Levrini, G., Lokas, S., Moeller, L., Rosich-Tell, B., Secchi, P., Torres, R., Davidson, M., & Snoeij, P. (2007). The radar mission for GMES operational land and sea services. *European Space Agency Bulletin, 131*, 10–17.

Baig, M. R. I., Shahfahad, Ahmad, I. A., Tayyab, M., Asgher, M. S., & Rahman, A. (2021). Coastal vulnerability mapping by integrating geospatial techniques and analytical hierarchy process (AHP) along the Vishakhapatnam Coastal Tract, Andhra Pradesh, India. *Journal of the Indian Society of Remote Sensing*, *49*(2), 215–231. https://doi.org/10.1007/s12524-020-01204-6

Behera, M. D., Prakash, J., Paramanik, S., Mudi, S., Dash, J., Varghese, R., Roy, P. S., Abhilash, P. C., Gupta, A. K., & Srivastava, P. K. (2022). Assessment of tropical cyclone amphan affected inundation areas using sentinel-1 satellite data. *Springer, 63*(1), 9–19. https://doi.org/10.1007/S42965-021-00187-W

Behera, R., Kar, A., Das, M. R., & Panda, P. P. (2019). GIS-based vulnerability mapping of the coastal stretch from Puri to Konark in Odisha using analytical hierarchy process. *Natural Hazards, 96*(2), 731–751. https://doi.org/10.1007/s11069-018-03566-0

Bera, R., & Maiti, R. (2021). An assessment of coastal vulnerability using geospatial techniques. *Environmental Earth Sciences, 80*(8), 306. https://doi.org/10.1007/s12665-021-09616-4

Bhagat, N. (2017). Flood frequency analysis using Gumbel's distribution method: A case study of lower Mahi Basin, India. *Journal of Water Resources and Ocean Science, 6*(4), 51. https://doi.org/10.11648/j.wros.20170604.11

Bhaskaran, P. K., Rao, A. D., & Murty, T. (2020). Tropical cyclone–induced storm surges and wind waves in the Bay of Bengal. *Wiley Online Library, 239–294*. https://doi.org/10.1002/9781119359203.CH17

Bhowmik, J., Irfanullah, H. M., & Selim, S. A. (2021). Empirical evidence from Bangladesh of assessing climate hazard-related loss and damage and state of adaptive capacity to address them. *Climate Risk Management, 31*, Article 100273. https://doi.org/10.1016/j.crm.2021.100273

Bhukya, S., Tiwari, M. K., & Patel, G. R. (2023). Assessment of spatiotemporal variation of agricultural and meteorological drought in Gujarat (India) using remote sensing and GIS. *Journal of the Indian Society of Remote Sensing, 51*(7), 1493–1510. https://doi.org/10.1007/s12524-023-01715-y

Bid, S., & Siddique, G. (2019). Identification of seasonal variation of water turbidity using NDTI method in Panchet Hill Dam, India. *Modeling Earth Systems and Environment, 5*(4), 1179–1200. https://doi.org/10.1007/s40808-019-00609-8

Biswas, B., Rahaman, A., & Barman, J. (2023). Comparative Assessment of FR and AHP models for landslide susceptibility mapping for Sikkim, India and preparation of suitable mitigation techniques. *Journal of the Geological Society of India, 99*(6), 791–801. https://doi.org/10.1007/S12594-023-2386-X

Bolliger, J., Schulte, L. A., Burrows, S. N., Sickley, T. A., & Mladenoff, D. J. (2004). Assessing ecological restoration potentials of Wisconsin (U.S.A.) using historical landscape reconstructions. *Restoration Ecology, 12*(1), 124–142. https://doi.org/10.1111/j.1526-100X.2004.00285.x

Borrows, P., & De Bruin, D. (2006). The management of riverine flood risk. *Irrigation and Drainage, 55*(1), S151–S157. https://doi.org/10.1002/ird.243

Chinmayi, H. K., Flynn, K. C., & Ashworth, A. J. (2024). Advancements in remote sensing techniques for earthquake engineering: A review. *Earthquake Research Advances, 100352*. https://doi.org/10.1016/j.eqrea.2024.100352

Clark, J. R. (1997). Coastal zone management for the new century. *Ocean & Coastal Management, 37*(2), 191–216. https://doi.org/10.1016/S0964-5691(97)00052-5

Danumah, J. H., Odai, S. N., Saley, B. M., Szarzynski, J., Thiel, M., Kwaku, A., Kouame, F. K., & Akpa, L. Y. (2016). Flood risk assessment and mapping in Abidjan district using multi-criteria analysis (AHP) model and geoinformation techniques, (cote d'ivoire). *Geoenvironmental Disasters, 3*(1). https://doi.org/10.1186/s40677-016-0044-y

Das, A., & Swain, P. K. (2024). Navigating the sea level rise: Exploring the interplay of climate change, sea level rise, and coastal communities in india. *Environmental Monitoring and Assessment, 196*, 1010. https://doi.org/10.1007/s10661-024-13191-z

Das, B., Dutta, D., & Ray, R. (2025). *Impact of tropical cyclone Yaas on coastal regions of Odisha and West Bengal: An assessment using Sentinel datasets*. Wiley Online Library. https://doi.org/10.1002/GJ.5153

Das, B., Sah, S., Reddy, K. V., Balaji, S. J., Desai, S., Mahajan, G., Singh, P., Saha, S., AR, U., Kumar, P., Sehgal, V. K., Singh, R. N., & Shetkar, D. A. (2025). Assessment of climate change vulnerability of agriculture in coastal districts of India using principal component analysis and entropy-based methods. *Theoretical and Applied Climatology, 156*(7), 409. https://doi.org/10.1007/s00704-025-05623-8

Das, R., & Samanta, G. (2023). Impact of floods and river-bank erosion on the riverine people in Manikchak Block of Malda District, West Bengal. *Environment, Development and Sustainability, 25*(11), 13595–13617. https://doi.org/10.1007/s10668-022-02648-1

Das, S., Roy, K., Mallick, B., & Ghosh, T. (2025). Delineation of flood vulnerable zones using AHP and GIS techniques in climatically vulnerable Sagar Island, Indian Sundarban Delta. *Discover Geoscience, 3*(1), 232. https://doi.org/10.1007/s44288-025-00348-7

Das, T. K., Haldar, S. K., Das Gupta, I., & Sen, S. (2014). River bank erosion induced human displacement and its consequences. *Living Reviews in Landscape Research, 8*. https://doi.org/10.12942/lrlr-2014-3

Das, T., Shahfahad, & Rahman, A. (2024). Assessing tropical cyclone risk for improving mitigation strategies in Coastal Odisha, India. *Environmental Science and Pollution Research, 31*, 53856–53876. https://doi.org/10.1007/s11356-024-33017-2

Das, T., Talukdar, S., Shahfahad, Baig, M. R. I., Hang, H. T., Siddiqui, A. M., & Rahman, A. (2024). Assessing vulnerability to cyclones in coastal Odisha using fuzzy logic integrated AHP: Towards effective risk management. *Springer*, *32*(3), 277–295. https://doi.org/10.1007/S41324-023-00556-8

Das, T., Talukdar, S., Shahfahad, Naikoo, M. W., Ahmed, I. A., Rahman, A., Islam, M. K., & Alam, E. (2024). Integration of fuzzy AHP and explainable AI for effective coastal risk management: A micro-scale risk analysis of tropical cyclones. *Progress in Disaster Science*, *23*. https://doi.org/10.1016/j.pdisas.2024.100357

de Brito, M. M., Evers, M., & Höllermann, B. (2017). Prioritization of flood vulnerability, coping capacity and exposure indicators through the Delphi technique: A case study in Taquari-Antas basin, Brazil. *International Journal of Disaster Risk Reduction, 24*, 119–128. https://doi.org/10.1016/j.ijdrr.2017.05.027

DeVries, B., Huang, C., Armston, J., Huang, W., Jones, J. W., & Lang, M. W. (2020). Rapid and robust monitoring of flood events using Sentinel-1 and Landsat data on the Google Earth Engine. *Remote Sensing of Environment, 240*. https://doi.org/10.1016/j.rse.2020.111664

Dewi, R. S., & Bijker, W. (2020). Dynamics of shoreline changes in the coastal region of Sayung, Indonesia. *Egyptian Journal of Remote Sensing and Space Science, 23*(2), 181–193. https://doi.org/10.1016/j.ejrs.2019.09.001

Dey, J., & Mazumder, S. (2023). Development of an integrated coastal vulnerability index and its application to the low-lying Mandarmani-Dadanpatrabar coastal sector, India. *Natural Hazards, 116*(3), 3243–3273. https://doi.org/10.1007/s11069-022-05805-x

Dhanalakshmi, S., Kankara, R. S., & Chenthamil Selvan, S. (2019). Impact assessment of sea level rise over coastal landforms: A case study of Cuddalore coast, south-east coast of India. *Environmental Earth Sciences, 78*(16), 494. https://doi.org/10.1007/s12665-019-8463-1

Dong, S., Yu, T., Farahmand, H., & Mostafavi, A. (2020). Bayesian modeling of flood control networks for failure cascade characterization and vulnerability assessment. *Computer-Aided Civil and Infrastructure Engineering, 35*(7), 668–684. https://doi.org/10.1111/mice.12527

Elboshy, B., Kanae, S., Gamaleldin, M., Ayad, H., Osaragi, T., & Elbarki, W. (2019). A framework for pluvial flood risk assessment in Alexandria considering the coping capacity. *Environment Systems and Decisions, 39*(1), 77–94. https://doi.org/10.1007/S10669-018-9684-7

Feagin, R. A., Mukherjee, N., Shanker, K., Baird, A. H., Cinner, J., Kerr, A. M., Koedam, N., Sridhar, A., Arthur, R., Jayatissa, L. P., Lo Seen, D., Menon, M., Rodriguez, S., Shamsuddoha, M., & Dahdouh-Guebas, F. (2010). Shelter from the storm? Use and misuse of coastal vegetation bioshields for managing natural disasters. *Conservation Letters, 3*(1), 1–11. https://doi.org/10.1111/j.1755-263X.2009.00087.x

Fernandez, P., Mourato, S., & Moreira, M. (2016). Social vulnerability assessment of flood risk using GIS-based multicriteria decision analysis. A case study of Vila Nova de Gaia. *Geomatics, Natural Hazards and Risk*, *7*(4), 1367–1389. https://doi.org/10.1080/19475705.2015.1052021

Giustarini, L., Chini, M., Hostache, R., Pappenberger, F., & Matgen, P. (2015). Flood hazard mapping combining hydrodynamic modeling and multi annual remote sensing data. *Remote Sensing, 7*(10), 14200–14226. https://doi.org/10.3390/rs71014200

Haldar, D., Nigam, R., Patnaik, C., Dutta, S., & Bhattacharya, B. (2016). Remote sensing-based assessment of impact of Phailin cyclone on rice in Odisha, India. *Springer, 14*(4), 451–461. https://doi.org/10.1007/S10333-015-0514-Y

Halder, B., Bandyopadhyay, J., Inácio, M., Banik, P., & Pereira, P. (2024). Impact analysis of cyclonic effects and landform change in part of Indian Sundarban using remote sensing and Google Earth Engine. *Environment, Development and Sustainability.* https://doi.org/10.1007/s10668-024-05520-6

Halder, S., & Bose, S. (2024). Sustainable flood hazard mapping with GLOF: A Google Earth Engine approach. *Natural Hazards Research*. https://doi.org/10.1016/j.nhres.2024.01.002

Hasan, M., Uddin, M. M., Mahbub-E-Kibria, A. S. M., Das, M. K., & Deb, D. (2024). A GIS-based integrated study assessing the socio-physical vulnerability of the south-central and south-western coastal regions of Bangladesh. *GeoJournal, 89*(4), 163. https://doi.org/10.1007/s10708-024-11185-5

Hastuti, A. W., Nagai, M., & Suniada, K. I. (2022). Coastal vulnerability assessment of Bali province, Indonesia using remote sensing and GIS approaches. *Remote Sensing, 14*(17), 4409. https://doi.org/10.3390/rs14174409

Hoque, M. A. A., Phinn, S., Roelfsema, C., & Childs, I. (2017). Tropical cyclone disaster management using remote sensing and spatial analysis: A review. *International Journal of Disaster Risk Reduction, 22*, 345–354. https://doi.org/10.1016/j.ijdrr.2017.02.008

Hoque, M. A. A., Phinn, S., Roelfsema, C., & Childs, I. (2018). Assessing tropical cyclone risks using geospatial techniques. *Applied Geography, 98*, 22–33. https://doi.org/10.1016/j.apgeog.2018.07.004

Hoque, M. A. A., Pradhan, B., Ahmed, N., & Roy, S. (2019). Tropical cyclone risk assessment using geospatial techniques for the eastern coastal region of Bangladesh. *Science of the Total Environment, 692*, 10–22. https://doi.org/10.1016/j.scitotenv.2019.07.132

Iqbal, S. (2017). Flood and erosion induced population displacements: A socio-economic case study in the Gangetic Riverine Tract at Malda District, West Bengal, India. *Journal of Human Ecology, 30*(3), 201–211. https://doi.org/10.1080/09709274.2010.11906290

Jacinth Jennifer, J., Saravanan, S., & Abijith, D. (2022). Integration of SAR and multi-spectral imagery in flood inundation mapping—A case study on Kerala floods 2018. *ISH Journal of Hydraulic Engineering, 28*(sup1), 480–490. https://doi.org/10.1080/09715010.2020.1791265

Jaman, T., Dharanirajan, K., & Shivaprasad Sharma, S. V. (2021). Assessment of impact of cyclone hazard on social vulnerability of Bhadrak District of Odisha State during Phailin Cyclone in 2013 and Titli Cyclone in 2018 using multi-criteria analysis and geospatial techniques. *International Journal of Disaster Risk Reduction, 53*, Article 101997. https://doi.org/10.1016/j.ijdrr.2020.101997

Jodhani, K. H., Gupta, N., Patel, D., Singh, S. K., Rathnayake, U., & Omar, P. J. (2025). Land transformation and its impact on temperature and vegetation in Gujarat. *Physics and Chemistry of the Earth, Parts A/B/C, 141*, 104190. https://doi.org/10.1016/j.pce.2025.104190

Joe, R. J. J., Pitchaimani, V. S., Gobinath, R., & Shyamala, G. (2025). Urban waterlogging vulnerability assess using SAR imagery and integrated terrain analysis. *Discover Sustainability, 6*(1), 54. https://doi.org/10.1007/s43621-025-00843-4

Joe, R. J. J., Pitchaimani, V. S., Mirra, T. V. N. S., & Karuppannan, S. (2025). Shoreline dynamics and anthropogenic influences on coastal erosion: A multi-temporal analysis for sustainable shoreline management along a southwest coastal district of India. *Environmental and Sustainability Indicators, 27*(March), Article 100744. https://doi.org/10.1016/j.indic.2025.100744

Johnson, L., Trammell, T., Bishop, T., Barth, J., Drzyzga, S., & Jantz, C. (2020). Squeezed from all sides: Urbanization, invasive species, and climate change threaten Riparian forest buffers. *Sustainability, 12*(4), 1448. https://doi.org/10.3390/su12041448

Kaliraj, S., Shunmugapriya, S., Lakshumanan, C., Suresh, D., Arun Prasad, K., & Srinivas, R. (2025). Flood risk zone mapping and future projections for the Thamirabarani river basin, Southern India: Insights from decadal rainfall trends and GIS-based analytical hierarchy process technique. *Natural Hazards, 121*(5), 5327–5361. https://doi.org/10.1007/s11069-024-07007-z

Kamil, E. A., Takaijudin, H., & Hashim, A. M. (2021). Mangroves as coastal bio-shield: A review of mangroves performance in wave attenuation. *Civil Engineering Journal (Iran), 7*(11), 1964–1981. https://doi.org/10.28991/cej-2021-03091772

Kantamaneni, K., Panneer, S., Krishnan, A., Shekhar, S., Bhat, L., R, A. K., & Rice, L. (2022). Appraisal of climate change and cyclone trends in Indian coastal states: A systematic approach towards climate action. *Arabian Journal of Geosciences, 15*(9), 814. https://doi.org/10.1007/s12517-022-10076-8

Kantamaneni, K., Sudha Rani, N. N. V., Rice, L., Sur, K., Thayaparan, M., Kulatunga, U., Rege, R., Yenneti, K., & Campos, L. C. (2019). A systematic review of coastal vulnerability assessment studies along Andhra Pradesh, India: A critical evaluation of data gathering, risk levels and mitigation strategies. *Water, 11*(2), 393. https://doi.org/10.3390/w11020393

Kara, B. C., & Hastaoğlu, K. Ö. (2026). AI-augmented synthetic aperture radar (SAR) for ground deformation: Introducing the adaptive regional AI system (ARAIS). *Environment and Earth Science, 85*, 77. https://doi.org/10.1007/s12665-025-12797-x

Karnata, S., & Kumari, N. (2025). Application of geospatial technologies and AI to detect and analyze the Shoreline change of Visakhapatnam, a coastal district (pp. 265–290). https://doi.org/10.1007/978-3-032-03714-5_12

Kumar, M. K., Dharanirajan, K., & Sabyasachi, S. (2021). Application of Gumbel's distribution method for flood frequency analysis of lower Ganga Basin (Farakka Barrage Station), West Bengal, india. *Disaster Advances, 14*(8), 51–58. https://doi.org/10.25303/148da5121

Kumar, R. (2019). Flood frequency analysis of the Rapti River Basin using Log Pearson Type-III and Gumbel Extreme Value-1 methods. *Journal of the Geological Society of India, 94*(5), 480–484. https://doi.org/10.1007/s12594-019-1344-0

Kumar, R., & Acharya, P. (2016). Flood hazard and risk assessment of 2014 floods in Kashmir Valley: A space-based multisensor approach. *Springer, 84*(1), 437–464. https://doi.org/10.1007/S11069-016-2428-4

Kumar, R., Rani, S., & Maharana, P. (2021). Assessing the impacts of Amphan cyclone over West Bengal, India: A multi-sensor approach. *Springer, 193*(5). https://doi.org/10.1007/S10661-021-09071-5

Kundzewicz, Z. W., Su, B., Wang, Y., Wang, G., Wang, G., Huang, J., & Jiang, T. (2019). Flood risk in a range of spatial perspectives—From global to local scales. *Natural Hazards and Earth System Sciences, 19*(7), 1319–1328. https://doi.org/10.5194/nhess-19-1319-2019

Lai, V. S., Wong, B. K., & Cheung, W. (2002). Group decision making in a multiple criteria environment: A case using the AHP in software selection. *European Journal of Operational Research, 137*(1), 134–144. https://doi.org/10.1016/S0377-2217(01)00084-4

Lal, P., Prakash, A., & Kumar, A. (2020). Google Earth Engine for concurrent flood monitoring in the lower basin of Indo-Gangetic-Brahmaputra plains. *Natural Hazards, 104*(2), 1947–1952. https://doi.org/10.1007/s11069-020-04233-z

Lalrindika, P., Zonunsanga, R., Vanlaltanpuia, V., & Rinawma, P. (2024). A geoinformatics and RUSLE model-based soil erosion assessment in a tropical mountainous area of Chite watershed, Mizoram, India. *Journal of Degraded and Mining Lands Management, 11*(3), 5875–5884. https://doi.org/10.15243/jdmlm.2024.113.5875

Lang, S., Füreder, P., Riedler, B., Wendt, L., Braun, A., Tiede, D., Schoepfer, E., Zeil, P., Spröhnle, K., Kulessa, K., Rogenhofer, E., Bäuerl, M., Öze, A., Schwendemann, G., & Hochschild, V. (2020). Earth observation tools and services to increase the effectiveness of humanitarian assistance. *European Journal of Remote Sensing, 53*(sup2), 67–85. https://doi.org/10.1080/22797254.2019.1684208

Li, L., Chen, Y., Yu, X., Liu, R., & Huang, C. (2015). Sub-pixel flood inundation mapping from multispectral remotely sensed images based on discrete particle swarm optimization. *ISPRS Journal of Photogrammetry and Remote Sensing, 101*, 10–21. https://doi.org/10.1016/j.isprsjprs.2014.11.006

Liu, H., Tatano, H., Kajitani, Y., & Yang, Y. (2022). Analysis of the influencing factors on industrial resilience to flood disasters using a semi-markov recovery model: A case study of the Heavy Rain Event of July 2018 in Japan. *International Journal of Disaster Risk Reduction, 82*, Article 103384. https://doi.org/10.1016/j.ijdrr.2022.103384

Lo, K. F. A., & Gunasiri, C. W. D. (2014). Impact of coastal land use change on shoreline dynamics in Yunlin County, Taiwan. *Environments—MDPI, 1*(2), 124–136. https://doi.org/10.3390/environments1020124

Mahapatra, M., Ramakrishnan, R., & Rajawat, A. S. (2015). Coastal vulnerability assessment of Gujarat coast to sea level rise using GIS techniques: A preliminary study. *Journal of Coastal Conservation, 19*(2), 241–256. https://doi.org/10.1007/s11852-015-0384-x

Mahmood, R., Roy, S., Hossain, M. B., & Zhang, L. (2025). Assessment of coastal vulnerability in the era of sea level rise along the coast of Bangladesh. *Natural Hazards, 121*(20), 24067–24099. https://doi.org/10.1007/s11069-025-07724-z

Majumder, S., Mukherjee, S., Sarkar, A., & Das, S. (2025). Identification of flood hazard-prone coastal zones using remote sensing tools: a study from the Sundarbans Deltaic Region of West Bengal, India. *Proceedings of the Indian National Science Academy*. https://doi.org/10.1007/s43538-025-00681-y

Malakar, K. D. (2020a). Flood frequency analysis using Gumbel's method: A case study of Lower Godavari River Division, India. *Journal of Scientific Computing, 2*(9), 33–51.

Malakar, K. D. (2020b). Rural livelihood and Mangrove degradation: A case study of Namkhana Block, West Bengal, India. *International Journal of Innovative Science and Research Technology, 5*(1), 721–726.

Malakar, K. D., Kuzur, G., Maity, D. K., Yadav, M., & Roy, S. (2022). A socio-ecological study of population, migration, urbanization, and socio-climate variation in Andhra Pradesh and Telangana, India. *Journal of Geography, Environment and Earth Science International, 26*(12), 1–33.

Malakar, K. D., Kumar, M., Anand, S., & Kuzur, G. (2023). Climate crisis and socio-ecological sustainability. In *Climate Change and Socio-Ecological Transformation. Advances in Geographical and Environmental Sciences*. Springer. https://doi.org/10.1007/978-981-99-4390-6_9

Malakar, K. D., Roy, S., & Kumar, M. (2025a). Coastal socio-ecological systems: GIScience in monitoring and management. In *Geospatial technologies in coastal ecologies monitoring and management. Advances in geographic information science*. Springer. https://doi.org/10.1007/978-3-031-92017-2_13

Malakar, K. D., Roy, S., Kumar, M. (2025b). Geospatial technologies in coastal ecologies monitoring and management on the Bengal Coast: Case studies and best practices. In *Geospatial technologies in coastal ecologies monitoring and management. Advances in geographic information science*. Springer. https://doi.org/10.1007/978-3-031-92017-2_16

Malakar, K. D., Roy, S., Kumar, M. (2025c). Monitoring and management of coastal ecologies in India: Community narratives through PGIS. In *Geospatial technologies in coastal ecologies monitoring and management. Advances in geographic information science*. Springer. https://doi.org/10.1007/978-3-031-92017-2_15

Mandal, K. K., Dharanirajan, K., Meena, M. L., Jaman, T., & Rana, S. (2024). Application of geospatial tools in the assessment of flood hazard impact on social vulnerability of Malda district, West Bengal, India. *Natural Hazards Research, 4*(3), 470–485. https://doi.org/10.1016/j.nhres.2023.11.008

Mandal, K. K., Sarkar, T., Alam, S., Dharanirajan, K., & Sharma, S. V. S. (2023). Application of Sentinel-1A SAR data for village level flood inundation mapping in Malda district, West Bengal, India. *Journal of Geographical Studies, 7*(1), 1–13. https://doi.org/10.21523/gcj5.22070101

Mandal, P., Maiti, A., Paul, S., Bhattacharya, S., & Paul, S. (2022). Mapping the multi-hazards risk index for coastal block of Sundarban, India using AHP and machine learning algorithms. *Tropical Cyclone Research and Review, 11*(4), 225–243. https://doi.org/10.1016/j.tcrr.2023.03.001

Mangukiya, N. K., & Sharma, A. (2022). Flood risk mapping for the lower Narmada basin in India: A machine learning and IoT-based framework. *Natural Hazards, 113*(2), 1285–1304. https://doi.org/10.1007/s11069-022-05347-2

Marfai, M. A., Almohammad, H., Dey, S., Susanto, B., & King, L. (2008). Coastal dynamic and shoreline mapping: Multi-sources spatial data analysis in Semarang Indonesia. *Environmental Monitoring and Assessment, 142*(1–3), 297–308. https://doi.org/10.1007/s10661-007-9929-2

McKenna, J., Cooper, A., & O'Hagan, A. M. (2008). Managing by principle: A critical analysis of the European principles of Integrated Coastal Zone Management (ICZM). *Marine Policy, 32*(6), 941–955. https://doi.org/10.1016/j.marpol.2008.02.005

Mishra, M., Kar, D., Debnath, M., Sahu, N., & Goswami, S. (2022). Rapid eco-physical impact assessment of tropical cyclones using geospatial technology: A case from severe cyclonic storms Amphan. *Springer, 110*(3), 2381–2395. https://doi.org/10.1007/S11069-021-05008-W

Misra, A., & Balaji, R. (2015). Decadal changes in the land use/land cover and shoreline along the coastal districts of southern Gujarat, India. *Environmental Monitoring and Assessment, 187*(7). https://doi.org/10.1007/s10661-015-4684-2

Moghimi, A., Singha, C., Fathi, M., Pirasteh, S., Mohammadzadeh, A., Varshosaz, M., Huang, J., & Li, H. (2024). Hybridizing genetic random forest and self-attention based CNN-LSTM algorithms for landslide susceptibility mapping in Darjiling and Kurseong, India. *Quaternary Science Advances, 14*. https://doi.org/10.1016/j.qsa.2024.100187

Mohanty, S., & Kumar, R. (2025). *GIS-based approach for vulnerability assessment of the Puri Coast, Odisha, India* (pp. 99–106). https://doi.org/10.1007/978-981-96-5043-9_9

Mohapatra, M., & Sharma, M. (2019). Cyclone warning services in India during recent years: A review. *Mausam, 70*(4), 635–666. https://doi.org/10.54302/mausam.v70i4.204

Mohapatra, M., Sharma, M., & Pattanaik, D. R. (2025). Evolution of cyclone monitoring and forecasting system in India: A review. *Mausam, 76*(1), 103–124. https://doi.org/10.54302/mausam.v76i1.6940

Moharrami, M., Javanbakht, M., & Attarchi, S. (2021). Automatic flood detection using Sentinel-1 images on the Google Earth Engine. *Environmental Monitoring and Assessment, 193*(5). https://doi.org/10.1007/S10661-021-09037-7

Mondal, D., & Pal, S. (2018). Monitoring dual-season hydrological dynamics of seasonally flooded wetlands in the lower reach of Mayurakshi River, Eastern India. *Geocarto International, 33*(3), 225–239. https://doi.org/10.1080/10106049.2016.1240720

Mondal, I., Ariful Hossain, S., Bandyopadhyay, J., Altuwaijri, H. A., Das, A., Jose, F., Saha, N., & Juliev, M. (2025). Potential effects of projected global sea level rise on Sundarbans mangrove wetland ecosystem: Insights from SLAMM and hybrid machine learning models. *Environment, Development and Sustainability, 27*(6), 14879–14912. https://doi.org/10.1007/s10668-025-06193-5

Mubialiwo, A., Abebe, A., Kawo, N. S., Ekolu, J., Nadarajah, S., & Onyutha, C. (2022). Hydrodynamic modelling of floods and estimating socio-economic impacts of floods in Ugandan River Malaba Sub-catchment. *Earth Systems and Environment*. https://doi.org/10.1007/s41748-021-00283-w

Mudi, S., Jaya, P., & Das, P. (2022). Flood hazard mapping in Assam using Sentinel-1 SAR data. In *Advances in geographic information science* (pp. 459–473). https://doi.org/10.1007/978-3-030-75197-5_20

Mukherjee, R., & Yadav, R. B. (2019). Flood frequency analysis of Ramganga River Basin in Western Gangetic Plain, India. National Geographical Journal of India, An International Peer-reviewed and refereed. *Journal of In, 65*(3). https://www.researchgate.net/publication/344507753

Murali R. M., Ankita M., & Vethamony P. (2018). A new insight to vulnerability of Central Odisha coast, India using analytical hierarchical process (AHP) based approach. *Journal of Coastal Conservation, 22*(4), 799–819. https://doi.org/10.1007/s11852-018-0610-4

Naren, A., & Maity, R. (2024). Coastal vulnerability considering combined effect of tropical cyclones and future sea level rise. *Journal of Coastal Conservation, 28*(3), 56. https://doi.org/10.1007/s11852-024-01055-0

Narendr, A., Vinay, S., Aithal, B. H., & Das, S. (2022). Multi-dimensional parametric coastal flood risk assessment at a regional scale using GIS. *Environment, Development and Sustainability, 24*(7), 9569–9597. https://doi.org/10.1007/s10668-021-01839-6

Natarajan, L., Sivagnanam, N., Usha, T., Chokkalingam, L., Sundar, S., Gowrappan, M., & Roy, P. D. (2021). Shoreline changes over last five decades and predictions for 2030 and 2040: A case study from Cuddalore, southeast coast of India. *Springer, 14*(3), 1315–1325. https://doi.org/10.1007/S12145-021-00668-5

Naz, A., & Chowdhury, A. (2022). Eco-engineering and mangrove restoration methods to stabilize earthen embankments and establishing bio-shield against natural disasters: A case study from Sundarban Ramsar Wetland, India (pp. 183–198). https://doi.org/10.1007/978-981-19-2738-6_10

Nghia, B. P. Q., Pal, I., Chollacoop, N., & Mukhopadhyay, A. (2022). Applying Google earth engine for flood mapping and monitoring in the downstream provinces of Mekong river. *Progress in Disaster Science, 14*. https://doi.org/10.1016/j.pdisas.2022.100235

Nguyen, H. H., McAlpine, C., Pullar, D., Leisz, S. J., & Galina, G. (2015). Drivers of coastal shoreline change: Case study of Hon Dat Coast, Kien

Giang, Vietnam. *Environmental Management, 55*(5), 1093–1108. https://doi.org/10.1007/s00267-015-0455-7

Pandey, A. C., Kaushik, K., & Parida, B. R. (2022). Google Earth Engine for large-scale flood mapping using SAR data and impact assessment on agriculture and population of Ganga-Brahmaputra Basin. *Sustainability (Switzerland), 14*(7). https://doi.org/10.3390/su14074210

Parida, B. R., Behera, S. N., Oinam, B., Patel, N. R., & Sahoo, R. N. (2018). Investigating the effects of episodic Super-cyclone 1999 and Phailin 2013 on hydro-meteorological parameters and agriculture: An application of remote sensing. *Remote Sensing Applications: Society and Environment, 10*, 128–137. https://doi.org/10.1016/j.rsase.2018.03.010

Parida, N. R., Parida, S., Samal, P., & Abisha, P. (2025). Geospatial and machine learning-based coastal vulnerability assessment of Puri, Odisha. *Journal of Coastal Conservation, 29*(6), 67. https://doi.org/10.1007/s11852-025-01149-3

Parthasarathy, K. S. S., & Deka, P. C. (2021). Remote sensing and GIS application in assessment of coastal vulnerability and shoreline changes: A review. *ISH Journal of Hydraulic Engineering, 27*(sup1), 588–600. https://doi.org/10.1080/09715010.2019.1603086

Passeri, D. L., Hagen, S. C., Bilskie, M. V., & Medeiros, S. C. (2015). On the significance of incorporating shoreline changes for evaluating coastal hydrodynamics under sea level rise scenarios. *Natural Hazards, 75*(2), 1599–1617. https://doi.org/10.1007/s11069-014-1386-y

Patel, K., Jain, R., Kalubarme, M. H., & Bhatt, T. (2024). Coastal erosion monitoring using multi-temporal remote sensing and sea surface temperature data in coastal districts of Gujarat state, India. *Geology, Ecology, and Landscapes, 8*(2), 194–207. https://doi.org/10.1080/24749508.2022.2127648

Patel, K., Jain, R., Patel, A. N., & Kalubarme, M. H. (2021). Shoreline change monitoring for coastal zone management using multi-temporal Landsat data in Mahi River estuary, Gujarat State. *Applied Geomatics, 13*(3), 333–347. https://doi.org/10.1007/s12518-021-00353-8

Patel, P., Mehta, D., & Sharma, N. (2023). Assessment of groundwater vulnerability using the GIS approach-based GOD method in Surat district of Gujarat state, India. *Water Practice and Technology, 18*(2), 285–294. https://doi.org/10.2166/wpt.2023.004

Pazhanivelan, S., Ragunath, K. P., Sudarmanian, N. S., Satheesh, S., & Sneka, K. (2024). Enhancing flood area mapping accuracy using advanced SAR

data processing. *International Journal of Environment and Climate Change, 14*(12), 783–799. https://doi.org/10.9734/ijecc/2024/v14i124662
Pednekar, G., & Siva Raju, S. (2020). Sea level rise and its socio-economic Impacts: A case study in Mumbai, India. In *Extreme weather events and human health* (pp. 155–169). Springer International Publishing. https://doi.org/10.1007/978-3-030-23773-8_11
Popa, M. C., Peptenatu, D., Draghici, C. C., & Diaconu, D. C. (2019). Flood hazard mapping using the flood and Flash-Flood Potential Index in the Buzau River catchment, Romania. *Water (Switzerland), 11*(10). https://doi.org/10.3390/w11102116
Prasad, R., & Singh, B. V. R. (2024). *Analysing land use and land cover change for sustainable coastal management: A case study of southern Coastal Gujarat, India* (pp. 67–85). https://doi.org/10.1007/978-3-031-67303-0_5
Raaijmakers, R., Krywkow, J., & van der Veen, A. (2008). Flood risk perceptions and spatial multi-criteria analysis: An exploratory research for hazard mitigation. *Natural Hazards, 46*(3), 307–322. https://doi.org/10.1007/s11069-007-9189-z
Rai, P. K., Chaubey, P. K., Mohan, K., & Singh, P. (2017). Geoinformatics for assessing the inferences of quantitative drainage morphometry of the Narmada Basin in India. *Applied Geomatics, 9*(3), 167–189. https://doi.org/10.1007/s12518-017-0191-1
Rajakumari, S., Minnu, A., & Sarunjith, K. J. (2022). Determination of vulnerable zones along Brahmapur coast, Odisha using AHP and GIS with validation against multiple cyclones. *Springer, 194*(4). https://doi.org/10.1007/S10661-022-09886-W
Ramkar, P., & Yadav, S. M. (2021). Flood risk index in data-scarce river basins using the AHP and GIS approach. *Natural Hazards, 109*(1), 1119–1140. https://doi.org/10.1007/s11069-021-04871-x
Rani, A., Gupta, S. K., Singh, S. K., Meraj, G., Kumar, P., Kanga, S., Đurin, B., & Dogančić, D. (2023). Predicting future land use utilizing economic and land surface parameters with ANN and Markov Chain models. *Earth, 4*(3), 728–751. https://doi.org/10.3390/earth4030039
Ranjan, A., Mandal, K. K., & Mallick, S. (2025). The land use and land cover changes, 1994–2024: Implications for livelihood options and employment opportunities in Dhanbad, India. *Spatial Information Research, 33*(1), 7–9. https://doi.org/10.1007/s41324-025-00605-4

Rao, A. D., Upadhaya, P., Ali, H., Pandey, S., & Warrier, V. (2020). Coastal inundation due to tropical cyclones along the east coast of India: An influence of climate change impact. *Springer, 101*(1), 39–57. https://doi.org/10.1007/S11069-020-03861-9

Raspini, F., Bianchini, S., Ciampalini, A., Del Soldato, M., Solari, L., Novali, F., Del Conte, S., Rucci, A., Ferretti, A., & Casagli, N. (2018). Continuous, semi-automatic monitoring of ground deformation using Sentinel-1 satellites. *Scientific Reports, 8*(1), 7253. https://doi.org/10.1038/s41598-018-25369-w

Rautela, K. S., Kumar, D., Gandhi, B. G. B. G. R., Kumar, A., & Dubey, A. K. (2023). Flood vulnerability assessment across Alaknanda River Basin using GIS-based combined analysis of geomorphometric approach and MCDM-AHP. *Journal of the Geological Society of India, 99*(11), 1604–1615. https://doi.org/10.1007/s12594-023-2512-9

Rehman, S., Sahana, M., Kumar, P., Ahmed, R., & Sajjad, H. (2021). Assessing hazards induced vulnerability in coastal districts of India using site-specific indicators: An integrated approach. *GeoJournal, 86*(5), 2245–2266. https://doi.org/10.1007/s10708-020-10187-3

Roy, P., Pal, S. C., Chakrabortty, R., Chowdhuri, I., Saha, A., & Shit, M. (2023). Effects of climate change and sea-level rise on coastal habitat: Vulnerability assessment, adaptation strategies and policy recommendations. *Journal of Environmental Management, 330*, Article 117187. https://doi.org/10.1016/j.jenvman.2022.117187

Saha, A. K., & Agrawal, S. (2020). Mapping and assessment of flood risk in Prayagraj district, India: a GIS and remote sensing study. *Nanotechnology for Environmental Engineering, 5*(2). https://doi.org/10.1007/s41204-020-00073-1

Sahana, M., Hong, H., Ahmed, R., Patel, P. P., Bhakat, P., & Sajjad, H. (2019). Assessing coastal island vulnerability in the Sundarban Biosphere Reserve, India, using geospatial technology. *Environmental Earth Sciences, 78*(10), 304. https://doi.org/10.1007/s12665-019-8293-1

Sahoo, B., & Bhaskaran, P. K. (2018). Multi-hazard risk assessment of coastal vulnerability from tropical cyclones—A GIS based approach for the Odisha coast. *Journal of Environmental Management, 206*, 1166–1178. https://doi.org/10.1016/j.jenvman.2017.10.075

Sahu, R. T., & Mehta, D. J. (2024). Impact of coastal inundation due to rise in sea level: A case study of Surat City, India. *Water Practice & Technology, 19*(5), 1753–1768. https://doi.org/10.2166/wpt.2024.116

Sai Priya, N. T., & Banerjee, A. (2024). *Urban growth patterns and spatial sustainability in coastal cities: The case of Visakhapatnam in India* (pp. 273–293. https://doi.org/10.1007/978-981-97-8370-0_12

Saini, I., Chandramouli, P., Samaddar, A., & Ghosh, S. (2013). Quantifying tropical cyclone cloud cover using Envisat retrievals–An example of a recent severe tropical cyclone', Thane. *Taylor & Francis, 34*(14), 4933–4950. https://doi.org/10.1080/01431161.2013.783249

Saleem Khan, A., Sabuj Kumar, M., Sudhir Chella, R., & Devdyuti, B. (2020). *Chennai city and coastal hazards: Addressing community-based adaptation through the lens of climate change and sea-level rise (CBACCS)* (pp. 777–798). https://doi.org/10.1007/978-3-030-37425-9_39

Samanta, R. K., Bhunia, G. S., Shit, P. K., & Pourghasemi, H. R. (2018). Flood susceptibility mapping using geospatial frequency ratio technique: A case study of Subarnarekha River Basin, India. *Modeling Earth Systems and Environment, 4*(1), 395–408. https://doi.org/10.1007/s40808-018-0427-z

Sanyal, J., & Lu, X. X. (2005). Remote sensing and GIS-based flood vulnerability assessment of human settlements: A case study of Gangetic West Bengal, India. *Hydrological Processes, 19*(18), 3699–3716. https://doi.org/10.1002/hyp.5852

Sar, N., Chatterjee, S., & Das Adhikari, M. (2015). Integrated remote sensing and GIS based spatial modelling through analytical hierarchy process (AHP) for water logging hazard, vulnerability and risk assessment in. *Springer, 1*(4). https://doi.org/10.1007/S40808-015-0039-9

Saravanan, S., Abijith, D., Kulithalai Shiyam Sundar, P., Reddy, N. M., Almohamad, H., Al Dughairi, A. A., Al-Mutiry, M., & Abdo, H. G. (2023). Multi-criterion analysis of cyclone risk along the coast of Tamil Nadu, India—A geospatial approach. *ISPRS International Journal of Geo-Information, 12*(8). https://doi.org/10.3390/ijgi12080341

Sarker, S., & Adnan, M. S. G. (2024). Evaluating multi-hazard risk associated with tropical cyclones using the fuzzy analytic hierarchy process model. *Natural Hazards Research, 4*(1), 97–109. https://doi.org/10.1016/j.nhres.2023.11.007

Sellaperumal, P., Kaliaperumal, R., Dhanaraju, M., Sudarmanian, N. S., Shanmugapriya, P., Satheesh, S., Singaram, M., Sivamurugan, A. P., Marimuthu, R., Rangasamy, B., & Tamilmounika, R. (2025). Time series analysis of Sentinel 1 A SAR data to retrieve annual rice area maps and long-term dynamics of start of season. *Scientific Reports, 15*(1), 1–14. https://doi.org/10.1038/s41598-025-91655-z

Shah, H., & Ramesh, R. (2022). Development-aligned mangrove conservation strategy for enhanced blue economy: A successful model from Gujarat, India. *Estuarine, Coastal and Shelf Science, 274*, Article 107929. https://doi.org/10.1016/j.ecss.2022.107929

Shen, X., Anagnostou, E. N., Allen, G. H., Robert Brakenridge, G., & Kettner, A. J. (2019). Near-real-time non-obstructed flood inundation mapping using synthetic aperture radar. *Remote Sensing of Environment*, *221*(August 2018), 302–315. https://doi.org/10.1016/j.rse.2018.11.008

Sivanpillai, R., Jacobs, K. M., Mattilio, C. M., & Piskorski, E. V. (2020). Rapid flood inundation mapping by differencing water indices from pre- and post-flood Landsat images. *Frontiers of Earth Science*. https://doi.org/10.1007/s11707-020-0818-0

Sorna, U. N., Saif, S. Bin, & Islam, F. A. S. (2025). Modeling relative sea-level rise through geodetic-hydrodynamic fusion: Quantifying multi-sectoral risks in the Bengal Delta. *International Journal of Applied and Natural Sciences*, *3*(3), 29–66. https://doi.org/10.61424/ijans.v3i3.541

Suparyanto dan Rosad (2015. (2020, August 6). 済無No Title No Title No Title. *Suparyanto Dan Rosad (2015*, *5*(3), 248–253. https://floodlist.com/asia/india-floods-jalpaiguri-west-bengal-october-2022

Swarna Latha, P., & Nageswara Rao, K. (2011). Use of geo-spatial information and technologies for analyzing morphological changes in the Greater Visakhapatnam coastal region, east coast of India. *Geocarto International, 26*(5), 391–411. https://doi.org/10.1080/10106049.2011.586733

Thenmozhi, M., Sujatha, M., Kavitha, M., Senthilraja, S., Babu, M., & Priya, V. (2024). Assessment of cyclone risk and case study of Gaja cyclone using GIS techniques and machine learning algorithms in coastal zone of Tamil Nadu, India. *Environmental Research, 246*. https://doi.org/10.1016/j.envres.2023.118089

Theocharidis, C., Prodromou, M., Doukanari, M., Kalogirou, E., Eliades, M., Kontoes, C., Hadjimitsis, D., & Neocleous, K. (2025). Integrated Coastal Vulnerability Index (ICVI) Assessment of Protaras Coast in Cyprus: Balancing tourism and coastal risks. *Geographies, 5*(1), 12. https://doi.org/10.3390/geographies5010012

Tiwari, V., Kumar, V., Matin, M. A., Thapa, A., Ellenburg, W. L., Gupta, N., & Thapa, S. (2020). Flood inundation mapping-Kerala 2018; Harnessing the power of SAR, automatic threshold detection method and Google Earth Engine. *PLoS ONE*, *15*(8 August). https://doi.org/10.1371/journal.pone.0237324

Tripathi, G., Pandey, A. C., Parida, B. R., & Kumar, A. (2020). Flood inundation mapping and impact assessment using multi-temporal optical and SAR satellite data: A case study of 2017 flood in Darbhanga District, Bihar, India. *Water Resources Management, 34*(6), 1871–1892. https://doi.org/10.1007/s11269-020-02534-3

Tripathy, P., & Malladi, T. (2022). Global flood mapper: A novel Google Earth Engine application for rapid flood mapping using Sentinel-1 SAR. *Natural Hazards, 114*(2), 1341–1363. https://doi.org/10.1007/s11069-022-05428-2

Vanama, V. S. K., Mandal, D., & Rao, Y. S. (2020). GEE4FLOOD: Rapid mapping of flood areas using temporal Sentinel-1 SAR images with Google Earth Engine cloud platform. *Journal of Applied Remote Sensing, 14*(03), 1. https://doi.org/10.1117/1.jrs.14.034505

Vanama, V. S. K., Musthafa, M., Khati, U., Gowtham, R., Singh, G., & Rao, Y. S. (2021). Inundation mapping of Kerala flood event in 2018 using ALOS-2 and temporal Sentinel-1 SAR images. *Current Science, 120*(5), 915–925. https://doi.org/10.18520/cs/v120/i5/915-925

Varikkodan, H., Balaji, S., Arjun, S., & Mandal, K. K. (2023). Tropical cyclone risk assessment of Port Blair, Andaman Islands, India by using numerical modelling and geospatial techniques. *Journal of Earth System Science, 132*(1). https://doi.org/10.1007/s12040-023-02055-8

Vasudevan, P., Sherly, M. A., & Chaudhuri, R. R. (2022). *Hydrological Risk Assessment of the coastal megacity Chennai using DRASTIC method and SWAT* (pp. 1–15). https://doi.org/10.1007/978-981-19-0304-5_1

Vekaria, D., Chander, S., Singh, R. P., & Dixit, S. (2022). A change detection approach to flood inundation mapping using multi-temporal Sentinel-1 SAR images, the Brahmaputra River, Assam (India): 2015–2020. *Journal of Earth System Science, 132*(1), 3. https://doi.org/10.1007/s12040-022-02020-x

Vincent, S. G. T., & Owens, K. A. (2021). Coastal wetlands of India: Threats and solutions. *Wetlands Ecology and Management, 29*(5), 633–639. https://doi.org/10.1007/s11273-021-09824-6

Yacouba, D., Guangdao, H., & Xingping, W. (2010). Applications of remote sensing in land use/land cover change detection in Puer and Simao counties, Yunnan Province. *Report and Opinion, 2*(9), 7–16.

Yang, Q., Liu, G., Li, H., Santagata, R., & Yang, Z. (2024). Understanding ecological restoration potential: The role of water resources and slope gradient limits. *Science of the Total Environment, 912*, Article 169001. https://doi.org/10.1016/j.scitotenv.2023.169001

14

Coastal-Climate Nexus and Participatory Interactions in the Indian Delta: Implications for Disaster Risk Reduction and Socio-ecological Sustainability

This chapter brings together a set of closely linked case studies that examine climate risk, social vulnerability, and community responses along the Indian delta coast, with a focus on coastal West Bengal. The purpose of the chapter is to show how climate change is experienced in everyday life across different coastal settings and how participatory approaches, supported by geospatial analysis, can contribute to disaster risk reduction and socio-ecological sustainability. Climate risk is treated here as a lived reality that shapes livelihoods, health, and relationships with the coastal environment. The case studies reflect a wide range of coastal conditions and social contexts. They include household-level livelihood and health stress in rural delta communities and tourism-based locations such as Digha. Other studies focus on mapping social vulnerability and traditional fishing grounds, as well as governance challenges related to interactions between mangrove ecosystems and coastal communities in the Sundarbans. The chapter also examines climate-induced migration from deltaic islands such as Sagar Island, patterns of shoreline change and coastal hazards, and the role of indigenous knowledge, community narratives, and climate justice in shaping resilience and adaptation. The casebook format allows these studies to be read

K. D. Malakar et al., *Coastal Ecologies at Risk*, Palgrave Studies in Environmental Sustainability, https://doi.org/10.1007/978-3-032-26081-9_14

together while respecting their local specificity. It highlights the value of combining community knowledge with spatial tools to understand how risks develop, how they are perceived, and how they are managed in practice. The chapter seeks to support more inclusive and grounded approaches to disaster risk reduction that reflect the realities of coastal life and provide lessons that are relevant for other climate-exposed delta regions.

Key Points of the Chapter

- Describes how climate risk affects livelihoods, health, and daily life in the Indian delta coast.
- Explains spatial and social differences in vulnerability across coastal communities of West Bengal.
- Examines the role of participatory approaches and local knowledge in understanding coastal risks.
- Highlights the value of geospatial analysis for mapping hazards, shoreline change, and vulnerability.
- Argues for innovative and inclusive risk reduction to support socio-ecological sustainability in delta regions.
- A collection of casebooks: Implications for Disaster Risk Reduction and Socio-ecological Sustainability;

  Theme 1: Livelihoods, Health, and Household-Level Climate Stress
  Theme 2: Social Vulnerability and Participatory Spatial Knowledge
  Theme 3: Conflict, Migration, and Participatory Governance
  Theme 4: Hazards, Adaptation, and Geospatial Modeling
  Theme 5: Community Narratives, Indigenous Knowledge, and Resilience
  Theme 6: Justice, Climate Gaps, and Participatory GIS.

14.1 Introduction

Coastal and deltaic regions in the Global South face growing pressure from climate change, environmental degradation, and recurring disasters. These regions are shaped by strong links between natural processes and human activity, which make them especially sensitive to climate variability and extreme events. Along the eastern coast of India, the Indian delta coast stands out as a highly dynamic and fragile socio-ecological region. Seasonal monsoon changes, cyclones, sea-level rise, shoreline erosion, salinity intrusion, and ecological transformation directly affect everyday life. Large populations depend on coastal and marine environments for fishing, agriculture, tourism, and related livelihoods, while health, culture, and social identity remain closely tied to these landscapes. In such conditions, disaster risk reduction and socio-ecological sustainability require approaches that are grounded in local contexts and informed by both spatial evidence and lived experience (Griggs & Reguero, 2021; Petzold & Scheffran, 2024; Roy et al., 2023). This section introduces a collection of casebooks that examine the coastal-climate nexus and participatory interactions across the Indian delta coast. The casebooks are designed to go beyond descriptive accounts. Each one serves as an analytical study that shows how climate stress, environmental change, and social vulnerability are experienced and addressed at different levels, including households, communities, ecosystems, and local institutions. By grouping diverse empirical studies within a shared framework, this chapter aims to connect geospatial analysis with participatory approaches and to demonstrate their relevance for disaster risk reduction and long-term socio-ecological sustainability.

Many existing studies of coastal risk rely on hazard maps, regional indicators, or planning tools developed at broader scales. These methods provide essential baseline information, but they often fail to capture how climate impacts differ across places and social groups. Changes in livelihoods, health conditions, migration patterns, and social relations are uneven and deeply shaped by local circumstances. This is particularly true in coastal regions where communities already face economic and environmental constraints (Laino et al., 2024; Viavattene et al., 2018; Zhang & Convertino, 2026). The casebook approach adopted in this chapter responds to this limitation by focusing on place-based evidence, community experience, and participatory forms of spatial knowledge, while maintaining methodological clarity and analytical consistency. Each casebook included in this section has three key objectives. The first is to document how climate-related risks appear in specific local settings, such as through livelihood disruption, declining ecosystem services, health stress, or displacement. The second is to show how participatory interactions, including participatory GIS, community mapping, field narratives, and indigenous knowledge, add depth to spatial analysis. The third objective is to draw out implications for disaster risk reduction and socio-ecological sustainability, so that research findings can inform planning practices, governance processes, and policy decisions.

The Indian delta coast, particularly in coastal West Bengal, offers a rich setting for examining climate-related risks and responses. The region includes ecologically sensitive environments such as the Sundarbans, rapidly changing coastal tourism areas like Digha, long-established marine fishing settlements, and low-lying deltaic islands exposed to flooding and erosion. These landscapes have been shaped by natural forces as well as long-term human intervention. As a result, risks emerge through overlapping environmental, social, and economic processes that cannot be fully understood through a single disciplinary approach. Climate change adds further stress by altering fish availability, increasing exposure to disease, degrading coastal ecosystems, and weakening livelihood security. At the same time, coastal communities respond through household-level coping practices, collective action, indigenous technical knowledge, and engagement with local institutions. Any meaningful discussion of disaster risk and sustainability must therefore account for both vulnerability and adaptive capacity. The casebooks presented in this chapter are structured to capture this balance through approaches that are spatially informed and socially grounded (Janmaimool et al., 2025; Malakar et al., 2020; Maulu et al., 2021).

To maintain clarity while accommodating diverse case material, the casebooks are organized into six themes. Each theme addresses a specific dimension of the coastal-climate nexus and shows how participatory interactions contribute to disaster risk reduction and sustainability.

Theme 1 focuses on livelihoods, health, and household-level climate stress. The casebooks under this theme examine rural coastal settlements and tourism-dependent areas. They show how changes in fisheries, temperature, salinity, and extreme events affect everyday health and livelihood security. These studies highlight that disaster risk develops over time and is closely linked to daily living conditions and access to resources. ***Theme 2*** addresses social vulnerability and participatory spatial knowledge. These casebooks combine GIS-based vulnerability assessments with participatory mapping of fishing grounds and community spaces. The findings demonstrate how vulnerability varies across locations and social groups, and how local spatial knowledge improves the relevance of risk mapping. And ***Theme 3*** focuses on conflict, migration, and participatory governance. Casebooks in this theme examine mangrove-human interactions and climate-induced migration from deltaic islands. They show how environmental change intensifies pressure on resources and exposes gaps in governance, while participatory approaches support locally appropriate responses.

Theme 4 examines hazards, adaptation, and geospatial analysis. These casebooks link local adaptation practices with spatial analysis of climate variability, shoreline change, and coastal vulnerability. The emphasis is on producing risk information that is technically sound and meaningful for local use. ***Theme 5*** centers on community narratives, indigenous knowledge, and resilience. Through field-based accounts and community assessments, these casebooks document how people interpret risk, remember past disasters, and apply indigenous knowledge in adaptation practices. This perspective highlights community agency rather than viewing vulnerability only as a limitation. And ***Theme 6*** addresses climate justice, adaptive gaps, and participatory GIS. These casebooks explore how social inequalities influence exposure to risk and access to adaptation support. Participatory GIS is used to bring marginal voices into decision-making and to support more equitable disaster risk reduction.

The casebooks presented in this chapter show that disaster risk reduction in coastal and deltaic regions cannot depend only on technical solutions or hazard-focused planning. Responses must recognize the close links between climate processes, ecosystems, and social systems. By combining geospatial analysis with participatory interactions, the casebooks provide grounded insights into how risks are shaped, understood, and managed at different scales. The chapter also highlights practical and policy-relevant lessons, including the importance of participatory risk mapping, livelihood-sensitive adaptation planning, conflict-aware conservation strategies, and governance approaches that address issues of justice and inclusion. In this way, the casebook collection contributes to wider discussions on socio-ecological sustainability and offers lessons that are relevant beyond the Indian delta coast. The remainder of Sect. 14.2 follows the six themes outlined above. Each casebook is presented as a distinct study while remaining connected to the wider framework of the chapter. Therefore, the demonstrate the value of case-based, participatory, and spatially informed research for strengthening disaster risk reduction and socio-ecological sustainability in coastal and deltaic regions.

14.2 A Collection of Casebooks on Coastal-Climate Nexus and Participatory Interactions in the Indian Delta Coast: Implications for Disaster Risk Reduction and Socio-ecological Sustainability

14.2.1 Theme 1: Livelihoods, Health, and Household-Level Climate Stress

Casebook 1

Rural Livelihoods, Health Vulnerabilities, and Climate Stress in Coastal Communities: A Case Study of West Bengal Delta.

(a) **Purpose**

The purpose of this casebook is to examine rural livelihoods, health vulnerabilities, and climate-related stress in a coastal area of the West Bengal delta. Research on rural lives, livelihoods, and poverty is important for explaining the social, economic, and environmental conditions that shape the well-being of rural populations, especially in regions that are highly exposed to climate variability (Malakar, 2020; Pingali et al.

2019; Jawaweera, 2010). Such research helps inform policies and development efforts that aim to promote growth and sustainability. This case study focuses on the rural area of Namkhana in West Bengal (Fig. 14.1), where livelihood activities, health conditions, and climatic factors are closely connected. Most households in this area depend on climate-sensitive resources for their daily living, which increases their exposure to environmental stress. The casebook examines how rural households experience climate-related pressures in everyday life and how these pressures influence livelihood security and health outcomes. It also considers how environmental conditions interact with social and economic factors to shape patterns of risk at both household and community levels. Primary data for this casebook were collected between 2023 and 2024 (N: 480). The analysis applies standard statistical methods commonly used in social science research to examine trends and relationships within the data. By documenting local experiences in a rural coastal setting, this casebook contributes to a clearer understanding of vulnerability and adaptation in the West Bengal delta and offers insights that are relevant to risk reduction and socio-ecological sustainability in other climate-exposed coastal regions.

(b) **Results and Discussion**

Rural Lives and Livelihoods

The study shows that agriculture, fishing, and forestry are the main sources of livelihood for households in rural Namkhana (Fig. 14.2). Agriculture supports about 31.1 percent of households and remains a key part of rural life. Paddy cultivation is the dominant agricultural activity, along with the production of vegetables and fruits. Farmers in the region face several challenges, including limited access to modern farming methods and equipment, inadequate irrigation facilities, and frequent exposure to natural hazards such as cyclones and floods. These constraints affect crop productivity and increase income insecurity.

Fishing is another major livelihood activity, involving approximately 38.09 percent of households. The coastal and estuarine environment

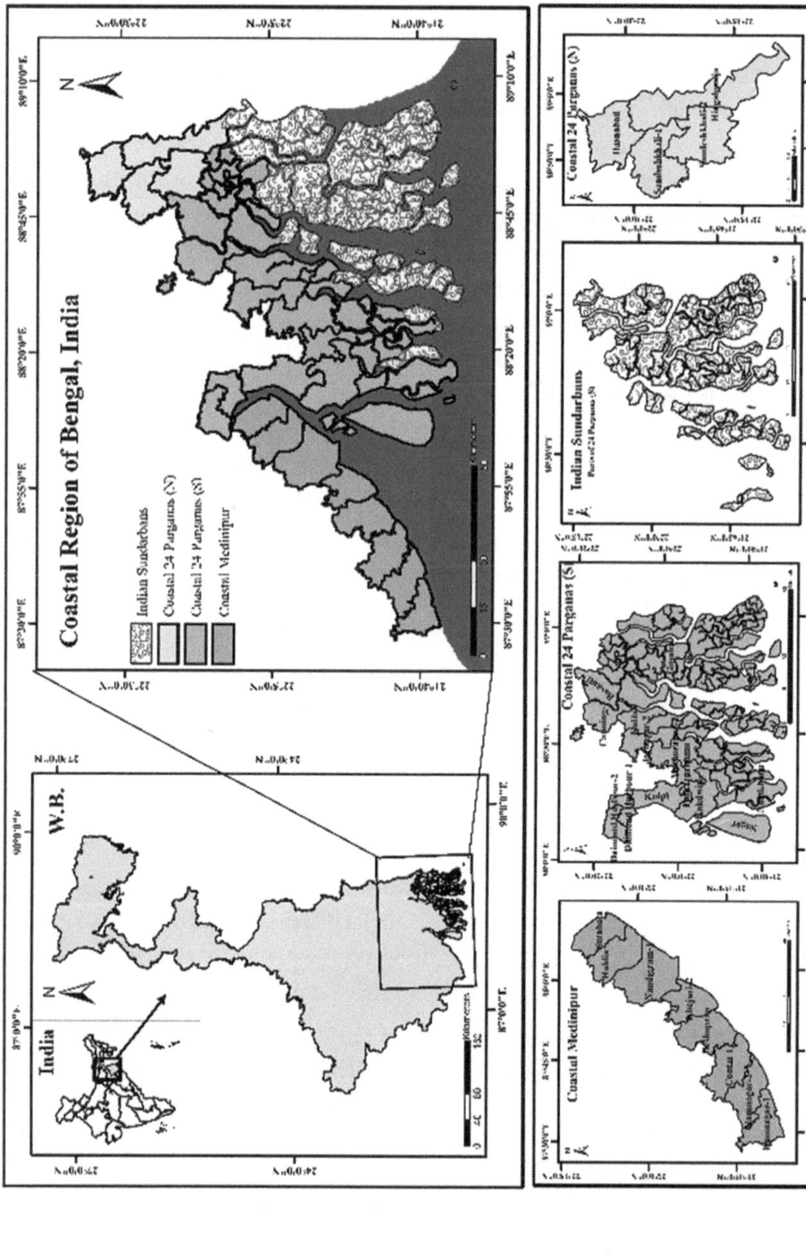

Fig. 14.1 Mapping study area: Coastal West Bengal (Indian delta coast) (prepared by the authors)

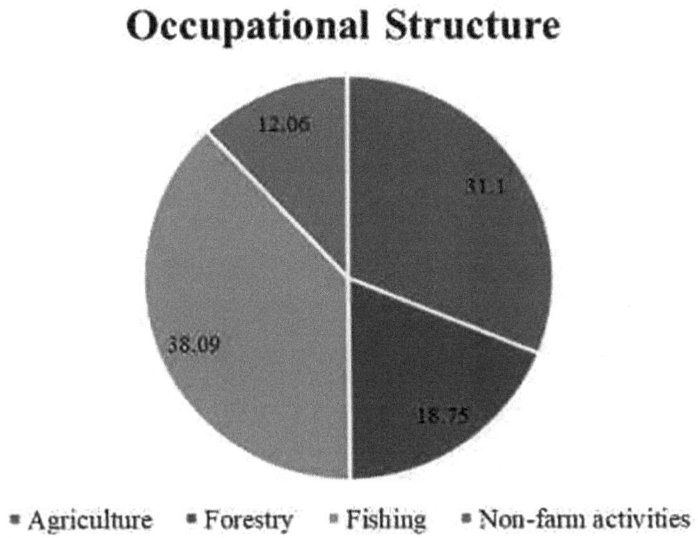

Fig. 14.2 Occupational structure of rural households in Namkhana (prepared by the authors)

provides a wide range of fishery resources, and fishing is largely carried out using traditional methods such as nets and traps. However, fishers are increasingly affected by declining fish stocks. Overfishing and destructive practices have reduced availability, while competition from large trawlers operating from outside the region has further limited the earnings of small-scale fishers.

Forestry provides livelihoods for about 18.75 percent of households and is closely linked to the mangrove ecosystem of the Sundarbans. This ecosystem plays an important role in coastal protection, carbon storage, and biodiversity conservation. Forest resources are managed through the involvement of both government agencies and local communities and are used for timber, fuelwood, and non-timber forest products. In addition to these primary activities, around 12.06 percent of households depend on non-farm sources of income, including handicrafts, small-scale trading, and wage labor.

Livelihood patterns in rural Namkhana are strongly shaped by local environmental conditions and natural resource availability. Persistent poverty, limited access to essential services such as healthcare and education, and environmental degradation continue to affect household well-being. Improving living conditions in the region requires development approaches that are sustainable and inclusive and that respond to the local social, economic, and environmental context (Fig. 14.2).

Health Vulnerabilities Due to Climate Change

The investigation also presents community views on changes in health problems linked to climate change in Namkhana and points to a general rise in health stress across different disease categories (Fig. 14.3). A majority of respondents, 59.17 percent, reported an increase in infectious diseases. At the same time, 29.38 percent were uncertain, while only a small proportion reported no change. No respondents reported a decrease, which suggests that infectious diseases are widely seen as becoming more common under changing climate conditions.

Reports of vector-borne diseases also indicate an increasing trend, with 41.67 percent of respondents noting a rise. However, uncertainty remains relatively high, as 23.75 percent were unsure and 19.79 percent

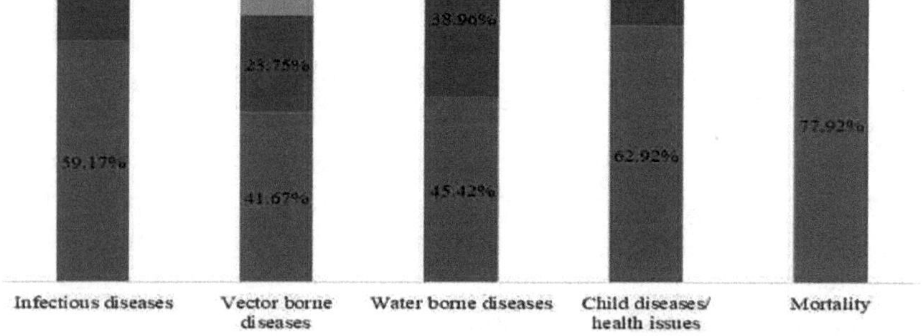

Fig. 14.3 Health problems/vulnerabilities due to climate change in Namkhana (prepared by the authors)

reported that they did not know. This pattern may reflect limited access to medical services or difficulties in clearly identifying vector related illnesses in rural areas. Water borne diseases show a similar pattern. About 45.42 percent of respondents reported an increase, while 38.96 percent expressed uncertainty. Seasonal flooding, waterlogging, and changes in water quality may influence these perceptions, even if people are not always able to directly link these conditions to specific diseases. Health problems among children appear to be especially affected. A large share of respondents, 62.92 percent, reported an increase in child health issues, indicating higher vulnerability among younger age groups. Perceptions of increased mortality are even more pronounced, with 77.92 percent reporting a rise. This reflects strong concern within the community about the overall health impacts of climate change. The findings point to growing health vulnerabilities in Namkhana. High levels of uncertainty across several categories also suggest gaps in health awareness, disease monitoring, and access to healthcare services in this climate-exposed coastal region.

Household Adaptation Strategies for the Effects of Climate Change

The finding describes the adaptation strategies adopted by households in Namkhana to manage climate-related challenges affecting fisheries, housing, water resources, land, and agriculture. The patterns show that households rely mainly on practical and low-cost measures, while long-term and technology-based adaptations remain limited.

In fish and shrimp farming, increasing the height of embankments is the most common response, adopted by 55.42 percent of households. This reflects the need to protect ponds from flooding and saline water intrusion. Placing nets around shrimp fields is also widely practiced by 41.04 percent of households to reduce loss during extreme events. Very few households reported using medicines, and none adopted practices such as liming or pond excavation, indicating limited technical knowledge and financial capacity in aquaculture adaptation.

During cyclones, household responses focus mainly on safety. About 40.21 percent of households shifted to cyclone shelters, while 20.63 percent migrated temporarily to safer locations. Nearly one-fifth reported forming community action teams, showing some level of collective effort. However, very few households had cyclone-resistant housing, highlighting weak long-term preparedness.

Salinity intrusion remains a major concern, but 54.38 percent of households reported taking no action. Some households practiced rainwater harvesting or dug ponds to secure freshwater. In response to bank and soil erosion, households mainly depended on earthen embankments, soil bunds, and stone structures, with mangrove replanting adopted by a smaller share. In agriculture, more than one-third of households reported no adaptation, while others adopted crop diversification or changes in cropping patterns. Overall, the table reflects uneven adaptation capacity and the need for stronger support to enhance climate resilience in Namkhana.

(c) **Implications for Innovative Risk Reduction and Socio-ecological Sustainability**

The findings from the discussions on rural livelihoods, health vulnerabilities, and household adaptation strategies point to important implications for innovative risk reduction and socio-ecological sustainability in coastal regions. In Namkhana, rural livelihoods depend heavily on climate-sensitive activities such as agriculture, fishing, and forest-based work. This dependence increases exposure to climate-related risks and shows the need for more diverse and resilient livelihood options. Promoting alternative income opportunities can help reduce vulnerability while also lowering pressure on natural resources. Health impacts linked to climate change further highlight the need to place public health at the center of disaster risk reduction. Rising levels of infectious diseases, water-related illnesses, and child health problems indicate that climate stress directly affects well-being. These patterns suggest that adaptation planning should include improved healthcare access, disease monitoring, and community health awareness. Strengthening health systems can reduce long-term risks and support household resilience. Household adaptation practices in Namkhana are largely short-term and focused on coping rather than prevention. Structural measures such as embankments and temporary relocation during cyclones remain the most common responses, while planned and ecosystem-based approaches are limited. At the same time, community actions and local coping practices show strong potential for collective risk reduction. Supporting these efforts through institutional backing, technical guidance, and recognition of indigenous knowledge can improve adaptation outcomes. Linking livelihood security, health resilience, and ecosystem management provides a practical pathway for innovative risk reduction and long-term socio-ecological sustainability in climate-exposed coastal areas.

Casebook 2

Climate Change, Coastal Ecology, and Tourism-Dependent Livelihoods: A Socio-ecological Case Study of Digha, Coastal West Bengal.

(a) **Purpose**

Climate change poses growing challenges for coastal ecosystems and tourism-based livelihoods, especially in regions where economic activities depend strongly on natural coastal environments. Rising sea levels, stronger storms, and long-term climate variability contribute to coastal erosion, beach degradation, and damage to sensitive ecosystems such as wetlands and mangrove forests. These environmental changes directly affect coastal tourism, as visitors are drawn to stable shorelines, healthy ecosystems, and attractive landscapes (Arabadzhyan et al. 2021; Becken, 2023; Karani & Failler, 2020; Moreno & Amelung, 2009).

Climate change also adds pressure to existing social and ecological problems in coastal areas. Pollution, overfishing, and habitat loss often intensify under changing environmental conditions, placing additional stress on fragile coastal ecosystems (Baloch et al., 2023; Gössling & Hall, 2006; Skrimizea & Parra, 2019). These combined stresses reduce ecological stability and biodiversity, which in turn affects the quality of tourism experiences. As environmental conditions decline, tourism-dependent livelihoods become more uncertain, particularly in communities with limited alternative sources of income.

This case study focuses on coastal Digha in West Bengal (Fig. 14.1) and examines the links between climate change, coastal ecology, and tourism-based livelihoods. The study is situated within the broader context of the coastal-climate nexus along the Indian delta coast. It explores how environmental change influences both ecological conditions and local economic activities. Primary survey data collected from coastal Digha were analyzed using standard statistical methods commonly used in social science research. By linking ecological change with livelihood outcomes, the study provides insights relevant to disaster risk reduction and socio-ecological sustainability in climate-exposed coastal tourism regions.

(b) **Results and Discussion**

Digha, located in the Indian state of West Bengal, is a major coastal tourism center known for its long sandy beaches, scenic shoreline, and rich marine environment. Tourism plays a key role in the local economy and provides livelihoods for a large section of the population. In recent years, however, climate change has begun to place increasing pressure on both the coastal environment and tourism-based livelihoods in the area.

One of the most visible impacts of climate change in Digha is sea-level rise, which has accelerated coastal erosion and reduced beach areas. These changes have altered the coastal landscape and affected local biodiversity. Beach erosion also poses a direct threat to tourism, as beaches are the main attraction for visitors. The gradual loss of beach space has created economic stress for local communities that depend on tourism-related activities. Climate change has also contributed to more frequent and intense extreme weather events, including cyclones and storm surges. These events endanger human life, damage infrastructure, and disturb coastal and marine ecosystems, increasing risks for the tourism sector.

Alongside climate-related pressures, Digha faces several socio-ecological challenges linked to the rapid growth of tourism. Pollution, overfishing, and habitat degradation have placed additional stress on coastal ecosystems. These pressures have weakened ecological health and reduced biodiversity, which affects the quality of tourism experiences. As environmental conditions continue to decline, the long-term sustainability of tourism-based livelihoods becomes uncertain.

Addressing these challenges requires an approach that recognizes the close links between social and ecological systems. Promoting environmentally responsible tourism, improving climate-resilient infrastructure, strengthening disaster preparedness, and encouraging active participation of local communities in coastal management are important steps. The case of Digha illustrates how climate change, coastal ecology, and tourism are closely connected and shows the need for integrated strategies that support disaster risk reduction and socio-ecological sustainability in coastal regions.

(c) **Implications for Innovative Risk Reduction and Socio-ecological Sustainability**

The discussion of Digha offers several insights for innovative risk reduction and socio-ecological sustainability in coastal tourism areas. Tourism in Digha depends strongly on beaches and coastal ecosystems, which are increasingly affected by sea-level rise, erosion, and extreme weather events. This situation shows that risk reduction efforts should not be limited to protecting buildings and infrastructure alone. Greater attention is needed for conserving natural coastal systems that support tourism and local livelihoods. Protecting beaches, dunes, and coastal vegetation can reduce exposure to hazards while helping sustain economic activity. The case also shows the importance of linking disaster risk reduction with coastal planning and tourism management. Repeated cyclones and storm surges expose weaknesses in preparedness, especially for small tourism businesses and informal workers. Improving early warning systems, strengthening access to safe shelters, and investing in climate-resilient infrastructure can reduce losses and improve safety. Measures to control pollution, regulate fishing pressure, and manage tourism growth are equally important for maintaining ecological health and reducing long-term risk. Community involvement plays a key role in building resilience. Local residents and tourism workers have direct experience of coastal change and can contribute to monitoring, conservation, and preparedness. Encouraging environmentally responsible tourism and involving local communities in coastal management can strengthen both social and ecological resilience. The experience of Digha shows that linking environmental protection, livelihood security, and risk reduction provides a practical pathway for long-term socio-ecological sustainability in climate-exposed coastal regions.

14.2.2 Theme 2: Social Vulnerability and Participatory Spatial Knowledge

Casebook 3

Mapping Social Vulnerability to Climate Risks in Coastal Bengal: A GIS-Based Community Assessment

(a) **Purpose**

Climate change presents an increasing challenge for coastal regions, where environmental pressures intersect with long-standing social and economic inequalities. Coastal communities are experiencing rising sea levels, more frequent and intense storms, and ongoing shoreline erosion, all of which affect livelihoods, settlement stability, and access to essential services (Hemani, 2014). These impacts are not evenly distributed. Socially and economically disadvantaged groups tend to experience greater exposure and fewer options for coping and recovery, largely due to limited resources and restricted access to institutional support (Shiiba et al., 2023; Johnson et al., 2023).

Communities with low incomes, fragile housing conditions, and inadequate infrastructure are particularly vulnerable to climate-related hazards. Indigenous populations and elderly residents often face heightened risks of displacement, health complications, and prolonged economic stress as climate impacts intensify (Jayawardhan, 2017; Shivanna, 2022; White et al., 2023;). Such vulnerabilities are shaped not only by physical location within hazard-prone coastal zones but also by social and institutional factors, including land tenure arrangements, governance practices, and unequal participation in decision-making processes. Climate risk in coastal areas therefore reflects a combination of environmental exposure and social disadvantage. Efforts to reduce vulnerability in coastal regions require approaches that extend beyond hazard-focused assessments. Social vulnerability needs to be examined as a multidimensional condition influenced by issues of equity, access, and local capacity. Engagement with coastal communities is essential in this process, as it allows for the inclusion of local knowledge, everyday experiences of risk, and community-defined priorities that are often absent from conventional planning frameworks (Carmen et al., 2022; Nunes, 2021). Inclusive and participatory decision-making can strengthen local adaptive capacity and contribute to more effective disaster risk reduction strategies.

In line with this perspective, this casebook examines social vulnerability to climate change in the coastal region of Bengal (Fig. 14.1) using a mixed-methods research approach that integrates geospatial analysis with community participation. Field-based case study work was carried out between April and June 2022 to identify key socio-economic, demographic, and institutional factors influencing vulnerability at the community level. The integration of spatial data with insights from local interactions provides a grounded understanding of how climate risks are experienced and managed in everyday coastal life. The casebook contributes to broader discussions on coastal resilience, disaster risk reduction, and socio-ecological sustainability in deltaic environments.

(b) **Results and Discussion**

Climate variability and long-term environmental change have increasingly influenced the livelihood conditions of coastal fishing communities. Evidence from field observations shows that fishers are keenly aware of changes in their surrounding marine environment and their effects on everyday fishing practices. Extreme weather events, altered monsoon behavior, irregular rainfall, shifts in sea surface salinity, seasonal variations, changes in water color, and the growing frequency of cyclones were consistently identified as key factors affecting fish availability and fishing outcomes (Fig. 14.4) (Pattanaik, 2021). Many respondents associated these environmental changes with a decline in traditional marine species, while others noted the appearance of new or unfamiliar species in

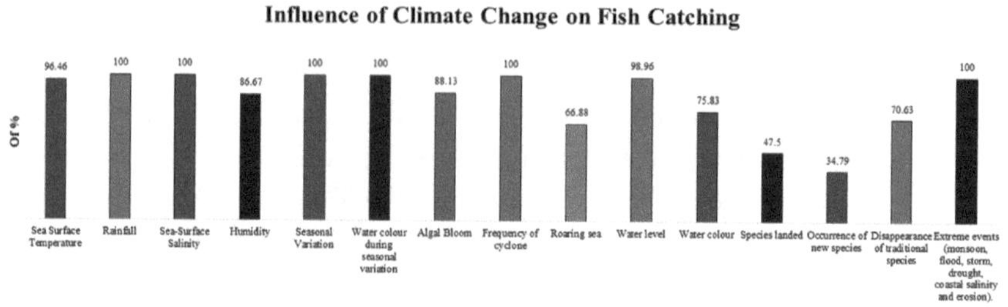

Fig. 14.4 Impacts of climate variability on fish catching activities in Coastal Bengal, India (prepared by the authors)

local fishing zones. Such shifts have created uncertainty around fishing-based livelihoods and raised concerns about long-term income security among coastal households.

Environmental stress has also played a significant role in shaping migration patterns within the study area. Community discussions indicated that several low-lying settlements in the Sundarban region have experienced repeated exposure to storms, coastal flooding, and saltwater intrusion. These conditions have gradually reduced opportunities for fishing and other daily income activities, contributing to food insecurity and unstable employment. As a response, some households relocated to relatively less exposed areas, including parts of Sagar Island. Migration in this context reflects a coping response to environmental pressure rather than a voluntary choice (Dasgupta, 2018).

Social and spatial inequalities strongly influence how climate impacts are distributed across coastal communities. Qualitative evidence suggests that vulnerability varies according to socio-economic conditions and geographic location. Marginalized groups residing in exposed areas of Sagar Island face greater hardship due to limited access to resources, fragile infrastructure, and restricted institutional support (Bera et al., 2022). These disparities widen the gap between communities with greater adaptive capacity and those that remain highly exposed to climate risks. Reducing this gap requires attention to social equity, inclusive governance practices, and meaningful participation of local communities in decision-making processes. Access to locally relevant climate and spatial information can further support informed adaptation at the community level.

Livelihood insecurity, displacement, and unequal exposure highlight the close relationship between climate stress and poverty in coastal regions. Households with limited economic resources and a strong dependence on climate-sensitive livelihoods are more likely to experience repeated losses from environmental hazards. These losses often weaken coping capacity and reduce opportunities for recovery, creating a cycle in which environmental degradation contributes to economic hardship, and poverty limits the ability to adapt to future risks. Addressing vulnerability in coastal areas therefore calls for integrated approaches that link disaster risk reduction with livelihood support and long-term social development.

(c) **Implications for Innovative Risk Reduction and Socio-ecological Sustainability**

Climate-related risks in coastal regions call for approaches that address both environmental change and the social conditions that shape vulnerability. Observations from fishing-dependent communities show that risk is closely connected to everyday livelihood practices, reliance on coastal ecosystems, and unequal access to resources and institutional support. Risk reduction efforts that focus only on physical hazards often fail to address these underlying factors, limiting their long-term effectiveness. Approaches that combine local experience with scientific assessment can lead to more relevant and practical disaster risk reduction strategies.

Socio-ecological sustainability in coastal areas depends on strengthening the relationship between communities and their surrounding environments. Conservation and restoration of coastal ecosystems such as mangroves and nearshore fisheries can reduce exposure to storms and erosion while supporting local livelihoods. Diversification of income sources, improved access to climate information, and community participation in planning processes can further enhance adaptive capacity and reduce dependence on climate-sensitive activities.

Social equity plays a central role in long-term sustainability. Marginalized groups often face greater exposure to climate hazards and fewer opportunities for recovery due to limited institutional support. Policies that link disaster risk reduction with livelihood security, social protection, and local capacity building can reduce vulnerability over time. Integrating risk reduction with socio-ecological sustainability supports resilient coastal development that benefits both communities and ecosystems.

Casebook 4

Participatory Mapping of Traditional Fishing Grounds in Coastal Bengal: Community Knowledge for Disaster Risk Reduction

(a) **Purpose**

Traditional fishing communities hold detailed knowledge of coastal and marine spaces that has developed through long-term engagement with their surrounding environment. Documenting this knowledge is important for fisheries planning, coastal management, and disaster risk reduction in climate-sensitive deltaic regions (André et al., 2022). Participatory mapping of fishing grounds allows fishers to identify spatial patterns of resource use, seasonal movements, and environmentally sensitive areas that are often missing from formal planning records (Dineshbabu et al., 2019). This approach supports inclusive decision-making and strengthens the connection between community knowledge and coastal governance. In the coastal region of Bengal, this casebook presents a case study that applies participatory mapping methods through direct engagement with local fishing communities. GPS-based spatial information was collected between September and November 2023 with the active participation of more than 120 community members. Fishers contributed location-specific details based on daily fishing practices and long-term experience of coastal waters. These data were mapped using Google Earth Pro as a participatory geographic information system tool to visualize and delineate traditional fishing areas. The mapped outputs provide insights into how livelihood spaces overlap with areas exposed to environmental risk. Integrating community-based spatial information into fisheries planning can support disaster preparedness, strengthen local adaptation strategies, and promote socio-ecological sustainability in the Indian deltaic coast.

(b) **Results and Discussion**

Spatial documentation of traditional fishing grounds provides a strong basis for participatory coastal governance and sustainable fisheries management. In the coastal region of Bengal, GPS data collected during fieldwork in 2023 were used to record the locations of fishing areas identified by traditional marine fishers. These spatial records were organized and mapped to improve understanding of how fishing activities are distributed across coastal and nearshore waters (Fig. 14.5). The mapping process also created space for dialogue between researchers and fishing communities on changing marine conditions and climate-related challenges.

Community participation was central to this process. Fishers shared place-based knowledge drawn from long-term experience of coastal environments, seasonal fishing practices, and changes in resource availability. Their involvement helped ensure that mapped fishing grounds reflected actual patterns of use rather than externally defined boundaries. Participation also supported discussion on sustainable fishing practices and local priorities for spatial management. Participatory mapping contributes to fisheries management by linking conservation objectives with the livelihood needs of coastal communities. Decision-making informed by community knowledge and spatial evidence strengthens local responsibility for marine resources and supports long-term sustainability of fishing-based livelihoods under changing environmental conditions.

(c) **Implications for Innovative Risk Reduction and Socio-ecological Sustainability**

Participatory documentation of traditional fishing grounds provides useful insights for improving risk reduction practices in coastal regions. Involving local fishers in mapping activities helps identify livelihood spaces that are closely connected to environmental hazards such as

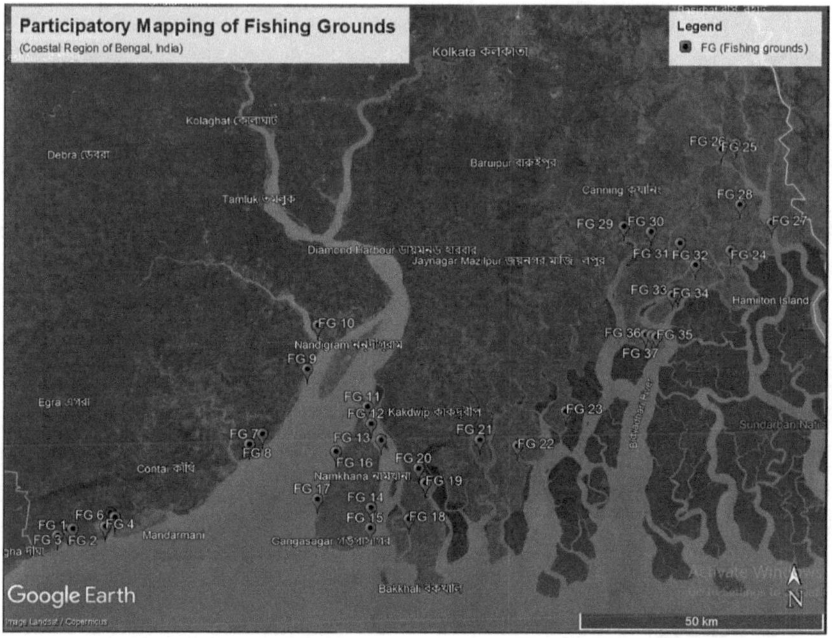

Fig. 14.5 Participatory mapping of fishing ground points in Coastal Bengal, India (prepared by the authors)

cyclones, coastal erosion, and changing marine conditions. These mapped areas reflect how climate-related risks are experienced in daily fishing activities rather than being defined only by technical hazard zones. Risk reduction strategies developed from this knowledge are more closely aligned with community needs and can support improved preparedness, early warning, and spatial planning. Socio-ecological sustainability in coastal areas depends on maintaining a balance between the use of marine resources and the protection of coastal ecosystems. Participatory mapping supports this balance by combining local ecological knowledge with spatial information. Fishers contribute insights on seasonal fishing patterns, movement routes, and environmentally sensitive areas, which can guide sustainable fishing practices and conservation efforts. Participation in mapping and decision-making processes also strengthens responsibility toward marine resources and supports adaptation to environmental change.

Inclusive planning processes further contribute to social equity by recognizing the knowledge and needs of fishing communities that are often excluded from formal management systems. Linking participatory risk reduction with livelihood support and ecosystem conservation can strengthen resilience and support long-term sustainability for both coastal communities and marine environments.

14.2.3 Theme 3: Conflict, Migration, and Participatory Governance

Casebook 5

Managing Mangrove-Human Conflicts in the Indian Sundarbans Using Participatory GIS Approaches

(a) **Purpose**

The Indian Sundarbans (Fig. 14.1) form one of the largest and most ecologically important mangrove regions in the world. As a UNESCO World Heritage Site, the area represents a highly sensitive socio-ecological setting where mangrove forests and long-established human settlements exist in close proximity (UNESCO, 1984). For generations, local communities have depended on these ecosystems for their livelihoods, particularly through fishing and other marine-based activities (Malakar et al., 2023a, 2023b). In recent decades, however, this close relationship has become increasingly fragile, leading to growing conflicts between conservation objectives and the everyday livelihood needs of local populations (Upadhyay et al., 2002).

Multiple environmental and socio-economic pressures have contributed to the intensification of these conflicts. Changes in land use, increasing coastal erosion, and climate-related stresses such as sea-level rise and extreme weather events have altered the physical landscape of the Sundarbans. At the same time, limited economic opportunities and persistent poverty have increased dependence on mangrove resources. This dependence has often resulted in overuse and encroachment, accelerating the degradation of mangrove cover and reducing the availability of ecosystem services that support both biodiversity and human well-being (Mitra, 2013; Sunkur et al., 2023).

Mangrove ecosystems in the Sundarbans perform several critical functions. They act as natural barriers against coastal hazards, reduce the impacts of storm surges, and support a wide range of plant and animal species. These forests also provide essential resources, including fish breeding habitats and materials used for fuel and shelter. The decline of mangrove ecosystems therefore has consequences that extend beyond environmental loss, increasing the vulnerability of coastal communities that rely heavily on these natural systems for survival.

Addressing mangrove-human conflicts in the Sundarbans requires approaches that account for both ecological protection and social realities. Participatory Geographic Information System methods offer a practical framework for linking spatial analysis with local knowledge. Through the direct involvement of community members in mapping resource use, identifying areas of conflict, and assessing environmental risks, participatory approaches help make management processes more inclusive and grounded in local conditions (Damastuti & de Groot, 2019).

This casebook examines the role of participatory spatial tools in supporting more balanced mangrove management practices in the Indian Sundarbans. Emphasis is placed on collaborative decision-making, sustainable livelihood planning, and adaptive land-use strategies that reflect local needs and environmental constraints. By aligning conservation efforts with community participation, the casebook highlights pathways toward strengthening the long-term resilience of both mangrove ecosystems and the people who depend on them.

This casebook is based on primary field data collected from selected coastal blocks of the Indian Sundarbans, including Hingalganj, Gosaba, Pathar Pratima, Kultali, and Namkhana in West Bengal. Fieldwork was conducted during August and September 2023 with the purpose of documenting interactions between mangrove ecosystems and local communities, as well as identifying areas of conflict and management concern. Participatory and qualitative research methods were adopted to ensure that local perspectives were clearly represented.

The field survey involved 25 respondents belonging to different age groups. Participants included individuals below 25 years of age, those between 25 and 35 years, and respondents above 35 years. This age diversity helped capture differences in livelihood practices, resource dependence, and perceptions of environmental change across generations. Ethical considerations were followed throughout the study, and the identities of respondents were kept confidential.

Data collection was carried out through household-level conversations and focus group discussions, which allowed for detailed engagement with community members. These interactions provided insights into daily resource use, conflict situations, and local views on mangrove management. The information collected from the field was examined using thematic analysis to identify key patterns and recurring issues.

Spatial information was supported using Google Maps and Google Earth for location identification and field verification. The integration of participatory methods with basic geospatial tools helped develop a grounded understanding of mangrove-human conflicts within the study area and supported the analysis of management opportunities.

(b) **Results and Discussion**

Conflicts

Field responses show that interactions between mangrove ecosystems and human activities in the Indian Sundarbans have led to several forms of conflict, mainly linked to strong livelihood dependence and growing environmental pressure. Local communities, especially traditional fishing groups, rely on mangroves for fishing, honey collection, fuelwood, and building materials. Continuous and unregulated extraction has gradually reduced mangrove cover, creating tension between livelihood needs and conservation efforts. The widespread use of mangrove wood for cooking, housing, and traditional medicine has contributed to habitat loss and weakened the natural protection these forests provide against storms and tidal surges. Shrimp farming and agricultural expansion have further increased pressure on mangrove areas through forest clearance, saline water intrusion, and soil degradation. Population growth and infrastructure development continue to encroach on mangrove landscapes, increasing competition for land and resources. Tourism activities, when poorly managed, have added to pollution and disturbance of sensitive habitats. Climate-related factors such as sea-level rise, coastal erosion, and extreme weather events have intensified these challenges by increasing ecological vulnerability and human risk.

Participatory Approaches to Conflict Management

Long-term solutions to conflicts between mangrove ecosystems and human activities depend on strong community involvement and participatory management practices that support both environmental protection and livelihood security. Consultations with local residents in the study area emphasize the value of community-based resource management, where people take part in monitoring mangrove use and follow controlled harvesting practices. This shared responsibility helps reduce pressure on mangrove resources while allowing communities to sustain their livelihoods. The use of alternative energy sources such as biogas, solar cooking, and fuelwood plantations outside mangrove areas can further reduce dependence on mangrove timber and limit deforestation (Anggraeni, 2024; Viswanathan, 2016).

Pressure on mangroves related to housing can be reduced by promoting alternative construction materials and discouraging the use of mangrove wood. Environmentally responsible shrimp farming practices, including integrated mangrove-based aquaculture, can help limit forest clearing while maintaining economic benefits. Agroforestry and improved agricultural practices can enhance soil quality, reduce erosion, and support food security. Planned settlement growth supported by clear land-use regulations can help protect sensitive mangrove zones. Sustainable tourism practices, along with waste management and visitor control, can reduce environmental disturbance. Climate adaptation measures such as mangrove restoration and strengthened coastal protection can further improve ecological stability and community resilience (VijayKumar et al., 2024; Sekhar et al., 2024).

(c) Implications for Innovative Risk Reduction and Socio-ecological Sustainability

This casebook discusses the implications of mangrove and human interactions in the Indian Sundarbans for innovative risk reduction and long-term socio-ecological sustainability. Field observations indicate that the degradation of mangrove ecosystems has increased environmental risks while also affecting the livelihood security of coastal communities. The loss of mangroves reduces natural protection against floods, coastal erosion, and cyclonic storms, placing both ecosystems and human settlements at greater risk. Addressing these challenges requires approaches that link environmental protection with social and economic needs.

Risk reduction in the Sundarbans is closely connected to the condition of mangrove ecosystems and the ability of local communities to respond to change. Community participation provides valuable insights into local vulnerabilities, seasonal resource pressures, and practical management options. Planned resource use, alternative livelihood opportunities, and appropriate land-use regulation can help reduce stress on mangrove systems while supporting daily livelihood activities. Restoration and protection of mangroves strengthen natural coastal defenses and support biodiversity, which lowers long-term exposure to environmental hazards.

Sustainable outcomes depend on approaches that recognize the close connection between ecological health and human well-being. Locally grounded and participatory practices encourage cooperation, strengthen adaptive capacity, and improve resilience. Such integrated efforts offer a clear pathway for reducing risk while promoting socio-ecological sustainability in the Sundarbans.

Casebook 6

Climate-Induced Migration and Adaptive Gaps in Deltaic Islands: A Case Study of Sagar Island

(a) Purpose

Sagar Island (Fig. 14.1), the largest island in the Sundarbans delta of West Bengal, is increasingly affected by climate-related environmental change. Rising sea levels, continuous coastal erosion, and the frequent occurrence of cyclones and storm surges have placed the island among the most vulnerable coastal regions of eastern India (Ghosh et al., 2024). These processes have gradually reshaped the island's physical environment, leading to the loss of cultivable land, damage to housing, and declining livelihood opportunities. Over time, such pressures have forced many residents to leave their original settlements and relocate within or outside the island. This process has resulted in the emergence of climate migrants, referring to people displaced primarily due to climate-induced loss of land, shelter, and income sources (Bera, 2020; Sinha, 2022).

The growing scale of climate-induced displacement on Sagar Island has also intensified the climate gap, which reflects unequal capacities among households and communities to cope with, adapt to, and recover from environmental stress. While some families are able to adjust through livelihood diversification or migration, many others remain exposed to repeated risks due to limited economic resources, insecure land tenure, and inadequate institutional support. This casebook examines how climate migrants on Sagar Island experience and manage this climate gap, with particular attention to their socio-economic vulnerabilities, everyday adaptation practices, and engagement with existing policy measures.

By focusing on the lived experiences of displaced households, the casebook contributes to wider discussions on climate justice in fragile coastal environments. It emphasizes the need for adaptation approaches that address immediate survival needs while also strengthening long-term resilience and social security. The analysis draws on qualitative fieldwork conducted across Sagar Island, based on narratives collected through 20 focus group discussions held between August and September 2023. Participatory methods were employed to capture community perspectives and to develop a grounded understanding of migration decisions, adaptation responses, and the challenges faced by climate migrants in a changing coastal landscape.

(b) **Results and Discussion**

Climate Inequality Among Climate Migrants

The concept of the climate gap draws attention to unequal levels of risk, vulnerability, and adaptive capacity among people displaced by environmental change. In coastal locations such as Sagar Island, climate-related migration has become a growing response to sea-level rise, continuing coastal erosion, and repeated exposure to cyclones and storm surges. These environmental pressures have steadily weakened local livelihood systems and reduced the ability of many households to remain in their original settlements. Migration in this context often occurs under compulsion rather than choice, leaving displaced families with limited resources and few alternatives for rebuilding their lives. After relocation, climate migrants commonly face persistent economic and social difficulties. The loss of agricultural land, fishing access, and other traditional sources of income undermines household security and increases dependence on informal and unstable work. Many migrants find it difficult to secure employment that reflects their skills or previous occupations, which further limits income stability. Housing conditions are often insecure, and access to essential services such as healthcare, education, and clean water remains uneven. These constraints restrict the capacity of migrants to adapt to new surroundings and increase their exposure to future climate risks.

The climate gap is shaped not only by material losses but also by social and institutional conditions. Displacement frequently disrupts social networks and cultural ties, contributing to stress, uncertainty, and challenges in integrating into host communities. At the institutional level, gaps in policy implementation, limited provision of climate-resilient infrastructure, and insufficient financial and administrative support reduce the effectiveness of adaptation efforts. These limitations often place climate migrants at the margins of development planning and disaster management systems. This casebook examines the climate gap as a multidimensional process operating across economic, social, and institutional domains. Attention is given to the everyday experiences of climate migrants and the broader structural conditions that influence their vulnerability and adaptive capacity. By drawing on both qualitative and quantitative perspectives, the casebook aims to provide a grounded understanding of climate-induced displacement and to support the development of adaptation strategies that are inclusive, locally relevant, and oriented toward long-term resilience.

Participatory Approaches to Addressing the Climate Gap

Managing the climate gap among displaced populations depends largely on meaningful involvement of affected communities in adaptation processes. In situations of climate-induced migration, strategies based only on top-down planning often fail to reflect local priorities and lived experiences. Community participation offers an alternative pathway by placing migrants at the center of decision-making and recognizing their knowledge, skills, and everyday practices as valuable inputs for adaptation.

Community-based adaptation provides a practical framework through which migrants and local actors jointly identify climate risks and develop responses that are suited to local social and environmental conditions. Activities such as training in climate-resilient agriculture and sustainable fishing support livelihood recovery and reduce economic insecurity. Participatory mapping and locally driven risk assessments help communities understand patterns of exposure and guide decisions related to relocation and site-specific adaptation. These practices encourage shared responsibility and strengthen confidence in collective action (Tanwattana and Toyoda, 2024; Igbokwe et al., 2024).

Engagement with local governance institutions remains a critical element in managing the climate gap. Platforms such as community climate action groups create space for dialogue between migrants and local authorities, allowing community concerns to inform policy priorities. Participatory budgeting processes further support inclusion by enabling communities to influence decisions on resource allocation for infrastructure development, healthcare services, and disaster preparedness. Such mechanisms contribute to greater transparency and improve institutional responsiveness to the needs of displaced populations (Restrepo-Mieth et al., 2023).

Social relationships and knowledge exchange also shape adaptive capacity in important ways. Peer learning within migrant and host communities supports the exchange of traditional ecological knowledge alongside scientific information, leading to locally grounded adaptation practices. Active participation of women and young people strengthens innovation and sustains engagement in adaptation efforts over time. These social processes help build trust and reinforce collective resilience under continuing climate stress (McMichael et al., 2024).

This casebook focuses on participatory approaches as a central component of climate gap management among climate migrants. Attention is given to the interactions between community involvement, institutional engagement, and knowledge sharing in shaping adaptation outcomes. By examining these processes, the casebook highlights pathways toward equitable and sustained adaptation in vulnerable coastal regions such as Sagar Island.

(c) **Implications for Innovative Risk Reduction and Socio-ecological Sustainability**

The climate gap highlights the need to rethink risk reduction approaches in ways that address both social vulnerability and environmental change. In vulnerable coastal settings such as Sagar Island, innovative risk reduction must move beyond emergency responses and focus on long-term socio-ecological sustainability. Strategies that integrate livelihood security, ecosystem protection, and social inclusion are essential for reducing repeated exposure to climate risks. Strengthening locally appropriate livelihoods, such as climate-resilient agriculture and sustainable fisheries, can reduce economic dependence on fragile environments while supporting household stability. Innovative risk reduction also depends on linking community knowledge with scientific planning. Participatory risk assessment, local early warning systems, and community-led monitoring can improve preparedness while ensuring that adaptation measures reflect everyday realities. Such approaches support trust, improve risk communication, and encourage shared responsibility between communities and institutions. At the same time, protecting coastal ecosystems such as mangroves and wetlands plays a critical role in reducing hazard impacts while sustaining biodiversity and ecosystem services. Socio-ecological sustainability requires policies that recognize migrants as active agents rather than passive recipients of support. Inclusive governance, secure access to resources, and long-term planning can reduce the climate gap while promoting resilience. These insights point toward adaptation pathways that strengthen both human well-being and ecological stability in climate-exposed coastal regions.

14.2.4 Theme 4: Hazards, Adaptation, and Geospatial Modeling

Casebook 7

Climate Variability, Natural Hazards, and Local Adaptation Strategies in Traditional Marine Fishing Communities of Coastal Bengal

(a) **Purpose**

Climate change and climate variability have become major concerns of the twenty-first century, affecting ecological systems, economic activities, and human livelihoods across the world (Adom, 2024). Coastal regions are especially vulnerable because they are directly exposed to sea-level rise, extreme weather events, and changes in ocean and atmospheric processes (Martyr-Koller et al., 2021). In India, the coastal districts of southern West Bengal, namely Medinipur East, South 24 Parganas, and North 24 Parganas, form an important region where long-term climatic change interacts with dense populations and livelihoods that depend heavily on natural resources, particularly agriculture and marine fisheries. Long-term climatic records show that climate variability in this coastal belt has been developing over more than a century. Analysis of data from 1901 to 2022 indicates statistically significant warming across all three districts. Average annual temperatures have increased steadily, with a cumulative rise of about 0.7 to 0.8 degrees Celsius during the study period. The increase in maximum temperature is strongest in Medinipur East, while North 24 Parganas shows higher values of atmospheric vapor pressure and potential evapotranspiration. These trends suggest increasing thermal stress and growing pressure on water availability. Rising evapotranspiration and vapor pressure point to higher atmospheric moisture and greater water demand, particularly for agriculture. Total rainfall does not show a statistically significant long-term trend, but the number of wet days has increased over time. This indicates a change in the timing and distribution of rainfall rather than in the total amount received. Such changes may lead to more frequent rainfall events occurring over shorter periods, increasing the risk of flooding, runoff, and soil erosion. Trends in cloud cover remain unclear, but the overall pattern indicates a gradual intensification of climatic stress in the coastal environment (Pattanaik, 2021).

The study area (Fig. 14.1) includes three coastal districts with different physical and socio-economic characteristics, but all are highly exposed to climate-related risks. Medinipur East is characterized by low-lying deltaic plains close to the Bay of Bengal and is highly prone to cyclones and coastal flooding. South 24 Parganas contains large parts of the Sundarbans mangrove region and is widely recognized for its ecological sensitivity and climate vulnerability. North 24 Parganas is relatively more urbanized but also includes extensive rural areas where livelihoods depend on traditional fishing and small-scale agriculture. High population density, widespread poverty, and strong dependence on natural resources increase the sensitivity of these districts to climatic change. This casebook is based on long-term climatic data obtained from the Climatic Research Unit and the Met Office, supported by relevant secondary sources (Harris et al. 2020). Trend analysis was carried out using the Mann–Kendall test and Sen's slope estimator, which are suitable for long time-series data with natural variability. The results provide a clear picture of long-term climatic changes in coastal West Bengal and offer a foundation for examining their effects on livelihoods, vulnerability, and local adaptation practices in the casebook that follows.

The present study focuses on connecting detailed climatic trend analysis with the responses of coastal communities living in the highly climate-sensitive coastal belt of West Bengal, India. It combines quantitative analysis of long-term climatic change with close examination of locally practiced adaptation strategies to explain how environmental changes are experienced and addressed at the community level. The findings are relevant for academic research as well as for policy formulation, particularly in relation to climate-resilient livelihoods, sustainable use of coastal resources, and inclusive approaches to adaptation planning. In this way, the study contributes to a clearer understanding of the links between climate change and development in the vulnerable coastal regions of Bengal.

(b) **Results and Discussion**

An analysis of climate variability and change in the three coastal districts of Bengal, namely Medinipur East, South 24 Parganas, and North 24 Parganas, over the period from 1901 to 2022 indicates a clear and steadily intensifying signal of climate change in the region (Fig. 14.6, 14.7, and 14.8). The assessment, based on the Mann–Kendall trend test and Sen's slope estimates, identifies widespread warming and moderate changes in hydro-climatic conditions across the study area.

All three districts show statistically significant increasing trends in average, maximum, and minimum temperatures, with p-values below 0.0001 and Kendall's tau values ranging from 0.35 to 0.45. Sen's slope estimates indicate that average annual temperatures have increased at rates ranging from 0.00576 °C per year in North 24 Parganas to 0.00627 °C per year in Medinipur East. Over the 121-year study period, this corresponds to a total temperature rise of approximately 0.7 to 0.8 °C. These results point to a persistent warming trend that is consistent with broader regional and global patterns of climate change.

Significant upward trends are also observed in potential evapotranspiration and vapor pressure across all three districts, with p-values at or below 0.0001. The rise in potential evapotranspiration, particularly evident in North 24 Parganas with a Sen's slope value of 0.00060, suggests increased water loss from soil and vegetation and a higher risk of agricultural water stress. Rising vapor pressure reflects increasing atmospheric moisture and further supports evidence of ongoing regional warming.

Rainfall trends show greater variability and do not indicate a clear long-term change. Although Sen's slope values are positive in all districts, suggesting slight increases in annual rainfall, the trends are not statistically significant. For instance, p-values of 0.1006 in Medinipur East and 0.6108 in North 24 Parganas indicate that a consistent long-term

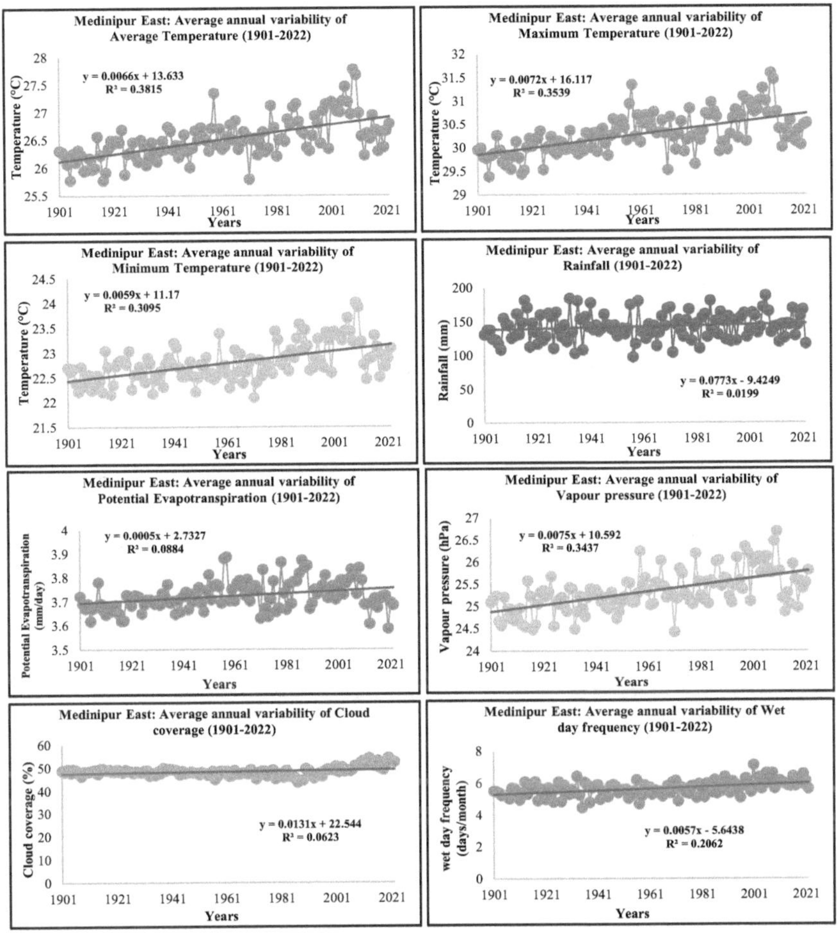

Fig. 14.6 Average annual variability of climatic variables (1901–2022) in East Medinipur

trend in total annual rainfall cannot be established. At the same time, substantial year-to-year variability remains evident across the districts.

Changes in rainfall characteristics are more apparent in the frequency of wet days. The number of days with measurable precipitation has increased significantly in all three districts, with p-values below 0.0001. This pattern indicates a shift in the timing and distribution of rainfall

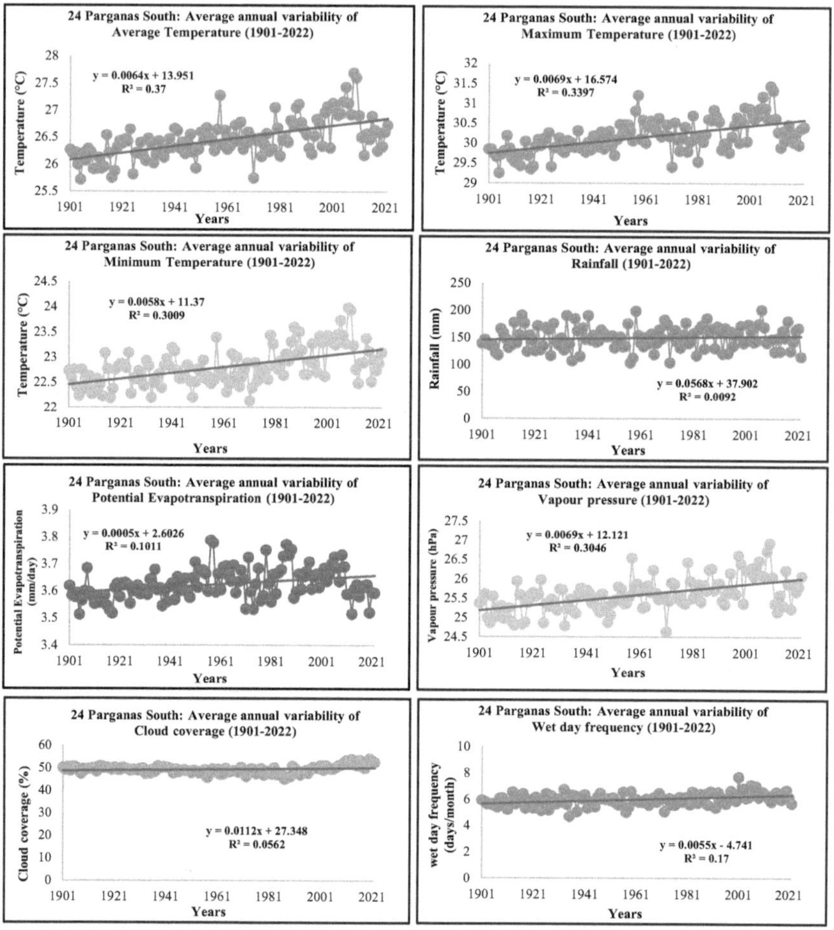

Fig. 14.7 Average annual variability of climatic variables (1901–2022) in 24 Parganas South

events, even though total annual rainfall has remained largely unchanged. Cloud cover shows a slight increasing tendency, but the associated trends are not statistically significant, suggesting that changes in cloudiness are weak or uncertain.

Comparison among the three districts highlights differences in the magnitude of climatic change. Medinipur East exhibits the strongest

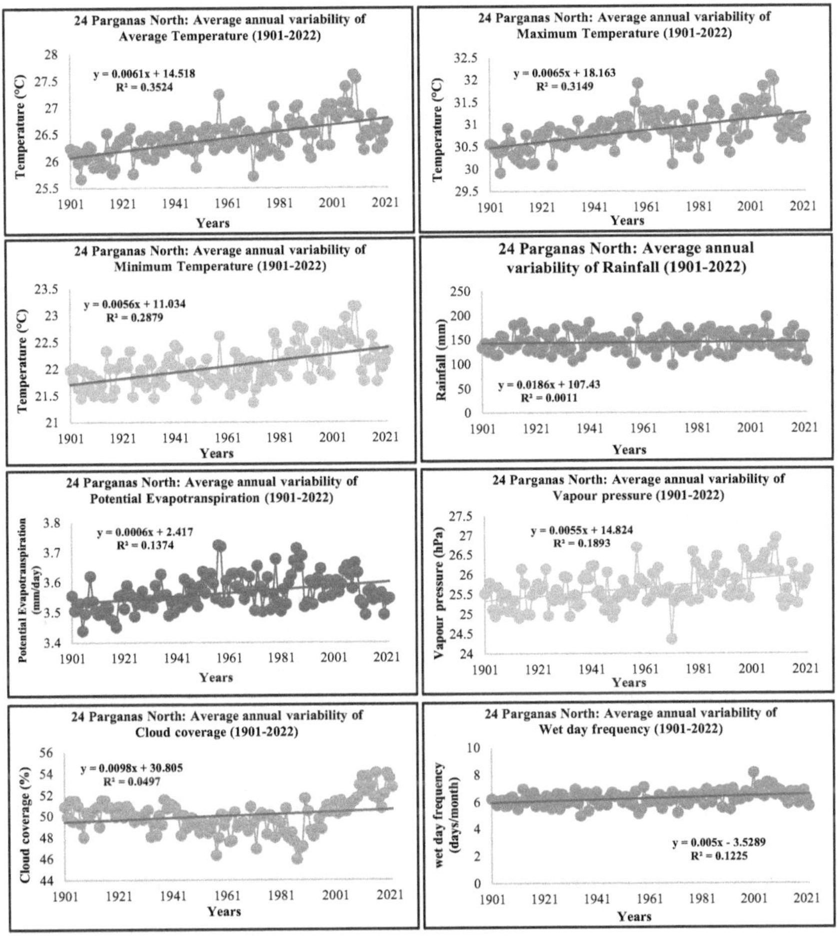

Fig. 14.8 Average annual variability of climatic variables (1901–2022) in 24 Parganas North

warming signal, particularly in maximum temperatures, with a Sen's slope of 0.00731 °C per year. North 24 Parganas records the highest average maximum temperature of 32.08 °C and also shows marked increases in vapor pressure and potential evapotranspiration, indicating heightened thermal and atmospheric moisture stress. South 24 Parganas displays trends similar to those observed in the other districts, although the overall magnitude of change is generally intermediate.

Natural Hazards and Local Adaptation Strategies

Traditional marine fishing communities along the coast of Bengal have long faced high levels of vulnerability due to climate variability and environmental change. Rising sea surface temperatures, irregular monsoon behavior, increasing cyclone activity, coastal erosion, and frequent storm surges have directly affected fishing conditions and household security. In response, these communities have developed a range of adaptive practices based on indigenous knowledge, socio-cultural traditions, and long-term experience of living with a dynamic coastal environment.

Changes in the timing and location of fishing activities represent one of the most visible adaptation strategies (Ojea et al., 2020). Fishers increasingly modify departure times to reduce exposure to unpredictable weather, rough seas, and sudden cyclonic disturbances. Decision-making now draws on a combination of traditional ecological knowledge and modern information sources, including weather forecasts accessed through mobile phones and radios. Observations of wind patterns, sea conditions, bird movement, and lunar phases continue to play an important role in assessing risk. At the same time, shifts in fish distribution linked to warming waters have forced fishers to move to deeper or more distant fishing grounds. This spatial adjustment increases fuel costs and significantly raises exposure to natural hazards such as high waves, strong currents, and extreme weather events.

Adaptation has also taken place through changes in fishing gear and equipment (Innes & Pascoe, 2010). Many fishers have modified nets to match changing fish size and species availability, with finer mesh nets increasingly used to target smaller and more abundant species. While this improves short-term catch security, it also raises concerns about long-term ecological sustainability. Some households have adopted mechanized or semi-mechanized boats to reach offshore fishing zones, marking a gradual shift away from fully artisanal practices. This transition, however, requires greater financial investment and exposes fishers to higher physical and economic risks, particularly during extreme weather events.

Livelihood diversification has emerged as an important coping strategy in the face of growing uncertainty (Deb & Haque, 2016; Huynh et al., 2021). Many fishing households supplement their income through small-scale agriculture, aquaculture such as brackish water shrimp farming, and wage labor during lean fishing periods. Women play a central role in sustaining household income through fish processing, net making, and small trading activities. These alternative income sources help reduce dependence on fishing alone and act as informal safety nets during periods of climate-related disruption or disaster.

Community-based disaster preparedness has become increasingly important as climate-related natural hazards intensify (Obregón et al., 2024). Coastal villages have improved cyclone awareness, made greater use of cyclone shelters, and actively engaged with early warning systems supported by government agencies and non-governmental organizations. Local institutions such as fishers' cooperatives and self-help groups play a key role in sharing information, coordinating evacuation and relief efforts, and supporting recovery after cyclones and flooding events. Despite these initiatives, many adaptation responses remain reactive and locally driven, with limited institutional support. Constraints such as restricted access to credit, inadequate coastal infrastructure, and policy gaps related to the rights and protection of small-scale fishers continue to limit long-term adaptive capacity.

The experience of traditional marine fishing communities in coastal Bengal highlights a form of adaptation that is grounded in everyday practice and shaped by repeated exposure to climate variability and natural hazards. Their responses demonstrate both resilience and vulnerability, underscoring the need for targeted policy support, strengthened institutional frameworks, and investment in climate-resilient infrastructure. Integrating local knowledge with formal planning processes is essential for reducing risk and safeguarding livelihoods in an increasingly hazardous coastal environment.

(c) **Implications for Innovative Risk Reduction and Socio-ecological Sustainability**

The observed patterns of climate variability and intensifying natural hazards in the coastal districts of West Bengal underline the need for risk reduction strategies that move beyond short-term coping and address long-term socio-ecological sustainability. Rising temperatures, changing rainfall characteristics, increased cyclone exposure, and coastal erosion are not isolated climatic events but interconnected processes that affect ecosystems, livelihoods, and social structures simultaneously. Risk reduction efforts therefore need to integrate climatic, ecological, and social dimensions rather than treating them separately.

Innovative risk reduction in this context lies in strengthening locally grounded adaptation while aligning it with scientific knowledge and institutional support. Traditional ecological knowledge, already used by fishing communities to assess weather and sea conditions, can be combined with improved early warning systems, localized climate services, and hazard-resilient infrastructure. Such integration can reduce exposure to extreme events while enhancing community preparedness. At the same time, promoting diversified and climate-resilient livelihoods reduces dependence on fragile coastal resources and helps spread risk across multiple income sources.

Socio-ecological sustainability also depends on maintaining the health of coastal ecosystems such as mangroves, estuaries, and nearshore fisheries, which provide natural protection against storms and support livelihoods. Policies that support ecosystem restoration, inclusive governance, and secure access to resources can strengthen both environmental resilience and social well-being. Together, these approaches offer a pathway toward reducing climate risk while sustaining the coupled human–environment systems of coastal Bengal.

Casebook 8

Monitoring Shoreline Change and Coastal Vulnerability Using Geospatial Technologies: Evidence from Coastal 24 Parganas, India

(a) **Purpose**

Coastal regions are among the most dynamic and sensitive environments in the world, shaped continuously by interactions between terrestrial and marine processes. These areas act as important transition zones that support diverse ecosystems and sustain human livelihoods such as agriculture, aquaculture, and marine fishing (Turner & Schaafsma, 2015). Because of their ecological and economic importance, coastal zones have long attracted human settlement and development. At the same time, this close interaction between natural systems and human activities increases the susceptibility of coastal environments to physical and ecological change.

In recent decades, coastal areas across the world have experienced growing pressure from both natural and human-induced factors. Sea-level rise, increasing frequency of extreme weather events, changes in sediment availability, and large-scale human interventions have contributed to accelerated shoreline instability and environmental degradation (Suntoro et al., 2025). These factors often interact with one another, intensifying coastal erosion and altering established geomorphic patterns.

Along the eastern coast of India, the deltaic Sundarban region located within the South and North 24 Parganas districts of West Bengal represents one of the most fragile coastal landscapes. This region is marked by highly dynamic shoreline conditions influenced by erosion, accretion, tidal processes, and riverine sediment transport. The Coastal South and North 24 Parganas (CS24) region forms a critical land-sea interface where geomorphic instability is further intensified by high population density, extensive embankment systems, and frequent exposure to climate-related hazards. Over the past three decades, significant changes in shoreline position have been recorded, indicating an urgent need for systematic analysis to support informed coastal management.

Shoreline change analysis has become an essential approach for examining long-term coastal morphodynamics. Advances in geospatial technologies and the availability of historical datasets allow for detailed assessment of spatial and temporal shoreline variations. Although erosion and accretion are natural components of coastal evolution, their scale and frequency in the CS24 region have been strongly influenced by human activities. Embankment construction, degradation of mangrove cover, and unplanned expansion of coastal settlements have altered natural sediment flows and reduced the adaptive capacity of the coastline. Climatic and hydro-geomorphological factors such as monsoonal rainfall, tidal fluctuations, and river-borne sediment transport continue to play a significant role in shaping shoreline behavior (Das & Swain, 2024; Kantamaneni et al., 2022).

This casebook adopts the perspective that coastal vulnerability emerges from the combined influence of natural processes and human interventions. A spatially explicit and interdisciplinary approach is applied to examine shoreline change in the CS24 region and to contribute to a clearer understanding of coastal dynamics. Such an approach is necessary for informing adaptive and sustainable coastal management strategies in regions facing increasing environmental and socio-economic pressures.

This casebook outlines the methodological framework used to examine shoreline change and coastal vulnerability in the Coastal 24 Parganas region over a thirty-year period from 1994 to 2024. The study uses a time-series geospatial approach to capture long-term shoreline behavior across this environmentally sensitive coastal setting. Landsat satellite imagery obtained from the USGS Earth Explorer platform serves as the primary data source. The long temporal record and consistent spatial resolution of Landsat imagery make it suitable for assessing gradual and episodic shoreline changes over extended periods.

Shoreline change analysis was conducted using the Digital Shoreline Analysis System (DSAS), an established extension for ArcGIS developed by the U.S. Geological Survey (Himmelstoss et al., 2024). DSAS supports the systematic measurement of shoreline movement through the generation of baselines and transects and the calculation of commonly used indicators such as Net Shoreline Movement (NSM), End Point Rate (EPR), and Weighted Linear Regression (WLR) (Thinh, 2022). For this study, a baseline was manually constructed along the entire coastline of the study area. From this baseline, 14,221 transects were generated at regular intervals and oriented perpendicular to the shoreline.

Shoreline positions were extracted using a combination of satellite image classification and manual digitization of the land–water interface. This approach helped improve the accuracy of shoreline delineation and reduce positional uncertainty. Among the available statistical techniques within DSAS, weighted linear regression was selected as the primary method for estimating shoreline change rates. This method reduces the influence of short-term variability and provides a clearer representation of long-term shoreline trends. Based on WLR results, each transect was classified as erosional, accretional, or stable. These classifications were then used to calculate descriptive statistics, including mean, maximum, and minimum rates of shoreline change, along with standard deviation values for erosional and accretional areas.

Field observations were incorporated to complement the geospatial analysis. A total of fifty field visits were carried out at selected locations within the study area, including Sagar Island, Namkhana, and Hingalganj. These visits focused on verifying shoreline conditions identified from satellite imagery and collecting qualitative information through discussions with local communities and relevant stakeholders. Observations from the field helped place the spatial patterns of shoreline change within a local context and improved confidence in the interpretation of results.

The integration of geospatial analysis with field-based observations strengthens the overall assessment of shoreline dynamics and coastal vulnerability. Satellite-derived data, geomorphic interpretation, and ground-level validation together provide a detailed picture of coastal change in one of India's most ecologically fragile and socio-economically vulnerable regions. The approach described in this casebook supports informed coastal planning and contributes to disaster risk reduction and climate change adaptation efforts in the Indian Sundarban delta.

(b) **Results and Discussion**

Shoreline change in the Coastal 24 Parganas (CS24) region from 1994 to 2024 shows strong spatial variation and continuous coastal adjustment (Fig. 14.9). Results obtained using the Digital Shoreline Analysis System (DSAS) and Weighted Linear Regression (WLR) indicate that large sections of the coastline have undergone noticeable morphological change during the study period. Of the 14,221 transects analyzed, 35.15 percent recorded erosion and 44.33 percent showed accretion, while only

0.16 percent remained relatively stable. This pattern reflects an active coastal system where sediment movement is ongoing and stable shoreline conditions are rare (Table 14.1).

The mean rate of shoreline change for the entire region indicates a slight net accretion of 0.456 meters per year, suggesting an overall tendency toward shoreline advance (Fig. 14.10). This average value, however, does not capture the sharp contrasts observed at the local scale. In several locations, shoreline advance reached rates as high as 49.3 meters per year, while in other areas erosion occurred at rates of up to minus 49.93 meters per year. These extreme values identify sections of the coastline where rapid change is taking place. Such areas face heightened risk, as intense erosion threatens settlements, infrastructure, and

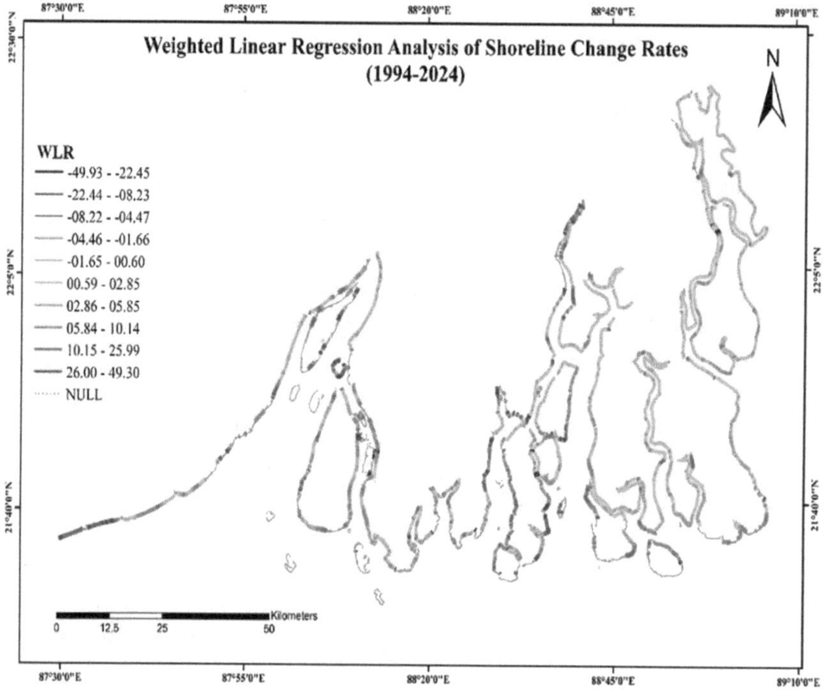

Fig. 14.9 Spatial distribution of shoreline change based on Weighted Linear Regression (WLR) along the Bengal coast (1994–2024)

Table 14.1 Summary of descriptive statistics from DSAS-based shoreline change and WLR results (1994–2024)

	CS24
Total Transects	14,221
Total Erosional Transects	4998
Total Accretional Transects	6304
Total Stable Transects	23
Erosional Transects (%)	35.14521
Accretional Transects (%)	44.32881
Stable Transects (%)	0.161733
Mean Shoreline Change (m/yr.)	0.456437
Max Shoreline Change (m/yr.)	49.3
Min Shoreline Change (m/yr.)	-49.93
Mean Erosion (m/yr.)	-1.01739
SD Erosion (m/yr.)	2.538276
Mean Accretion (m/yr.)	1.380874
SD Accretion (m/yr.)	2.94372

Note: CS24: Coastal 24 Parganas
Data calculated by the author using DSAS tools

coastal habitats, particularly in low-lying zones that experience frequent tidal flooding and storm surge events.

Erosion-affected transects recorded an average shoreline retreat rate of minus 1.02 meters per year, with a standard deviation of 2.54 meters, indicating considerable variation in erosion intensity across the region. Accretion-dominated transects showed a higher mean rate of shoreline

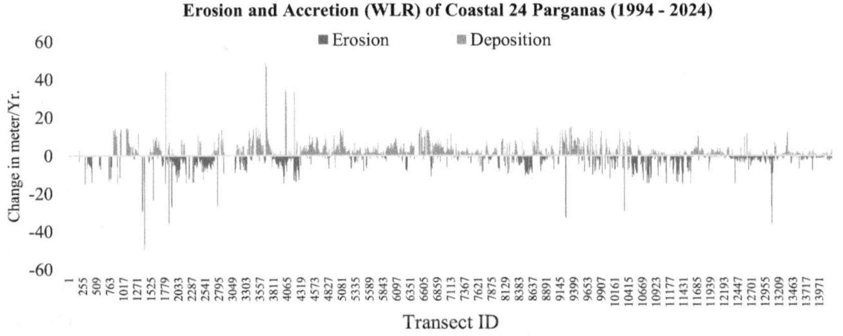

Fig. 14.10 Trends in erosion and accretion rates in the Coastal 24 Parganas region (1994–2024)

advance at 1.38 meters per year and a standard deviation of 2.94 meters. These results point to uneven sediment deposition patterns influenced by local coastal processes. The extremely small proportion of stable transects further highlights the absence of equilibrium conditions along much of the CS24 coastline.

Shoreline behavior in the region reflects the combined influence of natural processes and human activities. Sea-level rise, seasonal monsoon patterns, tidal dynamics, and changes in river-borne sediment supply continue to shape coastal morphology. Human interventions such as embankment construction, degradation of mangrove cover, and unplanned land use have further altered natural sediment movement and reduced the natural buffering capacity of the coastline. The findings presented in this casebook point to the need for adaptive and location-specific coastal management strategies that respond to ongoing shoreline change while addressing the social and ecological vulnerabilities of this densely populated and environmentally sensitive region.

(c) Implications for Innovative Risk Reduction and Socio-ecological Sustainability

Shoreline change patterns in the Coastal 24 Parganas region raise important concerns for coastal risk reduction and long-term socio-ecological sustainability (Malakar et al., 2025). The widespread presence of erosion and accretion, along with the very small number of stable shoreline sections, shows that coastal risks vary greatly across space and continue to change over time. This situation calls for a move away from uniform coastal protection measures toward strategies that respond to local shoreline conditions and levels of vulnerability. Areas experiencing rapid erosion require urgent attention, particularly where settlements, embankments, and essential infrastructure are increasingly exposed to tidal flooding and storm surge impacts.

The findings also point to the role of ecological measures in reducing coastal risk. Protection and restoration of mangrove forests can help weaken wave energy, support sediment stability, and strengthen the natural buffering capacity of the coastline. At the same time, better management of embankments and improved land-use planning are needed to limit further disturbance to natural sediment movement. Involving local communities in planning and decision-making can strengthen adaptation efforts by drawing on local knowledge and improving social resilience. Regular shoreline monitoring using geospatial techniques, supported by field observations, can improve the identification of emerging risks and guide adaptive management. This approach helps balance development pressures with ecological protection and supports more resilient livelihoods and sustainable coastal systems in this highly vulnerable region.

14.2.5 Theme 5: Community Narratives, Indigenous Knowledge, and Resilience

Casebook 9

Community Perspectives on Disaster Risk Reduction in the Coastal Sundarbans: Field Narratives and Participatory Insights

(a) **Purpose**

This casebook explores community perspectives on disaster risk reduction in the coastal Sundarbans of West Bengal, with a focus on how local residents perceive, experience, and respond to recurring coastal hazards. The region is frequently affected by cyclones, storm surges, tidal flooding, and shoreline erosion, which directly influence livelihoods, housing conditions, and access to essential services. Although formal risk reduction measures are in place, everyday coping practices and locally rooted knowledge continue to play an important role in shaping community resilience.

Field-based evidence for this casebook comes from qualitative research carried out in selected villages across the coastal blocks of South and North 24 Parganas (Fig. 14.1). Data collection involved participatory interviews with fifty respondents, informal group discussions with fifteen groups, and direct observation of daily practices among fishers, farmers, women's groups, and local leaders. Participants shared personal experiences of past disaster events, noted changes in the coastal environment, and discussed their views on existing protective measures such as embankments and cyclone shelters. Transect walks and site visits were conducted to document exposure to erosion, flooding, and embankment failure. These methods helped capture narratives rooted in lived experience and community priorities, while also supporting broader scientific assessments of coastal risk.

(b) **Results and Discussion**

Field narratives from the coastal Sundarbans indicate that local communities are keenly aware of growing environmental instability. Residents described visible shifts in shoreline position, higher tidal water levels, and an increase in embankment breaches. Many participants connected these changes to stronger cyclones and irregular monsoon patterns, while others pointed to the loss of mangrove cover and rising pressure on limited land as major sources of vulnerability. Fishing households reported declining catches and longer travel distances to reach productive fishing areas. Farming households noted increasing soil salinity, reduced crop yields, and frequent crop damage caused by tidal flooding.

Community discussions revealed both reliance on and dissatisfaction with embankments as protective structures. Embankments were widely seen as necessary for daily safety, especially during high tides and storm events. At the same time, repeated breaches and delayed repair work have weakened confidence in their long-term effectiveness. Temporary repairs carried out after disaster events were often described as inadequate, leading to repeated losses during later cyclones or flooding. Cyclone shelters were generally viewed as important, but concerns were raised about their distance from villages, overcrowding during emergencies, and limited accessibility for elderly residents and women.

Local coping practices remain central to how households manage disaster impacts. Families described practical actions such as raising house plinth levels, storing dry food and essential supplies, adjusting fishing schedules, and depending on social networks during crisis periods. Women often played a key role in managing household recovery and maintaining stability after disasters, although their participation in formal planning and decision-making processes remained limited. These observations highlight a clear gap between externally planned risk reduction measures and local priorities. Stronger communication, improved trust, and greater community involvement are essential for improving disaster preparedness and response in the coastal Sundarbans.

(c) Implications for Innovative Risk Reduction and Socio-ecological Sustainability

Community narratives from the coastal Sundarbans highlight the need for disaster risk reduction approaches that address physical, ecological, and social dimensions together. Long-term coastal safety cannot rely only on embankments, especially in areas where shoreline change is rapid and unpredictable. Greater focus on mangrove conservation and restoration can reduce wave impact, support sediment stability, and strengthen natural coastal protection. Local communities expressed strong support for such measures when they were connected to livelihood opportunities and community involvement.

Risk reduction strategies also need to be more inclusive and grounded in local contexts. Involving residents in embankment maintenance, cyclone shelter planning, and early warning communication can improve both effectiveness and trust. Participation of women in decision-making processes is particularly important, given their central role in maintaining household stability and supporting recovery after disasters. Clear communication between local institutions and community members can help align scientific assessments of risk with everyday experiences.

Continuous monitoring of coastal change, supported by regular engagement with local communities, can improve adaptive planning in a region where environmental conditions are constantly changing. Approaches that respect local knowledge while strengthening institutional support provide a practical path toward socio-ecological sustainability. These strategies can help protect livelihoods, reduce disaster-related losses, and preserve the ecological balance of the Sundarbans, one of India's most vulnerable and valuable coastal regions.

14.2.6 Theme 6: Justice, Climate Gaps, and Participatory GIS

Casebook 10

Climate Justice, Social Space, and Climate Gap in the Indian Delta: Evidence from Participatory GIS-Based Case Studies

(a) **Purpose**

Climate change is not only an environmental concern but also a social issue shaped by inequality, power relations, and access to resources. In the coastal belt of West Bengal in India, climate impacts are becoming increasingly visible through stronger cyclones, rising sea levels, saline intrusion, and frequent flooding. These changes place growing pressure on livelihoods, housing, and access to basic services. The impacts are unevenly distributed, with socially and economically marginalized communities facing the greatest risks despite contributing least to climate change (Mukherjee and Siddique, 2024). This imbalance reflects what is widely described as the climate gap, where differences in exposure and adaptive capacity deepen existing inequalities (Malakar et al., 2025).

Addressing the climate gap requires a shift away from top-down planning approaches toward processes that recognize local knowledge, lived experience, and spatial inequality. In this context, Participatory Geographic Information Systems (PGIS) provide a useful way to understand climate impacts in more inclusive and grounded ways. PGIS enables communities to map hazards, resources, and vulnerabilities while taking an active role in knowledge creation. This casebook examines climate impacts within the social and spatial realities of coastal Bengal through five interconnected case studies. These case studies show that PGIS functions not only as a mapping technique but also as a process that supports community engagement, participatory governance, and rights-based climate action (Malakar & Roy, 2024; Brown, 2012). Attention is given to how climate risks intersect with gender, caste, class, and institutional structures to shape uneven vulnerability and adaptive capacity.

The first case study focuses on climate vulnerability mapping and social inequality. Using PGIS, areas of high climate risk are mapped alongside social and economic indicators. The analysis shows that the most vulnerable zones often overlap with settlements of landless and low-income groups who have limited access to institutional support. Participatory mapping allows these communities to identify their own risks and express priorities for more inclusive adaptation planning. The second case study examines access to environmental resources among marginalized groups. Mapping exercises reveal uneven access to fisheries, agricultural land, and forest resources, shaped by political and institutional control as much as by environmental conditions. Community participation in this process supports shared understanding and opens space for more equitable resource management.

The third case study addresses gender, social space, and climate justice. Gender-sensitive PGIS methods map access to water points, cyclone shelters, agricultural fields, and decision-making spaces. The findings show that women often face greater constraints on mobility, safety, and participation in hazard-prone areas, increasing their vulnerability. Direct involvement of women in mapping activities helps bring these concerns into planning processes and supports gender-responsive adaptation strategies.

These case studies demonstrate how PGIS serves both analytical and practical purposes in climate justice efforts. The approach makes visible spatial patterns of inequality and exclusion that are often overlooked in conventional planning. It also enables communities to participate directly in shaping adaptation strategies and policy discussions. Participatory mapping encourages a sense of ownership and collective agency, which is essential for building long-term resilience.

This casebook positions PGIS as a link between spatial analysis and social action. By combining technical mapping with community knowledge, PGIS supports more just, inclusive, and locally grounded climate responses (de Carvalho and Giatti, 2018; Elwood, 2006). In the coastal region of Bengal, where climate pressures continue to intensify, such approaches are necessary to ensure that climate action aligns with principles of equity, human rights, and community resilience.

(b) **Results and Discussion**

b.1.Climate Vulnerability and Social Inequality

The coastal belt of Bengal along the Bay of Bengal is one of India's most climate-sensitive regions, frequently affected by cyclones, tidal surges, and saline water intrusion. These impacts have disrupted livelihoods and disproportionately affected marginal communities (Makwana & Patnaik, 2025). This case study examines the links between climate vulnerability and social inequality in selected villages of South 24 Parganas, West Bengal. Using a participatory mixed-method approach, the study draws on household surveys conducted in Kultali, Gosaba, and Patharpratima, along with participatory rural appraisal, focus group discussions, and community-based vulnerability mapping to capture both measurable risks and lived experiences.

Findings

Discussions with community members showed that climate stress in the coastal region is shaped by both environmental exposure and long-standing social inequalities. Participants described how frequent tidal surges and unpredictable cyclones have gradually reduced land security and weakened livelihoods. Elderly residents recalled the slow loss of fertile fields and homestead plots to saline water over recent decades. Women highlighted growing difficulties in accessing safe drinking water as local ponds become saline, increasing their daily workload. Marginalized groups, particularly landless laborers and Scheduled Caste households, reported delayed access to relief and recovery support, often leading to seasonal migration for insecure work. Younger participants expressed a sense of economic and cultural displacement as farming and fishing become less reliable. Community accounts also pointed to unequal access to early warnings and assistance, confirming that climate vulnerability in the region is uneven and closely linked to social position, access, and local power relations.

b.2 Marginalized Communities and Environmental Resources

The coastal belt of Bengal along the Bay of Bengal is an ecologically sensitive and socially vulnerable region that supports livelihoods through mangroves, fisheries, and fertile agricultural land (Pattanaik, 2021). Despite its natural richness, ongoing environmental change and unequal access to resources have increased marginalization among vulnerable communities (Malakar et al., 2023). This case study examines how access to environmental resources is shaped by social hierarchies, institutional arrangements, and changing ecological conditions in selected villages of the South 24 Parganas district of West Bengal.

The study uses a participatory mixed-methods approach that combines household surveys conducted with sixty households in Kultali, Namkhana, and Gosaba with qualitative methods such as participatory rural appraisal, resource mapping, focus group discussions, and key informant interviews. Quantitative data document patterns of land ownership and resource use, while qualitative methods explore lived experiences, power relations, and local governance practices. Integrating these data provides a grounded understanding of structural inequalities and everyday challenges related to access and control over environmental resources.

Findings

Survey findings show that access to productive resources in the study area is highly uneven. Only 27 percent of households reported owning cultivable land, while 41 percent depend on common pool resources such as forests and rivers for their livelihoods. Access to fisheries is particularly unequal. Better-off households often lease fishing ponds or control sections of creeks, while landless groups rely on daily wage labor or open access fishing that provides low and uncertain returns. Around 58 percent of respondents reported a decline in fish catches over the past decade, which they linked to rising salinity, illegal trawling, and siltation of water channels.

The collection of non-timber forest products such as honey, firewood, and crabs is regulated through permit systems that are often controlled by intermediaries or locally influential groups. Qualitative accounts provide deeper insight into these patterns. Women from landless households described growing dependence on mangrove forests for fuelwood, alongside increasing restrictions due to forest regulations. Small-scale fishers spoke of the difficulty of competing with commercial trawlers, often noting that little remains for them after larger operators extract their share. Members of marginalized caste groups reported limited participation in local decision-making, observing that access to resources tends to favor influential families. These unequal conditions increase livelihood insecurity and reduce the ability of poorer households to cope with environmental and economic stress.

b.3. Gender, Social Space, and Climate Justice

The coastal region of Bengal along the Bay of Bengal is exposed to a range of climate-related stresses, including cyclones, flooding, saline water intrusion, and land degradation. These environmental pressures interact with long-standing social hierarchies and shape how different groups experience vulnerability. Women are particularly affected due to gender-based divisions of labor, unequal access to resources, and limited participation in local decision-making processes. This case study examines the connections between social space, gender, and climate justice in selected villages of the South 24 Parganas district. It focuses on how climate impacts are experienced in everyday life and how women manage risk and responsibility within household and community settings.

The study uses a mixed-methods approach to examine the gendered dimensions of climate justice in the coastal region. Quantitative data were collected through household surveys conducted with sixty households across three villages, Mousuni, Kultali, and Gosaba. The surveys focused on water collection practices, livelihood activities, access to resources, and participation in adaptation planning. Qualitative data were gathered through participatory rural appraisal exercises, gender-segregated focus group discussions, oral testimonies, and participatory mapping of everyday spaces. This combination of methods helped capture both measurable gender differences and the lived experiences of women and other marginalized groups affected by climate-related environmental change.

A participatory research design guided the study to ensure that local perspectives were central to the analysis. Stratified household surveys provided information on gendered labor roles, dependence on natural resources, and involvement in decision-making. Qualitative methods such as participatory rural appraisal, focus group discussions, personal narratives, and mapping exercises documented daily experiences, perceptions of risk, and local power relations. Integrating these forms of evidence allowed for a grounded understanding of gendered climate vulnerability in coastal Bengal.

Findings

Quantitative findings show that women carry a major share of climate-related responsibilities within households. About 68 percent of women respondents reported being mainly responsible for collecting water and fuelwood, while 54 percent were involved in climate-sensitive livelihood activities such as paddy cultivation, crab collection, and small-scale fish processing. Access to safe drinking water emerged as a serious concern. Nearly 72 percent of women reported walking more than two kilometers each day during the dry season to collect potable water, a burden that increased further after major cyclonic events. Participation in formal decision-making remained limited, with only 19 percent of women reporting involvement in disaster planning or resource management meetings.

Qualitative narratives provided important insight into these patterns. Women described a steady reduction in their social and physical space as climate impacts disrupted livelihoods, daily mobility, personal safety, and health. Cyclone shelters were often reported to lack facilities that address women's needs, raising concerns about privacy and security during emergencies. Several young women shared experiences of leaving school to support household survival during periods of climate stress, highlighting the long-term nature of gendered vulnerability. Elderly women spoke about the decline of traditional coping practices, such as shared community ponds, which have become less reliable due to rising salinity and increasing competition over limited resources.

(c) **Implications for Innovative Risk Reduction and Socio-ecological Sustainability**

Findings from the three case studies show that risk reduction in the Indian delta cannot focus only on physical exposure to hazards. Climate risk is closely linked to social inequality, uneven access to resources, and gender-based vulnerability. Differences in caste, class, gender, and institutional access shape how communities experience and respond to climate stress. These patterns indicate the need to move away from uniform, infrastructure-centered solutions and toward approaches that respond to local social conditions and spatial variations in vulnerability.

Participatory GIS-based approaches provide useful tools for more inclusive and responsive risk reduction. Mapping exercises that combine hazard exposure with social and economic conditions help identify communities that face high levels of risk but receive limited institutional support. Such spatial information can support more targeted planning, improve transparency, and strengthen accountability in local governance. Mapping of environmental resources also highlights unequal control over land, fisheries, and forests, pointing to the importance of fair access and community-based management in sustaining local livelihoods.

Gender-focused findings show that climate impacts reshape everyday social space by increasing women's workloads, limiting mobility, and reducing participation in decision-making. Risk reduction strategies need to respond to these realities through gender-sensitive planning and by involving women more actively in local governance. Linking community knowledge with institutional support can strengthen adaptive capacity, reduce long-term risk, and support socio-ecological sustainability in the delta.

14.3 Conclusion and Arguments

This chapter brings together a range of case studies from the Indian delta coast to explain how climate risk, social vulnerability, and community responses are closely linked in everyday coastal life. The focus on coastal West Bengal highlights how environmental change affects people in different social and ecological settings. The case studies examine livelihood and health stress in rural delta communities as well as in tourism-based locations such as Digha. Other studies explore social vulnerability and traditional fishing practices, along with governance challenges related to interactions between mangrove ecosystems and coastal communities in the Sundarbans. The chapter also considers climate-induced migration from deltaic islands such as Sagar Island, shoreline change and coastal hazards, and the role of indigenous knowledge, community narratives, and climate justice in shaping resilience.

The arguments presented here show that disaster risk reduction in coastal and deltaic regions cannot depend only on technical assessments or hazard mapping. Risk develops through social, ecological, and institutional processes that vary across places and social groups. The case studies demonstrate that combining geospatial analysis with participatory approaches helps reveal these differences and supports more inclusive and innovative responses to risk. Such approaches strengthen local decision-making, improve the relevance of adaptation strategies, and recognize community knowledge as an essential resource. The chapter points to clear implications for innovative risk reduction and socio-ecological sustainability. Linking spatial tools with participatory processes supports livelihood-sensitive adaptation, conflict-aware conservation, and planning that takes equity and justice seriously. Although the chapter is grounded in the Indian delta coast, the insights gained from these case studies are relevant for other climate-exposed delta regions facing similar challenges.

References

Adom, P. K. (2024). The socioeconomic impact of climate change in developing countries over the next decades: A literature survey. *Heliyon., 10*(15), Article e35134. https://doi.org/10.1016/j.heliyon.2024.e35134

André, L. V., Van Wynsberge, S., Chinain, M., & Andréfouët, S. (2022). Benefits of collaboration between indigenous fishery management and data-driven spatial planning approaches: The case of a Polynesian traditional design (rāhui). *Fisheries Research, 256*, Article 106475. https://doi.org/10.1016/j.fishres.2022.106475

Anggraeni, I. D. (2024). The role of regulations in supporting the restoration of mangrove ecosystems for the achievement of sustainable development goals: A literature review. *Eduvest*, *4*(11), 10361–10378. https://doi.org/10.59188/eduvest.v4i11.43690

Arabadzhyan, A., Figini, P., García, C., González, M. M., Lam-González, Y. E., & León, C. J. (2021). Climate change, coastal tourism, and impact chains–a literature review. *CurrentIssuesinTourism,24*(16), 2233–2268. https://doi.org/10.1080/13683500.2020.1825351

Baloch, Q. B., Shah, S. N., Iqbal, N., Sheeraz, M., Asadullah, M., Mahar, S., & Khan, A. U. (2023). Impact of tourism development upon environmental sustainability: A suggested framework for sustainable ecotourism. *Environmental Science and Pollution Research, 30*(3), 5917–5930. https://doi.org/10.1007/s11356-022-22496-w

Becken, S. (2023). Climate change impacts on coastal tourism. CoastAdapt. (accessed 20 January 2023).Available at: https://coastadapt.com.au/sites/default/files/factsheets/T312_7_Coastal_Tourism.pdf

Bera, A. (2020). Review article on climate induced migration: Risk assessment and management: A case study of Sagar Island of Indian Sundarban. *Sgvu J Clim Change Water, 7*, 47–55.

Bera, A., Meraj, G., Kanga, S., Farooq, M., Singh, S. K., Sahu, N., & Kumar, P. (2022). Vulnerability and risk assessment to climate change in Sagar Island, India. *Water, 14*(5), 823. https://doi.org/10.3390/w14050823

Brown, G. (2012). An empirical evaluation of the spatial accuracy of public participation GIS (PPGIS) data. *Applied Geography, 34*, 289–294. https://doi.org/10.1016/j.apgeog.2011.12.004

Carmen, E., Fazey, I., Ross, H., et al. (2022). Building community resilience in a context of climate change: The role of social capital. *Ambio, 51*, 1371–1387. https://doi.org/10.1007/s13280-021-01678-9

Damastuti, E., & de Groot, R. (2019). Participatory ecosystem service mapping to enhance community-based mangrove rehabilitation and management in Demak, Indonesia. *Regional Environmental Change, 19*, 65–78. https://doi.org/10.1007/s10113-018-1378-7

Das, A., & Swain, P. K. (2024). Navigating the sea level rise: Exploring the interplay of climate change, sea level rise, and coastal communities in india. *Environmental Monitoring and Assessment, 196*, 1010. https://doi.org/10.1007/s10661-024-13191-z

Dasgupta, D. (2018). Homes, lives are swept away. The Straits Times. https://www.straitstimes.com/asia/south-asia/homes-lives-are-swept-away

de Carvalho, C. M., & Giatti, L. L. (2018). Participatory GIS for urban sustainability and resilience: A perspective of social learning and ecology of knowledge. In U. Azeiteiro, M. Akerman, W. Leal Filho, A. Setti & L. Brandli (Eds.), *Lifelong learning and education in healthy and sustainable cities*. World Sustainability Series. Springer. https://doi.org/10.1007/978-3-319-69474-0_2

Deb, A. K., & Haque, C. E. (2016). Livelihood diversification as a climate change coping strategy adopted by small-scale fishers of Bangladesh. *Climate Change Adaptation, Resilience and Hazards*, 345–368. https://doi.org/10.1007/978-3-319-39880-8_21

Dineshbabu, A. P., Thomas, S., Rohit, P., & Maheswarudu, G. (2019). Marine spatial planning for resource conservation, fisheries management and for ensuring fishermen security–global perspectives and Indian initiatives. *Current Science, 116*(4), 561–567.

Elwood, S. (2006). Participatory GIS and community planning: Restructuring technologies, social processes, and future research in PPGIS. In *Collaborative geographic information systems* (pp. 66–84). IGI Global Scientific Publishing. https://doi.org/10.4018/978-1-59140-845-1.ch004

Ghosh, A., Baidya, S., & Gupta, A.K. (2024). Climate change impact on Landuse and livelihood in Sundarbans: A case study of Sagar Island. In A. K. Gupta, A. Gupta & P. Acharya (Eds.), *Disaster risk and management under climate change. Disaster resilience and green growth*. Springer. https://doi.org/10.1007/978-981-99-4105-6_20

Gössling, S., & Hall, M. C. (2006). *Tourism and global environmental change: Ecological, social, economic and political interrelationships*. Routledge, Taylor & Francis. ISBN: 978-0-415-36132-3. https://library.oapen.org/bitstream/id/efa98d7b-4d06-4e48-b4d7-39b5eef61317/1006059.pdf

Griggs, G., & Reguero, B. G. (2021). Coastal adaptation to climate change and sea-level rise. *Water, 13*(16), 2151. https://doi.org/10.3390/w13162151

Harris, I., Osborn, T. J., Jones, P., et al. (2020). Version 4 of the CRU TS monthly high-resolution gridded multivariate climate dataset. *Sci Data, 7*, 109. https://doi.org/10.1038/s41597-020-0453-3

Hemani, C. (2014). Approaches to climate change adaptation of vulnerable coastal communities of India. In W. Leal Filho (Eds.), *Handbook of climate change adaptation*. Springer. https://doi.org/10.1007/978-3-642-40455-9_100-1

Himmelstoss, E. A., et al. (2024). Digital Shoreline analysis system version 6.0: U.S. Geological Survey software release, https://doi.org/10.5066/P13WIZ8M

Huynh, P. T., Le, N. D., Le, S. T., & Tran, T. N. (2021). Adaptive livelihood strategies among small-scale fishing households to climate change-related stressors in Central Coast Vietnam. *International Journal of Climate Change Strategies and Management, 13*(4/5), 492–510. https://doi.org/10.1108/IJCCSM-04-2020-0034

Igbokwe, J. C., Daraojimba, R. E., Okunade, B. A., Adewusi, O. E., A, O. B., & Adediran, F. E. (2024). Community engagement in local governance: A review of USA and African strategies. *World Journal of Advanced Research and Reviews, 21*(2), 105–112. https://doi.org/10.30574/wjarr.2024.21.2.0421

Innes, J. P., & Pascoe, S. (2010). A multi-criteria assessment of fishing gear impacts in demersal fisheries. *Journal of Environmental Management, 91*(4), 932–939. https://doi.org/10.1016/j.jenvman.2009.11.011

Janmaimool, P., Nunsunanon, S., & aphiphol Kriengsak, T. (2025). Aquaculture-related livelihood changes due to climate and non-climate stressors and well-being of residents in coastal communities. *Environmental Challenges, 20*, 101187. https://doi.org/10.1016/j.envc.2025.101187

Jawaweera, I. (2010). Livelihood and diversification in Rural Coastal Communities: Dependence on ecosystems services and possibilities for sustainable enterprising in Zanzibar, Tanzania. A thesis submitted to Stockholm University. Available at: https://www.diva-portal.org/smash/get/diva2:350389/FULLTEXT01.pdf

Jayawardhan, S. (2017). Vulnerability and climate change induced human displacement. *Consilience, 17*(1), 103–142.

Johnson, D., Blackett, P., Allison, A. E., & Broadbent, A. M. (2023). Measuring social vulnerability to climate change at the coast: Embracing complexity and context for more accurate and equitable analysis. *Water, 15*(19), 3408. https://doi.org/10.3390/w15193408

Kantamaneni, K., Panneer, S., Krishnan, A., et al. (2022). Appraisal of climate change and cyclone trends in Indian coastal states: A systematic approach towards climate action. *Arabian Journal of Geosciences, 15*, 814. https://doi.org/10.1007/s12517-022-10076-8

Karani, P., & Failler, P. (2020). Comparative coastal and marine tourism, climate change, and the blue economy in African Large Marine Ecosystems. *Environmental Development, 36*, Article 100572. https://doi.org/10.1016/j.envdev.2020.100572

Laino, E., Toledo, I., Aragonés, L., & Iglesias, G. (2024). A novel multi-hazard risk assessment framework for coastal cities under climate change. *Science of the Total Environment, 954*, Article 176638. https://doi.org/10.1016/j.scitotenv.2024.176638

Makwana, M., & Patnaik, U. (2025). Seasonal effects of extreme climate events and sea surface temperature indicators on the vulnerability of marine pelagic fisheries in the Bay of Bengal region. *Marine Environmental Research, 205*, Article 107009. https://doi.org/10.1016/j.marenvres.2025.107009

Malakar, K. D. (2020). Rural livelihood and Mangrove degradation: A case study of Namkhana Block, West Bengal, India. *International Journal of Innovative Science and Research Technology, 5*(1), 721–726.

Malakar, K. D., Kumar, M., Anand, S., & Kuzur, G. (2023a). Climate change and socio-ecological transformation: Applications and field narratives. In *Climate change and socio-ecological transformation; advances in geographical and environmental sciences* (pp. 295–374). Springer Nature Singapore. https://doi.org/10.1007/978-981-99-4390-6_7

Malakar, K. D., Kumar, M., Anand, S., & Kuzur, G. (2023b). Climate change and socio-ecological transformation: Applications and field narratives. In *Climate change and socio-ecological transformation. Advances in geographical and environmental sciences*. Springer. https://doi.org/10.1007/978-981-99-4390-6_7

Malakar, K. D., & Roy, S. (2024). *Mapping geospatial citizenship: The power of participatory GIS*. Springer. ISBN: 978-3-031-63106-1. https://doi.org/10.1007/978-3-031-63107-8

Malakar, K. D., Roy, S., & Kumar, M. (2025). Geospatial technologies in coastal ecologies monitoring and management on the Bengal Coast: Case studies and best practices. In *Geospatial technologies in coastal ecologies monitoring and management. Advances in geographic information science*. Springer. https://doi.org/10.1007/978-3-031-92017-2_16

Martyr-Koller, R., Thomas, A., Schleussner, C. F., Nauels, A., & Lissner, T. (2021). Loss and damage implications of sea-level rise on Small Island Developing States. *Current Opinion in Environmental Sustainability, 50*, 245–259. https://doi.org/10.1016/j.cosust.2021.05.001

Maulu, S., Hasimuna, O. J., Haambiya, L. H., Monde, C., Musuka, C. G., Makorwa, T. H., & Nsekanabo, J. D. (2021). Climate change effects on aquaculture production: Sustainability implications, mitigation, and adaptations. *Frontiers in Sustainable Food Systems, 5*, Article 609097. https://doi.org/10.3389/fsufs.2021.609097

McMichael, C., Yee, M., Cornish, G., Lutu, A., & McNamara, K. E. (2024). *Adaptive capacity in Pacific Islands: Responding to coastal and climatic change in Nagigi village, Fiji*. https://doi.org/10.31223/x53q6v

Mitra, A. (2013). *Sensitivity of mangrove ecosystem to changing climate*. Springer India: New Delhi. https://doi.org/10.1007/978-81-322-1509-7

Moreno, A., & Amelung, B. (2009). Climate change and coastal & marine tourism: Review and analysis. *Journal of Coastal Research, SI, 56*, 1140–1144.

Mukherjee, N., & Siddique, G. (2024). Impact of climate change in the Indian Sundarbans region. Springer. ISBN: 978-3-031-54237-4. https://doi.org/10.1007/978-3-031-54238-1

Nunes, A. R. (2021). Exploring the interactions between vulnerability, resilience and adaptation to extreme temperatures. *Natural Hazards, 109*, 2261–2293. https://doi.org/10.1007/s11069-021-04919-y

Obregón, C., Sokach, A., Neihapi, P., et al. (2024). Coastal fisheries and community-based support systems in post disaster contexts. *Maritime Studies, 23*, 35. https://doi.org/10.1007/s40152-024-00373-1

Ojea, E., Lester, S. E., & Salgueiro-Otero, D. (2020). Adaptation of fishing communities to climate-driven shifts in target species. *One Earth, 2*(6), 544–556. https://doi.org/10.1016/j.oneear.2020.05.012

Pattanaik, A. (2021). *Mangrove-dependent small-scale fisher (SSF) communities in the Sundarbans—vulnerable yet viable*. Thesis submitted to University of Waterloo.

Petzold, J., & Scheffran, J. (2024). Climate change and human security in coastal regions. *Cambridge Prisms: Coastal Futures, 2*, Article e5. https://doi.org/10.1017/cft.2024.2

Pingali, P., Aiyar, A., Abraham, M., & Rahman, A. (2019). Rural livelihood challenges: Moving out of agriculture. In *Transforming food systems for a rising India*. Palgrave Studies in Agricultural Economics and Food Policy. Palgrave Macmillan. https://doi.org/10.1007/978-3-030-14409-8_3

Restrepo-Mieth et al. 2023 Restrepo-Mieth, A., Perry, J., Garnick, J., & Weisberg, M. (2023). Community-based participatory climate action. *Global Sustainability (Cambridge University Press), 6*. https://doi.org/10.1017/sus.2023.12

Roy, P., Pal, S. C., Chakrabortty, R., Chowdhuri, I., Saha, A., & Shit, M. (2023). Effects of climate change and sea-level rise on coastal habitat: Vulnerability assessment, adaptation strategies and policy recommendations. *Journal of Environmental Management, 330*, Article 117187. https://doi.org/10.1016/j.jenvman.2022.117187

Sekhar, M., Lallawmkimi, M. C., Vijaya Rani, D., Upadhyay, L., Tiwari, P., Verma, U., & Rajput, A. (2024). Agroforestry practices and their impact on soil health and fertility: A review. *Journal of Experimental Agriculture International, 46*(9), 511–528. https://doi.org/10.9734/jeai/2024/v46i92849

Shiiba, N., Singh, P., Charan, D., et al. (2023). Climate change and coastal resiliency of Suva, Fiji: A holistic approach for measuring climate risk using the climate and ocean risk vulnerability index (CORVI). *Mitigation and Adaptation Strategies for Global Change, 28*, 9. https://doi.org/10.1007/s11027-022-10043-4

Shivanna, K. R. (2022). Climate change and its impact on biodiversity and human welfare. *Proceedings of the Indian National Science Academy, 88*, 160–171. https://doi.org/10.1007/s43538-022-00073-6

Sinha, D. (2022). Sundarbans' climate refugees face an uncertain future. DW (Nature and Environment, India). Retrieved 31 January 2025. Available at: https://www.dw.com/en/forgotten-people-sundarbans-climate-refugees-forced-to-move-again/a-61162969

Skrimizea, E., & Parra, C. (2019). Social-ecological dynamics and water stress in tourist islands: The case of Rhodes, Greece. *Journal of Sustainable Tourism, 27*(9), 1438–1456. https://doi.org/10.1080/09669582.2019.1630420

Sunkur, R., Kantamaneni, K., Bokhoree, C., & Ravan, S. (2023). Mangroves' role in supporting ecosystem-based techniques to reduce disaster risk and adapt to climate change: A review. *Journal of Sea Research, 196*, Article 102449. https://doi.org/10.1016/j.seares.2023.102449

Suntoro, A., Marbun, A. A. Y., & Sayekti, C. W. (2025). Climate change challenge: Coastal community adaptation towards achieving sustainable community resilience in Central Java, Indonesia. In *Climate change: Conflict and resilience in the age of Anthropocene*. Springer. https://doi.org/10.1007/978-3-031-85359-3_10

Tanwattana, P., & Toyoda, Y. (2024). Climate resilience action to policy: Embodied community-based climate risk resilience (CBCRR) in drought prone community in Thailand's economic corridor. *Climate Policy*, 1–18. https://doi.org/10.1080/14693062.2024.2430682

Thinh, N. A. (2022). Studying shoreline change in Ky Anh coastal area of Ha Tinh Province during 1989–2013 based on the digital shoreline analysis system (DSAS). In *Global changes and sustainable development in Asian emerging market economies* (p. 2). Springer. https://doi.org/10.1007/978-3-030-81443-4_33

Turner, R. K., & Schaafsma, M. (Eds.). (2015). *Coastal zones ecosystem services: From science to values and decision making*. Springer. ISBN: 978-3-319-17213-2. https://doi.org/10.1007/978-3-319-17214-9

UNESCO. (1984). Sundarbans National Park: World heritage. United Nations Educational, Scientific and Cultural Organization. Retrieved 6 November 2023. https://whc.unesco.org/en/list/452

Upadhyay, V. P., Ranjan, R., & Singh, J. S. (2002). Human–mangrove conflicts: The way out. *Current Science, 83*(11), 1328–1336. http://repository.ias.ac.in/72920/

Viavattene, C., Jiménez, J. A., Ferreira, O., Priest, S., Owen, D., & McCall, R. (2018). Selecting coastal hotspots to storm impacts at the regional scale: A Coastal Risk Assessment Framework. *Coastal Engineering, 134*, 33–47. https://doi.org/10.1016/j.coastaleng.2017.09.002

VijayKumar, R., Tiwari, P., Daniel, S., Kumar, K. R., Mishra, I., KS, A., & Shah, D. G. (2024). Agroforestry systems: A pathway to resilient and productive landscapes. *International Journal of Enviornment and Climate Change, 14*(12), 177–193. https://doi.org/10.9734/ijecc/2024/v14i124617

Viswanathan, P. (2016). Multifunctional benefits of community-based mangrove restoration in Gujarat: An analysis. In N. Ghosh, P. Mukhopadhyay, A. Shah & M. Panda (Eds.), *Nature, economy and society*. Springer. https://doi.org/10.1007/978-81-322-2404-4_12

White, B. P., Breakey, S., Brown, M. J., Smith, J. R., Tarbet, A., Nicholas, P. K., & Ros, A. M. V. (2023). Mental health impacts of climate change among vulnerable populations globally: An integrative review. *Annals of Global Health, 89*(1). https://doi.org/10.5334/aogh.4105

Zhang, J., & Convertino, M. (2026). Global mapping of potential coastal compound flood risk at 0.1° resolution. *Communications Earth & Environment, 7*, 83. https://doi.org/10.1038/s43247-025-03155-7

Glossary

Adaptive Capacity The ability of coastal communities and systems to adjust to climate stresses, reduce damage, and recover after disasters.

Blue Economy The sustainable use of ocean and coastal resources to support livelihoods, economic growth, and environmental health.

Climate Disasters The extreme climate-related events, including cyclones, floods, storm surges, and heatwaves, cause significant disruption to coastal environments and societies.

Climate-induced Disaster Risk The likelihood of losses resulting from the interaction between climate hazards, exposure, and vulnerability in coastal areas.

Coastal Ecologies The interconnected marine, estuarine, and terrestrial ecosystems sustain biodiversity, support livelihoods, and provide natural protection along coastal regions.

Coastal Infrastructural Resilience The ability of coastal infrastructure to resist damage, continue functioning, and recover quickly following climate-related events.

Coastal-Climate Nexus The interconnected relationship between coastal systems and climate processes that shapes risk, resilience, and sustainability.

K. D. Malakar et al., *Coastal Ecologies at Risk*, Palgrave Studies in Environmental Sustainability, https://doi.org/10.1007/978-3-032-26081-9

Community-based Adaptation (CbA) Adaptation approaches led by local communities that draw on local knowledge, participation, and context-specific solutions.

Disaster Risk Reduction (DRR) Policies and practices aimed at preventing new risks, reducing existing risks, and strengthening preparedness and resilience.

Early Warning Systems Integrated mechanisms that detect hazards, forecast risks, and communicate alerts to reduce loss of life and property.

Ecosystem-based Adaptation (EbA) Adaptation strategies that use natural ecosystems, such as mangroves and wetlands, to reduce climate risks and enhance resilience.

Exposure The location of people, infrastructure, livelihoods, or ecosystems in areas that are prone to climate-related hazards.

Geospatial Science The use of GIS, remote sensing, and spatial analysis to monitor hazards, vulnerabilities, and changes in coastal environments.

Governance Frameworks Institutional structures, policies, and legal arrangements that shape decision-making for coastal management and disaster resilience.

Integrated Coastal Zone Management (ICZM) A coordinated approach to managing coastal areas that balances environmental protection with social and economic development.

Participatory Disaster Risk Reduction Disaster risk reduction practices that incorporate community knowledge, local perceptions, and collective action.

Risk Assessment and Mapping The systematic analysis and spatial representation of hazards, vulnerabilities, and potential impacts to inform disaster risk reduction planning.

Risk Communication The process of sharing clear, timely, and reliable information to support informed decisions before, during, and after disasters.

Risk Perception The ways in which individuals and communities understand and interpret disaster risks based on experience, knowledge, and social context.

Socio-ecological Sustainability The long-term maintenance of healthy ecosystems and social well-being within interconnected coastal human–environment systems.

Socio-ecological Vulnerability The combined susceptibility of coastal communities and ecosystems to harm, shaped by exposure to hazards, sensitivity, and limited capacity to adapt.

Sustainable Coastal Development The development that meets current coastal needs while maintaining ecological balance and reducing future disaster risks.

Index

K. D. Malakar et al., *Coastal Ecologies at Risk*, Palgrave Studies in Environmental Sustainability, https://doi.org/10.1007/978-3-032-26081-9

FSC
www.fsc.org
MIX
Papier | Fördert
gute Waldnutzung
FSC® C083411

Kolibri 360 GmbH
Ferdinand-Jühlke-Straße 7
99095 Erfurt, Deutschland
produktsicherheit@kolibri360.de